Advances in Differential and Difference Equations with Applications 2020

Advances in Differential and Difference Equations with Applications 2020

Editor

Dumitru Baleanu

MDPI • Basel • Beijing • Wuhan • Barcelona • Belgrade • Manchester • Tokyo • Cluj • Tianjin

Editor
Dumitru Baleanu
Romania and Cankaya University
Turkey

Editorial Office
MDPI
St. Alban-Anlage 66
4052 Basel, Switzerland

This is a reprint of articles from the Special Issue published online in the open access journal *Mathematics* (ISSN 2227-7390) (available at: https://www.mdpi.com/journal/mathematics/special_issues/Advances_Differential_Difference_Equations_Applications_2020).

For citation purposes, cite each article independently as indicated on the article page online and as indicated below:

LastName, A.A.; LastName, B.B.; LastName, C.C. Article Title. *Journal Name* **Year**, *Article Number*, Page Range.

ISBN 978-3-03936-870-9 (Hbk)
ISBN 978-3-03936-871-6 (PDF)

© 2020 by the authors. Articles in this book are Open Access and distributed under the Creative Commons Attribution (CC BY) license, which allows users to download, copy and build upon published articles, as long as the author and publisher are properly credited, which ensures maximum dissemination and a wider impact of our publications.

The book as a whole is distributed by MDPI under the terms and conditions of the Creative Commons license CC BY-NC-ND.

Contents

About the Editor . . . . **vii**

Preface to "Advances in Differential and Difference Equations with Applications 2020" . . . . **ix**

Vasilii Zaitsev and Marina Zhuravleva
On Assignment of the Upper Bohl Exponent for Linear Time-Invariant Control Systems in a Hilbert Space by State Feedback
Reprinted from: *Mathematics* **2020**, *8*, 992, doi:10.3390/math8060992 . . . . **1**

María Pilar Velasco, David Usero, Salvador Jiménez, Luis Vázquez, José Luis Vázquez-Poletti and Mina Mortazavi
About Some Possible Implementations of the Fractional Calculus
Reprinted from: *Mathematics* **2020**, *8*, 893, doi:10.3390/math8060893 . . . . **21**

Vasilii Zaitsev and Inna Kim
Exponential Stabilization of Linear Time-Varying Differential Equations with Uncertain Coefficients by Linear Stationary Feedback
Reprinted from: *Mathematics* **2020**, *8*, 853, doi:10.3390/math8050853 . . . . **43**

Pedro Almenar and Lucas Jódar
The Sign of the Green Function of an n-th Order Linear Boundary Value Problem
Reprinted from: *Mathematics* **2020**, *8*, 673, doi:10.3390/math8050673 . . . . **59**

Eva María Ramos-Ábalos, Ramón Gutiérrez-Sánchez and Ahmed Nafidi
Powers of the Stochastic Gompertz and Lognormal Diffusion Processes, Statistical Inference and Simulation
Reprinted from: *Mathematics* **2020**, *8*, 588, doi:10.3390/math8040588 . . . . **81**

Osama Moaaz, Dimplekumar Chalishajar and Omar Bazighifan
Asymptotic Behavior of Solutions of the Third Order Nonlinear Mixed Type Neutral Differential Equations
Reprinted from: *Mathematics* **2020**, *8*, 485, doi:10.3390/math8040485 . . . . **95**

Kyung Won Hwang and Cheon Seoung Ryoo
Differential Equations Associated with Two Variable Degenerate Hermite Polynomials
Reprinted from: *Mathematics* **2020**, *8*, 228, doi:10.3390/math8020228 . . . . **109**

Anum Shafiq, Ilyas Khan, Ghulam Rasool, El-Sayed M. Sherif and Asiful H. Sheikh
Influence of Single- and Multi-Wall Carbon Nanotubes on Magnetohydrodynamic Stagnation Point Nanofluid Flow over Variable Thicker Surface with Concave and Convex Effects
Reprinted from: *Mathematics* **2020**, *8*, 104, doi:10.3390/math8010104 . . . . **127**

İbrahim Avci and Nazim Mahmudov
Numerical Solutions for Multi-Term Fractional Order DifferentialEquations with Fractional Taylor Operational Matrix of Fractional Integration
Reprinted from: *Mathematics* **2020**, *8*, 96, doi:10.3390/math8010096 . . . . **143**

Yonghyeon Jeon, Soyoon Bak and Sunyoung Bu
Reinterpretation of Multi-Stage Methods for Stiff Systems: A Comprehensive Review on Current Perspectives and Recommendations
Reprinted from: *Mathematics* **2019**, *7*, 1158, doi:10.3390/math7121158 . . . . **167**

Mouffak Benchohra, Noreddine Rezoug, Bessem Samet and Yong Zhou
Second Order Semilinear Volterra-Type Integro-Differential Equations with
Non-Instantaneous Impulses
Reprinted from: *Mathematics* 2019, 7, 1134, doi:10.3390/math7121134 . . . . **181**

Asifa Tassaddiq, Aasma Khalid, Muhammad Nawaz Naeem, Abdul Ghaffar, Faheem Khan, Samsul Ariffin Abdul Karim and Kottakkaran Sooppy Nisar
A New Scheme Using Cubic B-Spline to Solve Non-Linear Differential Equations Arising in
Visco-Elastic Flows and Hydrodynamic Stability Problems
Reprinted from: *Mathematics* 2019, 7, 1078, doi:10.3390/math7111078 . . . . **201**

Le Dinh Long, Nguyen Hoang Luc, Yong Zhou, and Can Nguyen
Identification of Source Term for the Time-Fractional Diffusion-Wave Equation by Fractional
Tikhonov Method
Reprinted from: *Mathematics* 2019, 7, 934, doi:10.3390/math7100934 . . . . **219**

Abdul Ghafoor, Sirajul Haq, Manzoor Hussain, Poom Kumam and Muhammad Asif Jan
Approximate Solutions of Time Fractional Diffusion Wave Models
Reprinted from: *Mathematics* 2019, 7, 923, doi:10.3390/math7100923 . . . . **243**

Mir Asma, W.A.M. Othman and Taseer Muhammad
Numerical Study for Darcy–Forchheimer Flow of Nanofluid due to a Rotating Disk with Binary
Chemical Reaction and Arrhenius Activation Energy
Reprinted from: *Mathematics* 2019, 7, 921, doi:10.3390/math7100921 . . . . **259**

Nichaphat Patanarapeelert and Thanin Sitthiwirattham
On Fractional Symmetric Hahn Calculus
Reprinted from: *Mathematics* 2019, 7, 873, doi:10.3390/math7100873 . . . . **275**

Dumitru Baleanu, Vladimir E. Fedorov, Dmitriy M. Gordievskikh and Kenan Taş
Approximate Controllability of Infinite-Dimensional Degenerate Fractional Order Systems in
the Sectorial Case
Reprinted from: *Mathematics* 2019, 7, 735, doi:10.3390/math7080735 . . . . **293**

Rajarama Mohan Jena, Snehashish Chakraverty and Dumitru Baleanu
On the Solution of an Imprecisely Defined Nonlinear Time-Fractional Dynamical Model
of Marriage
Reprinted from: *Mathematics* 2019, 7, 689, doi:10.3390/math7080689 . . . . **309**

Yong Zhou, Bashir Ahmad and Ahmed Alsaedi
Structure of Non-Oscillatory Solutions for Second Order Dynamic Equations on Time Scales
Reprinted from: *Mathematics* 2019, 7, 680, doi:10.3390/math7080680 . . . . **325**

About the Editor

Dumitru Baleanu is a professor at the Institute of Space Sciences, Magurele-Bucharest, Romania, and a visiting staff member at the Department of Mathematics, Cankaya University, Ankara, Turkey. Dumitru Baleanu got his Ph.D. from the Institute of Atomic Physics in 1996. His fields of interest include fractional dynamics and its applications, fractional differential equations and their applications, discrete mathematics, image processing, bio-informatics, mathematical biology, mathematical physics, soliton theory, Lie symmetry, dynamic systems on time scales, computational complexity, the wavelet method and its applications, quantization of systems with constraints, the Hamilton–Jacobi formalism, and geometries admitting generic and non-generic symmetries. Dumitru Baleanu is co-author of 15 books published by Springer, Elsevier, and World Scientific. His H index is 60, and he is a highly cited researcher in Mathematics and Engineering. Dumitru Baleanu won the 2019 Obada Prize. This prize recognizes and encourages innovative and interdisciplinary research that cuts across traditional boundaries and paradigms. It aims to foster universal values of excellence, creativity, justice, democracy, and progress, and to promote the scientific, technological, and humanistic achievements that advance and improve our world.

In addition, together with G.C. Wu, L.G. Zeng, X.C. Shi, and F. Wu, Dumitru Baleanu is the coauthor of Chinese Patent No ZL 2014 1 0033835.7 regarding chaotic maps and their important role in information encryption.

Preface to "Advances in Differential and Difference Equations with Applications 2020"

Differential and difference equations are extreme representations of complex dynamical systems.

During the last few decades, the theory of fractional differentiation has been successfully applied to the study of anomalous social and physical behaviors, where scaling power law of fractional order appears universal as an empirical description of such complex phenomena. Recently, the difference counterpart of fractional calculus has started to be intensively used for a better characterization of some real-world phenomena. Systems of delay differential equations have started to occupy a place of central importance in various areas of science, particularly in biological areas.

This book presents some19 original results regarding the theory and application of differential and difference equations which can be successfully used in dealing with real-world problems in various branches of science and engineering.

Dumitru Baleanu
Editor

Article

On Assignment of the Upper Bohl Exponent for Linear Time-Invariant Control Systems in a Hilbert Space by State Feedback

Vasilii Zaitsev *,† and Marina Zhuravleva †

Laboratory of Mathematical Control Theory, Udmurt State University, Izhevsk 426034, Russia; mrnzo@yandex.ru
* Correspondence: verba@udm.ru
† These authors contributed equally to this work.

Received: 18 May 2020; Accepted: 14 June 2020; Published: 17 June 2020

Abstract: We consider a linear continuous-time control system with time-invariant linear bounded operator coefficients in a Hilbert space. The controller in the system has the form of linear state feedback with a time-varying linear bounded gain operator function. We study the problem of arbitrary assignment for the upper Bohl exponent by state feedback control. We prove that if the open-loop system is exactly controllable then one can shift the upper Bohl exponent of the closed-loop system by any pregiven number with respect to the upper Bohl exponent of the free system. This implies arbitrary assignability of the upper Bohl exponent by linear state feedback. Finally, an illustrative example is presented.

Keywords: linear control system; Hilbert space; state feedback control; exact controllability; upper Bohl exponent

MSC: 34D08; 34A35; 93C05

---

1. Introduction

Consider a linear control system:

$$\dot{x}(t) = A(t)x(t) + B(t)u(t), \quad t \in \mathbb{R}. \tag{1}$$

Here $x \in \mathfrak{X}$ and $u \in \mathfrak{U}$ are the state and control vectors respectively, $\mathfrak{X}$ and $\mathfrak{U}$ are some finite-dimensional or infinite-dimensional Banach spaces. Suppose that the controller in system (1) has the form of linear static state feedback $u(t) = U(t)x(t)$. The closed-loop system has the form:

$$\dot{x}(t) = \big(A(t) + B(t)U(t)\big)x(t), \quad t \in \mathbb{R}. \tag{2}$$

Now we consider the elements of the gain operator $U(t)$ as controlling parameters. The problems of control over the asymptotic behavior of solutions to systems (2) by means of elements of gain operator $U(t)$ (in particular, the problem of stabilization for system (2)) belong to the classical problems of control theory. First results relate to stationary systems in finite-dimensional spaces. It was proved for complex [1] and real [2] finite-dimensional $(\mathfrak{X} = \mathbb{R}^n, \mathfrak{U} = \mathbb{R}^m)$ time-invariant $(A(t) \equiv A, B(t) \equiv B)$ systems that the condition of complete controllability of system (1) is necessary and sufficient for the arbitrary assignment of the eigenvalue spectrum $\lambda_1, \ldots, \lambda_n$ of the closed-loop system (2) by means of time-invariant $(U(t) \equiv U)$ feedback. This implies, in particular, stabilizability of (2) by means of $U(t) \equiv U$. First results for time-varying periodic systems in finite-dimensional spaces were obtained in [3]: It was proved that the complete controllability of system (1) is necessary and sufficient for the arbitrary assignment of the characteristic multipliers $\rho_1, \ldots, \rho_n$ of the closed-loop system (2) by means of periodic feedback. For time-varying non-periodic systems in finite-dimensional spaces, first results on stabilization were obtained in [4–6]. A transformation reducing system (2) to a canonical (block)-Frobenius form was used, which allows one to solve the eigenvalue assignment problem. However, rather restrictive conditions on the smoothness and boundedness of the coefficients of system (1) are required there. These conditions were weakened in [7] to the condition of uniform complete controllability in the sense of Kalman [8], and, on the basis of this property, sufficient conditions for exponential stabilization of system (2) were obtained. The proof of exponential stability is carried out using the second Lyapunov method (the Lyapunov function method).

In the framework of the first Lyapunov method of studying systems of differential equations in finite-dimensional spaces, a natural generalization of the concept of eigenvalue spectrum for non-stationary systems is the spectrum of Lyapunov exponents (see [9–11]). In addition to Lyapunov exponents, other Lyapunov invariants are known (that is, characteristics that do not change under the Lyapunov transformation, see [12]), which characterize the asymptotic behavior of solutions to a linear system of differential equations, for example, the Bohl exponents, the central (Vinograd) exponents, the exponential (Izobov) exponents, etc. In a series of studies [13–17], the results on arbitrary assignability of Lyapunov exponents and other Lyapunov invariants for system (2) in finite-dimensional spaces were proved, based on the property of uniform complete controllability in the sense of Kalman. In recent studies [18–23], these results have been partially extended to discrete-time systems. In finite-dimensional spaces, the Lyapunov exponents, the Bohl exponents, and other Lyapunov invariants were studied, for example in [24–26] for continuous-time systems and in [27–33] for discrete-time systems.

A large number of papers are devoted to stabilization problems of system (2) in infinite-dimensional spaces. We note here the studies [34–41]. Properties of the spectrum for systems in infinite-dimensional spaces were studied in [42–44].

In this paper, we studied the problem of arbitrary assignment of the upper Bohl exponent for continuous-time systems in an infinite-dimensional Hilbert space. The brief outline of the paper is as follows. In Section 2, some notations, definitions, and preliminary results are given and the concepts used throughout the paper are defined, as well as some basic theories, methods, and techniques. In Section 3, we analyze the problem of arbitrary assignment of the upper Bohl exponent by means of linear state feedback with a time-varying linear bounded gain operator function for linear time-invariant control system in a Hilbert space with bounded operator coefficients and prove that the property of exact controllability of the open-loop system is sufficient for arbitrary assignability of the upper Bohl exponent of the closed-loop system. Section 4 provides an illustrative example that emphasizes the theory. In Section 5, we revise the results obtained in the paper and also showcase future developments of the theory.

2. Notations, Definitions, and Preliminary Results

Let $\mathfrak{X}$ be a Banach space, $\mathfrak{X}^*$ be dual to $\mathfrak{X}$. By $\mathcal{L}(\mathfrak{X}_1, \mathfrak{X}_2)$ we denote a Banach space of linear bounded operators $A : \mathfrak{X}_1 \to \mathfrak{X}_2$. If $A \in \mathcal{L}(\mathfrak{X}_1, \mathfrak{X}_2)$, then $A^* \in \mathcal{L}(\mathfrak{X}_2^*, \mathfrak{X}_1^*)$ is its adjoint operator. By $I : \mathfrak{X} \to \mathfrak{X}$ denote the identity operator.

Consider a linear system of differential equations:

$$\dot{x}(t) = A(t)x(t), \quad t \in \mathbb{R}, \quad x \in \mathfrak{X}. \tag{3}$$

We suppose that the following conditions hold:

(a) $A(t) \in \mathcal{L}(\mathfrak{X}, \mathfrak{X})$ for any $t \in \mathbb{R}$;
(b) The function $\mathbb{R} \ni t \mapsto A(t) \in \mathcal{L}(\mathfrak{X}, \mathfrak{X})$ is piecewise continuous;
(c) $\sup\limits_{t \in \mathbb{R}} \|A(t)\| = a < +\infty$.

By a solution of system (3) we will understand, by definition, a solution of the integral equation:

$$x(t) = x_0 + \int_{t_0}^{t} A(s)x(s)\,ds, \qquad (4)$$

where

$$x(t_0) = x_0. \qquad (5)$$

Due to conditions imposed on $A(\cdot)$, a solution (4) of (3) is a continuous, piecewise continuously differentiable function and satisfies (3) almost everywhere ([45], Ch. III, Sect. 1.1, 1.2).

By $\Phi(t,\tau)$ denote the evolution operator of system (3) ([45], Ch. III, Sect. 1, p. 100) that is the solution of the operator system:

$$\frac{dX}{dt} = A(t)X, \quad X(\tau) = I.$$

By using the operator $\Phi(t,\tau)$, the solution of the initial value problem (3), (5) can be expressed by the formula $x(t) = \Phi(t,t_0)x_0$.

The evolution operator $\Phi(t,\tau)$ has the following properties ([45], Ch. III, Sect. 1, p. 101):
(A) $\Phi(t,t) = I$; (B) $\Phi(t,s)\Phi(s,\tau) = \Phi(t,\tau)$; (C) $\Phi(t,\tau) = [\Phi(\tau,t)]^{-1}$;

$$(D) \quad \exp\left(-\int_s^t \|A(\tau)\|\,d\tau\right) \leq \|\Phi^{\pm 1}(t,s)\| \leq \exp\left(\int_s^t \|A(\tau)\|\,d\tau\right), \quad s \leq t$$

(see [45], Ch. III, Sect. 2, (2.25)). It follows from property (D) and condition (c) that:

$$e^{-a(t-s)} \leq \|\Phi^{\pm 1}(t,s)\| \leq e^{a(t-s)}, \quad s \leq t. \qquad (6)$$

Definition 1. *The upper Bohl exponent ([45], Ch. III, Sect. 4) of system (3) is the number:*

$$\varkappa(A) = \varlimsup_{\tau,s \to +\infty} \frac{\ln\|\Phi(\tau+s,\tau)\|}{s}.$$

The upper Bohl exponent of system (3) characterizes asymptotic behavior of solutions of (3): The condition $\varkappa(A) < 0$ is necessary and sufficient for uniform exponential stability of all solutions to system (3). Due to the condition (c), the upper Bohl exponent of system (3) is finite ([45], Ch. III, Sect. 4, Theorem 4.3).

Let us apply the λ-transformation ([9], p. 249), ([45], Ch. III, Sect. 4, p. 124) to system (3) that is adding the disturbance λI to the operator $A(t)$ and consider the disturbed system:

$$\dot{z}(t) = (A(t) + \lambda I)z(t), \quad t \in \mathbb{R}, \quad z \in \mathfrak{X}. \qquad (7)$$

By $\Psi(t,\tau)$ denote the evolution operator of system (7).

Lemma 1. *For any $t, \tau \in \mathbb{R}$ the following equality holds:*

$$\Psi(t,\tau) = e^{\lambda(t-\tau)}\Phi(t,\tau). \qquad (8)$$

Proof. Calculating the derivative of the right-hand side of (8), we obtain:

$$\frac{d}{dt}\left(e^{\lambda(t-\tau)}\Phi(t,\tau)\right) = \lambda e^{\lambda(t-\tau)}\Phi(t,\tau) + e^{\lambda(t-\tau)}A(t)\Phi(t,\tau) = (A(t) + \lambda I)e^{\lambda(t-\tau)}\Phi(t,\tau). \qquad (9)$$

Next,

$$\left. e^{\lambda(t-\tau)}\Phi(t,\tau)\right|_{t=\tau} = I. \qquad (10)$$

It follows from (9) and (10) that $e^{\lambda(t-\tau)}\Phi(t,\tau)$ is the evolution operator of (7). Due to the uniqueness of the evolution operator, $e^{\lambda(t-\tau)}\Phi(t,\tau)$ coincides with $\Psi(t,\tau)$. □

Lemma 2. $\varkappa(A + \lambda I) = \varkappa(A) + \lambda.$

Proof. By using Lemma 1, we obtain:

$$\begin{aligned}
\varkappa(A + \lambda I) &= \varlimsup_{\tau,s \to +\infty} \frac{\ln \|\Psi(\tau+s,\tau)\|}{s} = \varlimsup_{\tau,s \to +\infty} \frac{\ln \|e^{\lambda s}\Phi(\tau+s,\tau)\|}{s} \\
&= \varlimsup_{\tau,s \to +\infty} \frac{\ln(e^{\lambda s}\|\Phi(\tau+s,\tau)\|)}{s} = \varlimsup_{\tau,s \to +\infty} \frac{\ln e^{\lambda s} + \ln \|\Phi(\tau+s,\tau)\|}{s} \\
&= \varlimsup_{\tau,s \to +\infty} \left(\frac{\lambda s}{s} + \frac{\ln \|\Phi(\tau+s,\tau)\|}{s}\right) = \varlimsup_{\tau,s \to +\infty} \left(\lambda + \frac{\ln \|\Phi(\tau+s,\tau)\|}{s}\right) \\
&= \lambda + \varlimsup_{\tau,s \to +\infty} \frac{\ln \|\Phi(\tau+s,\tau)\|}{s} = \lambda + \varkappa(A).
\end{aligned}$$

□

Let us consider another linear system of differential equations:

$$\dot{y}(t) = C(t)y(t), \quad t \in \mathbb{R}, \quad y \in \mathfrak{X}. \tag{11}$$

Suppose that the operator function $C(t)$ also satisfies conditions (a), (b), (c), i.e., $C(t) \in \mathcal{L}(\mathfrak{X},\mathfrak{X})$ $\forall t \in \mathbb{R}$, $C(\cdot)$ is piecewise continuous, and $\sup_{t \in \mathbb{R}} \|C(t)\| = c < +\infty$. By $\Theta(t,\tau)$ denote the evolution operator of system (11). Because of conditions imposed on $C(\cdot)$, we have the inequality:

$$e^{-c(t-s)} \leq \|\Theta^{\pm 1}(t,s)\| \leq e^{c(t-s)}, \quad s \leq t. \tag{12}$$

Definition 2. *Systems (3) and (11) are called kinematically similar on $\mathbb{R}$ ([45], Ch. IV, Sect. 2) if it is possible to establish between the totalities of all solutions of these systems a one-to-one correspondence:*

$$y(t) = L(t)x(t), \quad t \in \mathbb{R},$$

where $L(t)$ is a bounded linear operator function with a bounded inverse:

$$\|L(t)\| \leq d_1, \quad \|L^{-1}(t)\| \leq d_2, \quad t \in \mathbb{R}. \tag{13}$$

The following criterion holds (see [45], Ch. IV, Sect. 2, Lemma 2.1, (a)).

Lemma 3. *Systems (3) and (11) are kinematically similar on $\mathbb{R}$ if and only if there exists an operator function $\mathbb{R} \ni t \mapsto L(t) \in \mathcal{L}(\mathfrak{X},\mathfrak{X})$ satisfying (13) and such that the evolution operators of the systems are connected by the relation:*

$$\Theta(t,\tau)L(\tau) = L(t)\Phi(t,\tau). \tag{14}$$

Lemma 4 (see [45], Ch. IV, Sect. 2, Theorem 2.1). *If systems (3) and (11) are kinematically similar on $\mathbb{R}$, then $\varkappa(A) = \varkappa(C)$.*

Let us state sufficient conditions for kinematical similarity of systems (3) and (11) on $\mathbb{R}$ analogous to the corresponding conditions in a finite-dimensional space (see, e.g., [46]).

Lemma 5. *Suppose that the operator functions $A(t)$ and $C(t)$ satisfy conditions (a), (b), and (c), and there exists a sequence $\{t_i\}_{i \in \mathbb{Z}} \subset \mathbb{R}$ such that $0 < \rho_1 \leq t_{i+1} - t_i \leq \rho_2$ and $\Phi(t_{i+1},t_i) = \Theta(t_{i+1},t_i)$ for all $i \in \mathbb{Z}$. Then systems (3) and (11) are kinematically similar on $\mathbb{R}$.*

Proof. By using the group property (B) of evolution operators, we obtain for all $j > i$:

$$\Phi(t_j, t_i) = \Phi(t_j, t_{j-1}) \cdots \Phi(t_{i+1}, t_i) = \Theta(t_j, t_{j-1}) \cdots \Theta(t_{i+1}, t_i) = \Theta(t_j, t_i). \tag{15}$$

By (C), (15) holds for any $i, j \in \mathbb{Z}$. Let us construct the operator function:

$$L(t) = \Theta(t, t_0) \Phi(t_0, t). \tag{16}$$

By (15), we have $L(t_i) = I, i \in \mathbb{Z}$. Next, by (16), we have:

$$\Theta(t, \tau) L(\tau) = \Theta(t, \tau) \Theta(\tau, t_0) \Phi(t_0, \tau) = \Theta(t, t_0) \Phi(t_0, \tau),$$
$$L(t) \Phi(t, \tau) = \Theta(t, t_0) \Phi(t_0, t) \Phi(t, \tau) = \Theta(t, t_0) \Phi(t_0, \tau).$$

Hence, (14) is fulfilled. Let us prove that (13) is satisfied.

Let $t \in \mathbb{R}$ be an arbitrary number. Then, since $t_{i+1} - t_i \geq \rho_1$, there exists an $i_0 \in \mathbb{Z}$ such that $t \in [t_{i_0}, t_{i_0+1}]$. In this case, $t - t_{i_0} \leq \rho_2$. We have:

$$L(t) = \Theta(t, t_0) \Phi(t_0, t) = \Theta(t, t_{i_0}) \Theta(t_{i_0}, t_0) \Phi(t_0, t_{i_0}) \Phi(t_{i_0}, t)$$
$$= \Theta(t, t_{i_0}) L(t_{i_0}) \Phi(t_{i_0}, t) = \Theta(t, t_{i_0}) \Phi(t_{i_0}, t).$$

So, $L^{-1}(t) = \Phi(t, t_{i_0}) \Theta(t_{i_0}, t)$. Then, taking (6) and (12) into account, we obtain:

$$\|L(t)\| \leq \|\Theta(t, t_{i_0})\| \cdot \|\Phi(t_{i_0}, t)\| \leq e^{c(t - t_{i_0})} e^{a(t - t_{i_0})} \leq e^{(a+c)\rho_2} =: d_1,$$
$$\|L^{-1}(t)\| \leq \|\Phi(t, t_{i_0})\| \cdot \|\Theta(t_{i_0}, t)\| \leq e^{a(t - t_{i_0})} e^{c(t - t_{i_0})} \leq e^{(a+c)\rho_2} =: d_2.$$

Hence, (13) holds. By Lemma 3, the lemma is proved. □

Consider a linear control system:

$$\dot{x}(t) = A(t) x(t) + B(t) u(t), \quad t \in \mathbb{R}. \tag{17}$$

Here $x \in \mathfrak{X}$, $u \in \mathfrak{U}$; $\mathfrak{X}, \mathfrak{U}$ are Banach spaces; $A(t)$ satisfies conditions (a), (b), (c); $\forall t \in \mathbb{R}$ $B(t) \in \mathcal{L}(\mathfrak{U}, \mathfrak{X})$, the function $t \mapsto B(t)$ is piecewise continuous, and $\sup_{t \in \mathbb{R}} \|B(t)\| < +\infty$. Admissible controllers for (17) on some finite interval $[t_0, t_1]$ are functions $u(\cdot) \in L_p([t_0, t_1], \mathfrak{U})$, $p \geq 1$. For each admissible controller $u(\cdot)$, there is a unique solution of the initial value problem (17), (5) (see ([45], Ch. III, Sect. 1, (1.19)), [47]), determined by the formula:

$$x(t) = \Phi(t, t_0) x_0 + \int_{t_0}^{t} \Phi(t, \tau) B(\tau) u(\tau) \, d\tau.$$

Here $\Phi(t, \tau)$ is the evolution operator of the corresponding free system (3). We consider a control system (17) without imposing any geometric constraints on the control or on the state.

Definition 3 (see [47]). *System (17) is called exactly controllable on $[0, \vartheta]$ if for any $x_0, x_1 \in \mathfrak{X}$ there exists an admissible controller $u(t)$, $t \in [0, \vartheta]$, steering the solution of (17) from $x(0) = x_0$ to $x(\vartheta) = x_1$.*

Suppose that the controller in system (17) has the form of the linear state feedback:

$$u(t) = U(t) x(t), \tag{18}$$

where $U(t) \in \mathcal{L}(\mathfrak{X}, \mathfrak{U})$ $\forall t \in \mathbb{R}$, $U(\cdot)$ is piecewise continuous, and $\sup\limits_{t \in \mathbb{R}} \|U(t)\| < +\infty$. We say that the gain operator function $U(\cdot)$ satisfying these conditions is admissible. The closed-loop system has the form:

$$\dot{x}(t) = (A(t) + B(t)U(t))x(t). \tag{19}$$

By $\Phi_U(t, \tau)$ we denote the evolution operator of system (19).

Definition 4. *We say that system* (17) *admits a λ-transformation if there exists a constant $\sigma > 0$ such that, for any $\lambda \in \mathbb{R}$, there exists an admissible gain operator function $U(\cdot)$ ensuring that the evolution operator $\Phi_U(t, \tau)$ of system* (19) *satisfies the relation:*

$$\Phi_U((k+1)\sigma, k\sigma) = e^{\lambda \sigma} \Phi((k+1)\sigma, k\sigma) \tag{20}$$

for all $k \in \mathbb{Z}$.

This definition was given in [13] for systems (17) in finite-dimensional spaces (see also [48]). It is related to the definition of a λ-transformation of system (3).

Remark 1. *It follows from* (20) *that, for the evolution operator $\Phi_U(t, s)$ of system* (19)*, the relation $\Phi_U(k\sigma, \ell\sigma) = e^{\lambda(k-\ell)\sigma} \Phi(k\sigma, \ell\sigma)$ holds that is similar to* (8) *but is fulfilled on the set $\{k\sigma, k \in \mathbb{Z}\} \subset \mathbb{R}$.*

Theorem 1. *Suppose that system* (17) *admits a λ-transformation. Then, for any $\lambda \in \mathbb{R}$, there exists an admissible gain operator function $U(\cdot)$ such that the closed-loop system* (19) *and system* (7) *are kinematically similar on $\mathbb{R}$.*

Proof. It follows from (20) and (8) that, for all $k \in \mathbb{Z}$, the following equalities hold:

$$\Phi_U(t_{k+1}, t_k) = \Psi(t_{k+1}, t_k)$$

where $\Psi(t, s)$ is the evolution operator of system (7) and $t_k = k\sigma$. Now, applying Lemma 5 to systems (19) and (7), where $\rho_1 = \rho_2 = \sigma$, we obtain what is required. □

Definition 5. *We say that the upper Bohl exponent of system* (17) *is arbitrarily assignable by linear state feedback* (18) *if for any $\mu \in \mathbb{R}$ there exists an admissible gain operator function $U(\cdot)$ such that, for the closed-loop system* (19)*,*

$$\varkappa(A + BU) = \mu.$$

The corresponding definition in finite-dimensional spaces was given in [13] (see also [48]) for the upper (and lower) central (and Bohl) exponents.

3. Main Results

Consider a time-invariant control system (17):

$$\dot{x}(t) = Ax(t) + Bu(t), \quad t \in \mathbb{R}. \tag{21}$$

Here $x \in \mathfrak{X}$, $u \in \mathfrak{U}$; $\mathfrak{X}$ and $\mathfrak{U}$ are separable Hilbert spaces; $A \in \mathcal{L}(\mathfrak{X}, \mathfrak{X})$, $B \in \mathcal{L}(\mathfrak{U}, \mathfrak{X})$; $a := \|A\|$, $b := \|B\|$. For Hilbert spaces $\mathfrak{H}_1, \mathfrak{H}_2$, we suppose that, if $F \in \mathcal{L}(\mathfrak{H}_1, \mathfrak{H}_2)$, then $F^* \in \mathcal{L}(\mathfrak{H}_2, \mathfrak{H}_1)$, i.e., we identify $\mathfrak{H}_i^*$ with $\mathfrak{H}_i$. By $\langle \cdot, \cdot \rangle$ denote the scalar product (in the corresponding space). If $F^* = F \in \mathcal{L}(\mathfrak{X}, \mathfrak{X})$, then the inequality $F \geq \alpha I$ means, by definition, that $\langle Fx, x \rangle \geq \alpha \|x\|^2$ for all $x \in \mathfrak{X}$.

The evolution operator of the corresponding free system:

$$\dot{x}(t) = Ax(t), \quad t \in \mathbb{R},$$

has the form $\Phi(t, \tau) = \exp(A(t - \tau))$. Let us denote $\Phi(t) := \Phi(t, 0) = \exp(At)$.

Let us construct the *controllability gramian* $Q(\vartheta) : \mathfrak{X} \to \mathfrak{X}$, $\vartheta > 0$ (see ([49], Definition 4.1.3), ([50], Part IV, Ch. 2, Sect. 2.2, (2.9))):

$$Q(\vartheta)x = \int_0^\vartheta \Phi(s) BB^* \Phi^*(s) x \, ds. \tag{22}$$

We have $Q(\vartheta) \in \mathcal{L}(\mathfrak{X}, \mathfrak{X})$ (see [49], Lemma 4.1.4), $Q(\vartheta) = Q^*(\vartheta)$, and

$$\langle Q(\vartheta)x, x \rangle = \int_0^\vartheta \|B^* \Phi^*(s) x\|^2 \, ds \geq 0, \quad x \in \mathfrak{X}$$

(see [50], Part IV, Ch. 2, Sect. 2.2, (2.10)). By replacing s by $\vartheta - t$ in (22), we obtain that:

$$Q(\vartheta) = \int_0^\vartheta \Phi(\vartheta - t) BB^* \Phi^*(\vartheta - t) \, dt. \tag{23}$$

Lemma 6. $\|Q(\vartheta)\| \leq \vartheta e^{2a\vartheta} b^2$.

Proof. It follows from (D) that:

$$e^{-a\vartheta} \leq \|\Phi^{\pm 1}(t)\| \leq e^{a\vartheta}, \quad t \in [0, \vartheta]. \tag{24}$$

Moreover, since $\Phi(t) \in \mathcal{L}(\mathfrak{X}, \mathfrak{X})$, we have $\|\Phi^*(t)\| = \|\Phi(t)\|$ (see [49], Lemma A.3.41). Thus,

$$e^{-a\vartheta} \leq \left\|\left(\Phi^*(t)\right)^{\pm 1}\right\| \leq e^{a\vartheta}, \quad t \in [0, \vartheta]. \tag{25}$$

Similarly, $\|B^*\| = \|B\| = b$. By using (22), (24), and (25), we obtain:

$$\|Q(\vartheta)\| \leq \int_0^\vartheta \|\Phi(s) BB^* \Phi^*(s)\| \, ds \leq \int_0^\vartheta \|\Phi(s)\| \cdot \|B\| \cdot \|B^*\| \cdot \|\Phi^*(s)\| \, ds \leq \vartheta e^{2a\vartheta} b^2.$$

□

For $\vartheta > 0$, let us consider the operator $Q_0(\vartheta) : \mathfrak{X} \to \mathfrak{X}$ given by:

$$Q_0(\vartheta)x = \int_0^\vartheta \Phi(-t) BB^* \Phi^*(-t) x \, dt. \tag{26}$$

We have $Q_0(\vartheta) \in \mathcal{L}(\mathfrak{X}, \mathfrak{X})$, $Q_0^*(\vartheta) = Q_0(\vartheta)$, and $Q_0(\vartheta) \geq 0$. By (23), we have $Q(\vartheta) = \Phi(\vartheta) Q_0(\vartheta) \Phi^*(\vartheta)$.

Lemma 7. $\|Q_0(\vartheta)\| \leq \vartheta e^{2a\vartheta} b^2$.

The proof of Lemma 7 is similar to the proof of Lemma 6.

By ([49], Theorem 4.1.7), system (21) is exactly controllable on $[0, \vartheta]$ if and only if for some $\gamma > 0$ and all $x \in \mathfrak{X}$:

$$\langle Q(\vartheta)x, x \rangle \geq \gamma \|x\|^2. \tag{27}$$

Inequality (27) means that $Q(\vartheta) \geq \gamma I$.

Lemma 8. *System* (21) *is exactly controllable on* $[0, \vartheta]$ *if and only if, for some* $\gamma_1 > 0$,

$$Q_0(\vartheta) \geq \gamma_1 I. \tag{28}$$

Proof. By (23),

$$\langle Q(\vartheta)x, x \rangle = \int_0^\vartheta \langle \Phi(\vartheta)\Phi(-t)BB^*\Phi^*(-t)\Phi^*(\vartheta)x, x \rangle \, dt = \int_0^\vartheta \|B^*\Phi^*(-t)\Phi^*(\vartheta)x\|^2 dt$$

$$= \left| \Phi^*(\vartheta)x = y \right| = \int_0^\vartheta \|B^*\Phi^*(-t)y\|^2 dt = \langle Q_0(\vartheta)y, y \rangle. \tag{29}$$

($\Longrightarrow$). Suppose that system (21) is exactly controllable on $[0, \vartheta]$. Hence, for some $\gamma > 0$ and all $x \in \mathfrak{X}$, (27) holds. Set $\gamma_1 := \gamma e^{-2a\vartheta}$. Let $y \in \mathfrak{X}$ be an arbitrary element. Set $x := \left(\Phi^*(\vartheta)\right)^{-1} y$. Then $y = \Phi^*(\vartheta)x$. Hence, $\|y\| \leq \|\Phi^*(\vartheta)\| \cdot \|x\| \leq e^{a\vartheta}\|x\|$. Therefore, $\|x\| \geq e^{-a\vartheta}\|y\|$. By using (29) and (27), we obtain:

$$\langle Q_0(\vartheta)y, y \rangle = \langle Q(\vartheta)x, x \rangle \geq \gamma \|x\|^2 \geq \gamma e^{-2a\vartheta}\|y\|^2 = \gamma_1 \|y\|^2.$$

Hence, (28) holds.

($\Longleftarrow$). Suppose that (28) holds. Set $\gamma := \gamma_1 e^{-2a\vartheta}$. Let $x \in \mathfrak{X}$ be an arbitrary element. Set $y := \Phi^*(\vartheta)x$. Then $x = \left(\Phi^*(\vartheta)\right)^{-1} y$. Hence, $\|x\| \leq \left\|\left(\Phi^*(\vartheta)\right)^{-1}\right\| \cdot \|y\| \leq e^{a\vartheta}\|y\|$. Therefore, $\|y\| \geq e^{-a\vartheta}\|x\|$. By using (29) and (28), we obtain:

$$\langle Q(\vartheta)x, x \rangle = \langle Q_0(\vartheta)y, y \rangle \geq \gamma_1 \|y\|^2 \geq \gamma_1 e^{-2a\vartheta}\|x\|^2 = \gamma \|x\|^2.$$

Hence, (27) holds. Thus, system (21) is exactly controllable on $[0, \vartheta]$. □

Consider the operator control system:

$$\dot{Y}(t) = AY(t) + BU_1(t), \tag{30}$$

where $Y(t) : \mathfrak{X} \to \mathfrak{X}$, $U_1(t) : \mathfrak{X} \to \mathfrak{U}$, $t \in \mathbb{R}$.

Lemma 9. *Let system* (21) *be exactly controllable on* $[0, \vartheta]$ *for some* $\vartheta > 0$. *Then there exists* $\sigma(= 2\vartheta) > 0$ *such that for an arbitrary* $\lambda \in \mathbb{R}$ *there exists a continuous operator control function* $[0, \sigma] \ni t \mapsto U_1(t) \in \mathcal{L}(\mathfrak{X}, \mathfrak{U})$ *such that* $\|U_1(t)\| \leq \alpha_1$ *for some* $\alpha_1 \geq 0$ *for all* $t \in [0, \sigma]$, *steering the solution of* (30) *from:*

$$Y(0) = I \tag{31}$$

to

$$Y(\sigma) = e^{\lambda\sigma}\Phi(\sigma) \tag{32}$$

so that the operator solution $Y(t)$ *of* (30) *is a linear bounded operator function with a bounded inverse:*

$$\|Y(t)\| \leq \beta_1, \quad \|Y^{-1}(t)\| \leq \beta_2, \quad t \in [0, \sigma]. \tag{33}$$

Proof. Let system (21) be exactly controllable on $[0, \vartheta]$, $\vartheta > 0$. Set $\sigma := 2\vartheta$. Suppose that $\lambda \in \mathbb{R}$ is given. A solution of (30) with the initial condition (31) has the form:

$$Y(t) = \Phi(t) \cdot \left(I + \int_0^t \Phi(-s)BU_1(s) \, ds \right). \tag{34}$$

Condition (32) holds if and only if:

$$I + \int_0^\sigma \Phi(-s) B U_1(s)\, ds = e^{\lambda \sigma} I. \tag{35}$$

We will search for $U_1(t)$ in the form:

$$U_1(t) = B^* \Phi^*(-t) H, \tag{36}$$

where $H \in \mathcal{L}(\mathfrak{X}, \mathfrak{X})$. Then, it follows from (35) that:

$$I + Q_0(\sigma) H = e^{\lambda \sigma} I. \tag{37}$$

By definition (26) of $Q_0(\cdot)$, we have $Q_0(\sigma) \geq Q_0(\vartheta)$. By Lemma 8, $Q_0(\vartheta) \geq \gamma_1 I$ for some $\gamma_1 > 0$. Hence, $Q_0^{-1}(\sigma) \in \mathcal{L}(\mathfrak{X}, \mathfrak{X})$ and $\|Q_0^{-1}(\sigma)\| \leq \delta_1$ for some $\delta_1 > 0$. Finding H from (37), we obtain:

$$H = (e^{\lambda \sigma} - 1) Q_0^{-1}(\sigma). \tag{38}$$

Substituting (38) in (36), we obtain:

$$U_1(t) = B^* \Phi^*(-t) Q_0^{-1}(\sigma)(e^{\lambda \sigma} - 1), \quad t \in [0, \sigma]. \tag{39}$$

We have,

$$\|U_1(t)\| \leq \|B^*\| \cdot \|\Phi^*(-t)\| \cdot \|Q_0^{-1}(\sigma)\| \cdot |e^{\lambda \sigma} - 1| \leq b e^{a\sigma} \delta_1 |e^{\lambda \sigma} - 1| =: \alpha_1, \quad t \in [0, \sigma].$$

Substituting (39) in (34), we obtain:

$$Y(t) = \Phi(t) R(t) \tag{40}$$

where

$$R(t) = I + \int_0^t \Phi(-s) B B^* \Phi^*(-s)\, ds\, Q_0^{-1}(\sigma)(e^{\lambda \sigma} - 1). \tag{41}$$

We have, for all $t \in [0, \sigma]$,

$$\|R(t)\| \leq \|I\| + \int_0^t \|\Phi(-s) B B^* \Phi^*(-s)\|\, ds \cdot \|Q_0^{-1}(\sigma)\| \cdot |e^{\lambda \sigma} - 1| \leq 1 + \sigma e^{2a\sigma} b^2 \delta_1 |e^{\lambda \sigma} - 1|,$$

hence,

$$\|Y(t)\| \leq \|\Phi(t)\| \cdot \|R(t)\| \leq e^{a\sigma}(1 + \sigma e^{2a\sigma} b^2 \delta_1 |e^{\lambda \sigma} - 1|) =: \beta_1, \quad t \in [0, \sigma].$$

Thus, the first inequality in (33) holds.

Let us show that $R(t)$ has a bounded inverse for all $t \in [0, \sigma]$. Consider the operator

$$P(t) := R(t) Q_0(\sigma) = Q_0(\sigma) + (e^{\lambda \sigma} - 1) \int_0^t \Phi(-s) B B^* \Phi^*(-s)\, ds.$$

We have $P^*(t) = P(t), t \in [0, \sigma]$, and

$$P(t) = \int_0^\sigma \Phi(-s) B B^* \Phi^*(-s)\, ds - \int_0^t \Phi(-s) B B^* \Phi^*(-s)\, ds + e^{\lambda \sigma} \int_0^t \Phi(-s) B B^* \Phi^*(-s)\, ds$$

$$= \int_t^\sigma \Phi(-s) B B^* \Phi^*(-s)\, ds + e^{\lambda \sigma} \int_0^t \Phi(-s) B B^* \Phi^*(-s)\, ds =: P_1(t) + P_2(t).$$

We see that $P_i^*(t) = P_i(t)$ and $P_i(t) \geq 0$, $i = 1, 2$, $t \in [0, \sigma]$.

Let $t \in [0, \vartheta]$. Then,

$$P_1(t) = \int_t^\sigma \Phi(-s) BB^* \Phi^*(-s) \, ds$$

$$= \int_t^\vartheta \Phi(-s) BB^* \Phi^*(-s) \, ds + \int_\vartheta^{2\vartheta} \Phi(-s) BB^* \Phi^*(-s) \, ds =: P_3(t) + P_4(\vartheta).$$

We have $P_3^*(t) = P_3(t)$, $P_4^*(\vartheta) = P_4(\vartheta)$,

$$P_3(t) = \int_t^\vartheta \Phi(-s) BB^* \Phi^*(-s) \, ds \geq 0.$$

Next,

$$P_4(\vartheta) = \int_\vartheta^{2\vartheta} \Phi(-s) BB^* \Phi^*(-s) \, ds = \Phi^{-1}(\vartheta) \int_\vartheta^{2\vartheta} \Phi(\vartheta - s) BB^* \Phi^*(\vartheta - s) \, ds \left(\Phi^*(\vartheta)\right)^{-1}$$

$$= \Phi^{-1}(\vartheta) \int_0^\vartheta \Phi(-t) BB^* \Phi^*(-t) \, dt \left(\Phi^*(\vartheta)\right)^{-1} = \Phi^{-1}(\vartheta) Q_0(\vartheta) \left(\Phi^*(\vartheta)\right)^{-1}.$$

Since system (21) is exactly controllable on $[0, \vartheta]$, we have, by Lemma 8, $Q_0(\vartheta) \geq \gamma_1 I$. Therefore, $P_4(\vartheta) \geq \gamma_2 I$ for some $\gamma_2 \geq 0$ (namely, for $\gamma_2 := \gamma_1 e^{-2a\vartheta}$; the proof is similar to the proof of Lemma 8). So, we have $P_1(t) = P_3(t) + P_4(\vartheta) \geq \gamma_2 I$ for $t \in [0, \vartheta]$. Thus, $P(t) = P_1(t) + P_2(t) \geq \gamma_2 I$ for $t \in [0, \vartheta]$. Let $t \in [\vartheta, 2\vartheta]$. We have,

$$P_2(t) = e^{\lambda \sigma} \int_0^t \Phi(-s) BB^* \Phi^*(-s) \, ds \geq e^{\lambda \sigma} \int_0^\vartheta \Phi(-s) BB^* \Phi^*(-s) \, ds = e^{\lambda \sigma} Q_0(\vartheta) \geq e^{\lambda \sigma} \gamma_1 I.$$

So, $P(t) = P_1(t) + P_2(t) \geq e^{\lambda \sigma} \gamma_1 I$ for $t \in [\vartheta, 2\vartheta]$.

Thus, for all $t \in [0, \sigma]$, we have $P(t) \geq \gamma_3 I > 0$ where $\gamma_3 := \min\{\gamma_2, \gamma_1 e^{\lambda \sigma}\}$. Hence, there exists an inverse $P^{-1}(t) \in \mathcal{L}(\mathfrak{X}, \mathfrak{X})$ and $\|P^{-1}(t)\| \leq \delta_2$ for some $\delta_2 > 0$ for all $t \in [0, \sigma]$. Then $R(t) = P(t) Q_0^{-1}(\sigma)$ has a bounded inverse:

$$R^{-1}(t) = Q_0(\sigma) P^{-1}(t)$$

and, by using the estimation in Lemma 7, we obtain:

$$\|R^{-1}(t)\| \leq \|Q_0(\sigma)\| \cdot \|P^{-1}(t)\| \leq \sigma e^{2a\sigma} b^2 \delta_2 =: \delta_3.$$

By (40), we obtain:

$$Y^{-1}(t) = R^{-1}(t) \Phi^{-1}(t), \quad t \in [0, \sigma],$$

and

$$\|Y^{-1}(t)\| \leq \|R^{-1}(t)\| \cdot \|\Phi^{-1}(t)\| \leq \delta_3 e^{a\sigma} =: \beta_2, \quad t \in [0, \sigma].$$

□

Theorem 2. *Let system (21) be exactly controllable on $[0, \vartheta]$ for some $\vartheta > 0$. Then system (21) admits a λ-transformation.*

Proof. Let system (21) be exactly controllable on $[0, \vartheta]$, $\vartheta > 0$. Set: $\sigma := 2\vartheta$. Suppose that $\lambda \in \mathbb{R}$ is given. Let us construct the control function $U_1(t)$, $t \in [0, \sigma]$, in accordance with Lemma 9. Set

$$U_2(t) := U_1(t) Y^{-1}(t), \quad t \in [0, \sigma]. \tag{42}$$

Then $U_2(t) \in \mathcal{L}(\mathfrak{X}, \mathfrak{U})$, $t \in [0, \sigma]$, and

$$\|U_2(t)\| \leq \|U_1(t)\| \cdot \|Y^{-1}(t)\| \leq \alpha_1 \beta_2, \quad t \in [0, \sigma].$$

We have
$$U_1(t) = U_2(t)Y(t), \quad t \in [0, \sigma]. \tag{43}$$

Let us substitute (43) in (30). Then we obtain that the function $Y(t)$, $t \in [0, \sigma]$, defined by (40) is a solution of the system:
$$\dot{Y}(t) = (A + BU_2(t))Y(t) \tag{44}$$

satisfying the initial condition (31). Hence,
$$Y(t) = \Phi_{U_2}(t, 0), \quad t \in [0, \sigma], \tag{45}$$

where $\Phi_{U_2}(t, s)$, $t, s \in [0, \sigma]$, is the evolution operator of system (44). It follows from (45) and (32) that:
$$\Phi_{U_2}(\sigma, 0) = e^{\lambda \sigma} \Phi(\sigma). \tag{46}$$

Let us extend the function $U_2(t)$, $t \in [0, \sigma]$, onto $\mathbb{R}$ periodically with the period σ, i.e., construct the function:
$$U(t) \equiv U_2(t - k\sigma), \quad t \in [k\sigma, (k+1)\sigma), \quad k \in \mathbb{Z}. \tag{47}$$

Consider the system:
$$\dot{Y}(t) = (A + BU(t))Y(t), \quad t \in \mathbb{R}. \tag{48}$$

System (48) is σ-periodic. Therefore, the evolution operator $\Phi_U(t, s)$ of system (48) satisfies the following condition:
$$\Phi_U(t + k\sigma, s + k\sigma) = \Phi_U(t, s)$$

for all $t, s \in \mathbb{R}$ and $k \in \mathbb{Z}$. In particular,
$$\Phi_U((k+1)\sigma, k\sigma) = \Phi_U(\sigma, 0) = \Phi_{U_2}(\sigma, 0). \tag{49}$$

From (49), (46), and the equality $\Phi(\sigma) = \Phi((k+1)\sigma, k\sigma)$, it follows that:
$$\Phi_U((k+1)\sigma, k\sigma) = e^{\lambda \sigma} \Phi((k+1)\sigma, k\sigma). \tag{50}$$

Thus, the required equality is proved. The function $U(\cdot)$ is piecewise continuous and:
$$\sup_{t \in \mathbb{R}} \|U(t)\| = \sup_{t \in [0, \sigma)} \|U_2(t)\| \leq \alpha_1 \beta_2.$$

The theorem is proved. □

Corollary 1. *Let system (21) be exactly controllable on $[0, \vartheta]$ for some $\vartheta > 0$. Then, for any $\lambda \in \mathbb{R}$, there exists an admissible gain operator function $U(\cdot)$ such that the closed-loop system:*
$$\dot{x}(t) = (A + BU(t))x(t) \tag{51}$$

and system (7) are kinematically similar on $\mathbb{R}$.

Corollary 1 follows from Theorems 1 and 2.

Theorem 3. *Let system (21) be exactly controllable on $[0, \vartheta]$ for some $\vartheta > 0$. Then the upper Bohl exponent of system (21) is arbitrarily assignable by the linear state feedback (18).*

Proof. Let $\mu \in \mathbb{R}$ be given. Set,
$$\lambda := \mu - \varkappa(A). \tag{52}$$

For this λ, let us construct, by Corollary 1, an admissible gain operator function $U(\cdot)$ such that system (51) and system (7) are kinematically similar on $\mathbb{R}$. By Lemma 4, we have:

$$\varkappa(A + BU) = \varkappa(A + \lambda I). \tag{53}$$

By Lemma 2, we have:

$$\varkappa(A + \lambda I) = \varkappa(A) + \lambda. \tag{54}$$

From (53), (54), and (52), it follows that $\varkappa(A + BU) = \mu$ as required. □

4. Example

Let $\mathfrak{X} = \mathfrak{U} = \ell_2$ where ℓ_2 is the is the space of all sequences $x = (x_1, x_2, \ldots, x_n, \ldots)$ with the norm $\|x\| = \left(\sum\limits_{i=1}^{\infty} |x_i|^2 \right)^{1/2}$. The space ℓ_2 is a separable Hilbert space ([51], § 56). Consider a linear control system:

$$\dot{x}(t) = Ax(t) + Bu(t), \quad t \in \mathbb{R}, \quad x \in \mathfrak{X}, \quad u \in \mathfrak{U}, \tag{55}$$

where

$$A \colon (x_1, x_2, x_3, x_4, \ldots) \mapsto (-x_2, x_1, -x_4, x_3, \ldots), \tag{56}$$

$$B \colon (x_1, x_2, x_3, x_4, \ldots) \mapsto (x_1, 0, x_2, 0, x_3, 0, \ldots). \tag{57}$$

Considering elements of ℓ_2 as column-vectors with an infinite number of coordinates, one can identify the operators A and B with the following matrices with an infinite number of rows and columns:

$$A = \begin{bmatrix} 0 & -1 & 0 & 0 & \ldots \\ 1 & 0 & 0 & 0 & \ldots \\ 0 & 0 & 0 & -1 & \ldots \\ 0 & 0 & 1 & 0 & \ldots \\ \vdots & \vdots & \vdots & \vdots & \ddots \end{bmatrix}, \quad B = \begin{bmatrix} 1 & 0 & 0 & \ldots \\ 0 & 0 & 0 & \ldots \\ 0 & 1 & 0 & \ldots \\ 0 & 0 & 0 & \ldots \\ \vdots & \vdots & \vdots & \ddots \end{bmatrix}. \tag{58}$$

Consider the matrices $F = \begin{bmatrix} 0 & -1 \\ 1 & 0 \end{bmatrix}$, $G = \begin{bmatrix} 1 \\ 0 \end{bmatrix}$. One can write the matrices (58) in the following block-diagonal form:

$$A = \mathrm{diag}\,\{F, F, \ldots, F, \ldots\}, \quad B = \mathrm{diag}\,\{G, G, \ldots, G, \ldots\}. \tag{59}$$

We will use the following denotations for the matrices of the form (59):

$$A = \mathrm{diag}\,[F]_\infty, \quad B = \mathrm{diag}\,[G]_\infty.$$

Set $G_1 := FG$. Then $G_1 = \begin{bmatrix} 0 \\ 1 \end{bmatrix}$. Hence, $AB = \mathrm{diag}\,[G_1]_\infty$. Therefore, $\mathrm{span}\,\{B\mathfrak{U}, AB\mathfrak{U}\} = \mathfrak{X}$. It follows that system (55), (56), (57) is exactly controllable on $[0, \vartheta]$ for any $\vartheta > 0$. Let us take $\vartheta = \pi$. Let us show that the upper Bohl exponent of system (55), (56), (57) is arbitrarily assignable by linear state feedback (18).

Consider the system:

$$\dot{y}(t) = Fy(t), \quad t \in \mathbb{R}, \quad y \in \mathbb{R}^2. \tag{60}$$

The evolution operator $\Gamma(t,s)$ of system (60) has the form $\Gamma(t,s) = \Gamma(t-s)$ where $\Gamma(t) = \exp(Ft)$. Calculating the matrix exponent, we obtain that:

$$\Gamma(t) = \begin{bmatrix} \cos t & -\sin t \\ \sin t & \cos t \end{bmatrix}.$$

Let us construct the evolution operator $\Phi(t,s)$ of the free system:

$$\dot{x}(t) = Ax(t), \quad t \in \mathbb{R}.$$

We obtain $\Phi(t,s) = \Phi(t-s)$ where $\Phi(t) = \text{diag}\,[\Gamma(t)]_\infty$. Hence, $\Phi(\tau+s,\tau) = \text{diag}\,[\Gamma(s)]_\infty$. For any $y = \text{col}\,(y_1, y_2) \in \mathbb{R}^2$ we have:

$$\|\Gamma(s)y\|^2 = (y_1 \cos s - y_2 \sin s)^2 + (y_1 \sin s + y_2 \cos s)^2 = y_1^2 + y_2^2 = \|y\|^2.$$

From this, it follows that $\|\Phi(\tau+s,\tau)x\|^2 = \|x\|^2$ for all $\tau, s \in \mathbb{R}$ and $x \in \mathfrak{X}$. Hence, $\|\Phi(\tau+s,\tau)\| = 1$. So, $\ln \|\Phi(\tau+s,\tau)\| = 0$. Thus, $\varkappa(A) = 0$.

Let an arbitrary $\mu \in \mathbb{R}$ be given. Set $\lambda := \mu - \varkappa(A) = \mu$. Let us construct $U(\cdot)$, by Theorem 2, that ensures equality (50), and, hence, by Corollary 1, kinematic similarity of systems (51) and (7), and, thus, the equality $\varkappa(A+BU) = \lambda = \mu$. Set $\sigma := 2\vartheta = 2\pi$. We have:

$$\Phi(\sigma) = \Phi(2\pi) = \text{diag}\,[\Gamma(2\pi)]_\infty = \text{diag}\,[I_2]_\infty$$

where $I_2 = \begin{bmatrix} 1 & 0 \\ 0 & 1 \end{bmatrix}$, i.e., $\Phi(\sigma) = I \in \mathcal{L}(\mathfrak{X},\mathfrak{X})$. Next,

$$\Gamma(-s)GG^*\Gamma^*(-s) = \begin{bmatrix} \cos^2 s & -\cos s \sin s \\ -\cos s \sin s & \sin^2 s \end{bmatrix}.$$

Set $\Sigma(t) := \int_0^t \Gamma(-s)GG^*\Gamma^*(-s)\,ds$. Then,

$$\Sigma(t) = \begin{bmatrix} \dfrac{t}{2} + \dfrac{\sin 2t}{4} & \dfrac{\cos 2t - 1}{4} \\ \dfrac{\cos 2t - 1}{4} & \dfrac{t}{2} - \dfrac{\sin 2t}{4} \end{bmatrix},$$

and, hence, $\Sigma(\sigma) = \Sigma(2\pi) = \pi I_2$. We have:

$$Q_0(\sigma) = \int_0^\sigma \Phi(-s)BB^*\Phi^*(-s)\,ds = \int_0^\sigma \text{diag}\,[\Gamma(-s)]_\infty \text{diag}\,[GG^*]_\infty \text{diag}\,[\Gamma^*(s)]_\infty\,ds$$
$$= \int_0^\sigma \text{diag}\,[\Gamma(-s)GG^*\Gamma^*(s)]_\infty\,ds = \text{diag}\,[\Sigma(\sigma)]_\infty = \pi I \in \mathcal{L}(\mathfrak{X},\mathfrak{X}).$$

So, $Q_0^{-1}(\sigma) = \dfrac{1}{\pi}I \in \mathcal{L}(\mathfrak{X},\mathfrak{X})$, hence, by (38), $H = \dfrac{e^{2\pi\lambda}-1}{\pi}I \in \mathcal{L}(\mathfrak{X},\mathfrak{X})$. Denote,

$$\alpha := \dfrac{e^{2\pi\lambda}-1}{2\pi}. \tag{61}$$

By using (39), we obtain,

$$U_1(t) = \text{diag}\,[V_1(t)]_\infty, \quad \text{where} \quad V_1(t) = [2\alpha \cos t, -2\alpha \sin t]. \tag{62}$$

Next, by (41), we have $R(t) = \operatorname{diag}[K(t)]_\infty$, where $K(t) = I_2 + 2\alpha\Sigma(t)$. Set $S(t) := \Gamma(t)K(t)$. Multiplying $\Gamma(t)$ by $K(t)$, we obtain that:

$$S(t) = \begin{bmatrix} (1+\alpha t)\cos t + \alpha \sin t & -(1+\alpha t)\sin t \\ (1+\alpha t)\sin t & (1+\alpha t)\cos t - \alpha \sin t \end{bmatrix}. \tag{63}$$

By (40), we have,

$$Y(t) = \Phi(t)R(t) = \operatorname{diag}[\Gamma(t)]_\infty \operatorname{diag}[K(t)]_\infty = \operatorname{diag}[\Gamma(t)K(t)]_\infty = \operatorname{diag}[S(t)]_\infty. \tag{64}$$

Finding $\Delta(t) := \det S(t)$ from (63), we obtain that:

$$\Delta(t) = (1+\alpha t)^2 - \alpha^2 \sin^2 t.$$

It is easy to check that, for all $t \in [0,\sigma]$: $\Delta'(t) > 0$, if $\alpha > 0$; $\Delta'(t) < 0$, if $\alpha < 0$; and $\Delta'(t) = 0$, if $\alpha = 0$. Hence, for all $t \in [0,\sigma]$: if $\alpha > 0$, then $\Delta(t) \geq \Delta(0) = 1$; if $\alpha < 0$, then $\Delta(t) \geq \Delta(2\pi) = e^{4\pi\lambda} > 0$; if $\alpha = 0$, then $\Delta(t) = 1$. Thus, $\Delta(t)$ is separated from zero.

From (63), we obtain that

$$S^{-1}(t) = \frac{1}{\Delta(t)} \begin{bmatrix} (1+\alpha t)\cos t - \alpha \sin t & (1+\alpha t)\sin t \\ -(1+\alpha t)\sin t & (1+\alpha t)\cos t + \alpha \sin t \end{bmatrix}. \tag{65}$$

By (64), we have:

$$Y^{-1}(t) = \operatorname{diag}[S^{-1}(t)]_\infty. \tag{66}$$

Constructing $U_2(t)$ according to (42), by using (62), (66), and (65), we obtain:

$$U_2(t) = \operatorname{diag}[V_2(t)]_\infty, \quad V_2(t) = \frac{1}{\Delta(t)}\left[2\alpha(1+\alpha t) - 2\alpha^2 \sin t \cos t, -2\alpha^2 \sin^2 t\right]. \tag{67}$$

From (67) we obtain:

$$F + GV_2(t) = \begin{bmatrix} \dfrac{2\alpha(1+\alpha t) - 2\alpha^2 \sin t \cos t}{\Delta(t)} & \dfrac{-(1+\alpha t)^2 - \alpha^2 \sin^2 t}{\Delta(t)} \\ 1 & 0 \end{bmatrix}.$$

One can check that the matrix (63) satisfies the following matrix differential equation:

$$\dot{S}(t) = (F + GV_2(t))S(t), \quad t \in [0,\sigma]. \tag{68}$$

Next, by (67), we have:

$$A + BU_2(t) = \operatorname{diag}[F + GV_2(t)]_\infty. \tag{69}$$

Due to (68) and (69), the function (64) satisfies the system:

$$\dot{Y}(t) = (A + BU_2(t))Y(t), \quad t \in [0,\sigma],$$

and $Y(0) = I$. Hence, (45) and (46) holds. Constructing $U(t)$ according to (47), we obtain:

$$U(t) = \operatorname{diag}[V(t)]_\infty, \quad V(t) = \frac{1}{\Delta(t-2\pi k)}\left[2\alpha(1+\alpha(t-2\pi k)) - 2\alpha^2 \sin t \cos t, -2\alpha^2 \sin^2 t\right], \tag{70}$$
$$t \in [2\pi k, 2\pi(k+1)), \quad k \in \mathbb{Z}.$$

By Theorem 2, the gain operator function (70) with α defined by (61) ensures equality (50), kinematic similarity of systems:

$$\dot{x}(t) = (A + BU(t))x(t), \quad t \in \mathbb{R}, \tag{71}$$

and

$$\dot{x}(t) = (A + \lambda I)x(t), \quad t \in \mathbb{R}, \tag{72}$$

on $\mathbb{R}$, and the equality $\varkappa(A + BU) = \lambda = \mu$.

For numerical simulation, let us construct the projection of systems (71) and (72) into the space $\mathbb{R}^2 = \{(x_1, x_2), x_1, x_2 \in \mathbb{R}\}$. We obtain the systems

$$\dot{y}(t) = (F + GV(t))y(t), \quad t \in \mathbb{R}, \quad y \in \mathbb{R}^2, \tag{73}$$

and

$$\dot{y}(t) = (F + \lambda E)y(t), \quad t \in \mathbb{R}, \quad y \in \mathbb{R}^2. \tag{74}$$

Here E is the identity (2×2)-matrix. Systems (73) and (74) are kinematically similar, hence, since $\varkappa(F) = 0$, we have $\varkappa(F + GV) = \varkappa(F + \lambda E) = \lambda$. Let us take, for example, $\lambda = -1/4$. The equality $\varkappa(F + \lambda E) = -1/4$ means that system (74) (and (73)) is uniformly exponentially stable with the decay rate $1/4$.

Let $\Xi(t,s)$ denote the evolution matrix of system (73) and $\Omega(t,s)$ denote the evolution matrix of system (74). Let,

$$\Xi(t,0) =: \begin{bmatrix} \xi_{11}(t) & \xi_{12}(t) \\ \xi_{21}(t) & \xi_{22}(t) \end{bmatrix}, \quad \Omega(t,0) =: \begin{bmatrix} \omega_{11}(t) & \omega_{12}(t) \\ \omega_{21}(t) & \omega_{22}(t) \end{bmatrix}.$$

We have $\Xi(t,0)\big|_{t=0} = \Omega(t,0)\big|_{t=0} = E$. Let us construct the graphs of the functions $\xi_{ij}(t)$, $\omega_{ij}(t)$ (see, Figures 1–4).

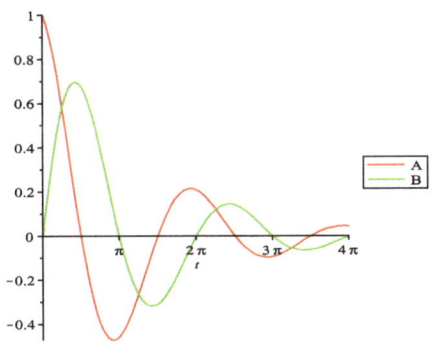

Figure 1. Graphs of the functions $A = \omega_{11}(t)$, $B = \omega_{21}(t)$.

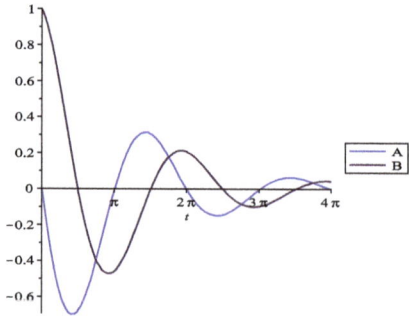

Figure 2. Graphs of the functions $A = \omega_{12}(t)$, $B = \omega_{22}(t)$.

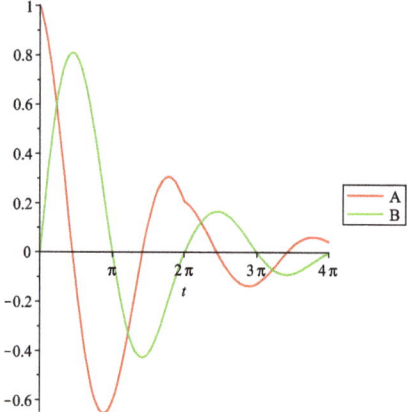

Figure 3. Graphs of the functions $A = \xi_{11}(t)$, $B = \xi_{21}(t)$.

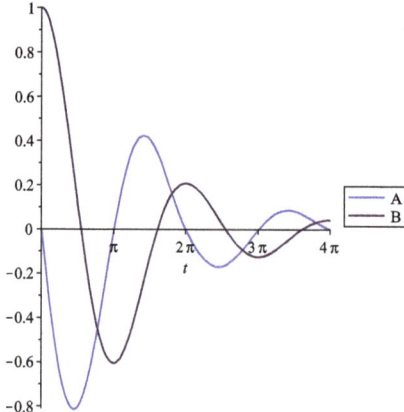

Figure 4. Graphs of the functions $A = \xi_{12}(t)$, $B = \xi_{22}(t)$.

One can see from system (73) and (74) (and from the graphs) that:

$$w_{11}(2\pi) = \xi_{11}(2\pi) = e^{-\pi/2} \approx 0.2079, \quad w_{12}(2\pi) = \xi_{12}(2\pi) = 0,$$
$$w_{21}(2\pi) = \xi_{21}(2\pi) = 0, \quad w_{22}(2\pi) = \xi_{11}(2\pi) = e^{-\pi/2} \approx 0.2079. \quad (75)$$

One can see also that the functions w_{ij} are smooth. Since the matrix of system (73) is piecewise continuous, the matrix function $\Xi(t,0)$ is piecewise smooth and its derivative can be discontinuous at the points $t = 2\pi k$. Calculating the one-sided limits from (73), we obtain that:

$$\dot{\xi}_{11}(t)\Big|_{t=2\pi-0} = (e^{-\pi/2} - 1)/\pi \approx -0.2521, \quad \dot{\xi}_{11}(t)\Big|_{t=2\pi+0} = (e^{-\pi/2} - 1)e^{-\pi/2}/\pi \approx -0.0524,$$

$$\dot{\xi}_{12}(t)\Big|_{t=2\pi-0} = \dot{\xi}_{12}(t)\Big|_{t=2\pi+0} = -e^{-\pi/2}, \quad \dot{\xi}_{21}(t)\Big|_{t=2\pi-0} = \dot{\xi}_{21}(t)\Big|_{t=2\pi+0} = e^{-\pi/2},$$

$$\dot{\xi}_{22}(t)\Big|_{t=2\pi-0} = \dot{\xi}_{22}(t)\Big|_{t=2\pi+0} = 0,$$

i.e., only the function $\xi_{11}(t)$ has the discontinuous derivative at the point $t = 2\pi$. This is confirmed by the graphs.

It follows from (75) that:

$$\Xi(2\pi,0) = \Omega(2\pi,0) = \begin{bmatrix} e^{-\pi/2} & 0 \\ 0 & e^{-\pi/2} \end{bmatrix} \approx \begin{bmatrix} 0.2079 & 0 \\ 0 & 0.2079 \end{bmatrix}.$$

By periodicity, we have:

$$\Xi(4\pi,0) = \Omega(4\pi,0) = \begin{bmatrix} e^{-\pi} & 0 \\ 0 & e^{-\pi} \end{bmatrix} \approx \begin{bmatrix} 0.0432 & 0 \\ 0 & 0.0432 \end{bmatrix},$$

and so on, $\Xi(2\pi k, 0) = \Omega(2\pi k, 0) = e^{-\pi k/2} E$, $k \in \mathbb{Z}$. The graphs confirm asymptotic equivalence of the behavior of solutions of systems (73) and (74).

Remark 2. *The advantage of the developed method is that it allows us to establish the exact asymptotics (i.e., exact equality $\varkappa(A + BU) = \mu$) for the closed-loop system, in contrast to, e.g., [35], from which one can only obtain the inequality $\Lambda(A + BU) \leq \varkappa$ for the upper Lyapunov exponent Λ of the closed-loop system. The problem of exact assignment of the upper Bohl exponent for a system in infinite-dimensional space in the presented formulation has not been previously investigated. Moreover, the developed method allows us to assign exact values for other asymptotic invariants of the closed-loop system (central exponents, exponential exponents etc.). A disadvantage is that the analytical expressions for the controller (and for solutions of the closed-loop system) can be complicated, in contrast to the stabilization problem [35]. This method can be applied to any system with the property of exact controllability. The choice of matrices in the example in a rather simple form was made for illustrative purposes because in this case the analytical expressions for the controller and for the solutions of the closed-loop system is not very complicated.*

5. Conclusions

For a linear time-invariant control system in a Hilbert space with bounded operator coefficients, we examined the problem of arbitrary assignment of the upper Bohl exponent by means of linear state feedback with a time-varying linear bounded gain operator function. We have proved that the property of exact controllability of the open-loop system is sufficient for arbitrary assignability of the upper Bohl exponent of the closed-loop system. We plan to extend these results to systems without necessarily bounded operator A but generating a C_0-continuous semigroup. We plan to prove similar results for systems with dynamic output feedback. Further development of these results may be their extension to systems with periodic coefficients and with arbitrary time-varying non-periodic coefficients, to systems in general Banach spaces, or to systems with discrete time. We expect to apply the results to specific systems, for example, to systems with delays, considering them as abstract systems of differential equations in an infinite-dimensional space.

Author Contributions: Conceptualization, V.Z.; methodology, V.Z.; formal analysis, V.Z. and M.Z.; investigation, V.Z. and M.Z.; writing–original draft preparation, V.Z. and M.Z.; writing–review and editing, V.Z. and M.Z.; visualization, V.Z. and M.Z.; supervision, V.Z.; project administration, V.Z.; funding acquisition, V.Z. All authors have read and agreed to the published version of the manuscript.

Funding: This research was funded by the Ministry of Science and Higher Education of the Russian Federation in the framework of state assignment No. 075-00232-20-01, Project 0827-2020-0010 "Development of the theory and methods of control and stabilization of dynamical systems".

Conflicts of Interest: The authors declare no conflict of interest.

References

1. Popov, V.M. Hyperstability and optimality of automatic systems with several control functions. *Revue Roumaine des Sciences Techniques–Serie Electrotechnique et Energetique* **1964**, *9*, 629–690.
2. Wonham, W.M. On pole assignment in multi-input controllable linear systems. *IEEE Trans. Autom. Control.* **1967**, *12*, 660–665. [CrossRef]
3. Brunovsky, P. Controllability and linear closed-loop controls in linear periodic systems. *J. Differ. Equ.* **1969**, *6*, 296–313. 10.1016/0022-0396(69)90019-9. [CrossRef]
4. Silverman, L.M. Transformation of time-variable systems to canonical (phase-variable) form. *IEEE Trans. Autom. Control.* **1966**, *11*, 300–303. [CrossRef]
5. Wolovich, W.A. On the stabilization of controllable system. *IEEE Trans. Autom. Control.* **1968**, *13*, 569–572. [CrossRef]
6. Silverman, L.M.; Anderson, B.D.O. Controllability, observability and stability of linear systems. *SIAM J. Control* **1968**, *6*, 121–130. [CrossRef]
7. Ikeda, M.; Maeda, H.; Kodama, S. Stabilization of linear systems. *SIAM J. Control* **1972**, *10*, 716–729. [CrossRef]
8. Kalman, R.E. Contribution to the theory of optimal control. *Boletin de la Sociedad Matematica Mexicana* **1960**, *5*, 102–119.
9. Bylov, B.F.; Vinograd, R.E.; Grobman, D.M.; Nemytskii, V.V. *Theory of Lyapunov Exponents*; Nauka: Moscow, Russia, 1966.
10. Demidovich, B.P. *Lectures on the Mathematical Stability Theory*; Nauka: Moscow, Russia, 1967.
11. Barreira, L. *Lyapunov Exponents*; Birkhäuser: Cham, Switzerland, 2017. [CrossRef]
12. Makarov, E.K.; Popova, S.N. *Controllability of Asymptotic Invariants of Time-Dependent Linear Systems*; Belaruskaya Navuka: Minsk, Belarus, 2012.
13. Makarov, E.K.; Popova, S.N. Global controllability of central exponents of linear systems. *Russ. Math.* **1999**, *43*, 56–63.
14. Makarov, E.K.; Popova, S.N. The global controllability of a complete set of Lyapunov invariants for two-dimensional linear systems. *Differ. Equ.* **1999**, *35*, 97–107.
15. Popova, S.N. Global controllability of the complete set of Lyapunov invariants of periodic systems. *Differ. Equ.* **2003**, *39*, 1713–1723. [CrossRef]
16. Popova, S.N. Global reducibility of linear control systems to systems of scalar type. *Differ. Equ.* **2004**, *40*, 43–49. [CrossRef]
17. Popova, S.N. On the global controllability of Lyapunov exponents of linear systems. *Differ. Equ.* **2007**, *43*, 1072–1078. [CrossRef]
18. Babiarz, A.; Banshchikova, I.; Czornik, A.; Makarov, E.; Niezabitowski, M.; Popova, S. On assignability of Lyapunov spectrum of discrete linear time-varying system with control. In Proceedings of the 2016 21st International Conference on Methods and Models in Automation and Robotics (MMAR), Miedzyzdroje, Poland, 29 August–1 September 2016. [CrossRef]
19. Babiarz, A.; Czornik, A.; Makarov, E.; Niezabitowski, M.; Popova, S. Pole placement theorem for discrete time-varying linear systems. *SIAM J. Control Optim.* **2017**, *55*, 671–692. [CrossRef]

20. Popova S. Assignability of certain Lyapunov invariants for linear discrete-time systems. *IFAC-PapersOnLine* **2018**, *51*, 40–45. [CrossRef]
21. Babiarz, A.; Banshchikova, I.; Czornik, A.; Makarov, E.; Niezabitowski, M.; Popova, S. Necessary and sufficient conditions for assignability of the Lyapunov spectrum of discrete linear time-varying systems. *IEEE Trans. Autom. Control* **2018**, *63*, 3825–3837. [CrossRef]
22. Babiarz, A.; Banshchikova, I.; Czornik, A.; Makarov, E.; Niezabitowski, M.; Popova, S. Proportional local assignability of Lyapunov spectrum of linear discrete time-varying systems. *SIAM J. Control Optim.* **2019**, *57*, 1355–1377. [CrossRef]
23. Babiarz, A.; Banshchikova, I.; Czornik, A.; Makarov, E.; Niezabitowski, M.; Popova, S. Assignability of Lyapunov spectrum for discrete linear time-varying systems. *Springer Proc. Math. Stat.* **2020**, *312*, 133–147. [CrossRef]
24. Barabanov, E.A.; Konyukh, A.V. Bohl exponents of linear differential systems. *Mem. Differ. Equ. Math. Phys.* **2001**, *24*, 151–158.
25. Bykov, V.V. Bohl exponents and Baire classes of functions. *J. Math. Sci.* **2015**, *210*, 168–185. [CrossRef]
26. Bykov, V.V. Baire classes of Lyapunov invariants. *Sb. Math.* **2017**, *208*, 620–643. [CrossRef]
27. Barabanov, E.; Vaidzelevich, A.; Czornik, A.; Niezabitowski, M. Exact boundaries of mobility of the Lyapunov, Perron and Bohl exponents of linear difference systems under small perturbations of the coefficient matrices. *AIP Conf. Proc.* **2016**, *1738*, 130007. [CrossRef]
28. Czornik, A.; Konyukh, A.; Niezabitowski, M. Which functions may be the upper Bohl function of the diagonal discrete linear time-varying systems? *J. Math. Anal. Appl.* **2017**, *452*, 1420–1433. [CrossRef]
29. Babiarz, A.; Czornik, A.; Konyukh, A.; Niezabitowski, M. What types of functions may be the lower Bohl function of a diagonal discrete linear time-varying systems? *J. Frankl. Inst.* **2017**, *354*, 5131–5144. [CrossRef]
30. Babiarz, A.; Czornik, A.; Niezabitowski, M. On the assignability of regularity coefficients and central exponents of discrete linear time-varying systems. *IFAC-PapersOnLine* **2019**, *52*, 64–69. [CrossRef]
31. Babiarz, A.; Czornik, A.; Niezabitowski, M. Relations between Bohl exponents and general exponent of discrete linear time-varying systems. *J. Differ. Equ. Appl.* **2019**, *25*, 560–572. [CrossRef]
32. Czornik, A.; Konyukh, A.; Konyukh, I.; Niezabitowski, M.; Orwat, J. On Lyapunov and upper Bohl exponents of diagonal discrete linear time-varying systems. *IEEE Trans. Autom. Control* **2019**, *64*, 5171–5174. [CrossRef]
33. Pinto, N.; Robledo, G. Some relations between Bohl exponents and the exponential dichotomy spectrum. *J. Differ. Equ. Appl.* **2019**, *25*, 573–582. [CrossRef]
34. Datko, R. A linear control problem in an abstract Hilbert space. *J. Differ. Equ. Appl.* **1971**, *9*, 346–359. [CrossRef]
35. Slemrod, M. The linear stabilization problem in Hilbert space. *J. Funct. Anal.* **1972**, *11*, 334–345. [CrossRef]
36. Russell, D.L. Controllability and stabilizability theory for linear partial differential equations: Recent progress and open questions. *SIAM Rev.* **1978**, *20*, 639–739. [CrossRef]
37. Pritchard, A.J.; Zabczyk, J. Stability and stabilizability of infinite-dimensional systems. *SIAM Rev.* **1981**, *23*, 25–52. [CrossRef]
38. Maciej Przyłuski, K.; Rolewicz, S. On stability of linear time-varying infinite-dimensional discrete-time systems. *Syst. Control Lett.* **1984**, *4*, 307–315. [CrossRef]
39. Logemann, H. Stability and stabilizability of linear infinite-dimensional discrete-time systems. *IMA J. Math. Control Inf.* **1992**, *9*, 255–263. [CrossRef]
40. Luo, Z.-H.; Guo, B.-Z.; Morgu, O. *Stability and Stabilization of Infinite Dimensional Systems with Applications*; Springer: London, UK, 1999. [CrossRef]
41. Raymond, J.-P. Stabilizability of infinite dimensional systems by finite dimensional control. *Comput. Methods Appl. Math.* **2019**, *19*, 267–282. [CrossRef]
42. Barreira, L.; Valls, C. Stability of the Lyapunov exponents under perturbations. *Ann. Funct. Anal.* **2017**, *8*, 398–410. [CrossRef]
43. Barreira, L.; Dragičević, D.; Valls, C. Nonuniform spectrum on Banach spaces. *Adv. Math.* **2017**, *321*, 547–591. [CrossRef]
44. Barreira, L.; Dragičević, D.; Valls, C. Spectrum for compact operators on Banach spaces. *J. Math. Soc. Jpn.* **2019**, *71*, 1–17. [CrossRef]
45. Daleckii, J.L.; Krein, M.G. *Stability of Solutions of Differential Equations in Banach Space*; American Mathematical Society: Providence, RI, USA, 1974.

46. Makarov, E.K. On the discreteness of asymptotic invariants of linear differential systems. *Differ. Equ.* **1998**, *34*, 1323–1331.
47. Triggiani, R. Controllability and observability in Banach space with bounded operators. *SIAM J. Control* **1975**, *13*, 462–491. [CrossRef]
48. Zaitsev, V.A. Lyapunov reducibility and stabilization of nonstationary systems with an observer. *Differ. Equ.* **2010**, *46*, 437–447. [CrossRef]
49. Curtain, R.F.; Zwart, H. *An Introduction to Infinite-Dimensional Linear Systems Theory*; Springer: New York, NY, USA, 1995. [CrossRef]
50. Zabczyk, J. *Mathematical Control Theory*; Birkhäuser: Boston, MA, USA, 2008. [CrossRef]
51. Kolmogorov, A.N.; Fomin, S.V. *Elements of the Theory of Functions and Functional Analysis*; Graylock Press: Rochester, NY, USA, 1957.

© 2020 by the authors. Licensee MDPI, Basel, Switzerland. This article is an open access article distributed under the terms and conditions of the Creative Commons Attribution (CC BY) license (http://creativecommons.org/licenses/by/4.0/).

Article

About Some Possible Implementations of the Fractional Calculus

María Pilar Velasco [1,*,†], David Usero [2,†], Salvador Jiménez [1,†], Luis Vázquez [2,†], José Luis Vázquez-Poletti [2,†] and Mina Mortazavi [3,†]

[1] Department of Applied Mathematics to the Information and Communications Technologies, Universidad Politécnica de Madrid, 28040 Madrid, Spain; s.jimenez@upm.es
[2] Instituto de Matemática Interdisciplinar, Universidad Complutense de Madrid, 28040 Madrid, Spain; umdavid@mat.ucm.es (D.U.); lvazquez@fdi.ucm.es (L.V.); jlvazquez@fdi.ucm.es (J.L.V.-P.)
[3] Department of Applied Mathematics, School of Mathematical Sciences, Ferdowsi University of Mashhad, Mashhad 9177948974, Iran; minamortazavi5@gmail.com
* Correspondence: mp.velasco@upm.es
† These authors contributed equally to this work.

Received: 6 May 2020; Accepted: 26 May 2020; Published: 2 June 2020

Abstract: We present a partial panoramic view of possible contexts and applications of the fractional calculus. In this context, we show some different applications of fractional calculus to different models in ordinary differential equation (ODE) and partial differential equation (PDE) formulations ranging from the basic equations of mechanics to diffusion and Dirac equations.

Keywords: fractional calculus; fractional differential equations; nonlocal effects

1. From Elementary Mathematical Analysis to Fractional Derivatives

A first reason to justify the use of fractional operators is the need to introduce memory terms into differential models in a natural form. In this sense, we can consider the classical Calculus Fundamental Theorem with the introduction of a convolution kernel F associated to a function g leads to the natural form of introducing a memory term by changing the convolution kernel of the integral:

$$F(x) = F(a) + \int_a^x K(g(x) - g(s)) \cdot f(s) ds \qquad (1)$$

Now F is the generalized primitive of the function f and the convolution kernel K is a memory term that could be different in each specific problem, changing the definition of fractional operator consequently. This generalization of the integral can be considered as a base to construct possible definitions for fractional integrals.

From these considerations, the fractional calculus emerges in the mathematical world as the study of integral and derivative operators of non-integer orders on domains of real or complex functions. Several definitions of fractional derivatives D^α have been developed progressively with the objective to generalize the concept of ordinary derivative D, such that for $\alpha = 1$ the ordinary operator can be recovered [1–3].

Fractional calculus is a powerful mathematical tool that allows to create intermediate-order parameters equations and offers modeling scenarios where fundamental mathematical questions converge and appropriate numerical algorithms can de developed. A lot of fractional operators have been defined in the literature; however, not all of them can be used in each real-world application. In this context, we appreciate very much the enthusiastic and clarifier paper of D. Baleanu and A. Fernandez [4] and M.D. Ortigueira and J.A.T. Machado [5–7].

1.1. From Factorial to the Gamma Function

The first definitions of fractional operators are related to the use of the Gamma function is a function that generalizes the definition of the factorial to non-positive numbers. Its definition is:

$$\Gamma(z) = \int_0^\infty s^{z-1} e^{-s} ds \tag{2}$$

for any complex number z with positive real part.

Using integration by parts in Equation (2), a fundamental property of the Gamma function is obtained:

$$\Gamma(z) = (z-1)\Gamma(z-1), \tag{3}$$

which allows to give the Gamma function of a positive integer number as

$$\Gamma(n) = (n-1)!. \tag{4}$$

In this context, the Gamma function is a generalization of the concept of factorial.

In 1738, Euler introduces the first generalization of ordinary derivative, verifying that the fractional derivation made sense for the potential function x^a. And in 1819, Lacroix starts from the m-order derivative of the function x^n, with m and n positive integer numbers

$$\frac{d^m}{dx^m} x^n = \frac{n!}{(n-m)!} x^{n-m}, \tag{5}$$

to determine the $1/2$ order derivative of the function x^a, using the generalization of the factorial function by the Gamma function:

$$\frac{d^{1/2}}{dx^{1/2}} x^a = \frac{\Gamma(a+1)}{\Gamma\left(a+\frac{1}{2}\right)} x^{a-\frac{1}{2}}, \tag{6}$$

such that for $a=1$:

$$\frac{d^{1/2}}{dx^{1/2}} x = \frac{\sqrt{x}}{\Gamma\left(\frac{3}{2}\right)} = \frac{2\sqrt{x}}{\sqrt{\pi}}. \tag{7}$$

This is the result that will be obtained with the called Riemann–Lioville fractional derivative.

This concept can be generalized to any order and the following relation between the ordinary and fractional case with the Riemann–Liouville fractional derivative is obtained:

$$\frac{d^n}{dx^n} x^m = \frac{m!}{(m-n)!} x^{m-n} \Rightarrow \frac{d^\alpha}{dx^\alpha} x^\mu = \frac{\Gamma(\mu+1)}{\Gamma(\mu-\alpha+1)} x^{\mu-\alpha}. \tag{8}$$

Later, some provisional definitions of fractional operators were introduced by Fourier, Abel, Liouville and Riemman, without much success. Until, in 1870, N. Ya. Sonine started from the Cauchy formula for repeated integration:

$$({}_aI_x^n f)(x) = \int_a^x dx_1 \int_a^{x_1} dx_2 \cdots \int_a^{x_{n-1}} f(t)dt = \frac{1}{(n-1)!} \int_a^x (x-t)^{n-1} f(t) dt \tag{9}$$

and, using the generalization of the factorial function by the Gamma function, he obtained the actual definition of the fractional integral of Riemann–Liouville:

$$({}_aI_x^\alpha f)(x) = \frac{1}{\Gamma(\alpha)} \int_a^x (x-t)^{\alpha-1} f(t) dt, \quad \Re(\alpha) > 0 \tag{10}$$

although in 1884 Laurent formulated it definitively.

1.2. Some Definitions of Fractional Integrals and Derivatives

An important definition of fractional integral and derivative corresponds to Riemann–Liouville:

- Left-side Riemann–Liouville Fractional Integral of order $\alpha > 0$:

$$_aI_x^\alpha \phi(x) = \frac{1}{\Gamma(\alpha)} \int_a^x (x-t)^{\alpha-1} \phi(t)dt, \qquad x > a. \tag{11}$$

- Right-side Riemann–Liouville Fractional Integral of order $\alpha < 0$:

$$_xI_b^\alpha \phi(x) = \frac{1}{\Gamma(\alpha)} \int_x^b (x-t)^{\alpha-1} \phi(t)dt, \qquad x < b. \tag{12}$$

- Left-side Riemann–Liouville Fractional Derivative of order $\alpha > 0$:

$$_aD_x^\alpha \phi(x) = \frac{1}{\Gamma(n-\alpha)} \left(\frac{\partial}{\partial x}\right)^n \int_a^x (x-t)^{\alpha-(n-1)} \phi(t)dt = D^n(_aI_x^{n-\alpha}f)(x), \qquad x > a. \tag{13}$$

- Right-side Riemann–Liouville Fractional Derivative of order $\alpha < 0$:

$$_xD_b^\alpha \phi(x) = \frac{1}{\Gamma(n-\alpha)} \left(-\frac{\partial}{\partial x}\right)^n \int_x^b (x-t)^{\alpha-(n-1)} \phi(t)dt = (-D)^n(_xI_b^{n-\alpha}f)(x), \qquad x < b. \tag{14}$$

In all cases $n \in \mathbb{N}$, such that $0 \leq n-1 < \alpha < n$.

These operators recover the classical operators for the parameter $\alpha = 1$ and the algebra of these operators is different to the classical operators:

- Let $f \in L_p(a,b)$ $(1 \leq p \leq \infty)$ and $Re(\alpha), Re(\beta) > 0$. Then:

$$(_aI_x^\alpha {}_aI_x^\beta f)(x) = (_aI_x^{\alpha+\beta} f)(x) \tag{15}$$

in $[a,b]$.

- Let $f \in L_1(a,b)$, $\alpha, \beta > 0$, such that $n-1 < \alpha \leq n$, $m-1 < \beta \leq m$ $(n,m \in \mathbb{N})$ and $\alpha + \beta < n$, $f_{m-\alpha} = {}_aI_x^{m-\alpha}f \in AC^m([a,b])$. Then:

$$(_aD_x^\alpha {}_aD_x^\beta f)(x) = (_aD_x^{\alpha+\beta}f)(x) - \sum_{j=1}^m (_aD_x^{\beta-j}f)(a+) \frac{(x-a)^{-j-\alpha}}{\Gamma(1-j-\alpha)}. \tag{16}$$

- Let $f \in L_1(a,b)$, $\alpha \geq \beta > 0$. Then:

$$(_aD_x^\alpha {}_aI_x^\beta f)(x) = (_aI_x^{\beta-\alpha}f)(x), \quad \alpha \leq \beta \tag{17}$$
$$(_aD_x^\alpha {}_aI_x^\beta f)(x) = (_aD_x^{\alpha-\beta}f)(x), \quad \alpha \geq \beta \tag{18}$$

Related to the integrals of Riemann–Liouville, the definition of Caputo fractional derivative appears:

- Left-side Caputo Fractional Derivative of order $\alpha > 0$:

$$_a^C D_x^\alpha \phi(x) = {}_aD_x^\alpha \left(\phi(x) - \sum_{k=0}^{n-1} \frac{\phi^{(k)}(a)}{k!}(x-a)^k \right)$$

$$= \frac{1}{\Gamma(n-\alpha)} \int_a^x \frac{\phi^{(n)}(t)}{(x-t)^{\alpha+1-n}} dt, \qquad x > a. \tag{19}$$

- Right-side Caputo Fractional Derivative of order $\alpha < 0$:

$$_x^C D_b^\alpha \phi(x) = \frac{(-1)^n}{\Gamma(n-\alpha)} \int_x^b \frac{\phi^{(n)}(t)}{(x-t)^{\alpha+1-n}} dt, \quad x < b. \tag{20}$$

We have, as before, $n \in \mathbb{N}$ such that $0 \leq n-1 < \alpha < n$, and now the $n+1$ derivatives of function ϕ must be continuous and bounded in $[a,b]$.

The following identity established the relation between the Riemann–Liouville and Caputo fractional derivatives, for f a suitable function (for instance, f n-derivable):

$$(_aD_x^\alpha f)(x) = (_a^C D_x^\alpha f)(x) + \sum_{j=0}^{n-1} \frac{f^{(j)}(a)}{\Gamma(1+j-\alpha)}(x-a)^{j-\alpha}. \tag{21}$$

The extension of Riemann–Liouville fractional operators to infinity intervals leads to the Liouville fractional operators:

- Left-side Liouville Fractional Integral of order $\alpha > 0$:

$$_+^L I_x^\alpha \phi(x) = \frac{1}{\Gamma(\alpha)} \int_{-\infty}^x (x-t)^{\alpha-1} \phi(t) dt, \quad x \in \mathbb{R}. \tag{22}$$

- Right-side Liouville Fractional Integral of order $\alpha < 0$:

$$_x^L I_-^\alpha \phi(x) = \frac{1}{\Gamma(\alpha)} \int_x^\infty (x-t)^{\alpha-1} \phi(t) dt, \quad x \in \mathbb{R}. \tag{23}$$

- Left-side Liouville Fractional Derivative of order $\alpha > 0$:

$$_+^L D_x^\alpha \phi(x) = \frac{1}{\Gamma(n-\alpha)} \left(\frac{\partial}{\partial x}\right)^n \int_{-\infty}^x (x-t)^{\alpha-(n-1)} \phi(t) dt = D^n(_+ I_x^{n-\alpha} f)(x), \quad x \in \mathbb{R}. \tag{24}$$

- Right-side Liouville Fractional Derivative of order $\alpha < 0$:

$$_x^L D_-^\alpha \phi(x) = \frac{1}{\Gamma(n-\alpha)} \left(-\frac{\partial}{\partial x}\right)^n \int_x^\infty (x-t)^{\alpha-(n-1)} \phi(t) dt = (-D)^n(_x I_-^{n-\alpha} f)(x), \quad x \in \mathbb{R}. \tag{25}$$

In all cases $n \in \mathbb{N}$, such that $0 \leq n-1 < \alpha < n$.

1.3. Mittag-Leffler Functions

Special functions related to the eigenfunctions of fractional operators are the Mittag-Leffler functions. They appear in the solution of many fractional differential equations. The Mittag-Leffler functions are generalizations of the exponential function and they was introduced by the mathematician G.M. Mittag-Leffler in 1903:

$$\begin{aligned} E_\alpha(t) &= \sum_{k=0}^\infty \frac{t^k}{\Gamma(\alpha k + 1)} \quad (\alpha > 0, \alpha \in \mathbb{R}), \\ E_{\alpha,\beta}(t) &= \sum_{k=0}^\infty \frac{t^k}{\Gamma(\alpha k + \beta)} \quad (\alpha, \beta > 0, \alpha, \beta \in \mathbb{R}). \end{aligned} \tag{26}$$

Here, we have some elementary properties of the Mittag-Leffer function. For some values of the parameters α, β, Mittag-Leffer functions return known classical functions, for example:

$$E_1(t) = e^t, \tag{27}$$

$$E_2(t) = \cosh(\sqrt{t}). \tag{28}$$

The relevance of the Mittag-Lefler functions is their behavior as generalized exponential functions associated to the Riemann–Liouville and Caputo fractional derivatives:

$$_0D_t^\alpha t^{\alpha-1} E_{\alpha,\alpha}(\lambda t^\alpha) = \lambda t^{\alpha-1} E_{\alpha,\alpha}(\lambda t^\alpha), \tag{29}$$

and

$$_0^C D_t^\alpha E_\alpha(\lambda t^\alpha) = \lambda E_\alpha(\lambda t^\alpha). \tag{30}$$

On the other hand, a very useful extension of the Mittag-Leffer function is related to extend the fractional derivatives defined by using a Mittag-Leffler kernel which is non-local and non-singular ([8,9]).

$$^{ABR}D_{a+}^\alpha f(t) = \frac{B(\alpha)}{1-\alpha} \frac{d}{dt} \int_a^t f(x) E_\alpha\left(\frac{-\alpha}{1-\alpha}(t-x)^\alpha\right) dx, \tag{31}$$

$$^{ABC}D_{a+}^\alpha f(t) = \frac{B(\alpha)}{1-\alpha} \int_a^t f'(x) E_\alpha\left(\frac{-\alpha}{1-\alpha}(t-x)^\alpha\right) dx \tag{32}$$

valid for $0 < \alpha < 1$, with $B(\alpha)$ being a normalisation function. This new definition has many applications at the same time that satisfies the extensions of the product rule and chain rule.

Futhermore, these functions have many uses, for instance, they allow to address fractal kinetics from basic functions ([10]) or they can be used as a simplification tool combined to exponent/powerlaw mathematical congruence ([11]).

1.4. Some Ideas for Numerical Integration

It is simple to extend some classical methods from integer to non-integer orders. For instance, the most basic first order explicit method for numerical integration of ordinary differential equations with a given initial value, known as Euler methods.

In classical dynamical models, symplectic integration schemes preserve the flow of the hamiltonian while other classical integrators as Runge–Kutta schemes do not necessarily conserve it. This is immediately translated into the conservation of the first integral of motion and long term stability of the scheme. In fractional dynamics, energy is not conserved. Despite this, using a symplectic scheme in fractional mechanics ensures that the observed instabilities will be certainly due to the fractional operators. Long-term stability is inherited in this fractional mapping as is shown below.

Let us consider the following initial value problem with Caputo fractional derivative:

$$_0^C D_t^\alpha y(t) = y^\alpha, \quad t \in [0, T], \quad 0 < \alpha < 1 \tag{33}$$

$$y(0) = y_0 \tag{34}$$

where y^α is a Lipschitz function with constant L.

By using a truncated Taylor series for the Caputo fractional derivative, with step $t_n = nh$ with $h = \frac{T}{N}$ y $n = 0, 1, 2, \ldots, N$, we obtain:

$$y(t+h) = y(t) + h^\alpha \frac{_0^C D_t^\alpha y(t)}{\Gamma(\alpha)} \tag{35}$$

and then the fractional Euler method for this initial value problem is:

$$y_{n+1} = y_n + \frac{h^\alpha}{\Gamma(\alpha)} f(t_n, y_n) \tag{36}$$

with convergence order h^α.

Other more complex generalizations of classical methods are possible. For instance, fractional Hamiltonian–Jacobi methods. The equations of motions for a one-dimensional Hamiltonian system $H = \frac{1}{2}p^2 + V(x)$ with unit mass, are defined as

$$\dot{x} = \frac{\partial H}{\partial p} = p, \quad \dot{p} = -\frac{\partial H}{\partial x} = -V'(x), \tag{37}$$

that is associated to the second order equation $\ddot{x} + V'(x) = 0$. This system can be generalized by using Caputo time-fractional derivatives

$$\begin{cases} {}_0^C D_t^\alpha x = p, \\ {}_0^C D_t^\alpha p = -V'(x), \end{cases} \tag{38}$$

where $0 < \alpha \leq 1$.

And the system Equation (38) with initial condition $x(0) = x_0$, $p(0) = p_0$ is equivalent to

$$\begin{cases} x(t) = x(0) + {}_0 I_t^\alpha p(t) \\ p(t) = p(0) - {}_0 I_t^\alpha V'(x(t)) \end{cases} \tag{39}$$

For the numeric solution of system Equation (39) we have developed a map (see [12])

$$\begin{cases} p_n = p_0 - \dfrac{(\Delta t)^\alpha}{\Gamma(\alpha+1)} \sum_{k=0}^{n-1} V'(x_k)[(n-k)^\alpha - (n-k-1)^\alpha] \\ x_n = x_0 + \dfrac{(\Delta t)^\alpha}{\Gamma(\alpha+1)} \sum_{k=0}^{n-1} p_{k+1}[(n-k)^\alpha - (n-k-1)^\alpha] \end{cases} \tag{40}$$

When $\alpha = 1$, this is equivalent to a second order symplectic integrator $p_n = p_{n-1} - \Delta t\, V'(x_{n-1})$, $x_n = x_{n-1} + \Delta t\, p_n$, and mapping Equation (38) provides an orbit (x_n, p_n) approaching the exact orbit at $t = n\Delta t$ when $\Delta t \to 0$. Futhermore, the term p_{k+1} can be replaced by p_k in order to return the second order Euler scheme which is not symplectic as $\Delta t \to 0$.

The orbit at step n depends on all the previous states up to the initial one due to the memory kernel of the fractional integral and then we have an infinite dimensional mapping. The computational complexity of the orbit up to (x_n, p_n) is of order n^2 whereas it is of order n for $\alpha = 1$.

This map has been tested taking $\alpha = 1$ with standard models and using different potentials which solutions are known. In particular for the harmonic oscillator and initial conditions $x_0 = 1$, $p_0 = 0$ and $\Delta t = 0.01$, solution has been compared with $x(t) = \cos(t)$, $p(t) = \sin(t)$ with error smaller than 0.5% after 10,000 steps.

The map Equation (40) has been used to simulate numerical solutions to significant non-linear fractional generalized Hamiltonian problems with a potential $V(x)$, where the explicit solution cannot be found. For instance, standard academic cases like free particle motion ($V = 0$) and a uniformly accelerated particle ($V = kx$) [13], or the simple oscillator $V = \frac{1}{2}\omega^2 x^2$, the double well potential $V = \frac{1}{4}x^4 - \frac{1}{2}x^2$ (see Figure 1) and the pendulum $V = \cos(x)$ [12]. In this pioneer numerical work, it was observed that the fractional derivative introduces a damping effect which can be either algebraic or exponential depending on the time scales of the system. It is considered in a more general context in [10] or in [11].

Riemann–Liouville time fractional problem could be integrated in this way, changing Riemann–Liouville derivative by Caputo's and applying a similar mapping.

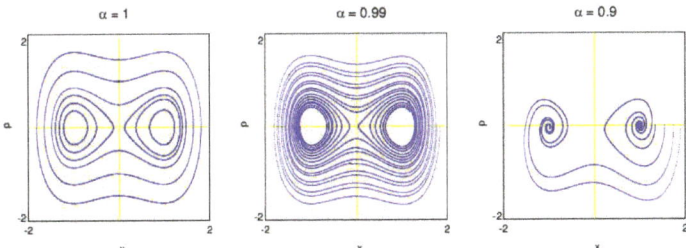

Figure 1. Phase portrait for the nonlinear oscillator with double well potential $V = \frac{1}{4}x^4 - \frac{1}{2}x^2$ (corresponding to the time interval $0 \leq t \leq 20$).

Non-homogeneous systems can be studied through the generalizations of this map. For instance, by the introduction of an external force $f(t)$ to simulate a forced-damped oscillator [14]:

$$\begin{cases} {}^C_0D^\alpha_t x = p, \\ {}^C_0D^\alpha_t p = -\omega_0^2 x + f(t), \end{cases} \qquad (41)$$

and this change means the introduction of an extra force term $f_n = f(n\Delta t)$ in the first equation of (40).

In particular, the system evolves to a limit cycle for an harmonic forcing ($f(t) = A_0 \cos(\omega t)$), similar to the classic case with the forced-damped oscillator. Varying the forcing frequency ω a resonance motion is reproduced, with amplitude

$$A_{res} = \frac{A_0}{2\omega_0} \left| \frac{1}{(i\omega)^\alpha - i\omega} \right| \qquad (42)$$

(see Figure 2).

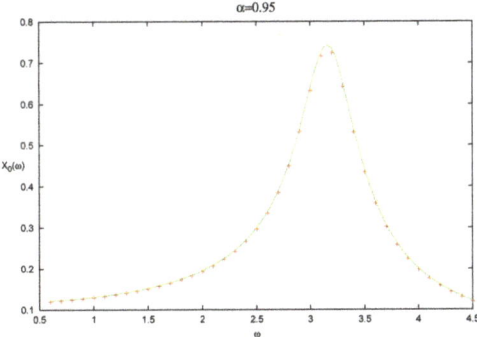

Figure 2. Plot of the amplitude of the limit cycle of $x(t)$ for different forcing frequencies ω versus the theoretical amplitude Equation (42) with $\alpha = 0.95$.

2. Variational Problems and Euler–Lagrange Equations

2.1. Nonlocal, Fractional Calculus of Variations

Classical mechanics can be viewed as founded on Hamilton's principle of stationary action. It is an elegant theory based on a very simple axiom. Some authors, appealed both by this and by the potential effectiveness of fractional modeling, have developed a fractional mechanics.

The guiding idea is to keep the same axiom (Hamilton's principle) and allow to have fractional derivatives as variables. Apparently, the founding papers are due to Riewe [15,16]. He uses fractional derivatives and builds the corresponding Euler–Lagrange equations in a systematic way. He also gives the corresponding Hamilton equations. As an application, he provides a formulation that includes a dissipative force. His presentation is not without some limitation.

Agrawall and other authors have generalized from them a Lagrangian and a Hamiltonian formalism [17–19] that also includes constraints.

From these references we see that right and left fractional derivatives appear in the Lagrangian and in the fractional Euler–Lagrange equations. The implications that this poses to the causality have not been dealt with.

We propose a slightly more general formalism that allows systems with different nonlocal terms in the Lagrangian, including fractional integrals and fractional derivatives of Caputo and Riemann–Liouville kind.

2.1.1. Positions

Let us call t, the time, the independent variable, $x(t)$ a function and y some linear functional that depends on x over the whole range of times through a relation yet to be defined:

$$y(t) = G[t, x]. \tag{43}$$

We consider a (Lagrangian) function of these, $L(t, x, y)$, and we look for the corresponding E-L equation for the action $\int L$ using the ideas of the calculus of variation. For instance, following Agrawal [17], we consider $x^*(t)$ and $y^*(t)$ the functions that make the action stationary and write $x(t) = x^*(t) + \varepsilon \eta(t)$. This implies, due to the assumed linearity of G that

$$y(t) = y^*(t) + \varepsilon G[t, \eta(s)]. \tag{44}$$

With this, the action becomes a function of ε with derivative:

$$\frac{d}{d\varepsilon} \int_0^T L(t, x, y)\, dt = \int_0^T \left(\frac{\partial L(t, x, y)}{\partial x} \eta(t) + \frac{\partial L(t, x, y)}{\partial y} G[t, \eta(s)] \right) dt = 0. \tag{45}$$

The next step is to write the integrand in terms of $\eta(t)$ and declare each factor to be zero. For this, we need to suppose a specific form to the functional G.

Let us suppose a nonlocal dependence (in time) of y on x of the form

$$y(t) = G[t, x(s)] := \int_0^T K(t,s)\, x(s)\, ds, \tag{46}$$

where $K(t,s)$ is some kernel, independent of both x and y. With this we have from Equation (45)

$$\int_0^T \left(\frac{\partial L(t,x,y)}{\partial x} \eta(t) + \frac{\partial L(t,x,y)}{\partial y} \int_0^T K(t,s) \eta(s)\, ds \right) dt = 0$$

$$\iff \int_0^T \int_0^T \left(\frac{\partial L(t,x,y)}{\partial x} \eta(s) \delta(t-s) + \frac{\partial L(t,x,y)}{\partial y} K(t,s) \eta(s) \right) ds\, dt = 0$$

$$\iff \int_0^T \int_0^T \left(\frac{\partial L(t,x,y)}{\partial x} \delta(t-s) + \frac{\partial L(t,x,y)}{\partial y} K(t,s) \right) dt\, \eta(s) ds = 0$$

$$\implies \int_0^T \left(\frac{\partial L(t,x,y)}{\partial x} \delta(t-s) + \frac{\partial L(t,x,y)}{\partial y} K(t,s) \right) dt = 0$$

$$\iff \frac{\partial L(s,x,y)}{\partial x} + \int_0^T \frac{\partial L(t,x,y)}{\partial y} K(t,s)\, dt = 0. \tag{47}$$

The use of Dirac's delta is justified if we understand that L is zero for t outside $[0, T]$. Besides, for all this manipulation to make sense, we need the kernel K and both partials of L to be integrable and $\partial L/\partial x$ to have a finite L^2-norm. If we choose a singular kernel, as below, we will see that some other conditions might be necessary.

Equation (47) is the corresponding E-L equation, we may exchange in it the name of the variables t and s and we finally have as necessary condition for the stationary action:

$$\frac{\partial L(t,x,y)}{\partial x} + \int_0^T \frac{\partial L(s,x,y)}{\partial y} K(s,t)\,ds = 0. \tag{48}$$

If we compare the integral with Equation (46) we see that the variables in the kernel are interchanged.

This is an important feature and it implies that the equation at a given time t involves values of x from both the past and the future. We will see this more clearly below when we deal with fractional derivatives.

2.1.2. Velocities

Let us suppose now that, instead of depending on x, $y(t)$ depends linearly on $\dot{x}(s)$ through

$$y(t) = G[t,\dot{x}(s)] := \int_0^T K(t,s)\,\dot{x}(s)\,ds. \tag{49}$$

We repeat what we did previously: we consider $x^*(t)$, $y^*(t)$ and write $x(t) = x^*(t) + \varepsilon\eta(t)$, which supposes now that

$$y(t) = y^*(t) + \varepsilon G[t,\dot{\eta}(s)]. \tag{50}$$

With this, the action becomes again a function of ε with the condition of the stationary trajectories given by:

$$\frac{d}{d\varepsilon}\int_0^T L(t,x,y)\,dt = \int_0^T \left(\frac{\partial L(t,x,y)}{\partial x}\eta(t) + \frac{\partial L(t,x,y)}{\partial y}G[t,\dot{\eta}(s)]\right)dt = 0. \tag{51}$$

Substituting the value of G, we have

$$\int_0^T \left(\frac{\partial L(t,x,y)}{\partial x}\eta(t) + \frac{\partial L(t,x,y)}{\partial y}\int_0^T K(t,s)\dot{\eta}(s)\,ds\right)dt = 0$$

$$\iff \int_0^T \left(\frac{\partial L(t,x,y)}{\partial x}\eta(t) - \frac{\partial L(t,x,y)}{\partial y}\int_0^T \frac{\partial K(t,s)}{\partial s}\eta(s)\,ds\right)dt = 0$$

$$\iff \int_0^T \int_0^T \left(\frac{\partial L(t,x,y)}{\partial x}\eta(s)\delta(t-s) - \frac{\partial L(t,x,y)}{\partial y}\frac{\partial K(t,s)}{\partial s}\eta(s)\right)ds\,dt = 0$$

$$\iff \int_0^T \int_0^T \left(\frac{\partial L(t,x,y)}{\partial x}\delta(t-s) - \frac{\partial L(t,x,y)}{\partial y}\frac{\partial K(t,s)}{\partial s}\right)dt\,\eta(s)\,ds = 0$$

$$\implies \int_0^T \left(\frac{\partial L(t,x,y)}{\partial x}\delta(t-s) - \frac{\partial L(t,x,y)}{\partial y}\frac{\partial K(t,s)}{\partial s}\right)dt = 0. \tag{52}$$

From here we get as Euler–Lagrange equation

$$\frac{\partial L(s,x,y)}{\partial x} - \int_0^T \frac{\partial L(t,x,y)}{\partial y}\frac{\partial K(t,s)}{\partial s}\,dt = 0$$

$$\iff \frac{\partial L(t,x,y)}{\partial x} - \frac{\partial}{\partial t}\int_0^T \frac{\partial L(s,x,y)}{\partial y} K(s,t)\,ds = 0, \tag{53}$$

where we have interchanged in the last step the names of the variables s and t. As in the previous case, where y depended only on positions, the evolution of the system at time t depends on both past and future values.

2.1.3. Extension of the Velocities Case

We may also consider the following possibility: y depends on x as given by Equation (46), and L depends on x and y but also on $\dot{y}$.

Since the procedure is linear, we may consider just that L depends on x and $\dot{y}$. The necessary condition for stationary trajectories becomes:

$$\int_0^T \left(\frac{\partial L(t,x,y')}{\partial x} \eta(t) + \frac{\partial L(t,x,y')}{\partial y'} \frac{d}{dt} \int_0^T K(t,s)\eta(s)\,ds \right) dt = 0. \tag{54}$$

Operating similarly as before, we have:

$$\Longrightarrow \frac{\partial L(s,x,y')}{\partial x} + \int_0^T \frac{\partial L(t,x,y')}{\partial y'} \frac{\partial K(t,s)}{\partial t} dt = 0$$

$$\Longrightarrow \frac{\partial L(s,x,y')}{\partial x} + \left[\frac{\partial L(t,x,y')}{\partial y'} K(t,s)\right]_{t=0}^{t=T} - \int_0^T \frac{\partial}{\partial t} \frac{\partial L(t,x,y')}{\partial y'} K(t,s) dt = 0$$

$$\Longrightarrow \frac{\partial L(s,x,y')}{\partial x} + \frac{\partial L(T,x,y')}{\partial y'} K(T,s) - \frac{\partial L(0,x,y')}{\partial y'} K(0,s)$$

$$- \int_0^T \frac{\partial}{\partial t} \frac{\partial L(t,x,y')}{\partial y'} K(t,s) dt = 0. \tag{55}$$

Or, otherwise:

$$\Longrightarrow \frac{\partial L(s,x,y')}{\partial x} - \int_0^T \frac{\partial}{\partial t} \frac{\partial L(t,x,y')}{\partial y'} K(t,s) dt = 0, \tag{56}$$

if we consider that L and its derivatives are all zero at the boundaries. As before, we may rewrite this as:

$$\frac{\partial L(t,x,\dot{y})}{\partial x} + \frac{\partial L(T,x,\dot{y})}{\partial \dot{y}} K(T,t) - \frac{\partial L(0,x,\dot{y})}{\partial \dot{y}} K(0,t)$$

$$- \int_0^T \frac{\partial}{\partial s} \frac{\partial L(s,x,y')}{\partial y'} K(s,t) ds = 0, \tag{57}$$

or, as:

$$\frac{\partial L(t,x,\dot{y})}{\partial x} - \int_0^T \frac{\partial}{\partial s} \frac{\partial L(s,x,y')}{\partial y'} K(s,t) ds = 0. \tag{58}$$

We have, once more, this dependency on values from the past and from the future.

2.1.4. General Case

We may consider a Lagrangian with three variables, for instance, y_1 depending on x, $\dot{y}_1$, and y_2 depending on $\dot{x}$, through two linear functionals, as above, with kernels K_1 and K_2, respectively:

$$y_1(t) = \int_0^T K_1(t,s)\, x(s)\, ds, \qquad y_2(t) = \int_0^T K_2(t,s)\, \dot{x}(s)\, ds. \tag{59}$$

In that case, due to the linearity of all the previous manipulations, the Euler–Lagrange equation will just have a contribution from each and be of the form:

$$\frac{\partial L(t,x,y_1,y_2,\dot{y}_1)}{\partial x} + \int_0^T \frac{\partial L(s,x,y_1,y_2,y_1')}{\partial y_1} K_1(s,t)\, ds$$

$$- \int_0^T \frac{\partial}{\partial s} \frac{\partial L(s,x,y_1,y_2,y')}{\partial y_1'} K_1(s,t)\, ds$$

$$- \frac{\partial}{\partial t} \int_0^T \frac{\partial L(s,x,y_1,y_2,y_1')}{\partial y_2} K_2(s,t)\, ds = 0. \tag{60}$$

This reminds us of the original calculus of variations (i.e., local and nonfractional) with higher order formulation. The case, for instance, of a Lagrangian that depends on x, $\dot{x}$ and $\ddot{x}$.

2.1.5. Fractional Integrals

We may obtain fractional integrals for y given by Equation (46) choosing, for instance, the following kernel:

$$K_{I^+}(t,s) = \frac{1}{\Gamma(\alpha)} \frac{\Theta(t-s)}{(t-s)^{1-\alpha}}, \tag{61}$$

where $\alpha > 0$ and Θ is the heavyside function:

$$\Theta(s) = \begin{cases} 0, & \text{if } s \leq 0, \\ 1, & \text{if } s > 0. \end{cases} \tag{62}$$

Kernel Equation (61) gives us the left-sided Riemann–Liouville integral with lower boundary 0:

$$y(t) = {}_{0^+}I_t^\alpha x(t) := \frac{1}{\Gamma(\alpha)} \int_0^t \frac{x(s)}{(t-s)^{1-\alpha}}, \tag{63}$$

while interchanging the variables and considering $K_{I^-}(t,s) = K_{I^+}(s,t)$, we obtain the right-sided integral with upper boundary T:

$$y(t) = {}_{T^-}I_t^\alpha x(t) := \frac{1}{\Gamma(\alpha)} \int_t^T \frac{x(s)}{(s-t)^{1-\alpha}}. \tag{64}$$

For $y(t) = {}_{0^+}I_t^\alpha x(t)$, the Euler–Lagrange Equation (48) becomes:

$$\frac{\partial L(t,x,y)}{\partial x} + \frac{1}{\Gamma(\alpha)} \int_t^T \frac{\partial L(s,x,y)}{\partial y} \frac{1}{(s-t)^{1-\alpha}} ds = 0 \tag{65}$$

$$\iff \frac{\partial L(t,x,y)}{\partial x} + {}_{T^-}I_t^\alpha \left(\frac{\partial L(t,x,y)}{\partial y} \right) = 0, \tag{66}$$

while in the second case, where $y(t) = {}_{T^-}I_t^\alpha x(t)$, we obtain:

$$\frac{\partial L(t,x,y)}{\partial x} + {}_{0^+}I_t^\alpha \left(\frac{\partial L(t,x,y)}{\partial y} \right) = 0. \tag{67}$$

We see that, independently of the choice we consider, the Euler–Lagrange equation for the system has both kinds of integrals: left-sided and the right-sided, one as the variable y, the other applied to the partial derivative of L with respect to y. This supposes that the evolution equation at any intermediate time $t \in (0, T)$, has elements from both the past and the future.

2.1.6. Fractional Derivatives

Caputo: In order to have y to represent a fractional derivative, we may consider the dependence on velocities and use in Equation (49) the kernel

$$K_{C^+}(t,s) = \frac{1}{\Gamma(1-\alpha)} \frac{\Theta(t-s)}{(t-s)^\alpha}, \quad \alpha \in (0,1). \tag{68}$$

We have that $y(t)$ is the (left) fractional Caputo derivative of order α:

$$y(t) = \frac{1}{\Gamma(1-\alpha)} \int_0^T \frac{\Theta(t-s)}{(t-s)^\alpha} \dot{x}(s)\, ds$$

$$= \frac{1}{\Gamma(1-\alpha)} \int_0^t \frac{\dot{x}(s)}{(t-s)^\alpha}\, ds = {}_0^C D_t^\alpha x(t). \tag{69}$$

The Euler–Lagrange equation is in this case

$$\frac{\partial L(t,x,y)}{\partial x} - \frac{1}{\Gamma(1-\alpha)} \frac{\partial}{\partial t} \int_t^T \frac{\partial L(s,x,y)}{\partial y} \frac{1}{(s-t)^\alpha}\, ds = 0$$

$$\iff \frac{\partial L(t,x,y)}{\partial x} + {}_{T-}^{RL} D_t^\alpha \left(\frac{\partial L(t,x,y)}{\partial y} \right) = 0. \tag{70}$$

Conversely, if we choose in Equation (49) the kernel

$$K_{C-}(t,s) = -K_{C+}(s,t) = \frac{-1}{\Gamma(1-\alpha)} \frac{\Theta(s-t)}{(s-t)^\alpha}, \tag{71}$$

we have that y is the right Caputo derivative of order α

$$y(t) = \frac{-1}{\Gamma(1-\alpha)} \int_t^T \frac{\dot{x}(s)}{(s-t)^\alpha}\, ds = {}_{T-}^C D_t^\alpha x(t), \tag{72}$$

and the corresponding Euler–Lagrange Equation (53) is now

$$\frac{\partial L(t,x,y)}{\partial x} + \frac{1}{\Gamma(1-\alpha)} \frac{\partial}{\partial t} \int_0^t \frac{\partial L(s,x,y)}{\partial y} \frac{1}{(t-s)^\alpha}\, ds = 0$$

$$\iff \frac{\partial L(t,x,y)}{\partial x} + {}_{0+}^{RL} D_t^\alpha \left(\frac{\partial L(t,x,y)}{\partial y} \right) = 0. \tag{73}$$

We see that for both kernels we have in the Euler–Lagrange equations derivatives form both sides, as in the previous case.

In both circumstances we have an equation that involves values from the past and from the future.

Example 1. $L(t,x,y) = \frac{1}{2} y^2 - U(x)$. We have:

$$\frac{\partial L(t,x,y)}{\partial x} = -U'(x), \quad \frac{\partial L(t,x,y)}{\partial y} = y,$$

and the Euler–Lagrange equation is just:

$${}_{T-}^{RL} D_t^\alpha \left({}_{0+}^C D_t^\alpha x(t) \right) = U'(x)$$

$$\iff \frac{1}{\Gamma^2(1-\alpha)} \frac{\partial}{\partial t} \int_t^T \frac{1}{(s-t)^\alpha} \left(\int_0^s \frac{x'(u)}{(s-u)^\alpha}\, du \right) ds = -U'(x),$$

if we take the left-derivative for y, and is:

$${}_{0+}^{RL} D_t^\alpha \left({}_{T-}^C D_t^\alpha x(t) \right) = U'(x),$$

if we chose the right-derivative. By the way, the sign that seemed to be missing ("the force is minus the derivative of the potential") is included inside the right-derivative in both cases, as we have seen right above for one of them.

Riemann–Liouville: we have to use the extension of the velocities case and consider in Equation (46) the kernel:

$$K_{RL^+}(t,s) = \frac{1}{\Gamma(1-\alpha)} \frac{\Theta(t-s)}{(t-s)^\alpha}, \quad \alpha \in (0,1), \qquad (74)$$

since it yields:

$$\dot{y} = \frac{1}{\Gamma(1-\alpha)} \frac{\partial}{\partial t} \int_0^t \frac{x(s)}{(t-s)^\alpha} ds = {}^{RL}_{0^+}D_t^\alpha x(t). \qquad (75)$$

The corresponding Euler–Lagrange equation is:

$$\frac{\partial L(t,x,\dot{y})}{\partial x} - \int_0^T \frac{\partial}{\partial s}\frac{\partial L(s,x,y')}{\partial y'} K_{RL^+}(s,t)ds = 0$$

$$\Longleftrightarrow \frac{\partial L(t,x,\dot{y})}{\partial x} - \frac{1}{\Gamma(1-\alpha)}\int_t^T \frac{\partial}{\partial s}\frac{\partial L(s,x,y')}{\partial y'} \frac{1}{(s-t)^\alpha} ds = 0$$

$$\Longleftrightarrow \frac{\partial L(t,x,\dot{y})}{\partial x} + {}^C_{T^-}D_t^\alpha\left(\frac{\partial L(s,x,y')}{\partial y'}\right) = 0. \qquad (76)$$

Conversely, if we consider the kernel:

$$K_{RL^-}(t,s) = -K_{RL^+}(s,t) = \frac{-1}{\Gamma(1-\alpha)}\frac{\Theta(s-t)}{(s-t)^\alpha}, \quad \alpha \in (0,1), \qquad (77)$$

we obtain:

$$\dot{y} = \frac{-1}{\Gamma(1-\alpha)}\frac{\partial}{\partial t}\int_t^T \frac{x(s)}{(s-t)^\alpha} ds = {}^{RL}_{T^-}D_t^\alpha x(t), \qquad (78)$$

and the Euler–Lagrange equation:

$$\frac{\partial L(t,x,\dot{y})}{\partial x} + {}^C_{0^+}D_t^\alpha\left(\frac{\partial L(s,x,y')}{\partial y'}\right) = 0. \qquad (79)$$

As always, we obtain integrals that cover both ranges: from 0 to t and from t to T.

We also see that the result is very similar that for the Caputo derivative, with an exchange of the role of the Caputo and the Riemann–Liouville derivatives.

2.2. Momenta and Hamilton Formalism

For each variable y we may define a momentum associated to it, in analogy to classical mechanics, as

$$p(t) = \frac{\partial L(t,x,y)}{\partial y}, \qquad (80)$$

or

$$p(t) = \frac{\partial L(t,x,\dot{y})}{\partial \dot{y}}, \qquad (81)$$

and we can express the previous Euler–Lagrange equations as

$$\frac{\partial L(t,x,y)}{\partial x} + \int_0^T K(s,t)\, p(s)\, ds = 0, \qquad (82)$$

when y depends on positions, and when depending on velocities (first case)

$$\frac{\partial L(t,x,y)}{\partial x} - \frac{\partial}{\partial t}\int_0^T K(s,t)\, p(s)\, ds = 0. \qquad (83)$$

or (second case)

$$\frac{\partial L(t,x,\dot{y})}{\partial x} - \int_0^T K(s,t)\frac{\partial p(s)}{\partial s} ds = 0. \qquad (84)$$

Using the four kernels considered before and the corresponding Euler–Lagrange equations we may give the Hamiltonian approach to the systems.

For instance, when y depends on velocities, the momentum is given by Equation (80), we have with kernel Equation (68):

$$\begin{cases} {}^C_{0^+}D^\alpha_t x(t) = y(t), \\ {}^{RL}_{T^-}D^\alpha_t p(t) = -\dfrac{\partial L(t,x,y)}{\partial x}, \end{cases} \tag{85}$$

and with kernel Equation (71):

$$\begin{cases} {}^C_{T^-}D^\alpha_t x(t) = y(t), \\ {}^{RL}_{0^+}D^\alpha_t p(t) = -\dfrac{\partial L(t,x,y)}{\partial x}. \end{cases} \tag{86}$$

When y depends on positions, the Lagrangian depends on $\dot{y}$ and the momentum is Equation (81), we obtain with kernel Equation (74):

$$\begin{cases} {}^{RL}_{0^+}D^\alpha_t x(t) = \dot{y}(t), \\ {}^C_{T^-}D^\alpha_t p(t) = -\dfrac{\partial L(t,x,\dot{y})}{\partial x}, \end{cases} \tag{87}$$

and with kernel (77):

$$\begin{cases} {}^{RL}_{T^-}D^\alpha_t x(t) = \dot{y}(t), \\ {}^C_{0^+}D^\alpha_t p(t) = -\dfrac{\partial L(t,x,\dot{y})}{\partial x}. \end{cases} \tag{88}$$

Whenever from Equation (80) we can express y (or $\dot{y}$) as a function of x and p, i.e., we can provide a Legendre transformation, the previous systems of equations correspond to the Hamilton equations.

As for the general case presented in Section 2.1.4, we may consider several functions y in the Lagrangian, and each will give rise to its corresponding momentum and a corresponding evolution equation for it.

2.3. Interpretation

We have seen that nonlocality (as considered above) plus Hamilton's principle implies that the equations are not causal, in the sense that the evolution at a given time t depends on values from all times and not just from the present (or the past).

It would indicate that if we want evolution equations that involve values just from the past, as in all the heuristic models, we cannot use the framework of the calculus of variations: the motion is no longer along paths that make an action stationary. This is irrelevant for both mathematicians and engineers, but it is irksome for physicists (such a pretty theory...).

But the same can be seen to happen in classical mechanics: as long as the solution is unique for any initial problem at a given time, the solution can be ran backwards or forwards in time. We can express our solution using any time we like: from $t = 0$ we may predict the values for any $t > 0$, but from $t = t_{\max}$ we can give the values for any $t < t_{\max}$.

But, in that case, we can always use only values from the past to get our solution while in this framework this is no longer possible.

We may change our idea of causality (at least inside the "ideal" world described by mechanics) and allow that, as long as the solution can be determined for all times, it is irrelevant what values are involved: the behaviour of the system is given. This whenever we can establish existence and unicity of the solution which is, in general, still an open problem for many fractional equations.

3. New Mathematical Scenarios: New Families of Functions and Equations

As mathematical tool, fractional operators establish important relations between transform integrals and special functions. So, the combined use of integral representations, exponential operational rules and special polynomials provides a powerful tool in the formalism of fractional calculus ([20,21]). Furthermore, fractional operators allow to elude singularities and reduce linear ordinary equations with variable coefficients. As a consequence, an extension of the classical integral representation of the related special functions can be obtained by using fractional operators ([22,23]).

From a applied point of view, fractional calculus offers a modeling scenario where fundamental mathematical questions converge and appropriate numerical algorithms can de developed. For these reasons, fractional calculus has many applications in different areas [24].

The main contribution of the fractional calculus is the consideration of intermediate-order dimensions through integrals and derivatives of arbitrary order ([25,26]). This has allowed to get a better modeling in different applications, for instance to model biomedical and biological phenomena ([27]). A large number of models considering long-range dependence and systems with memory are constructed with integro-differential and fractional equations.

In classical physics, many fundamental equations are based on similar laws:

- Hooke's Law: $F(t) = kx(t)$;
- Newtonian Fluid: $F(t) = k\frac{dx}{dt}(t)$;
- Newton's second law: $F(t) = k\frac{d^2x}{dt^2}(t)$.

As an interpolation of these equations, a fractional approach gives the possibility to look for intermediate or mixed behaviours:

$$F(t) = k\frac{d^\alpha x}{dt^\alpha}(t) \tag{89}$$

Some other contexts are the diffusion processes associated to the basic diffusion equation:

$$\frac{\partial u}{\partial t} = \frac{\partial^2 u}{\partial x^2}, \tag{90}$$

as we show in Table 1:

Table 1. Contexts of diffusion.

Law	Darcy: $\vec{q} = -K\,\text{Grad}\,h$	Fourier: $\vec{Q} = -\kappa\,\text{Grad}\,T$	Fick: $\vec{f} = -D\,\text{Grad}\,C$	Ohm: $\vec{j} = -\sigma\,\text{Grad}\,V$
Flux	Subterranean Water: q	Heat: Q	Solute: f	Charge: j
Potential	Hydrostatic Charge: h	Temperature: T	Concentration: C	Voltage: V
Medium's Property	Hydraulic Conductivity: K	Thermal Conductivity: κ	Diffusion Coefficient: D	Electric Conductivity: σ

The diffusion equation can be generalized through the fractional operators that allow to make a natural interpolation among equations, starting with the first order wave equation and ending with the second order wave equation:

$$\text{First order wave equation (hyperbolic):}\ \frac{\partial u}{\partial t} = \frac{\partial u}{\partial x} \tag{91}$$

$$\text{Interpolation:}\ \frac{\partial u}{\partial t} = \frac{\partial^\alpha u}{\partial x^\alpha} \tag{92}$$

$$\text{Diffusion equation (parabolic):}\ \frac{\partial u}{\partial t} = \frac{\partial^2 u}{\partial x^2} \tag{93}$$

$$\text{Interpolation: } \frac{\partial^\alpha u}{\partial t^\alpha} = \frac{\partial^2 u}{\partial x^2} \tag{94}$$

$$\text{Wave equation (hyperbolic): } \frac{\partial^2 u}{\partial t^2} = \frac{\partial^2 u}{\partial x^2} \tag{95}$$

Another fractional approach associated to the previous one is the use of Dirac-type fractional equations [28–30].

$$A\frac{\partial \psi}{\partial t} + B\frac{\partial \psi}{\partial x} = 0 \longrightarrow \quad A\frac{\partial^\alpha \psi}{\partial t^\alpha} + B\frac{\partial \psi}{\partial x} = 0 \quad \longrightarrow A\frac{\partial^{1/2}\psi}{\partial t^{1/2}} + B\frac{\partial \psi}{\partial x} = 0$$

$$\psi = \begin{pmatrix} \varphi \\ \zeta \end{pmatrix}$$

Pauli algebra $A^2 = I$
$B^2 = I$
$\{A, B\} = 0$

$$\frac{\partial^2 u}{\partial t^2} - \frac{\partial^2 u}{\partial x^2} = 0 \longrightarrow \quad \begin{array}{c} \gamma = 2\alpha \\ \frac{\partial^\gamma u}{\partial t^\gamma} - \frac{\partial^2 u}{\partial x^2} = 0 \end{array} \longrightarrow \frac{\partial u}{\partial t} - \frac{\partial^2 u}{\partial x^2} = 0$$

In this way, the following equation,

$$A\frac{\partial^{1/2}\psi}{\partial t^{1/2}} + B\frac{\partial \psi}{\partial x} = 0, \tag{96}$$

describes two coupled diffusion processes or a diffusion process with internal degrees of freedom. Depending on the chosen representation of the Pauli algebra, that A and B must verify, we obtain a system of equations coupled or decoupled:

$$A_1 = \begin{pmatrix} 0 & 1 \\ 1 & 0 \end{pmatrix} \quad B_1 = \begin{pmatrix} 0 & 1 \\ -1 & 0 \end{pmatrix} \implies \begin{cases} \partial_t^\alpha \varphi = \varphi \\ \partial_t^\alpha \zeta = -\zeta \end{cases} \tag{97}$$

$$A_2 = \begin{pmatrix} 1 & 0 \\ 0 & -1 \end{pmatrix} \quad B_2 = \begin{pmatrix} 0 & 1 \\ -1 & 0 \end{pmatrix} \implies \begin{cases} \partial_t^\alpha \varphi = \zeta \\ \partial_t^\alpha \zeta = \varphi \end{cases} \tag{98}$$

$$A\frac{\partial^\alpha \psi}{\partial t^\alpha} + B\frac{\partial \psi}{\partial x} = 0 \xrightarrow{\gamma = 2\alpha} \frac{\partial^\gamma u}{\partial t^\gamma} - \frac{\partial^2 u}{\partial x^2} = 0$$

In the study of the temporal inversion ($t \to -t$), we have the invariance of the fractional Dirac equation for the values $0 < \alpha < 1$:

$$\alpha = \frac{1}{3}, \frac{2}{3}, \frac{1}{5}, \frac{2}{5}, \frac{3}{5}, \frac{4}{5}, \frac{1}{7}, \frac{2}{7}, \ldots, \frac{6}{7}, \frac{1}{9}, \ldots$$

Some other fractional differential equations are obtained by considering the root $1/3$ of both the wave and diffusion equations:

$$\text{Wave equation: } P\partial_t^{2/3}\varphi + Q\partial_x^{2/3}\varphi = 0 \tag{99}$$

$$\text{Diffusion equation: } P\partial_t^{1/3}\varphi + Q\partial_x^{2/3}\varphi = 0 \tag{100}$$

where

$$P^3 = I \quad Q^3 = -I \quad PPQ + PQP + QPP = 0 \quad QQP + QPQ + PQQ = 0 \tag{101}$$

A possible realization is in terms of the 3×3 matrices associated to the Silvester algebra:

$$P = \begin{pmatrix} 0 & 0 & 1 \\ \omega^2 & 0 & 0 \\ 0 & \omega & 0 \end{pmatrix}, \quad Q = \Omega \begin{pmatrix} 0 & 0 & 1 \\ \omega & 0 & 0 \\ 0 & \omega^2 & 0 \end{pmatrix}, \tag{102}$$

with ω a cubic root of the unity and Ω a cubic root of the negative unity. In this case, φ has three components.

4. Nonlocal Phenomena in Space and/or Time. Applications

We use the term non-locality if what happens in a spatial point or at a given time depends on an average over an interval that contains that value. Thus, the non-local effects in space correspond to long-range interactions (many spatial scales), while the non-local effects in time suppose memory or delay effects (many temporal scales).

These phenomena are associated to integral or integro-differential equations, which appear in multiple contexts:

- Potential theory: Newton and Coulomb laws of the inverse of the square of the distance.
- Problems in geophysics: three-dimensional maps of the Earth's inside.
- Problems in electricity and magnetism.
- Hereditary phenomena in physics (materials with memory: hysteresis) and biology (ecological processes: accumulation of metals).
- Problems of evolution of populations.
- Problems of radiation.
- Optimization, control systems.
- Communication theory.
- Mathematical economy.

These different phenomena can be described by fractional differential equations, and it sets out two fundamental questions:

1. Are the models with space and/or time fractional derivatives consistent with the fundamental laws and symmetries of Nature?
2. How can the fractional differentiation order be experimentally observed and how does a fractional derivative emerge from models without fractional derivatives?

As example to answer the first question, it is interesting to remark that, for instance, the fractional diffusion equation with some kind of time fractional derivative,

$$\text{Interpolation: } \frac{\partial^\alpha u}{\partial t^\alpha} = \frac{\partial^2 u}{\partial x^2}, \tag{103}$$

verifies the second law of thermodynamics only if the following the generalized Fourier law is satisfied:

$$\frac{\partial^{\alpha-1} u}{\partial x^{\alpha-1}} \frac{\partial u}{\partial x} > 0. \tag{104}$$

Not all fractional operators satisfy the condition aforementioned, so it might be the key to choose the convenient fractional derivative or to apply restrictions to the initial and boundary conditions of the problem. Let us define $\rho = \alpha - 1$. When $\rho = 1$, the condition Equation (104) is trivial; but when $0 < \rho < 1$, this issue is a complex problem with different solutions according to the selected fractional operator and conditions [31].

4.1. Application of Fractional Calculus to Model Atmospheric Effects of Absorption

The time-fractional Cauchy problem is well-known:

$$^C_0 D_t^\alpha u(t,x) - \lambda_+^L D_x^\beta u(t,x) = 0, \quad t > 0, x \in \mathbb{R}, 0 < \alpha \le 1, \beta > 0 \tag{105}$$

$$\lim_{x \to \pm\infty} u(t,x) = 0, \quad u(0+,x) = g(x). \tag{106}$$

and its solution in the space of functions with Laplace and Fourier transforms, $LF = L(R^+) \times F(R)$, is defined by

$$u(t,x) = \frac{1}{2\pi} \int_{-\infty}^{\infty} G(k) E_\alpha(\lambda(-ik)^\beta t^\alpha) e^{-ikx} dk, \tag{107}$$

where the Mittag-Leffler function is evaluated on the complex plane and $G(k)$ represents the Fourier transform of $g(x)$.

For instance, for $\beta = 1$ and $g(x) = e^{-\mu|x|}$, $\mu > 0$:

$$u(t,x) = e^{-\mu|x|} E_\alpha(-\mu\lambda t^\alpha) \tag{108}$$

The fundamental solution of the problem is obtained for $g(x) = \delta(x)$, $G(k) = 1$ and, in this case, the moments for $\beta = 1$ have the following expression:

$$<x^n> = \int_{-\infty}^{\infty} x^n u(t,x) dx = (-\lambda t^\alpha)^n \frac{\Gamma(n+1)}{\Gamma(\alpha n + 1)}, \quad n = 0, 1, 2, \ldots \tag{109}$$

When replacing t by the wave-length of the radiation, λ, the moment for $n = 1$ and an appropriate constant β returns the Angstrom law,

$$\tau = \frac{\beta}{\lambda^\alpha}, \tag{110}$$

that is used to model the coefficient of molecular scattering τ for the absorption of the incoming energy inside the Martian atmosphere due to the dust [32,33]. The parameters α and β would be fixed in function of the Martian dust features. So, this relation shows a possible application of fractional calculus to model the dynamic of the Martian atmosphere. Deep studies, by using cloud computing, on this issue have been developed in [34–36].

In the context of the spatial exploration, other new original application of the fractional calculus analysis is the prediction and identification of dust devils and correlations between wind and seismic signals in a Martian meteorological payload packet. We recently started this project on the basis of our previous experience in the missions to Mars and to apply in ExoMars22 (initially ExoMars20 but now delayed due to the Covid-19) [37–40].

4.2. Chaos in a Fractional Duffing'S Equation

Duffing's equation has been a model for many studies on chaotic systems. It considers a simple but complex system where chaos can appear depending on the values of the parameters. The mathematical model is a time-forced, dissipative, second order nonlinear differential equation that can be viewed as a perturbed Hamiltonian or Lagrangian system [41]. Different possible potentials can be considered but the fundamental equation corresponds to a model for a long and slender vibrating beam set between two permanent magnets, subjected to an external sinusoidal force.

Duffing's equation shows many paradigmatic features of chaotic systems, in a somewhat simple frame, in the Theory of Dynamical Systems. The fractional counterpart we have chosen may possess similar relevant behaviours of other more general fractional models. This is the basis for this study. Besides, although the presence of chaotic behaviour in fractional Duffing's equations has been documented, many questions remain open.

It is possible to extend Duffing's equation into a fractional one in many ways, either as a second order differential equation, or as a system of simultaneous two first order equations, with some or all of these derivatives replaced by fractional ones. Different authors have, thus, considered different fractional equations, playing also with the fractional order of derivation [42,43]. In our case, we have chosen to replace only the first order derivative by a fractional Caputo derivative. The equation we consider is

$$\ddot{x} + \gamma D_t^\alpha x - x + x^3 = f_0 \cos(\omega t), \tag{111}$$

where $\alpha \in (0,1)$ and $D_t^\alpha x$ stands for the Caputo fractional derivative with lower limit 0 of order α. This equation has the advantage of a regular solution (at least C^2) whose existence can be ensured [44]. This is not merely a mathematical model since it can be viewed as the same mechanical device represented by Duffing's equation but immersed in a viscous medium. For some values of the parameters, a strange attractor is obtained, quite similar to the one that appears in the classical (i.e., non fractional) Duffing's equation.

We have studied the controlability of the chaotic regime in the presence of both harmonic and nonharmonic external perturbations, considering geometrical resonances for the second case. Using resonant Jacobi functions, we obtain conditions for external, additional, drivings that ensure chaos-free responses in our model [44].

We have also characterized the chaotic behaviour in our fractional model computing the maximum but, also, all the other Lyapunov characteristic exponents. We have used, as a reference, the fiduciary orbit technique and we have built a perturbative approach with a local equation that allows to estimate all the exponents [45].

The results show that a chaotic regime exists for the fractional Duffing's equation but also a regular regime with a long transient time. This regular regime, in practice, can be assimilated in many cases to a chaotic regime for quite long transient times, although the solutions for even longer times tend to regular, almost-periodic limiting curves.

5. Conclusions

We deeply thank the enlightening research papers of D. Baleanu and A. Fernandez [4] and M.D. Ortigueira and J.A.T. Machado ([5–7]). These works are a source of inspiration in order to continue the studies about fractional calculus and its applications to real world phenomena. In the light of the results of these works, in this paper we present new scenarios of discussion that might complement the previous studies. The main novelties of this work, can be summarized in the following items:

- The effect of the nonlocality, associated to the structure itself of the fractional derivative, manifests that in a solution we can observe the coexistence of two decays exponential and polynomial according to the time scale we consider (Section 1.4).
- Up to our best knowledge, the obtained results in Section 2 by using the Dirac delta are new. We show that the loss of causality in fractional mechanics is not specific to choosing any particular fractional derivative but is intrinsic to the formulation of nonlocal dependence with more general kernels. We present an approach, using a Dirac delta formulation, that simplifies the, otherwise, more cumbersome computations.
- Concerning Section 3, there are many remarkable issues. We show explicitly the use of the fractional calculus as an instrument to create new equations as interpolation among other classical ones very well known. An important issue is the interpolation between the parabolic and the hyperbolic dynamics. In this case, we have challenging dynamics attending to the behaviours under discrete symmetries as the time and space inversion.

Dirac obtained his famous equation by considering the square root of the Klein–Gordon equation. It is related to the basic idea of evolution depending only on the initial configuration of the system. At the same time, Dirac introduced the concept of internal degrees of freedom: the spin of a particle. In this contribution, we apply the above idea of Dirac to the square root of the classical heat equation and we obtain a fractional diffusion equation with internal degrees of freedom.

We extend the idea of considering a general root equation of a given one, and we obtain a connection between the Silvester algebra and the fractional calculus.

- In Section 4, we show one example where a fractional diffusion equation does not satisfy the second law of Thermodynamics, and we consider the use of the fractional calculus to model the dust dynamics with the associated electromagnetic interaction in Earth and Mars atmospheres.

On the other hand, these developments set out interesting questions and discussions about the adjustment and the reliability of the mathematical models to the dynamic of the real processes; for instance, the objectivity in the descriptions of the models. In this sense, the following items are remarkable:

- In physics, the laws must have the same form in all the inertial reference systems, otherwise we could distinguish an inertial reference system from other one by internal experiments.
- The above statement implies that we must have a suitable dictionary to relate the measurements in one system to other one.
- For inertial systems we have the Galileo and the Lorentz transforms (dictionaries).
- Einstein generalized the Galileo transform to the Lorentz transform in order to take into account that two inertial systems cannot distinguish each other by either internal mechanic or electromagnetic experiments (special relativity).
- Einstein generalized the above statements to the accelerated reference systems and created the general relativity.
- As a consequence, given an evolution fractional differential equation should be analysed its behaviour under Galileo and Lorentz transformations, as well as the discrete symmetries of the time and space inversion. This preliminary analysis will enlighten, for instance, the reversibility and causality issues associated to the equation.
- Concerning the velocity issue, we consider it in the general sense as the variation of a quantity with time.

Author Contributions: Investigation, M.P.V., D.U., S.J., L.V., J.L.V.-P. and M.M.; Methodology, M.P.V., D.U., S.J., L.V., J.L.V.-P. and M.M.; Writing—original draft, M.P.V., D.U., S.J., L.V., J.L.V.-P. and M.M.; Writing—review and editing, M.P.V., D.U., S.J., L.V., J.L.V.-P. and M.M. All authors have read and agreed to the published version of the manuscript.

Funding: This research has been carried out partially in the framework of the IN-TIME project, funded by the European Commission under the Horizon 2020 Marie Sklodowska-Curie actions Research and Innovation Staff Exchange (RISE) (Grant Agreement 823934). Futhermore, this research was funded by Ministerio de Economía, Industria y Competitividad of Spanish Government (ESP2016-79135-R).

Acknowledgments: Authors thank the collaboration of the research groups Finish Meteorological Institute (Ari-Matti Harri) and Space Research Institute of Russian Academy of Sciences (Oleg Korablev).

Conflicts of Interest: The authors declare no conflict of interest.

References

1. Samko, S.G.; Kilbas, A.A.; Marichev, O.I. *Fractional Integrals and Derivatives: Theory and Applications*; Taylor & Francis: London, UK, 2002.
2. Podlubny, I. *Fractional Differential Equations*; Academic Press: San Diego, CA, USA, 1999.
3. Kilbas, A.A.; Srivastava, H.M.; Trujillo, J.J. *Theory and Applications of Fractional Differential Equations*; Elsevier: Amsterdam, The Netherlands, 2006.
4. Baleanu, D.; Fernández, A. On fractional operators and their classifications. *Mathematics* **2019**, *7*, 830. [CrossRef]
5. Ortigueira, M.D.; Machado, J.A.T. Which Derivative? *Fractal Fract.* **2017**, *1*, 3. [CrossRef]
6. Ortigueira, M.D.; Machado, J.A.T. On fractional vectorial calculus. *Bull. Pol. Acad. Tech.* **2018**, *66*, 389–402.

7. Ortigueira, M.D.; Machado, J.A.T. Fractional Derivatives: The Perspective of System Theory. *Mathematics* **2019**, *7*, 150. [CrossRef]
8. Atangana, A.; Baleanu, D. New fractional derivatives with non-local and non-singular kernel. Theory and application to heat transfer model. *Therm. Sci.* **2016**, *20*, 763–769. [CrossRef]
9. Baleanu, D.; Fernandez, A. On some new properties of fractional derivatives with Mittag-Leffler kernel. *Commun. Nonlinear Sci. Numer. Simul.* **2018**, *59*, 444–462. [CrossRef]
10. Sopasakis, P.; Sarimveis, H.; Macheras, P.; Dokoumetzidis, A. Fractional calculus in pharmacokinetics. *J. Pharmacokinet. Pharmacodyn.* **2018**, *45*, 107–125. [CrossRef]
11. Ionescu, C.M. A computationally efficient Hill curve adaptation strategy during continuous monitoring of dose-effect relation in anaesthesia. *Nonlinear Dyn.* **2018**, *92*, 843–852. [CrossRef]
12. Turchetti, G.; Usero, D.; Vázquez, L. Hamiltonian systems with fractional time derivative. *Tamsui Oxf. J. Math. Sci.* **2002**, *18*, 31–44.
13. Usero, D. Propagación de Ondas no Lineales en Medios Heterogéneos. Ph.D. Thesis, Universidad Complutense de Madrid, Madrid, Spain, 2004.
14. Usero, D.; Vázquez, L. Fractional derivative: A new formulation for damped systems. *Localization Energy Transf. Nonlinear Syst.* **2003**, 296–303. [CrossRef]
15. Riewe, F. Nonconservative Lagrangian and Hamiltonian mechanics. *Phys. Rev. E* **1996**, *53*, 1890–1899. [CrossRef] [PubMed]
16. Riewe, F. Mechanics with fractional derivatives. *Phys. Rev. E* **1997**, *55*, 3581–3592. [CrossRef]
17. Agrawall, O.P. Formulation of Euler–Lagrange equations for fractional variational problems. *J. Math. Anal. Appl.* **2001**, *272*, 368–379. [CrossRef]
18. Muslih, S.I.; Baleanu, D. Hamiltonian formulation of systems with linear velocities within Riemann–Liouville fractional derivatives. *J. Math. Anal. Appl.* **2005**, *304*, 599–606. [CrossRef]
19. Rabei, E.M.; Nawafleh, K.H.I.; Hijjawi, R.S.; Muslih, S.I.; Baleanu, D. The Hamilton formalism with fractional derivatives. *J. Math. Anal. Appl.* **2007**, *327*, 891–897. [CrossRef]
20. Dattoli, G.; Cesarano, C.; Ricci, P.E.; Vázquez, L. Fractional operators, integral representations and special polynomials. *Int. J. Appl. Math.* **2003**, *10*, 131–139.
21. Dattoli, G.; Cesarano, C.; Ricci, P.E.; Vázquez, L. Special polynomials and fractional calculus. *Math. Comput. Model.* **2003**, *37*, 729–733. [CrossRef]
22. Rodríguez-Germá, L.; Trujillo, J.J.; Velasco, M.P. Fractional calculus framework to avoid singularities of differential equations. *Fract. Cal. Appl. Anal.* **2008**, *11*, 431–441.
23. Rivero, M.; Rodríguez-Germá, L.; Trujillo, J.J.; Velasco, M.P. Fractional operators and some special functions. *Comput. Math. Appl.* **2010**, *59*, 1822–1834. [CrossRef]
24. Magin, R.L. *Fractional Calculus in Bioengineering*; Begell House Publishers: Danbury, CT, USA, 2006.
25. Rocco, A.; West, B.J. Fractional calculus and the evolution of fractal phenomena. *Physica A* **1999**, *265*, 535–546. [CrossRef]
26. West, B.J.; Bologna, M.; Grigolini, P. *Physics of Fractal Operators*; Springer Verlag: New York, NY, USA, 2003.
27. Ionescu, C.; Lopes, A.; Copot, D.; Machado, J.A.T.; Bates, J.H.T. The role of fractional calculus in modeling biological phenomena: A review. *Commun. Nonlinear Sci. Numer. Simul.* **2017**, *51*, 141–159. [CrossRef]
28. Larsson, M.; Balatsky, A. Paul Dirac and the Nobel Prize in Physics. *Phys. Today* **2019**, *72*, 46–52. [CrossRef]
29. Vázquez, L. Fractional Diffusion Equations with Internal Degrees of Freedom. *J. Comp. Math.* **2003**, *21*, 491–494.
30. Pierantozzi, T.; Vázquez, L. An Interpolation between the Wave and Diffusion Equations through the Fractional Evolution Equations Dirac Like. *J. Math. Phys.* **2005**, *46*, 113521. [CrossRef]
31. Vázquez, L.; Trujillo, J.J.; Velasco, M.P. Fractional heat equation and the second law of thermodynamics. *Fract. Cal. Appl. Anal.* **2011**, *14*, 334–342. [CrossRef]
32. Angstrom, A. On the atmospheric transmission of sun radiation and on dust in the air. *Geogr. Ann.* **1929**, *11*, 156–166.
33. Córdoba-Jabonero, C.; Vázquez, L. Characterization of atmospheric aerosols by an in-situ photometris technique in planetary environments. *SPIE* **2005**, *4878*, 54–58.
34. Jiménez, S.; Usero, D.; Vázquez, L.; Velasco, M.P. Fractional diffusion models for the atmosphere of Mars. *Fractal Fract.* **2018**, *2*, 1. [CrossRef]

35. Vázquez, L.; Velasco, M.P.; Vázquez-Poletti, J.L.; Llorente, I.M.; Usero, D.; Jiménez, S. Modeling and simulation of the atmospheric dust dynamic: Fractional Calculus and Cloud Computing. *Int. J. Numer. Anal. Model.* **2018**, *15*, 74–85.
36. Vázquez-Poletti, J.L.; Llorente, I.M.; Velasco, M.P.; Vicente-Retortillo, A.; Aguirre, C.; Caro-Carretero, R.; Valero, F.; Vázquez, L. Martian Computing Clouds: A Two Use Case Study. In Proceedings of the Seventh Moscow Solar System Symposium (7M-S3), Moscow, Russia, 10–14 October 2016.
37. De Lucas, E.; Miguel, M.J.; Mozos, D.; Vázquez, L. Martian dust devils detector over FPGA. *Geosci. Instrum. Method Data Syst.* **2012**, *1*, 23–31. [CrossRef]
38. Aguirre, C.; Franzese, G.; Esposito, F.; Vázquez, L.; Caro-Carretero, R.; Vilela-Mendes, R.; Ramírez-Nicolás, M.; Cozzolino, F.; Popa, C.I. Signal-adapted tomography as a tool for dust devil detection. *Aeolian Res.* **2017**, *29*, 12–22. [CrossRef]
39. Gómez-Elvira, J.; Armiens, C.; Castañer, L.; Domínguez, M.; Genzer, M.; Gómez, F.; Haberle, R.; Harri, A.M.; Jiménez, V.; Kahanpaa, H.; et al. REMS: The Environmental Sensor Suite for the Mars Science Laboratory Rover. *Space Sci. Rev.* **2012**, *170*, 583–640 [CrossRef]
40. Domínguez-Pumar, M.; Pérez, E.; Ramón, M.; Jiménez, V.; Bermejo, S.; Pons-Nin, J. Acceleration of the Measurement Time of Thermopiles Using Sigma-delta Control. *Sensors* **2019**, *19*, 3159. [CrossRef] [PubMed]
41. Guckenheimer, J.; Holmes, P.H. *Nonlinear Oscillations, Dynamical Systems, and Bifurcations of Vector Fields*; Springer-Verlag: New York, NY, USA, 1986.
42. Gao, X.; Yu, J. Chaos in the fractional order periodically forced complex Duffing's oscillators. *Chaos Solitons Fractals* **2005**, *24*, 1097–1104. [CrossRef]
43. Sheu, L.J.; Chen, H.K.; Chen, J.H.; Tam, L.M. Chaotic dynamics of the fractionally damped Duffing equation. *Chaos Solitons Fractals* **2007**, *32*, 1459–1468. [CrossRef]
44. Jiménez, S.; González, J.A.; Vázquez, L. Fractional Duffing's equation and geometrical resonance. *Int. J. Bifurcation Chaos* **2013**, *23*, 1350089-1–1350089-13. [CrossRef]
45. Jiménez, S.; Zufiria, P.J. Chaos in a fractional Duffing's equation. *Dyn. Syst. Differ. Equ. Appl.* **2015**, *10*, 660–669.

© 2020 by the authors. Licensee MDPI, Basel, Switzerland. This article is an open access article distributed under the terms and conditions of the Creative Commons Attribution (CC BY) license (http://creativecommons.org/licenses/by/4.0/).

Article

Exponential Stabilization of Linear Time-Varying Differential Equations with Uncertain Coefficients by Linear Stationary Feedback

Vasilii Zaitsev *,† and Inna Kim †

Laboratory of Mathematical Control Theory, Udmurt State University, 426034 Izhevsk, Russia; kimingeral@gmail.com
* Correspondence: verba@udm.ru
† These authors contributed equally to this work.

Received: 6 May 2020; Accepted: 22 May 2020; Published: 24 May 2020

Abstract: We consider a control system defined by a linear time-varying differential equation of n-th order with uncertain bounded coefficients. The problem of exponential stabilization of the system with an arbitrary given decay rate by linear static state or output feedback with constant gain coefficients is studied. We prove that every system is exponentially stabilizable with any pregiven decay rate by linear time-invariant static state feedback. The proof is based on the Levin's theorem on sufficient conditions for absolute non-oscillatory stability of solutions to a linear differential equation. We obtain sufficient conditions of exponential stabilization with any pregiven decay rate for a linear differential equation with uncertain bounded coefficients by linear time-invariant static output feedback. Illustrative examples are considered.

Keywords: linear differential equation; exponential stability; linear output feedback; stabilization; uncertain system

MSC: 34D20; 93C05; 93D15; 93D23

---

1. Introduction

Consider a control system defined by an ordinary differential equation with time-varying coefficients of n-th order

$$x^{(n)} + p_1(t)x^{(n-1)} + \ldots + p_n(t)x = u, \tag{1}$$

where $x \in \mathbb{R}$ is the state variable, $u \in \mathbb{R}$ is the control input, $t \in \mathbb{R}_+ := [0, +\infty)$. We suppose that the functions $p_i(t)$ are measurable but exact values of these functions at time moments t are unknown, we know only that the functions are bounded on $\mathbb{R}_+$ and lower and upper bounds (α_i and β_i) are known:

$$\alpha_i \le p_i(t) \le \beta_i, \quad t \in \mathbb{R}_+, \quad i = \overline{1,n}. \tag{2}$$

Functions $p_i(t)$ can be arbitrary, in particular, they can vary fast or slowly. Denote $\mathbf{x} = (x, \dot{x}, \ldots, x^{(n-1)})$. We consider a problem of feedback stabilization for system (1). One needs to construct a function $u(t, \mathbf{x})$, $u(t, \mathbf{0}) = 0$, such that, for system (1) closed-loop by $u = u(t, \mathbf{x})$, the zero solution is exponentially stable and has a given decay rate. The stated problem essentially relates to the problems of robust stabilization.

Let us assume that $p_i(t)$ are time-invariant (and hence, are known), i.e., $p_i(t) \equiv p_i(= \alpha_i = \beta_i)$. In that case, the stabilization problem is trivial. In fact, we construct

$$v_i = p_i - \phi_i, \tag{3}$$

where $\phi_i \in \mathbb{R}$, $i = \overline{1,n}$, are chosen such that the polynomial

$$\lambda^n + \phi_1 \lambda^{n-1} + \ldots + \phi_n \tag{4}$$

is stable (i.e., $\operatorname{Re} \lambda_j < -\theta < 0$ for all roots λ_j, $j = \overline{1,n}$, of (4)). Then system (1) closed-loop by the control

$$u(\mathbf{x}) = v_1 x^{(n-1)} + \ldots + v_n x \tag{5}$$

has the form

$$x^{(n)} + \phi_1 x^{(n-1)} + \ldots + \phi_n x = 0, \tag{6}$$

and the zero (and hence, every) solution of (6) is exponentially stable.

Now, assume that $p_i(t)$ are time-varying. Then we can not construct the control by using (3) because $p_i(t)$ are unknown. Let the feedback control law have the form (5), where v_i are constant. The closed-loop system has the form

$$x^{(n)} + (p_1(t) - v_1) x^{(n-1)} + \ldots + (p_n(t) - v_n) x = 0. \tag{7}$$

We study the following problem: *construct constants $v_1, \ldots, v_n \in \mathbb{R}$ such that all solutions of (7) are exponentially stable with a given decay of rate*. This problem is non-trivial due to the following reasons. For studying this problem, we need use some sufficient conditions for exponential stability of linear time-varying systems. The problem of obtaining some sufficient conditions for (asymptotic, exponential) stability of linear time-varying systems

$$\dot{x} = A(t)x, \quad t \in \mathbb{R}_+, \quad x \in \mathbb{R}^n, \tag{8}$$

is one of the important and difficult problems in the theory of differential equations and control theory [1]. In contrast to systems with constant coefficients ($A(t) \equiv A$), the condition $\operatorname{Re} \lambda_j < 0$, $j = \overline{1,n}$, fulfilled for the eigenvalues of the matrix of the system (8) is neither a sufficient nor a necessary condition for the asymptotic stability of the system (8) (see, e.g., [2], ([3], §9)). Some sufficient conditions for asymptotic and exponential stability of linear time-varying systems (8) and linear time-varying differential equations

$$x^{(n)} + q_1(t) x^{(n-1)} + \ldots + q_n(t) x = 0 \tag{9}$$

were obtained in [1–11]. The following theorem take place.

Theorem 1. *Suppose the functions $q_i(t)$ are measurable and bounded on $\mathbb{R}_+$ and the following inequalities hold:*

$$0 < \sigma_i \le q_i(t) \le \omega_i, \quad t \in \mathbb{R}_+, \quad i = \overline{1,n}. \tag{10}$$

Let the polynomial

$$P_1(\lambda) = \lambda^n + \omega_1 \lambda^{n-1} + \sigma_2 \lambda^{n-2} + \omega_3 \lambda^{n-3} + \ldots, \tag{11}$$
$$P_2(\lambda) = \lambda^n + \sigma_1 \lambda^{n-1} + \omega_2 \lambda^{n-2} + \sigma_3 \lambda^{n-3} + \ldots \tag{12}$$

have only real roots. Then all solutions of (9) are exponentially tends to 0 as $t \to +\infty$.

Theorem 1 was proved by A.Yu. Levin in [6]. Note that these roots (of the polynomials (11) and (12)) are negative necessarily due to positivity of σ_i, ω_i, $i = \overline{1,n}$. Next, it follows from the proof of Theorem 1 [6] that every solution $x(t)$ of (9) along with its derivatives up to $(n-1)$-th order has the form $O(e^{-\nu_n t})$ as $t \to +\infty$, where $-\nu_n < 0$ is the largest of the roots of polynomials (11), (12).

By using standard replacement $y_1 = x, y_2 = x', \ldots, y_n = x^{(n-1)}$, one can rewrite the control system (1), (5) in the form

$$\dot{y} = A(t)y + Bu, \tag{13}$$
$$u = Vy. \tag{14}$$

Here $A(t)$ is the companion matrix for the polynomial with the coefficients $p_i(t)$, $B = \text{col}[0, \ldots, 0, 1]$, $V = [v_n, \ldots, v_1]$.

A large number of papers are devoted to the problems of robust asymptotic stability and stabilization for linear systems. We note here the famous works [12–18] and recent works [19–22]. The problems of stabilization of uncertain linear systems using linear matrix inequalities were studied in [23–33].

Uncertain systems (13), (14) were studied in [34–37] and in other works of A.H. Gelig and I.E. Zuber. In particular, it follows from results of [34] that system (13) is exponentially stabilizable by feedback control (14). This result is supplemented and developed in this paper. The difference between this result and the results obtained in the work is as follows. Firstly, we achieve exponential stabilization of (7) not only with some decay rate as it follows from [34] but with an arbitrary pregiven decay rate. Secondly, in contrast to [34], which uses the Second Lyapunov Method (Method of Lyapunov Function), we apply, in some sense, the First Lyapunov Method (which uses the roots of characteristic polynomial) and non-oscillation theory. Thirdly, we extend these stabilization results to systems with static output feedback control.

In this work, using Theorem 1, we prove results on exponential stabilization with any pregiven decay rate by linear stationary static state or output feedback for a control system defined by a linear time-varying differential equation of the n-th order with uncertain coefficients.

2. Preliminary Results

Theorem 2. *For any $\eta > 0$ for any $n \in \mathbb{N}$ there exist polynomials*

$$f(\lambda) = \lambda^n + \delta_1 \lambda^{n-1} + \gamma_2 \lambda^{n-2} + \delta_3 \lambda^{n-3} + \ldots, \tag{15}$$
$$g(\lambda) = \lambda^n + \gamma_1 \lambda^{n-1} + \delta_2 \lambda^{n-2} + \gamma_3 \lambda^{n-3} + \ldots \tag{16}$$

such that the following properties hold:

(i) $0 < \gamma_i \leq \delta_i - 1, i = 1, \ldots, n$;

(ii) the roots $-a_i$, $i = 1, \ldots, n$, of $f(\lambda)$ and the roots $-b_i$, $i = 1, \ldots, n$, of $g(\lambda)$ are real (and hence, negative);

(iii) the following inequalities hold:

$$0 > -\eta \geq -a_1 > -b_1 > -b_2 > -a_2 > -a_3 > -b_3 > \ldots > -a_{2\ell-1} > -b_{2\ell-1} > -b_{2\ell} > -a_{2\ell} \tag{17}$$
(if n is even and $n = 2\ell$);

$$0 > -\eta \geq -b_1 > -a_1 > -a_2 > -b_2 > -b_3 > -a_3 > \ldots > -a_{2\ell} > -b_{2\ell} > -b_{2\ell+1} > -a_{2\ell+1} \tag{18}$$
(if n is odd and $n = 2\ell + 1$).

Proof. At first, suppose that the theorem is proved for any $\eta \geq 1$. Let us construct, for $\eta = 1$, the polynomials (15), (16) providing properties (*i*), (*ii*), (*iii*), and denote them by $f_1(\lambda), g_1(\lambda)$. Now, let $\eta \in (0,1)$. Then, let us set $f(\lambda) := f_1(\lambda), g(\lambda) := g_1(\lambda)$. Hence, conditions (*i*), (*ii*) are satisfied. Since $-\eta > -1$, condition (*iii*) holds as well. Thus, without loss of generality, one can assume that $\eta \geq 1$.

Let us give the proof by induction on n. The statements that we have to prove are different for odd and even numbers n: for even n, we need to ensure inequalities (17), in addition to (*i*) and (*ii*), and for odd n, we need to ensure inequalities (18). Therefore, the induction base as well as the induction hypothesis and the induction step should depend on whether the number n is even or odd. That is why we should check the induction base for $n = 1$ and $n = 2$.

Let $n = 1$. For any $\eta \geq 1$, we set $\gamma_1 := \eta$, $\delta_1 := \eta + 1$. Then the polynomials $f(\lambda) = \lambda + \delta_1$ and $g(\lambda) = \lambda + \gamma_1$ have the roots $-a_1 = -\delta_1$ and $-b_1 = -\gamma_1$ respectively. Obviously, conditions (i), (ii), and inequalities (18) are satisfied.

Let $n = 2$. For any $\eta \geq 1$, we set

$$a_1 := \eta, \quad a_2 := 5\eta, \quad b_1 := 2\eta, \quad b_2 := 3\eta, \tag{19}$$

$$f(\lambda) := (\lambda + a_1)(\lambda + a_2), \quad g(\lambda) := (\lambda + b_1)(\lambda + b_2). \tag{20}$$

Then

$$\delta_1 = 6\eta, \quad \gamma_1 = 5\eta, \quad \delta_2 = 6\eta^2, \quad \gamma_2 = 5\eta^2. \tag{21}$$

By (19), (20), condition (ii) and inequality (17) are satisfied. By (21) and the inequality $\eta \geq 1$, condition (i) is satisfied. The induction base is proved.

Let us put forward the induction hypothesis. Suppose that the assertion of the theorem is true for $n = k$. Then, let us prove that the assertion of the theorem is true for $n = k+1$. We will carry out the induction step for even and odd k separately.

By the induction hypothesis, there exist polynomials

$$f(\lambda) = \lambda^k + \delta_1 \lambda^{k-1} + \gamma_2 \lambda^{k-2} + \ldots, \tag{22}$$

$$g(\lambda) = \lambda^k + \gamma_1 \lambda^{k-1} + \delta_2 \lambda^{k-2} + \ldots \tag{23}$$

such that

$$0 < \gamma_i \leq \delta_i - 1, \quad i = \overline{1,k}, \tag{24}$$

$$f(\lambda) = \prod_{i=1}^{k}(\lambda + a_i), \quad g(\lambda) = \prod_{i=1}^{k}(\lambda + b_i), \quad a_i, b_i \in \mathbb{R}, \quad a_i, b_i > 0, \quad i = \overline{1,k}, \tag{25}$$

$$0 > -\eta \geq -a_1 > -b_1 > -b_2 > -a_2 > \ldots > -a_{2\ell-1} > -b_{2\ell-1} > -b_{2\ell} > -a_{2\ell} \quad \text{(if } k = 2\ell\text{)}, \tag{26}$$

$$0 > -\eta \geq -b_1 > -a_1 > -a_2 > -b_2 > \ldots > -a_{2\ell} > -b_{2\ell} > -b_{2\ell+1} > -a_{2\ell+1} \quad \text{(if } k = 2\ell + 1\text{)}. \tag{27}$$

Let us prove that there exist polynomials

$$F(\lambda) = \lambda^{k+1} + \Delta_1 \lambda^k + \Gamma_2 \lambda^{k-1} + \Delta_3 \lambda^{k-2} + \ldots, \tag{28}$$

$$G(\lambda) = \lambda^{k+1} + \Gamma_1 \lambda^k + \Delta_2 \lambda^{k-1} + \Gamma_3 \lambda^{k-2} + \ldots \tag{29}$$

such that

$$0 < \Gamma_i \leq \Delta_i - 1, \quad i = \overline{1,k+1}, \tag{30}$$

$$F(\lambda) = \prod_{i=1}^{k+1}(\lambda + A_i), \quad G(\lambda) = \prod_{i=1}^{k+1}(\lambda + B_i), \quad A_i, B_i \in \mathbb{R}, \quad A_i, B_i > 0, \quad i = \overline{1,k+1}, \tag{31}$$

$$0 > -\eta \geq -B_1 > -A_1 > -A_2 > -B_2 > \ldots > -A_{2\ell} > -B_{2\ell} > -B_{2\ell+1} > -A_{2\ell+1} \quad \text{(if } k = 2\ell\text{)}, \tag{32}$$

$$0 > -\eta \geq -A_1 > -B_1 > -B_2 > -A_2 > \ldots > -A_{2\ell+1} > -B_{2\ell+1} > -B_{2\ell+2} > -A_{2\ell+2} \quad \text{(if } k = 2\ell + 1\text{)}. \tag{33}$$

We assume that $\delta_0 := 1$, $\gamma_0 := 1$. Set

$$C_1 := \max_{i=\overline{1,\ell}} \left\{ \frac{\delta_{2i-1} - \gamma_{2i-1} + 1}{\delta_{2i-2}}, \frac{1}{\delta_{2\ell}} \right\}, \quad C_2 := \max_{j=\overline{1,\ell}} \frac{\delta_{2j} - \gamma_{2j} + 1}{\delta_{2j-1}}, \quad N := \max_{j=\overline{1,\ell}} \frac{\gamma_{2j-1}}{\delta_{2j-1}} \quad (34)$$

for the case if $k = 2\ell$, and

$$C_1 := \max_{i=\overline{1,\ell+1}} \frac{\delta_{2i-1} - \gamma_{2i-1} + 1}{\delta_{2i-2}}, \quad C_2 := \max_{j=\overline{1,\ell}} \left\{ \frac{\delta_{2j} - \gamma_{2j} + 1}{\delta_{2j-1}}, \frac{1}{\delta_{2\ell+1}} \right\}, \quad N := \max_{j=\overline{1,\ell+1}} \frac{\gamma_{2j-1}}{\delta_{2j-1}} \quad (35)$$

for the case if $k = 2\ell + 1$. Then $C_1 > 0$, $C_2 > 0$, $0 < N < 1$. Consider lines

$$y = x + C_1, \quad x = Ny + C_2. \quad (36)$$

They intersect at the point $M_0(x_0, y_0)$ with the coordinates $x_0 = \dfrac{C_1 N + C_2}{1 - N} > 0$, $y_0 = \dfrac{C_1 + C_2}{1 - N} > 0$.
Consider the set $\Omega_0 = \{(x,y) \in \mathbb{R}^2 : y \geq x + C_1, x \geq Ny + C_2\}$. The set Ω_0 is a cone, with a vertex at the point M_0, located in the first quadrant of the xOy-plane and bounded by half-lines (36) where $x \geq x_0$. The ray $m = \{(x,y) \in \mathbb{R}^2 : x - x_0 = \dfrac{1+N}{2}(y - y_0), x \geq x_0\}$ is contained in Ω_0. Consider the inequality system

$$\begin{cases} y \geq x + C_1, \\ x \geq Ny + C_2, \\ x > a_k. \end{cases} \quad (37)$$

The solution of system (37) is the set $\Omega_1 = \Omega_0 \cap \{x > a_k\}$. The set Ω_1 is non-empty. In particular, the point $M_1(\hat{x}, \hat{y})$ lying on the ray m with $\hat{x} = \max\{x_0 + 1, a_k + 1\}$ is contained in Ω_1. Calculating $\hat{y}$, we obtain that $\hat{y} = \dfrac{2}{1+N} \max\{1, a_k - x_0 + 1\} + y_0$.
Set

$$A_i := b_i, \quad B_i := a_i, \quad i = \overline{1,k}, \quad (38)$$

$$A_{k+1} := \hat{y}, \quad B_{k+1} := \hat{x}, \quad (39)$$

$$F(\lambda) := \prod_{i=1}^{k+1}(\lambda + A_i), \quad G(\lambda) := \prod_{i=1}^{k+1}(\lambda + B_i). \quad (40)$$

Then condition (31) is satisfied. Next, since $\hat{x} > a_k$, it follows that

$$B_k < B_{k+1}. \quad (41)$$

Next, since $(\hat{x}, \hat{y})$ is a solution of (37), we have

$$A_{k+1} = \hat{y} \geq \hat{x} + C_1 > \hat{x} = B_{k+1}. \quad (42)$$

Thus, it follows from inequalities (41), (42), equalities (38) and induction hypothesis (26), (27) that inequalities (32) are satisfied if $k = 2\ell$, and inequalities (33) are satisfied if $k = 2\ell + 1$.

Let us prove inequalities (30). From the definition (40) of the polynomials $F(\lambda)$, $G(\lambda)$ and equalities (38), (25) we obtain that

$$F(\lambda) = g(\lambda)(\lambda + A_{k+1}), \quad G(\lambda) = f(\lambda)(\lambda + B_{k+1}). \quad (43)$$

Substituting (22), (23) and (28), (29) into (43) and opening the brackets, we obtain equalities

$$\Delta_{2i-1} = A_{k+1}\delta_{2i-2} + \gamma_{2i-1}, \qquad \Gamma_{2i-1} = B_{k+1}\gamma_{2i-2} + \delta_{2i-1}, \qquad i = \overline{1,\ell},$$
$$\Delta_{2\ell+1} = A_{k+1}\delta_{2\ell}, \qquad \Gamma_{2\ell+1} = B_{k+1}\gamma_{2\ell},$$
$$\Delta_{2j} = B_{k+1}\delta_{2j-1} + \gamma_{2j}, \qquad \Gamma_{2j} = A_{k+1}\gamma_{2j-1} + \delta_{2j}, \qquad j = \overline{1,\ell},$$

for the case if $k = 2\ell$, and equalities

$$\Delta_{2i-1} = A_{k+1}\delta_{2i-2} + \gamma_{2i-1}, \qquad \Gamma_{2i-1} = B_{k+1}\gamma_{2i-2} + \delta_{2i-1}, \qquad i = \overline{1,\ell+1},$$
$$\Delta_{2j} = B_{k+1}\delta_{2j-1} + \gamma_{2j}, \qquad \Gamma_{2j} = A_{k+1}\gamma_{2j-1} + \delta_{2j}, \qquad j = \overline{1,\ell},$$
$$\Delta_{2\ell+2} = B_{k+1}\delta_{2\ell+1}, \qquad \Gamma_{2\ell+2} = A_{k+1}\gamma_{2\ell+1},$$

for the case if $k = 2\ell + 1$. The inequalities $\Gamma_i > 0$, $i = \overline{1,k+1}$, are satisfied due to inequalities (24) and the inequalities $A_{k+1} > 0$, $B_{k+1} > 0$. The inequalities

$$\Gamma_i \leq \Delta_i - 1, \quad i = \overline{1, k+1}, \tag{44}$$

are equivalent to the inequality system

$$\begin{cases} \gamma_{2i-1} + A_{k+1}\delta_{2i-2} \geq B_{k+1}\gamma_{2i-2} + \delta_{2i-1} + 1, & i = \overline{1,\ell}, \\ A_{k+1}\delta_{2\ell} \geq B_{k+1}\gamma_{2\ell} + 1, \\ \gamma_{2j} + B_{k+1}\delta_{2j-1} \geq A_{k+1}\gamma_{2j-1} + \delta_{2j} + 1, & j = \overline{1,\ell}, \end{cases} \tag{45}$$

for the case if $k = 2\ell$, and are equivalent to the inequality system

$$\begin{cases} \gamma_{2i-1} + A_{k+1}\delta_{2i-2} \geq B_{k+1}\gamma_{2i-2} + \delta_{2i-1} + 1, & i = \overline{1,\ell+1}, \\ \gamma_{2j} + B_{k+1}\delta_{2j-1} \geq A_{k+1}\gamma_{2j-1} + \delta_{2j} + 1, & j = \overline{1,\ell}, \\ B_{k+1}\delta_{2\ell+1} \geq A_{k+1}\gamma_{2\ell+1} + 1, \end{cases} \tag{46}$$

for the case if $k = 2\ell + 1$. System (45) is equivalent to the inequality system

$$\begin{cases} A_{k+1} \geq B_{k+1}\dfrac{\gamma_{2i-2}}{\delta_{2i-2}} + \dfrac{\delta_{2i-1} - \gamma_{2i-1} + 1}{\delta_{2i-2}}, & i = \overline{1,\ell}, \\ A_{k+1} \geq B_{k+1}\dfrac{\gamma_{2\ell}}{\delta_{2\ell}} + \dfrac{1}{\delta_{2\ell}}, \\ B_{k+1} \geq A_{k+1}\dfrac{\gamma_{2j-1}}{\delta_{2j-1}} + \dfrac{\delta_{2j} - \gamma_{2j} + 1}{\delta_{2j-1}}, & j = \overline{1,\ell}. \end{cases} \tag{47}$$

System (46) is equivalent to the inequality system

$$\begin{cases} A_{k+1} \geq B_{k+1}\dfrac{\gamma_{2i-2}}{\delta_{2i-2}} + \dfrac{\delta_{2i-1} - \gamma_{2i-1} + 1}{\delta_{2i-2}}, & i = \overline{1,\ell+1}, \\ B_{k+1} \geq A_{k+1}\dfrac{\gamma_{2j-1}}{\delta_{2j-1}} + \dfrac{\delta_{2j} - \gamma_{2j} + 1}{\delta_{2j-1}}, & j = \overline{1,\ell}, \\ B_{k+1} \geq A_{k+1}\dfrac{\gamma_{2\ell+1}}{\delta_{2\ell+1}} + \dfrac{1}{\delta_{2\ell+1}}. \end{cases} \tag{48}$$

For the case if $k = 2\ell$, the following inequalities hold:

$$\dfrac{\gamma_{2i}}{\delta_{2i}} \leq 1, \quad i = \overline{0,\ell}; \qquad \dfrac{\gamma_{2j-1}}{\delta_{2j-1}} \leq N, \quad j = \overline{1,\ell}.$$

For the case if $k = 2\ell + 1$, the following inequalities hold:

$$\frac{\gamma_{2i}}{\delta_{2i}} \leq 1, \quad i = \overline{0, \ell}; \quad \frac{\gamma_{2j-1}}{\delta_{2j-1}} \leq N, \quad j = \overline{1, \ell+1}.$$

Thus, it follows from definitions (34), (35) that to satisfy inequalities (47) (for the case if $k = 2\ell$) and inequalities (48) (for the case if $k = 2\ell + 1$) it is sufficient to satisfy inequalities

$$\begin{cases} A_{k+1} \geq B_{k+1} + C_1 \\ B_{k+1} \geq N A_{k+1} + C_2. \end{cases} \quad (49)$$

By (39), inequalities (49) hold because $(\hat{x}, \hat{y}) \in \Omega_0$. Therefore, inequalities (44) are satisfied. Hence, (30) are satisfied. Thus, the induction step is proved. The theorem is proved. □

3. Time-Invariant Stabilization by Static State Feedback

Definition 1. *We say that system (1) is exponentially stabilizable with the decay rate $\theta > 0$ by linear stationary static state feedback (5) if there exist constants $v_1, \ldots, v_n \in \mathbb{R}$ such that every solution $x(t)$ of the closed-loop system (7) is exponentially stable with the decay rate θ, i.e., $x(t)$ along with its derivatives up to $(n-1)$-th order has the form $O(e^{-\theta t})$ as $t \to +\infty$.*

Theorem 3. *System (1) is exponentially stabilizable with an arbitrary pregiven decay rate $\theta > 0$ by linear stationary static state feedback (5).*

Proof. Let an arbitrary $\theta > 0$ be given. Denote $\rho_i := \beta_i - \alpha_i$, $i = \overline{1, n}$, where α_i, β_i are from (2). We have $\rho_i \geq 0$, $i = \overline{1, n}$. We set $L := \max\{1, \rho_1, \sqrt{\rho_2}, \ldots, \sqrt[n]{\rho_n}\}$. Then

$$L \geq 1 > 0, \quad L \geq \rho_1, \quad L^2 \geq \rho_2, \quad \ldots, \quad L^n \geq \rho_n. \quad (50)$$

Set $\eta := \theta/L$. Then $\eta > 0$. Let us construct the polynomials (15), (16) according to Theorem 2 so that properties (i), (ii), (iii) are satisfied. Then the roots $-a_i$ and $-b_i$ ($i = \overline{1, n}$) of the polynomials $f(\lambda)$ and $g(\lambda)$ are real and the following inequalities hold:

$$-a_i \leq -\eta, \quad -b_i \leq -\eta, \quad i = \overline{1, n}. \quad (51)$$

Let us construct the polynomials $P_1(\lambda)$, $P_2(\lambda)$ by formulas (11), (12) where $\omega_i = \delta_i L^i$, $\sigma_i = \gamma_i L^i$, $i = \overline{1, n}$. Then $P_1(\lambda)$ and $P_2(\lambda)$ have the roots $-c_i := -a_i L$ and $-d_i := -b_i L$ ($i = \overline{1, n}$) respectively. These roots are real and by virtue of (51) the following inequalities hold:

$$-c_i \leq -\theta, \quad -d_i \leq -\theta, \quad i = \overline{1, n}. \quad (52)$$

We set $v_i := \alpha_i - \gamma_i L^i$, $i = \overline{1, n}$, in (5) and consider the closed-loop system (7). System (7) has the form (9) where $q_i(t) = p_i(t) - v_i$, $i = \overline{1, n}$. Taking into account inequalities (2), (50) and property (i), for every $i = \overline{1, n}$ for all $t \in \mathbb{R}_+$, we have

$$0 < \sigma_i = \gamma_i L^i = \alpha_i - \alpha_i + \gamma_i L^i \leq p_i(t) - v_i =: q_i(t) \leq$$
$$\leq \beta_i - \alpha_i + \gamma_i L^i = \rho_i + \gamma_i L^i \leq L^i (1 + \gamma_i) \leq \delta_i L^i = \omega_i.$$

Thus, inequalities (10) hold. Applying Theorem 1 and inequalities (52), we obtain that the closed-loop system (7) is exponentially stable with the decay rate θ. The theorem is proved. □

Example 1. Let $n = 2$. Consider a control system (1):

$$x'' + p_1(t)x' + p_2(t)x = u, \quad t \in \mathbb{R}_+, \quad x \in \mathbb{R}, \quad u \in \mathbb{R}. \tag{53}$$

Suppose that $p_1(t), p_2(t)$ satisfy conditions $\alpha_1 \leq p_1(t) \leq \beta_1, \alpha_2 \leq p_2(t) \leq \beta_2, t \in \mathbb{R}_+$. Suppose, for simplicity, that $\rho_1 := \beta_1 - \alpha_1 \leq 1, \rho_2 := \beta_2 - \alpha_2 \leq 1$ (one can achieve this by replacing time $\tilde{x}(t) = x(\mu t)$). Let $\theta > 0$ be an arbitrary number. One needs to construct the controller $u = u(\mathbf{x})$ in (53) where

$$u(\mathbf{x}) = v_1 x' + v_2 x \tag{54}$$

with constant numbers v_1, v_2 such that the closed-loop system

$$x'' + (p_1(t) - v_1)x' + (p_2(t) - v_2)x = 0 \tag{55}$$

is exponentially stable with the decay rate θ. Without loss of generality, we suppose that $\theta \geq 1$. For constructing (54) we use the proof of Theorem 3. We have $L = 1$. Set $\eta := \theta$. Then $\eta \geq 1$. Let us construct the polynomials (15), (16) according to Theorem 2: $f(\lambda) := \lambda^2 + 6\eta\lambda + 5\eta^2, g(\lambda) := \lambda^2 + 5\eta\lambda + 6\eta^2$. Then $\gamma_1 = 5\eta$, $\gamma_2 = 5\eta^2, \delta_1 = 6\eta, \delta_2 = 6\eta^2$. Due to $\eta \geq 1$, condition (i) holds. Next, the equalities $P_1(\lambda) = f(\lambda)$, $P_2(\lambda) = g(\lambda)$ hold. The gain coefficients constructed by Theorem 3 have the form

$$v_1 = \alpha_1 - 5\theta, \quad v_2 = \alpha_2 - 5\theta^2. \tag{56}$$

Let us substitute (56) into (54). The closed-loop system (55) take the form

$$x'' + (s_1(t) + 5\theta)x' + (s_2(t) + 5\theta^2)x = 0, \quad t \in \mathbb{R}_+. \tag{57}$$

Here

$$0 \leq s_1(t) := p_1(t) - \alpha_1 \leq \beta_1 - \alpha_1 = \rho_1 \leq 1 = L,$$
$$0 \leq s_2(t) := p_2(t) - \alpha_2 \leq \beta_2 - \alpha_2 = \rho_2 \leq 1 = L^2.$$

All solutions of (57) are exponentially stable with the decay rate θ. Let us check it. The substitution $z_1 = x, z_2 = x'$ reduces Equation (57) to the system

$$\dot{z} = A(t)z, \quad t \in \mathbb{R}_+,$$

$$z = \begin{bmatrix} z_1 \\ z_2 \end{bmatrix}, \quad A(t) = \begin{bmatrix} 0 & 1 \\ -(s_2(t) + 5\theta^2) & -(s_1(t) + 5\theta) \end{bmatrix}. \tag{58}$$

Let us show that system (58) is exponentially stable with the decay rate θ. The substitution

$$z(t) = e^{-\theta t} y(t). \tag{59}$$

reduce system (58) to the system

$$\dot{y} = B(t)y, \quad t \in \mathbb{R}_+,$$

$$y = \begin{bmatrix} y_1 \\ y_2 \end{bmatrix}, \quad B(t) = \begin{bmatrix} \theta & 1 \\ -(s_2(t) + 5\theta^2) & -(s_1(t) + 4\theta) \end{bmatrix}. \tag{60}$$

Let us show that system (60) is Lyapunov stable. Set $S = \begin{bmatrix} 7\theta^2 & 2\theta \\ 2\theta & 1 \end{bmatrix}$. Then $S > 0$ in the sense of quadratic forms. Next, we have

$$B^T(t)S + SB(t) = \begin{bmatrix} -6\theta^3 - 4\theta s_2(t) & -4\theta^2 - 2\theta s_1(t) - s_2(t) \\ -4\theta^2 - 2\theta s_1(t) - s_2(t) & -4\theta - 2s_1(t) \end{bmatrix}. \tag{61}$$

Here and throughout, T is the transposition. Let us find the principal minors of (61). We obtain

$$\Delta_1 = -2\theta(3\theta^2 + 2s_2(t)) < 0, \quad \Delta_2 = -4\theta - 2s_1(t) < 0,$$
$$\Delta_{1,2} = \det(B^T(t)S + SB(t)) = 8\theta^4 - 4\theta^3 s_1(t) + 8\theta^2 s_2(t) - 4\theta^2 s_1^2(t) + 4\theta s_1(t)s_2(t) - s_2^2(t).$$

We have

$$8\theta^4 - 4\theta^3 s_1(t) - 4\theta^2 s_1^2(t) = 4\theta^3(\theta - s_1(t)) + 4\theta^2(\theta^2 - s_1^2(t)) \geq 0,$$
$$8\theta^2 s_2(t) - s_2^2(t) = s_2(t)(8\theta^2 - s_2(t)) \geq 0.$$

Hence $\Delta_{1,2} \geq 0$. Thus, (61) is negative-semidefinite. Therefore, system (60) is stable. Hence, all solutions of (60) are bounded as $t \to +\infty$. Then, by (59), $\|z(t)\| = O(e^{-\theta t})$, $t \to +\infty$, as required.

As an example of numerical simulation, consider system (53) with $p_1(t) = \frac{t}{1+t^2}$, $p_2(t) = -\frac{1}{1+t^2}$:

$$x'' + \frac{t}{1+t^2}x' - \frac{1}{1+t^2}x = u. \tag{62}$$

We have $\alpha_1 := -1/2 \leq p_1(t) \leq 1/2 =: \beta_1$, $\alpha_2 := -1 \leq p_1(t) \leq 0 =: \beta_2$, $\rho_1 := \beta_1 - \alpha_1 = 1$, $\rho_2 := \beta_2 - \alpha_2 = 1$. The free system (i.e., system (62) with $u = 0$) has a general solution

$$x(t) = C_1 t + C_2 \sqrt{t^2 + 1}$$

and, obviously, is unstable. Let us set $\theta := 1$, $\eta := \theta = 1$. The gain coefficients (56) have the form

$$v_1 = \alpha_1 - 5\theta = -11/2, \quad v_2 = \alpha_2 - 5\theta^2 = -6.$$

The closed-loop system (57) take the form

$$x'' + \left(\frac{11}{2} + \frac{t}{1+t^2}\right)x' + \left(6 - \frac{1}{1+t^2}\right)x = 0. \tag{63}$$

System (63) is exponentially stable with the decay rate $\theta = 1$. Some graphs of the solutions to system (63) are shown in Figure 1.

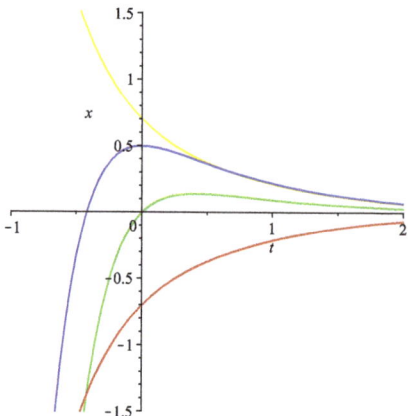

Figure 1. Graphs of the solutions to (63).

4. Time-Invariant Stabilization by Static Output Feedback

Consider a linear control system defined by a linear differential equation of n-th order with time-varying uncertain coefficients satisfying (2); the input is a stationary linear combination of m variables and their derivatives of order $\leq n - p$; the output is a k-dimensional vector of stationary linear combinations of the state x and its derivatives of order $\leq p - 1$:

$$x^{(n)} + \sum_{i=1}^{n} p_i(t) x^{(n-i)} = \sum_{\tau=1}^{m} \sum_{l=p}^{n} b_{l\tau} w_\tau^{(n-l)}, \quad x \in \mathbb{R}, \quad b_{l\tau} \in \mathbb{R}, \quad t \in \mathbb{R}_+, \tag{64}$$

$$y_j = \sum_{\nu=1}^{p} c_{\nu j} x^{(\nu-1)}, \quad j = \overline{1,k}, \quad c_{\nu j} \in \mathbb{R}, \tag{65}$$

$w = \mathrm{col}(w_1, \ldots, w_m) \in \mathbb{R}^m$ is an input vector; $y = \mathrm{col}(y_1, \ldots, y_k) \in \mathbb{R}^k$ is an output vector. Let the control in (64), (65) have the form of linear static output feedback

$$w = Uy. \tag{66}$$

We suppose that the gain matrix U is time-invariant. The closed-loop system has the form

$$x^{(n)} + q_1(t) x^{(n-1)} + \ldots + q_n(t) x = 0, \quad t \in \mathbb{R}_+, \tag{67}$$

where the coefficients $q_i(t)$ of (67) depends on $p_i(t)$, $b_{l\tau}$, $c_{\nu j}$, U. On the basis of system (64), (65), we construct the $n \times m$-matrix $B = \{b_{l\tau}\}$, $l = \overline{1,n}$, $\tau = \overline{1,m}$, and the $n \times k$-matrix $C = \{c_{\nu j}\}$, $\nu = \overline{1,n}$, $j = \overline{1,k}$, where $b_{l\tau} = 0$ for $l < p$ and $c_{\nu j} = 0$ for $\nu > p$. Denote by J the matrix whose entries of the first superdiagonal are equal to unity and whose remaining entries are zero; we set $J^0 := I$. By $\mathrm{Sp}\, Q$ denote the trace of a matrix Q.

Definition 2. *We say that system (64), (65) is exponentially stabilizable with the decay rate $\theta > 0$ by linear stationary static output feedback (66) if there exists a constant $m \times k$-matrix U such that every solution $x(t)$ of the closed-loop system (67) is exponentially stable with the decay rate θ.*

Theorem 4. *Suppose that linear stationary output feedback (66) bring system (64), (65) to the closed system (67). Then the coefficients $q_i(t)$, $i = \overline{1,n}$, of (67) satisfy the equalities*

$$q_i(t) = p_i(t) - r_i,$$

where
$$r_i = \text{Sp}(C^T J^{i-1} BU), \quad i = \overline{1, n}. \tag{68}$$

The proof of Theorem 4 is identical to the proof of Theorem 1 [38].

Let us introduce the mapping vec that unwraps an $n \times m$-matrix $H = \{h_{ij}\}$ row-by-row into the column vector $\text{vec } H = \text{col}(h_{11}, h_{12}, \ldots, h_{1m}, \ldots, h_{n1}, \ldots, h_{nm})$. For any $k \times m$-matrices X, Y, the obvious equality holds:
$$\text{Sp}(XY^T) = (\text{vec } X)^T \cdot (\text{vec } Y). \tag{69}$$

Let us construct the $k \times m$-matrices
$$C^T J^0 B, \ C^T JB, \ \ldots, \ C^T J^{n-1} B \tag{70}$$

and the $mk \times n$-matrix
$$P = [\text{vec}(C^T J^0 B), \ldots, \text{vec}(C^T J^{n-1} B)].$$

Denote $r = \text{col}(r_1, \ldots, r_n) \in \mathbb{R}^n$, $\psi = \text{vec}(U^T)$. Equalities (68) represent a linear system of n equations with respect to the coefficients of the matrix U. Taking into account (69), one can rewrite system (68) in the form
$$P^T \psi = r. \tag{71}$$

Suppose that matrices (70) are linearly independent. Then rank $P = n$. Hence, the system of linear equations (71) is solvable for any vector $r \in \mathbb{R}^n$. In particular, system (71) has the solution $\psi = P(P^T P)^{-1} r$.

By Theorem 3, for any pregiven $\theta > 0$ there exists a constant vector $r = \text{col}(r_1, \ldots, r_n)$ such that system (67) with $q_i(t) = p_i(t) - r_i$ is exponentially stable with the decay rate θ. Resolving system (71) for that r with respect to ψ and constructing U by the formula $U = (\text{vec}^{-1} \psi)^T$, we find the gain matrix of feedback (66) exponentially stabilizing system (64), (65) with the decay rate θ. Thus, the following theorem is proved.

Theorem 5. *System (64), (65) is exponentially stabilizable with an arbitrary pregiven decay rate $\theta > 0$ by linear stationary static output feedback (66) if matrices (70) are linearly independent.*

Example 2. Let $n = 3$. Consider a control system
$$x''' + p(t)x = w_1' + w_1 - w_2' + w_2, \quad t \in \mathbb{R}_+, \quad x \in \mathbb{R}, \quad w = \text{col}(w_1, w_2) \in \mathbb{R}^2, \tag{72}$$
$$y_1 = x - x', \quad y_2 = x + x', \quad y = \text{col}(y_1, y_2) \in \mathbb{R}^2. \tag{73}$$

System (72), (73) has the form (64), (65) where $n = 3$, $m = k = p = 2$. Suppose that $p(t)$ is an arbitrary measurable function satisfying the condition $0 \le p(t) \le 1$. Let $\theta > 0$ be an arbitrary number. One needs to construct feedback control (66), where $U = \{u_{ij}\}_{i,j=1}^2$, with constant u_{ij}, $i, j = 1, 2$, providing exponential stability of the closed-loop system with the decay rate θ. Without loss of generality, we suppose that $\theta \ge 1$. By Theorem 4, the closed-loop system has the form
$$x''' - r_1 x'' - r_2 x' + (p(t) - r_3) x = 0, \tag{74}$$

where r_i have the form (68), and
$$B = \begin{bmatrix} 0 & 0 \\ 1 & -1 \\ 1 & 1 \end{bmatrix}, \quad C = \begin{bmatrix} 1 & 1 \\ -1 & 1 \\ 0 & 0 \end{bmatrix}.$$

At first, let us construct a constant vector $r = \text{col}(r_1, r_2, r_3)$, providing exponential stability of (74). For constructing r we use the proof of Theorem 3. We have $\alpha_1 = \beta_1 = 0$, $\alpha_2 = \beta_2 = 0$, $\alpha_3 = 0$, $\beta_3 = 1$. Then $\rho_1 = 0$, $\rho_2 = 0$, $\rho_3 = 1$, $L = 1$. Set $\eta := \theta$. Using the proof of Theorem 2, we construct the polynomials (15), (16) such that properties (i), (ii), (iii) are satisfied:

$$f(\lambda) := (\lambda + 2\eta)(\lambda + 3\eta)(\lambda + 14\eta) = \lambda^3 + 19\eta\lambda^2 + 76\eta^2\lambda + 84\eta^3,$$
$$g(\lambda) := (\lambda + \eta)(\lambda + 5\eta)(\lambda + 12\eta) = \lambda^3 + 18\eta\lambda^2 + 77\eta^2\lambda + 60\eta^3.$$

Then $\gamma_1 = 18\eta$, $\gamma_2 = 76\eta^2$, $\gamma_3 = 60\eta^3$, $\delta_1 = 19\eta$, $\delta_2 = 77\eta^2$, $\delta_3 = 84\eta^3$. Conditions (i), (ii), (iii) hold. Coefficients r_1, r_2, r_3 have the form

$$r_1 = -18\theta, \quad r_2 = -76\theta^2, \quad r_3 = -60\theta^3. \tag{75}$$

Let us substitute (75) into (74). The closed-loop system (74) take the form

$$x''' + 18\theta x'' + 76\theta^2 x' + (p(t) + 60\theta^3)x = 0. \tag{76}$$

All solutions of (76) are exponentially stable with the decay rate θ. Let us check it. The substitution $z_1 = x$, $z_2 = x'$, $z_3 = x''$ reduces Equation (76) to the system

$$\dot{z} = A(t)z, \quad t \in \mathbb{R}_+, \tag{77}$$

$$z = \begin{bmatrix} z_1 \\ z_2 \\ z_3 \end{bmatrix}, \quad A(t) = \begin{bmatrix} 0 & 1 & 0 \\ 0 & 0 & 1 \\ -(p(t) + 60\theta^3) & -76\theta^2 & -18\theta \end{bmatrix}.$$

Let us show that the system (77) is exponentially stable with the decay rate θ. The substitution

$$z(t) = e^{-\theta t} y(t). \tag{78}$$

reduce the system (77) to the system

$$\dot{y} = B(t)y, \quad t \in \mathbb{R}_+, \tag{79}$$

$$y = \begin{bmatrix} y_1 \\ y_2 \\ y_3 \end{bmatrix}, \quad B(t) = \begin{bmatrix} \theta & 1 & 0 \\ 0 & \theta & 1 \\ -(p(t) + 60\theta^3) & -76\theta^2 & -17\theta \end{bmatrix}.$$

Let us show that system (79) is Lyapunov stable. Set $S = \begin{bmatrix} 9000\theta^4 & 2580\theta^3 & 150\theta^2 \\ 2580\theta^3 & 804\theta^2 & 46\theta \\ 150\theta^2 & 46\theta & 3 \end{bmatrix}$. Let us find the successive principal minors s_i, $i = 1, 2, 3$, of S. We have $s_1 = 9000\theta^4 > 0$, $s_2 = \det \begin{bmatrix} 9000\theta^4 & 2580\theta^3 \\ 2580\theta^3 & 804\theta^2 \end{bmatrix} = 579{,}600\theta^6 > 0$, $s_3 = \det S = 208{,}800\theta^6 > 0$. Then $S > 0$ in the sense of quadratic forms. Next, we have

$$B^T(t)S + SB(t) = \begin{bmatrix} -300\theta^2 p(t) & -46\theta p(t) & -3p(t) \\ -46\theta p(t) & -224\theta^3 & -10\theta^2 \\ -3p(t) & -10\theta^2 & -10\theta \end{bmatrix}. \tag{80}$$

Let us find the principal minors of (80). We obtain

$$\Delta_1 = -300\theta^2 p(t) \leq 0, \quad \Delta_2 = -224\theta^3 < 0, \quad \Delta_3 = -10\theta < 0,$$
$$\Delta_{1,2} = 67{,}200\theta^5 p(t) - 2116\theta^2 p^2(t) = 4\theta^2 p(t)(16{,}800\theta^3 - 529p(t)) \geq 0,$$
$$\Delta_{1,3} = 3000\theta^3 p(t) - 9p^2(t) = 3p(t)(1000\theta^3 - 3p(t)) \geq 0, \quad \Delta_{2,3} = 2140\theta^4 > 0,$$
$$\Delta_{1,2,3} = \det(B^T(t)S + SB(t)) = -642{,}000\theta^6 p(t) + 20{,}416\theta^3 p^2(t) = -16\theta^3 p(t)(40{,}125\theta^3 - 1276p(t)) \leq 0.$$

Hence, (80) is negative-semidefinite. Thus, the system (79) is stable. Hence, all solutions of (79) are bounded as $t \to +\infty$. Then, by (78), $\|z(t)\| = O(e^{-\theta t})$, $t \to +\infty$, as required.

Next, let us construct matrices (70) and P. We obtain $P = \begin{bmatrix} -1 & 0 & 1 \\ 1 & -2 & 1 \\ 1 & 2 & 1 \\ -1 & 0 & 1 \end{bmatrix}$. Obviously, rank $P = 3$ and matrices (70) are linearly independent. Resolving system (71) where r_i has the form (75), we obtain

$$\psi = \text{col}\,[9\theta/2 - 15\theta^3, -9\theta/2 + 19\theta^2 - 15\theta^3, -9\theta/2 - 19\theta^2 - 15\theta^3, 9\theta/2 - 15\theta^3].$$

Thus, the gain matrix has the form

$$U = \begin{bmatrix} 9\theta/2 - 15\theta^3 & -9\theta/2 - 19\theta^2 - 15\theta^3 \\ -9\theta/2 + 19\theta^2 - 15\theta^3 & 9\theta/2 - 15\theta^3 \end{bmatrix}. \tag{81}$$

We obtain that feedback (66) with the matrix (81) exponentially stabilizes the system (72), (73) with the decay rate θ.

As an example of numerical simulation, consider system (72), (73) where

$$\widehat{p}(t) = \begin{cases} 1, & t \in [0,1), \\ 0, & t \in [1,2), \end{cases} \quad p(t) = \widehat{p}(t - 2k), \quad t \in [2k, 2(k+1)), \quad k \in \mathbb{Z}.$$

We have $0 \leq p(t) \leq 1$. The function $p(t)$ is ω-periodic with the period $\omega = 2$. The free system

$$x''' + p(t)x = 0, \quad x \in \mathbb{R}, \tag{82}$$

is equivalent to the system of differential equations

$$\dot{z} = \begin{bmatrix} 0 & 1 & 0 \\ 0 & 0 & 1 \\ -p(t) & 0 & 0 \end{bmatrix} z, \quad z \in \mathbb{R}^3. \tag{83}$$

System (83) is ω-periodic. Since system (83) is piecewise constant, the monodromy matrix $\Phi(\omega)$ for system (83) can be found explicitly. Calculating approximately eigenvalues λ_1, λ_2, and λ_3 of $\Phi(\omega)$, we obtain $\lambda_{1,2} \approx 0.418 \pm 2.167i$, $\lambda_3 \approx 0.205$. Hence, $|\lambda_1| = |\lambda_2| > 1$. Thus, system (83) (and hence, Equation (82)) is unstable. Let us set $\theta := 1$, $\eta := \theta = 1$. The gain matrix (81) has the form

$$U = \begin{bmatrix} -21/2 & -77/2 \\ -1/2 & -21/2 \end{bmatrix}.$$

The closed-loop system (76) take the form

$$x''' + 18x'' + 76x' + (p(t) + 60)x = 0. \tag{84}$$

System (84) is exponentially stable with the decay rate $\theta = 1$. Some graphs of the solutions to system (84) are shown in Figure 2.

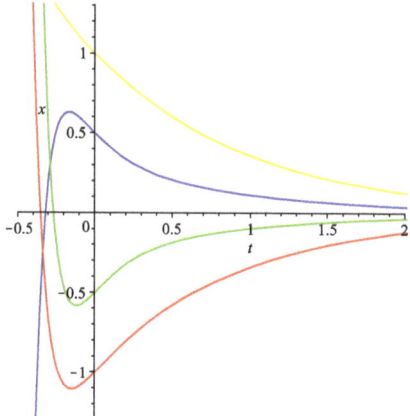

Figure 2. Graphs of the solutions to (84).

5. Conclusions

We examined the problem of exponential stabilization with any pregiven decay rate for a linear time-varying differential equations with uncertain bounded coefficients by means of stationary linear static feedback. We have received sufficient conditions for the solvability of this problem by state and output feedback. For this purpose, the first Lyapunov method and the Levin theorem on non-oscillatory absolute stability were used. We plan to extend these results to systems of differential equation including systems with delays. A further development of these results may be their extension to systems (64), (65), (66), when $b_{l\tau}$ and (or) c_{vj} depend on t. So far this question remains open.

Author Contributions: All authors contributed equally to this manuscript. All authors have read and agreed to the published version of the manuscript.

Funding: This research was funded by the Ministry of Science and Higher Education of the Russian Federation in the framework of state assignment No. 075-00232-20-01, project 0827-2020-0010 "Development of the theory and methods of control and stabilization of dynamical systems" and by the Russian Foundation for Basic Research (project 20–01–00293).

Acknowledgments: The research was performed using computing resources of the collective use center of IMM UB RAS "Supercomputer center of IMM UB RAS".

Conflicts of Interest: The authors declare no conflict of interest.

References

1. Aeyels, D.; Peuteman, J. Uniform asymptotic stability of linear time-varying systems. In *Open Problems in Mathematical Systems and Control Theory*; Blondel, V., Sontag, E.D., Vidyasagar, M., Willems, J.C., Eds; Springer: London, UK, 1999; pp. 1–5. [CrossRef]
2. Ilchmann, A.; Owens, D.H.; Prätzel-Wolters, D. Sufficient conditions for stability of linear time-varying systems. *Syst. Control Lett.* **1987**, *9*, 157–163. [CrossRef]
3. Bylov, B.F.; Vinograd, R.E.; Grobman, D.M.; Nemytskii, V.V. *Theory of Lyapunov Exponents*; Nauka: Moscow, Russia, 1966.
4. Demidovich, B.P. *Lectures on the Mathematical Stability Theory*; Nauka: Moscow, Russia, 1967.
5. Zhu, J.J. A necessary and sufficient stability criterion for linear time-varying systems. In Proceedings of the 28th Southeastern Symposium on System Theory, Baton Rouge, Louisiana, USA, 31 March–2 April 1996; pp. 115–119. [CrossRef]
6. Levin, A.Y. Absolute nonoscillatory stability and related questions. *St. Petersburg Math. J.* **1993**, *4*, 149–161.
7. Ragusa, M.A. Necessary and sufficient condition for a VMO function. *Appl. Math. Comput.* **2012**, *218*, 11952–11958. [CrossRef]

8. Zhou, B. On asymptotic stability of linear time-varying systems. *Automatica* **2016**, *68*, 266–276. [CrossRef]
9. Wan, J.-M. Explicit solution and stability of linear time-varying differential state space systems. *Int. J. Control Autom. Syst.* **2017**, *15*, 1553–1560. [CrossRef]
10. Vrabel, R. A note on uniform exponential stability of linear periodic time-varying systems. *IEEE Trans. Autom. Control* **2020**, *65*, 1647–1651. [CrossRef]
11. Zhou, B.; Tian, Y.; Lam, J. On construction of Lyapunov functions for scalar linear time-varying systems. *Syst. Control Lett.* **2020**, *135*, 104591. [CrossRef]
12. Kharitonov, V.L. The asymptotic stability of the equilibrium state of a family of systems of linear differential equations. *Differ. Uravn.* **1978**, *14*, 2086–2088.
13. Petersen, I.R. A stabilization algorithm for a class of uncertain linear systems. *Syst. Control Lett.* **1987**, *8*, 351–357. [CrossRef]
14. Zhou, K.; Khargonekar, P.P. Robust stabilization of linear systems with norm-bounded time-varying uncertainty. *Syst. Control Lett.* **1988**, *10*, 17–20. [CrossRef]
15. Khargonekar, P.P.; Petersen, I.R.; Zhou, K. Robust stabilization of uncertain linear systems: Quadratic stabilizability and H^∞ control theory. *IEEE Trans. Autom. Control* **1990**, *35*, 356–361. [CrossRef]
16. Xie, L.; de Souza, C.E. Robust H_∞ control for linear systems with norm-bounded time-varying uncertainty. *IEEE Trans. Autom. Control* **1992**, *37*, 1188–1191. [CrossRef]
17. Zhabko, A.P.; Kharitonov, V.L. Necessary and sufficient conditions for the stability of a linear family of polynomials. *Autom. Remote Control* **1994**, *55*, 1496–1503.
18. Kharitonov, V.L. Robust stability analysis of time delay systems: A survey. *Annu. Rev. Control* **1999**, *23*, 185–196. [CrossRef]
19. Sadabadi, M.S.; Peaucelle, D. From static output feedback to structured robust static output feedback: A survey. *Annu. Rev. Control* **2016**, *42*, 11–26. [CrossRef]
20. Blanchini, F.; Colaneri, P. Uncertain systems: Time-varying versus time-invariant uncertainties. In *Uncertainty in Complex Networked Systems. Systems and Control: Foundations and Applications*; Başar, T., Ed.; Birkhäuser: Cham, Switzerland, 2018. [CrossRef]
21. Carniato, L.A.; Carniato, A.A.; Teixeira, M.C.M.; Cardim, R.; Mainardi Junior, E.I.; Assunção E. Output control of continuous-time uncertain switched linear systems via switched static output feedback. *Int. J. Control* **2018**, *93*, 1127–1146. [CrossRef]
22. Gu, D.-K.; Liu, G.-P.; Duan, G.-R. Robust stability of uncertain second-order linear time-varying systems. *J. Frankl. Instit.* **2019**, *356*, 9881–9906. [CrossRef]
23. Barmish, B.R. Necessary and sufficient conditions for quadratic stabilizability of an uncertain system. *J. Optim. Theory Appl.* **1985**, *46*, 399–408. [CrossRef]
24. Xie, L.; Shishkin, S.; Fu, M. Piecewise Lyapunov functions for robust stability of linear time-varying systems. *Syst. Control Lett.* **1997**, *31*, 165–171. [CrossRef]
25. Ramos, D.C.W.; Peres, P.L.D. An LMI approach to compute robust stability domains for uncertain linear systems. In Proceedings of the 2001 American Control Conference, Arlington, VA, USA, 25–27 June 2001. [CrossRef]
26. Montagner, V.F.; Peres, P.L.D. A new LMI condition for the robust stability of linear time varying systems. In Proceedings of the 42nd IEEE Conference on Decision and Control, Maui, HI, USA, 9–12 December 2003. [CrossRef]
27. Bliman, P.A. A convex approach to robust stability for linear systems with uncertain scalar parameters. *SIAM J. Control Optim.* **2004**, *42*, 2016–2042. [CrossRef]
28. Geromel, J.C.; Colaneri, P. Robust stability of time varying polytopic systems. *Syst. Control Lett.* **2006**, *55*, 81–85. [CrossRef]
29. Chesi, G.; Garulli, A.; Tesi, A.; Vicino, A. Robust stability of time-varying polytopic systems via parameter-dependent homogeneous Lyapunov functions. *Automatica* **2007**, *43*, 309–316. [CrossRef]
30. Hu, T.; Blanchini, F. Non-conservative matrix inequality conditions for stability/stabilizability of linear differential inclusions. *Automatica* **2010**, *46*, 190–196. [CrossRef]
31. Chesi, G. Sufficient and necessary LMI conditions for robust stability of rationally time-varying uncertain systems. *IEEE Trans. Autom. Control* **2013**, *58*, 1546–1551. [CrossRef]

32. Gritli, H.; Belghith, S. New LMI conditions for static output feedback control of continuous-time linear systems with parametric uncertainties. In Proceedings of the 2018 European Control Conference (ECC), Limassol, Cyprus, 12–15 June 2018. [CrossRef]
33. Gritli, H.; Belghith, S.; Zemouche, A. LMI-based design of robust static output feedback controller for uncertain linear continuous systems. In Proceedings of the 2019 International Conference on Advanced Systems and Emergent Technologies (IC_ASET), Hammamet, Tunisia, 19–22 March 2019. [CrossRef]
34. Gelig, A.H.; Zuber, I.E. Invariant stabilization of classes of uncertain systems with delays. *Autom. Remote Control* **2011**, *72*, 1941–1950. [CrossRef]
35. Zakharenkov, M.; Zuber, I.; Gelig, A. Stabilization of new classes of uncertain systems. *IFAC-PapersOnLine* **2015**, *48*, 1024–1027. [CrossRef]
36. Gelig, A.H.; Zuber, I.E.; Zakharenkov, M.S. New classes of stabilizable uncertain systems. *Autom. Remote Control* **2016**, *77*, 1768–1780. [CrossRef]
37. Gelig, A.K.; Zuber, I.E. Multidimensional output stabilization of a certain class of uncertain systems. *Autom. Remote Control* **2018**, *79*, 1545–1557. [CrossRef]
38. Zaitsev, V.A. Modal control of a linear differential equation with incomplete feedback. *Differ. Equ.* **2003**, *39*, 145–148. [CrossRef]

© 2020 by the authors. Licensee MDPI, Basel, Switzerland. This article is an open access article distributed under the terms and conditions of the Creative Commons Attribution (CC BY) license (http://creativecommons.org/licenses/by/4.0/).

Article

The Sign of the Green Function of an n-th Order Linear Boundary Value Problem

Pedro Almenar Belenguer [1,*,†] **and Lucas Jódar** [2,†]

1. Vodafone Spain, Avda. América 115, 28042 Madrid, Spain
2. Instituto Universitario de Matemática Multidisciplinar, Universitat Politècnica de València, Camino de Vera s/n, 46022 Valencia, Spain; ljodar@imm.upv.es
* Correspondence: pedro.almenar1@vodafone.com
† These authors contributed equally to this work.

Received: 28 March 2020; Accepted: 23 April 2020; Published: 29 April 2020

Abstract: This paper provides results on the sign of the Green function (and its partial derivatives) of an n-th order boundary value problem subject to a wide set of homogeneous two-point boundary conditions. The dependence of the absolute value of the Green function and some of its partial derivatives with respect to the extremes where the boundary conditions are set is also assessed.

Keywords: n-th order linear differential equation; two-point boundary value problem; Green function

MSC: 34B05; 34B27; 34C10; 34C11

1. Introduction

Let J be a compact interval in $\mathbb{R}$ and let us consider the real disfocal differential operator $L: C^n(J) \to C(J)$ defined by

$$Ly = y^{(n)}(x) + a_{n-1}(x)y^{(n-1)}(x) + \cdots + a_0(x)y(x), \quad x \in J, \tag{1}$$

where $a_j(x) \in C(J)$, $0 \le j \le n-1$. Following Eloe and Ridenhour [1], let Ω_l be the set whose members are collections of l different ordered integer indices i such that $0 \le i \le n-1$, let $k \in \mathbb{N}$ be such that $1 \le k \le n-1$, let $\alpha \in \Omega_k$ be the set $\{\alpha_1, \ldots, \alpha_k\}$ and $\beta \in \Omega_{n-k}$ be the set $\{\beta_1, \ldots, \beta_{n-k}\}$, both associated to the homogeneous boundary conditions

$$y^{(\alpha_i)}(a) = 0, \quad i = 1, 2, \ldots, k, \quad \alpha_i \in \alpha, \tag{2}$$

$$y^{(\beta_i)}(b) = 0, \quad i = 1, 2, \ldots, n-k, \quad \beta_i \in \beta, \tag{3}$$

where $[a, b] \subset J$. Throughout this paper we will impose the condition that, for any integer m such that $1 \le m \le n$, at least m terms of the sequence $\alpha_1, \ldots, \alpha_k, \beta_1, \ldots, \beta_{n-k}$ are less than m. Due to their resemblance with the conditions defined by Butler and Erbe in [2], we will call them admissible boundary conditions (note that (2) and (3) are not exactly the same boundary conditions defined by Butler and Erbe since the latter applied to the so-called quasiderivatives of $y(x)$ and not to derivatives). In particular, if for every integer m such that $1 \le m \le p+1$, exactly m terms of the sequence $\alpha_1, \ldots, \alpha_k, \beta_1, \ldots, \beta_{n-k}$ are less than m, we will say that the boundary conditions are p-alternate. In the case $p = n - 1$ we will call the boundary conditions strongly admissible. The admissible conditions cover well known cases like conjugate boundary conditions ($\alpha_1 = 0, \alpha_2 = 1, \ldots, \alpha_k = k-1$ and $\beta_1 = 0, \beta_2 = 1, \ldots, \beta_{n-k} = n-k-1$), focal boundary conditions (right focal with $\alpha_1 = 0, \alpha_2 = 1, \ldots, \alpha_k = k-1$ and $\beta_1 = k, \beta_2 = k+1, \ldots, \beta_{n-k} = n-1$ or left focal with $\alpha_1 = n-k, \alpha_2 = n-k+1, \ldots, \alpha_k = n-1$ and

$\beta_1 = 0, \beta_2 = 1, \ldots, \beta_{n-k} = n-k-1$) and many other. The focal boundary conditions are also strongly admissible (or $(n-1)$-alternate).

The purpose of this paper will be to provide results on the sign of $G(x,t)$, the Green function associated to the problem

$$Ly = 0, \quad x \in (a,b),$$
$$y^{(\alpha_i)}(a) = 0, \; \alpha_i \in \alpha; \quad y^{(\beta_i)}(b) = 0, \; \beta_i \in \beta, \tag{4}$$

as well as some of its partial derivatives with regards to x, both in the interval (a,b) and at the extremes a and b. We will also analyze the dependence of the absolute value of $G(x,t)$ and its derivatives with respect to the extremes a and b. In this sense, this paper represents an extension of the work by Eloe and Ridenhour [1] which in turn extended previous results from Peterson [3,4], Elias [5] and Peterson and Ridenhour [6]. Note that the disfocality of L on $[a,b]$, according to Nehari [7], implies that $y(x) \equiv 0$ is the only solution of $Ly = 0$ satisfying $y^{(i)}(x_i) = 0$, $i = 0, 1, 2, \ldots, n-1$, with $x_i \in [a,b]$, and also guarantees the existence of the Green function of (4).

It is well known (see for instance [8], Chapter 3) that problems of the type

$$Ly = f, \quad x \in (a,b),$$
$$y^{(\alpha_i)}(a) = 0, \; \alpha_i \in \alpha; \quad y^{(\beta_i)}(b) = 0, \; \beta_i \in \beta, \tag{5}$$

with $f \in C[a,b]$ being an input function, have a solution given by $y(x) = \int_a^b G(x,t)f(t)dt$. Therefore, the knowledge of the sign of $G(x,t)$ and its derivatives can provide information on the sign of the solution $y(x)$ and these same derivatives, at least when f does not change sign on (a,b). This was already used by Eloe and Ridenhour in [1] to show that a clamped beam is stiffer that a simply supported beam. Likewise, the evolution of $G(x,t)$ as a or b vary can also provide insights on the dependence of the value of $y(x)$ on these extremes and can allow comparing the effect of a longer separation of the extremes when the same input function f is applied to a system modeled by (5).

The knowledge about the sign of $G(x,t)$ is also useful to find information about the eigenvalues and eigenfunctions of the general problem

$$Ly = \lambda \sum_{l=0}^{\mu} c_l(x) y^{(l)}(x), \quad x \in (a,b),$$
$$y^{(\alpha_i)}(a) = 0, \; \alpha_i \in \alpha; \quad y^{(\beta_i)}(b) = 0, \; \beta_i \in \beta, \tag{6}$$

with $\mu \leq n-1$, $c_l(x) \in C(J)$ for $0 \leq l \leq \mu$. These problems are tackled by converting them in the equivalent integral problem

$$My(x) = \frac{1}{\lambda} y(x), \quad x \in [a,b], \tag{7}$$

where M is the operator $M: C^\mu[a,b] \to C^n[a,b]$ defined by

$$My(x) = \int_a^b G(x,t) \sum_{l=0}^{\mu} c_l(t) y^{(l)}(t) dt, \quad x \in [a,b]. \tag{8}$$

If the partial derivative of $G(x,t)$ of the highest order whose sign is constant on (a,b) is not lower than μ, it is possible to define a cone P associated to that partial derivative such that $MP \subset P$ and, with the help of the cone theory elaborated by Krein and Rutman [9] and Krasnosel'skii [10], prove that there exists a solution of (7) associated to the smallest eigenvalue λ. Moreover, it is possible to determine some properties of λ and even compare the values of λ for different boundary conditions.

Refs. [11–17] are examples that follow this approach. In all these, therefore, the knowledge of the sign of the derivatives of $G(x,t)$ is critical.

The non-linear version of (6), namely

$$Ly = \lambda f(y,x), \quad x \in (a,b), \tag{9}$$

subject to different homogeneous, mixed or integral boundary conditions (see for instance [18,19]), is also addressed usually by converting it in the integral problem

$$\frac{1}{\lambda}y = \int_a^b G(x,t)f(y(t),t)dt, \quad x \in (a,b). \tag{10}$$

In most of these problems, the information about the sign of the Green function is relevant to apply other tools (fixed-point theorems, upper and lower solutions method, fixed-point index theory, etc.) to determine the existence of a solution. In some of them, the knowledge of the sign of the partial derivatives can help to achieve the same goal ([18,20,21]).

As for a physical applicability, problems of the type (5), (6) and (9) appear in many situations, like the study of the deflections of beams, both straight ones with non-homogeneous cross-sections in free vibration (which are subject to the fourth-order linear Euler-Bernoulli equation) and curved ones with different shapes. An account of these and other applications can be found in [22], Chapter IV.

Throughout the paper we will use the terms $G(\alpha,\beta,x,t)$ and $G_{a,b}(x,t)$ (and further $G_{a,b}(\alpha,\beta,x,t)$) when we want to highlight the dependence of the Green function of (4) on the boundary conditions (α,β) and the extremes a, b, respectively. That will be particularly useful when we manipulate Green functions subject to different boundary conditions or different extremes. We will denote by $H(x,t)$ and $I(x,t)$ the partial derivatives of $G(\alpha,\beta,x,t)$ with respect to the extreme b and a, respectively, that is

$$H(x,t) = \frac{\partial G(x,t)}{\partial b}, \quad I(x,t) = \frac{\partial G(x,t)}{\partial a}, \quad (x,t) \in [a,b] \times [a,b]. \tag{11}$$

We will say that a, b are interior to A, B if $A \leq a < b \leq B$ and $A < a$ or $b < B$. We will use the expression $card\{D\}$ to denote the number of elements (or cardinal) of the set D.

Likewise, if we assume that y is a function with $(n-1)th$ derivative in $[a,b]$, we will make use of the following nomenclature associated to (α,β):

- $K(\alpha,\beta)$ is the minimum derivative of $y(x)$ for which the boundary conditions (α,β) specify that $y^{(i)}(a) = 0$ or $y^{(i)}(b) = 0$ for $i = K(\alpha,\beta) + 1, \ldots, n-1$, with $K(\alpha,\beta) = n-1$ if both $y^{(n-1)}(a) \neq 0$ and $y^{(n-1)}(b) \neq 0$.
- $m(\alpha,i)$ is the number of derivatives of y of order equal or higher than i which the boundary conditions α do not specify to be zero at a.
- $n(\beta,i)$ is the number of derivatives of y of order higher than i which the boundary conditions β do specify to be zero at b.
- $\alpha_A \in \alpha$ is the greatest index such that $y^{(j)}(a) = 0$ for $\alpha_1 \leq j \leq \alpha_A$ and $y^{(j)}(b) \neq 0$ for $\alpha_1 \leq j \leq \alpha_A - 1$, and $\beta_B \in \beta$ is the greatest index such that $y^{(j)}(b) = 0$ for $\beta_1 \leq j \leq \beta_B$ and $y^{(j)}(a) \neq 0$ for $\beta_1 \leq j \leq \beta_B - 1$. Note that if $\beta_B \notin \alpha$ then the boundary conditions are p-alternate with $p > \beta_B$, whereas if $\beta_B \in \alpha$ and $\beta_B > 0$ then the boundary conditions are $(\beta_B - 1)$-alternate.
- $S(\alpha)$ is the sum of all indices of α. Likewise, $S(\beta)$ is the sum of all indices of β.

To make these definitions clear, let us use some examples. Let us assume that $n = 8$, $k = 4$, $\alpha = \{0,1,2,5\}$ and $\beta = \{3,4,5,7\}$. Then $\alpha_A = 2$ (since $3 \notin \alpha$), $\beta_B = 5$ (since $6 \notin \beta$ but also $5 \in \alpha$), $K(\alpha,\beta) = 6$ (since $6 \notin \alpha \cup \beta$ and $7 \in \beta$), $S(\alpha) = 0 + 1 + 2 + 5 = 8$ and $S(\beta) = 3 + 4 + 5 + 7 = 19$. Likewise, let us assume that $n = 7$, $k = 2$, $\alpha = \{3,5\}$ and $\beta = \{0,1,2,4,5\}$. Then $\alpha_A = 3$, $\beta_B = 2$, $K(\alpha,\beta) = 6$, $S(\alpha) = 8$ and $S(\beta) = 12$.

As for the organization of the paper, Section 2 will provide the main results of the paper. Concretely, in the Section 2.1 we will tackle the general case of admissible boundary conditions, in the Section 2.2 we will prove some additional results associated to p-alternate boundary conditions and in the Section 2.3 we will cover the strongly admissible boundary conditions. Finally in Section 3 we will elaborate some conclusions.

2. Results

2.1. The Sign of the Green Function and Its Derivatives on the Admissible Case

In this subsection, we will prove some basic results concerning the sign of the Green function of the problem (4) and its derivatives, as well as comparisons of their absolute values when the extremes a and b vary. To this end, it is interesting to recall a couple of results from Eloe and Ridenhour, which we will state (modified slightly using our notations) for completeness.

Theorem 1 (Theorem 3.3 of [1]). 1. If $\alpha_1 = 0$, then for $i = 0, \ldots, \beta_1$,

$$(-1)^{n-k} \frac{\partial^i G(x,t)}{\partial x^i} > 0, \quad (x,t) \in (a,b) \times (a,b). \tag{12}$$

2. If $\beta_1 = 0$, then for $i = 0, \ldots, \alpha_1$,

$$(-1)^{n-k+i} \frac{\partial^i G(x,t)}{\partial x^i} > 0, \quad (x,t) \in (a,b) \times (a,b). \tag{13}$$

Theorem 2 (Theorem 3.4 of [1]). Let us suppose that $\max(\alpha_k, \beta_{n-k}) < n - 1$, and that a_1, b_1 are extremes interior to a_2, b_2, with $[a_2, b_2] \subset J$.

1. If $\alpha_1 = 0$, then for $i = 0, \ldots, \beta_1$,

$$(-1)^{n-k} \frac{\partial^i G_{a_2,b_2}(x,t)}{\partial x^i} > (-1)^{n-k} \frac{\partial^i G_{a_1,b_1}(x,t)}{\partial x^i} > 0, \quad (x,t) \in (a_1,b_1) \times (a_1,b_1). \tag{14}$$

2. If $\beta_1 = 0$, then for $i = 0, \ldots, \alpha_1$,

$$(-1)^{n-k+i} \frac{\partial^i G_{a_2,b_2}(x,t)}{\partial x^i} > (-1)^{n-k+i} \frac{\partial^i G_{a_1,b_1}(x,t)}{\partial x^i} > 0, \quad (x,t) \in (a_1,b_1) \times (a_1,b_1). \tag{15}$$

These theorems, although of considerable scope, unfortunately, do not yield information on the sign of all the partial derivatives of $G(x,t)$ at the extremes a and b, whose knowledge is necessary for the application of cone theory to the eigenvalue problem (6) mentioned in the Introduction, as well as for the analysis of the strongly admissible case (see Section 2.3). Likewise, they do not cover the dependence of $G(x,t)$ with the extremes a and b when either α_k or β_{n-k} are equal to $n-1$. These shortcomings and the lack of explicit proofs of these theorems in [1] (the reader is left to obtain them following the techniques devised by the authors in previous sections of that paper) lead us to dedicate this subsection to reproduce what we suppose were the steps used by Eloe and Ridenhour to obtain Theorems 1 and 2 as well as to prove the missing results (see Remark 2 for some examples of the latter).

We will start with a Lemma that can be considered an extension of [1], Lemma 2.3 to the problem (4). As Eloe and Ridenhour pointed out, [1], Lemma 2.3 was in essence proved by Peterson and Ridenhour in [6] for the case $(\alpha_1, \ldots, \alpha_k) = (0, \ldots, k-1)$.

Lemma 1. *Let us assume that L is disfocal on $[a,b]$ and that $y(x)$ is a nontrivial solution of $Ly = 0$ which satisfies the $n-1$ homogeneous boundary conditions*

$$y^{(\alpha_i)}(a) = 0, \quad \alpha_i \in \alpha, \quad \alpha \in \Omega_{k-1},$$
$$y^{(\beta_i)}(b) = 0, \quad \beta_i \in \beta, \quad \beta \in \Omega_{n-k}. \qquad (16)$$

Let us also assume that

$$\text{card}\{i: 0 \le i \le j-1, y^{(i)}(a) = 0\} + \text{card}\{i: 0 \le i \le j-1, y^{(i)}(b) = 0\} \ge j, \quad j = 1, \ldots, K(\alpha, \beta). \qquad (17)$$

Then $y(x)$ is essentially unique (to within the norm) and satisfies

1. *Neither $y(x)$ nor any of its derivatives vanish at a or b on derivatives lower than $K(\alpha, \beta) + 1$ and different from those of (16), that is*

$$y^{(i)}(a) \neq 0, \quad i = 0, \ldots, K(\alpha, \beta), \quad i \notin \alpha,$$
$$y^{(i)}(b) \neq 0, \quad i = 0, \ldots, K(\alpha, \beta), \quad i \notin \beta. \qquad (18)$$

2. $y^{(i)}(x) \neq 0$, $x \in (a,b)$, $0 \le i \le \max(\alpha_1, \beta_1)$. *Moreover, if (α, β) are p-alternate, $y^{(i)}(x) \neq 0$, $x \in (a,b)$, $0 \le i \le p+1$.*
3. *If $y^{(i)}(a) = 0, 0 \le i \le K(\alpha, \beta) - 1$, there exists an $\epsilon > 0$ such that $y^{(i)}(x)y^{(i+1)}(x) > 0$, $x \in (a, a+\epsilon)$.*
4. *If $y^{(i)}(a) \neq 0, 0 \le i \le K(\alpha, \beta) - 1$, there exists an $\epsilon > 0$ such that $y^{(i)}(x)y^{(i+1)}(x) < 0$, $x \in (a, a+\epsilon)$.*
5. *If $y^{(i)}(b) = 0, 0 \le i \le K(\alpha, \beta) - 1$, there exists an $\epsilon > 0$ such that $y^{(i)}(x)y^{(i+1)}(x) < 0$, $x \in (b-\epsilon, b)$.*
6. *If $y^{(i)}(b) \neq 0, 0 \le i \le K(\alpha, \beta) - 1$, there exists an $\epsilon > 0$ such that $y^{(i)}(x)y^{(i+1)}(x) > 0$, $x \in (b-\epsilon, b)$.*

Proof. Following the argumentation of [6], let us denote by l_j, r_j the following values

$$l_j = \text{card}\{i: 0 \le i \le j-1, y^{(i)}(a) = 0\}, \quad 0 \le j \le n,$$

$$r_j = \text{card}\{i: 0 \le i \le j-1, y^{(i)}(b) = 0\}, \quad 0 \le j \le n.$$

We will show by induction that the number of zeroes of $y^{(j)}(x)$ in the interval (a,b) (let us name it $z_j(a,b)$) is at least $l_j + r_j - j$. For $j = 0$ it is straightforward, so let us assume that the hypothesis holds for $j-1$, that is,

$$z_{j-1}(a,b) \ge l_{j-1} + r_{j-1} - j + 1.$$

If we consider the possible zeroes of $y^{(j-1)}(x)$ at a or b, Rolle's theorem mandates that

$$z_j(a,b) \ge z_{j-1}(a,b) + l_j - l_{j-1} + r_j - r_{j-1} - 1$$

$$\ge l_{j-1} + r_{j-1} - j + 1 + l_j - l_{j-1} + r_j - r_{j-1} - 1 = l_j + r_j - j.$$

From the definition of l_j, r_j, this result also implies that the number of zeroes of $y^{(j)}(x)$ in $[a,b]$ (let us name it $z_j[a,b]$), satisfies

$$z_j[a,b] \ge l_{j+1} + r_{j+1} - j.$$

With this in mind it is immediate to see that the condition (17) translates into

$$z_j[a,b] \ge 1, \quad j = 0, \ldots, K(\alpha, \beta) - 1, \qquad (19)$$

whereas the definition of $K(\alpha, \beta)$ implies

$$z_j[a,b] \ge 1, \quad j = K(\alpha, \beta) + 1, \ldots, n-1. \qquad (20)$$

The key insight for the rest of the proof is that any additional zero of $y^{(i)}(x)$ on $[a,b]$ for $i = 0, \ldots, K(\alpha, \beta)$ not forced by the homogeneous boundary conditions nor by Rolle's theorem will imply, again by Rolle's theorem, that $z_{K(\alpha,\beta)}[a,b] = 1$ which together with (19) and (20) give

$$z_j[a,b] \geq 1, \quad j = 0, \ldots, n-1.$$

Since L is disfocal on $[a,b]$ by hypothesis, such an additional zero will mean $y \equiv 0$. This proves properties 1 and 2 (the p-alternate condition grants that only one homogeneous boundary condition -at either a or b- is set in each derivative up to p-th one, so these boundary conditions cannot force, at least via Rolle's theorem, any zeroes in (a,b) in the derivatives up to the $(p+1)$-th one) and also the fact that y is essentially unique to within the norm (if there were two different solutions y_1 and y_2 one could create a non trivial linear combination y_3 of these two with a zero of $y_3^{(K(\alpha,\beta))}$ in $[a,b]$).

As for property 3, if $i + 1 \leq K(\alpha, \beta)$ then the number of zeroes of $y^{(i+1)}(x)$ on (a,b) must be finite (otherwise from Rolle's theorem we would end up with a zero of $y^{(K(\alpha,\beta))}(x)$ on (a,b) and the disfocality of L on $[a,b]$ would force $y \equiv 0$) and there must be an $\epsilon > 0$ such that $y^{(i+1)}(x) \neq 0$ on $(a, a + \epsilon)$. Since

$$y^{(i)}(x) = y^{(i)}(a) + \int_a^x y^{(i+1)}(s)ds = \int_a^x y^{(i+1)}(s)ds,$$

it must follow that $y^{(i)}(x) y^{(i+1)}(x) > 0$ on $(a, a + \epsilon)$.

To prove property 4, let $x_i \in (a,b]$ be such that $y^{(i)}(x_i) = 0$ and $y^{(i)}(x) \neq 0$ on $[a, x_i)$ (the existence of x_i is granted by (19)). There cannot be any zeroes of $y^{(i+1)}(x)$ on (a, x_i) since, by the previous argumentation, this would imply again a zero of $y^{(K(\alpha,\beta))}(x)$ on (a,b) and therefore $y \equiv 0$. As

$$-y^{(i)}(x) = y^{(i)}(x_i) - y^{(i)}(x) = \int_x^{x_i} y^{(i+1)}(s)ds,$$

one gets to $y^{(i)}(x) y^{(i+1)}(x) < 0$ on (a, x_i).

The proof of properties 5 and 6 is similar to that of properties 3 and 4, respectively. □

Remark 1. *It is important to stress that the results 3–6 of the previous Lemma only apply if $i \leq K(\alpha, \beta) - 1$. If $y^{(K(\alpha,\beta))}(x) \neq 0$ on $[a,b]$ we cannot deduce anything about the zeroes of higher derivatives of $y(x)$ on $[a,b]$, as the disfocality condition would already not be met in $y^{(K(\alpha,\beta))}$.*

The next Theorem extends [1], Lemma 2.4 and Theorem 2.1 to the problem (4).

Theorem 3. *Let us assume that the boundary conditions (α, β), with $\alpha \in \Omega_k$ and $\beta \in \Omega_{n-k}$, are admissible. Then one has*

$$(-1)^{m(\alpha,i)} \frac{\partial^i G(\alpha, \beta, a, t)}{\partial x^i} > 0, \ 0 \leq i \leq n-1, \ i \notin \alpha, \tag{21}$$

and

$$(-1)^{n(\beta,i)} \frac{\partial^i G(\alpha, \beta, b, t)}{\partial x^i} > 0, \ 0 \leq i \leq n-1, \ i \notin \beta. \tag{22}$$

In addition:

1. *If $\alpha_1 = 0$ then*

$$(-1)^{n-k} \frac{\partial^i G(\alpha, \beta, x, t)}{\partial x^i} > 0, \ 0 \leq i \leq \beta_1, \ x \in (a,b). \tag{23}$$

2. *If $\beta_1 = 0$ then*

$$(-1)^{n-k+i} \frac{\partial^i G(\alpha, \beta, x, t)}{\partial x^i} > 0, \ 0 \leq i \leq \alpha_1, \ x \in (a,b). \tag{24}$$

Proof. Let us note first that the admissibility of the boundary conditions imposes that $\alpha_1 = 0$ or $\beta_1 = 0$.

We will focus initially on the case $\alpha_1 = 0$, for which we will follow a similar approach as that used in [1], Lemma 2.4. Thus, as a starting point, let us fix $t \in [a,b]$ and let us consider the boundary conditions (α, β) with $\alpha = \{0, \ldots, k-1\}$, which (as it is straightforward to show) are always admissible regardless of the value of k and β. From [1], Lemma 2.4 one has (22) and from [1], Theorem 2.1 one gets (23) and

$$(-1)^{n-k} \frac{\partial^k G(\alpha, \beta, a, t)}{\partial x^k} > 0. \tag{25}$$

If $k < n - 1$, we can pick new boundary conditions (α', β') with $\alpha' = \{0, \ldots, k\}$ and $\beta' = \beta \setminus \beta_{n-k}$ (that is $\beta' = \{\beta_1, \ldots, \beta_{n-k-1}\}$, for which [1], Theorem 2.1 gives again

$$(-1)^{n-k-1} \frac{\partial^{k+1} G(\alpha', \beta', a, t)}{\partial x^{k+1}} > 0. \tag{26}$$

We can build the function $g_1(x) = G(\alpha', \beta', x, t) - G(\alpha, \beta, x, t)$, which is n-times continuously differentiable (the difference of the Green functions compensate the discontinuity of their $(n-1)$-th partial derivatives with regards to x at $x = t$) and satisfies

$$Lg_1 = 0, \quad x \in (a,b);$$

$$g_1^{(j)}(a) = 0, \ 0 \leq j \leq k-1; \quad g_1^{(\beta_j)}(b) = 0, \ \beta_j \in \beta \setminus \beta_{n-k};$$

$$g_1^{(k)}(a) = -\frac{\partial^k G(\alpha, \beta, a, t)}{\partial x^k}; \quad g_1^{(\beta_{n-k})}(b) = \frac{\partial^{\beta_{n-k}} G(\alpha', \beta', b, t)}{\partial x^{\beta_{n-k}}}. \tag{27}$$

From (25) and (27) it follows

$$(-1)^{n-k} g_1^{(k)}(a) < 0. \tag{28}$$

The boundary conditions of g_1 are (α, β'). It is straighforward to prove that $K(\alpha, \beta') = n - 1$ and that g_1 satisfies the hypothesis (17) of Lemma 1 for $1, \ldots, n-1$. In consequence, one can apply properties 1 and 4 of Lemma 1 to g_1 and, taking (28) into account, get to

$$(-1)^{n-k} g_1^{(k+1)}(a) > 0.$$

From here and (26) one has

$$(-1)^{n-k-1} \frac{\partial^{k+1} G(\alpha, \beta, a, t)}{\partial x^{k+1}} = (-1)^{n-k-1} \frac{\partial^{k+1} G(\alpha', \beta', a, t)}{\partial x^{k+1}} - (-1)^{n-k-1} g_1^{(k+1)}(a) > 0. \tag{29}$$

This argument can be repeated recursively to obtain

$$(-1)^{n-i} \frac{\partial^i G(\alpha, \beta, a, t)}{\partial x^i} > 0, \ k \leq i \leq n-1, \tag{30}$$

which is (21).

Next, we will proceed by induction over $S(\alpha)$. Thus, let us consider admissible (but not strongly admissible) boundary conditions (α, β) with $\alpha \in \Omega_k$ and $\beta \in \Omega_{n-k}$, and let us define new conditions (α', β) by taking α and replacing the homogeneous boundary condition α_i by $\alpha_k + 1$ (that is, α' specifies $y^{(\alpha_k+1)}(a) = 0$ instead of $y^{(\alpha_i)}(a) = 0$). Let us assume that (α', β) are also admissible.

The function $g_2(x) = G(\alpha', \beta, x, t) - G(\alpha, \beta, x, t)$ is n-times continuously differentiable and satisfies

$$Lg_2 = 0, \quad x \in (a,b);$$

$$g_2^{(\alpha_j)}(a) = 0, \ \alpha_j \in \alpha, \ \alpha_j \neq \alpha_i; \quad g_2^{(\beta_j)}(b) = 0, \ \beta_j \in \beta;$$

$$g_2^{(\alpha_i)}(a) = \frac{\partial^{\alpha_i} G(\alpha', \beta, a, t)}{\partial x^{\alpha_i}}; \quad g_2^{(\alpha_k+1)}(a) = -\frac{\partial^{\alpha_k+1} G(\alpha, \beta, a, t)}{\partial x^{\alpha_k+1}}. \tag{31}$$

Let (α'', β) be the homogeneous boundary conditions satisfied by g_2, with $\alpha'' \in \Omega_{k-1}$. We will prove now that

$$K(\alpha'', \beta) = \max(K(\alpha', \beta), \alpha_k + 1), \tag{32}$$

and that g_2 complies with the hypotheses of Lemma 1 for $K(\alpha'', \beta)$.

If $K(\alpha', \beta) > \alpha_k + 1$ then $K(\alpha'', \beta) = K(\alpha', \beta)$ as the only difference between (α', β) and (α'', β) is precisely $\alpha_k + 1$. In that case

$$\operatorname{card}\{\beta_l \in \beta, j < \beta_l \le n - 1\} \le n - 1 - j - 1 = n - j - 2$$

for $\alpha_k + 1 \le j < K(\alpha'', \beta)$, since $K(\alpha'', \beta) \notin \beta$ as per the definition of $K(\alpha'', \beta)$. Following the nomenclature of Lemma 1 and noting that

$$l_n(\alpha', \beta) + r_n(\alpha', \beta) = n,$$

it follows

$$l_{j+1}(\alpha', \beta) + r_{j+1}(\alpha', \beta) \ge n - (n - j - 2) = j + 2, \quad \alpha_k + 1 \le j < K(\alpha'', \beta),$$

which in turn means

$$l_{j+1}(\alpha'', \beta) + r_{j+1}(\alpha'', \beta) \ge j + 1, \quad \alpha_k + 1 \le j < K(\alpha'', \beta). \tag{33}$$

Since

$$l_{j+1}(\alpha'', \beta) + r_{j+1}(\alpha'', \beta) = l_{j+1}(\alpha', \beta) + r_{j+1}(\alpha', \beta) \ge j + 1, \quad j \le \alpha_k, \tag{34}$$

due to the admissibility of (α', β), from (33) and (34) it follows that the condition (17) holds for (α'', β) and $1, \ldots, K(\alpha'', \beta)$.

On the other hand, if $K(\alpha', \beta) < \alpha_k + 1$, since (α', β) are admissible there cannot be an order above $K(\alpha', \beta)$ which belongs to α' and β at the same time, which implies that the number of boundary conditions above $K(\alpha', \beta)$ is limited by

$$\operatorname{card}\{\alpha_l \in \alpha', K(\alpha', \beta) < \alpha_l \le n - 1\} + \operatorname{card}\{\beta_l \in \beta, K(\alpha', \beta) < \beta_l \le n - 1\} = n - 1 - K(\alpha', \beta),$$

and therefore

$$\operatorname{card}\{\alpha_l \in \alpha'', K(\alpha', \beta) < \alpha_l \le n - 1\} + \operatorname{card}\{\beta_l \in \beta, K(\alpha', \beta) < \beta_l \le n - 1\} = n - 2 - K(\alpha', \beta).$$

This means that there must exist an index l with $K(\alpha', \beta) + 1 \le l \le n - 1$ such that $l \notin \alpha'' \cup \beta$. That index l must obviously be $K(\alpha'', \beta)$. As the only difference between (α', β) and (α'', β) is precisely $\alpha_k + 1$, it follows that $K(\alpha'', \beta) = \alpha_k + 1$. The admissibility of (α', β) grants that (α'', β) fulfils the condition (17) of Lemma 1 for $1, \ldots, K(\alpha'', \beta)$, also in this case $K(\alpha'', \beta) = \alpha_k + 1$.

Moving on, from the induction hypothesis we know that

$$(-1)^{m(\alpha, \alpha_k + 1)} \frac{\partial^{\alpha_k + 1} G(\alpha, \beta, a, t)}{\partial x^{\alpha_k + 1}} > 0, \tag{35}$$

which together with (31) gives

$$(-1)^{m(\alpha, \alpha_k + 1)} g_2^{(\alpha_k + 1)}(a) < 0. \tag{36}$$

Since the number of derivatives of g_2 between $g_2^{(\alpha_i)}$ and $g_2^{(\alpha_k+1)}$ which are not specified to be zero at a is $m(\alpha', \alpha_i) - m(\alpha, \alpha_k + 1) + 1$, applying properties Properties 3 and 4 of Lemma 1 to $g_2(x)$ one gets

$$(-1)^{m(\alpha', \alpha_i)} g_2^{(\alpha_i)}(a) > 0, \tag{37}$$

that is

$$(-1)^{m(\alpha', \alpha_i)} \frac{\partial^{\alpha_i} G(\alpha', \beta, a, t)}{\partial x^{\alpha_i}} > 0. \tag{38}$$

In a similar manner, for $x \in (a, a + \epsilon)$

$$(-1)^{m(\alpha', j)} g_2^{(j)}(x) > 0, \; j \leq \alpha_i, \tag{39}$$

and since $m(\alpha', j) = m(\alpha, j)$ for $j < \alpha_i$ from the induction hypothesis one obtains

$$(-1)^{m(\alpha', j)} \frac{\partial^j G(\alpha', \beta, a, t)}{\partial x^j} = (-1)^{m(\alpha', j)} g_2^{(j)}(a) + (-1)^{m(\alpha, j)} \frac{\partial^j G(\alpha, \beta, a, t)}{\partial x^j} > 0, \; j \notin \alpha, \; j < \alpha_i. \tag{40}$$

Equations (38) and (40) prove (21) for $j \leq \alpha_i$.

Before addressing (21) for $\alpha_i > j$, which will require a different function g_3, let us focus on (23) and (22), in this order. Thus, from (39), the definition of $m(\alpha', j)$ and property property 2 of Lemma 1 it follows

$$(-1)^{n-k} g_2^{(i)}(x) > 0, \; 0 \leq i \leq \beta_1, \; x \in (a, b). \tag{41}$$

Since by the induction hypothesis

$$(-1)^{n-k} \frac{\partial^i G(\alpha, \beta, x, t)}{\partial x^i} > 0, \; 0 \leq i \leq \beta_1, \; x \in (a, b), \tag{42}$$

from (41) and (42) one gets to

$$(-1)^{n-k} \frac{\partial^i G(\alpha', \beta, x, t)}{\partial x^i} = (-1)^{n-k} g_2^{(i)}(x) + (-1)^{n-k} \frac{\partial^i G(\alpha, \beta, x, t)}{\partial x^i} > 0, \; 0 \leq i \leq \beta_1, \; x \in (a, b), \tag{43}$$

which is (23).

On the other hand, (41) also implies $(-1)^{n-k} g_2^{(\beta_1)}(x) = (-1)^{n(\beta, \beta_1)+1} g_2^{(\beta_1)}(x) > 0$ for $x \in (b - \epsilon, b)$. Applying properties properties 1, 5 and 6 of Lemma 1, one has

$$(-1)^{n(\beta, j)} g_2^{(j)}(b) > 0, \; 0 \leq j \leq K(\alpha'', \beta), \; j \notin \beta. \tag{44}$$

Since the induction hypothesis on b implies

$$(-1)^{n(\beta, j)} \frac{\partial^j G(\alpha, \beta, b, t)}{\partial x^j} > 0, \; 0 \leq j \leq n - 1, \; j \notin \beta, \tag{45}$$

from (44) and (45) we get to

$$(-1)^{n(\beta, j)} \frac{\partial^j G(\alpha', \beta, b, t)}{\partial x^j} > 0, \; 0 \leq j \leq K(\alpha'', \beta), \; j \notin \beta, \tag{46}$$

or rather

$$(-1)^{n(\beta, j)} \frac{\partial^j G(\alpha', \beta, b, t)}{\partial x^j} > 0, \; 0 \leq j \leq \max(K(\alpha', \beta), \alpha_k + 1), \; j \notin \beta, \tag{47}$$

if we consider (32). The extension of (47) to (22) is straightforward since if $\max(K(\alpha', \beta), \alpha_k + 1) < n - 1$ then $\{\max(K(\alpha', \beta), \alpha_k + 1) + 1, \ldots, n - 1\} \subset \beta$.

Let us move on to prove (21) for $j > \alpha_i$. For that let us consider the boundary conditions $(\hat{\alpha}, \hat{\beta})$, defined by $\hat{\alpha} = \alpha' \cup \{\alpha_i\}$ (or in another way, $\hat{\alpha} = \alpha \cup \{\alpha_k + 1\}$), $\hat{\alpha} \in \Omega_{k+1}$ and $\hat{\beta} = \beta \backslash \beta_{n-k}$, $\hat{\beta} \in \Omega_{n-k-1}$. $(\hat{\alpha}, \hat{\beta})$ are admissible since:

1. If $\beta_{n-k} \geq \alpha_i$, the property is straightforward as (α', β) are also admissible.
2. If $\beta_{n-k} < \alpha_i$, then (reusing the nomenclature of Lemma 1) one has $l_{j+1}(\hat{\alpha}, \hat{\beta}) + r_{j+1}(\hat{\alpha}, \hat{\beta}) = n$ for $j = \alpha_k + 1, \ldots, n-1$ and in particular $l_{\alpha_k+2}(\hat{\alpha}, \hat{\beta}) + r_{\alpha_k+2}(\hat{\alpha}, \hat{\beta}) = n$ which in turn implies (note $\alpha_k + 1 < n$)

$$l_{j+1}(\hat{\alpha}, \hat{\beta}) + r_{j+1}(\hat{\alpha}, \hat{\beta}) \geq n - (\alpha_k + 1 - j) = n - \alpha_k + j - 1 \geq j + 1,$$

for $\beta_{n-k} \leq j \leq \alpha_k + 1$. As there is no change in the boundary conditions associated to derivatives of order lower than β_{n-k}, this proves the admissibility of $(\hat{\alpha}, \hat{\beta})$.

Thus, let us define the function $g_3(x) = G(\alpha', \beta, x, t) - G(\hat{\alpha}, \hat{\beta}, x, t)$, which is n-times continuously differentiable and satisfies

$$Lg_3 = 0, \quad x \in (a, b);$$

$$g_3^{(\alpha_j)}(a) = 0, \ \alpha_j \in \alpha, \ \alpha_j \neq \alpha_i; \quad g_3^{(\beta_j)}(b) = 0, \ \beta_j \in \beta \backslash \beta_{n-k};$$

$$g_3^{(\alpha_i)}(a) = \frac{\partial^{\alpha_i} G(\alpha', \beta, a, t)}{\partial x^{\alpha_i}}; \quad g_3^{(\beta_{n-k})}(b) = -\frac{\partial^{\beta_{n-k}} G(\hat{\alpha}, \hat{\beta}, b, t)}{\partial x^{\beta_{n-k}}}. \tag{48}$$

From (38) and (48) it follows

$$(-1)^{m(\alpha', \alpha_i)} g_3^{(\alpha_i)}(a) > 0. \tag{49}$$

The boundary conditions for g_3 are $(\alpha', \hat{\beta})$. We will prove now that

$$K(\alpha', \hat{\beta}) = \max(K(\alpha', \beta), \beta_{n-k}), \tag{50}$$

and that $(\alpha', \hat{\beta})$ satisfy the condition (17) of Lemma 1 for $1, \ldots, K(\alpha', \hat{\beta})$.

If $K(\alpha', \beta) > \beta_{n-k}$ then $K(\alpha', \hat{\beta}) = K(\alpha', \beta)$ as the only difference between $(\alpha', \hat{\beta})$ and (α', β) is precisely β_{n-k}. In that case we can follow a similar reasoning as before to state

$$\text{card}\{\alpha_l \in \alpha', \ j < \alpha_l \leq n-1\} \leq n - 1 - j - 1 = n - j - 2,$$

for $\beta_{n-k} \leq j < K(\alpha', \beta)$, so, using again the nomenclature of Lemma 1 for (α', β)

$$l_{j+1}(\alpha', \beta) + r_{j+1}(\alpha', \beta) \geq n - (n - j - 2) = j + 2, \ \beta_{n-k} \leq j < K(\alpha', \beta) = K(\alpha', \hat{\beta}).$$

That in turn implies

$$l_{j+1}(\alpha', \hat{\beta}) + r_{j+1}(\alpha', \hat{\beta}) \geq j + 1, \ \beta_{n-k} \leq j < K(\alpha', \hat{\beta}),$$

or

$$l_j(\alpha', \hat{\beta}) + r_j(\alpha', \hat{\beta}) \geq j, \ \beta_{n-k} + 1 \leq j \leq K(\alpha', \hat{\beta}). \tag{51}$$

Since

$$l_j(\alpha', \hat{\beta}) + r_j(\alpha', \hat{\beta}) = l_j(\alpha', \beta) + r_j(\alpha', \beta) \geq j, \ j \leq \beta_{n-k}, \tag{52}$$

from (51) and (52) it follows that $(\alpha', \hat{\beta})$ satisfy the condition (17) for $1, \ldots, K(\alpha', \hat{\beta})$ when $K(\alpha', \hat{\beta}) = K(\alpha', \beta)$.

On the other hand, if $K(\alpha',\beta) < \beta_{n-k}$, since (α',β) are admissible, there cannot be an order above $K(\alpha',\beta)$ which belongs to α' and β at the same time, which implies that the number of boundary conditions above $K(\alpha',\beta)$ is limited by

$$\operatorname{card}\{\alpha_l \in \alpha',\ K(\alpha',\beta)+1 \leq \alpha_l \leq n-1\} + \operatorname{card}\{\beta_l \in \beta,\ K(\alpha',\beta)+1 \leq \beta_l \leq n-1\} = n-1-K(\alpha',\beta),$$

and therefore

$$\operatorname{card}\{\alpha_l \in \alpha',\ K(\alpha',\beta)+1 \leq \alpha_l \leq n-1\} + \operatorname{card}\{\beta_l \in \hat{\beta},\ K(\alpha',\beta)+1 \leq \beta_l \leq n-1\} = n-2-K(\alpha',\beta).$$

This means that there must exist an index l with $K(\alpha',\beta)+1 \leq l \leq n-1$ such that $l \notin \alpha' \cup \hat{\beta}$. That index l must obviously be $K(\alpha',\hat{\beta})$. As the only difference between (α',β) and $(\alpha',\hat{\beta})$ is precisely β_{n-k}, it follows that $K(\alpha',\hat{\beta}) = \beta_{n-k}$. The admissibility of (α',β) grants that $(\alpha',\hat{\beta})$ fulfils the condition (17) of Lemma 1 for $1,\ldots,K(\alpha',\hat{\beta})$, also in this case $K(\alpha',\hat{\beta}) = \beta_{n-k}$.

Since $K(\alpha',\hat{\beta}) \geq \beta_{n-k}$, in all cases where $K(\alpha',\hat{\beta}) \leq \alpha_i$, $\frac{\partial^j G(\alpha',\hat{\beta},a,t)}{\partial x^j} = 0$ for $j = \alpha_i+1,\ldots,n-1$, eliminating the need for proving (21) in these scenarios. In the rest of the cases we can apply properties 3 and 4 of Lemma 1 to g_3 and (49) to yield

$$(-1)^{m(\alpha',j)} g_3^{(j)}(a) > 0, \quad \alpha_i < j \leq K(\alpha',\hat{\beta}),\ j \notin \alpha. \tag{53}$$

Due to the definition of $\hat{\beta}$, we can apply in this case induction over $S(\beta)$ and assume

$$(-1)^{m(\hat{\alpha},j)} \frac{\partial^j G(\hat{\alpha},\hat{\beta},a,t)}{\partial x^j} > 0, \quad \alpha_i < j \leq n-1,\ j \notin \alpha. \tag{54}$$

From (53) and (54), and the fact that $m(\hat{\alpha},j) = m(\alpha',j)$ for $\alpha_i < j \leq n-1$, one finally gets to

$$(-1)^{m(\alpha',j)} \frac{\partial^j G(\alpha',\beta,a,t)}{\partial x^j} = (-1)^{m(\alpha',j)} g_3^{(j)}(a) + (-1)^{m(\alpha',j)} \frac{\partial^j G(\hat{\alpha},\hat{\beta},a,t)}{\partial x^j}$$

$$= (-1)^{m(\alpha',j)} g_3^{(j)}(a) + (-1)^{m(\hat{\alpha},j)} \frac{\partial^j G(\hat{\alpha},\hat{\beta},a,t)}{\partial x^j} > 0, \quad \alpha_i < j \leq K(\alpha',\hat{\beta}),\ j \notin \alpha, \tag{55}$$

or, taking (50) into account

$$(-1)^{m(\alpha',j)} \frac{\partial^j G(\alpha',\beta,a,t)}{\partial x^j} > 0, \quad \alpha_i < j \leq \max(K(\alpha',\beta),\beta_{n-k}),\ j \notin \alpha. \tag{56}$$

The extension of (56) to (21) is straightforward as if $\max(K(\alpha',\beta),\beta_{n-k}) < n-1$ then $\frac{\partial^j G(\alpha',\beta,a,t)}{\partial x^j} = 0$ for $j = \max(K(\alpha',\beta),\beta_{n-k})+1,\ldots,n-1$. This completes the proof of the case $\alpha_1 = 0$.

Let us focus now on the case $\alpha_1 > 0$, $\beta_1 = 0$. For that we will consider the function

$$G'(\beta,\alpha,x,t) = (-1)^n G(\alpha,\beta,b+a-x,b+a-t), \tag{57}$$

which as one can readily show (see e.g., [8], Chapter 3, page 105) is the Green function of the problem

$$L'G' = 0,\quad (x,t) \in \{(a,t) \cup (t,b)\} \times (a,b),$$

$$\frac{\partial^{\beta_j} G'(\beta,\alpha,a,t)}{\partial x^{\beta_j}} = 0,\ j=1,\ldots,n-k;\quad \frac{\partial^{\alpha_j} G'(\beta,\alpha,b,t)}{\partial x^{\alpha_j}} = 0,\ j=1,\ldots,k; \tag{58}$$

with L' defined as

$$L'y = y^{(n)}(x) - a_{n-1}(b+a-x)y^{(n-1)}(x) + \cdots + (-1)^n a_0(b+a-x)y(x). \tag{59}$$

Since $\beta_1 = 0$ is a boundary condition applied at a, G' satisfies the hypotheses of the first part of this theorem. Thus, from (21), (22) and (23) one gets to

$$(-1)^{m(\beta,i)}\frac{\partial^i G'(\beta,\alpha,a,t)}{\partial x^i} > 0, \ 0 \leq i \leq n-1, \ i \notin \beta, \tag{60}$$

$$(-1)^{n(\alpha,i)}\frac{\partial^i G'(\beta,\alpha,b,t)}{\partial x^i} > 0, \ 0 \leq i \leq n-1, \ i \notin \alpha, \tag{61}$$

and

$$(-1)^k \frac{\partial^i G'(\beta,\alpha,x,t)}{\partial x^i} > 0, \ 0 \leq i \leq \alpha_1, \ x \in (a,b). \tag{62}$$

(60), (61), (62) and the relationship

$$(-1)^{n-j}\frac{\partial^j G'(\beta,\alpha,b+a-x,b+a-t)}{\partial x^j} = \frac{\partial^j G(\alpha,\beta,x,t)}{\partial x^j}, \ 0 \leq j \leq n-1, \tag{63}$$

finally yield

$$(-1)^{n(\alpha,i)+n-i}\frac{\partial^i G(\alpha,\beta,a,t)}{\partial x^i} > 0, \ 0 \leq i \leq n-1, \ i \notin \alpha, \tag{64}$$

$$(-1)^{m(\beta,i)+n-i}\frac{\partial^i G(\alpha,\beta,b,t)}{\partial x^i} > 0, \ 0 \leq i \leq n-1, \ i \notin \beta, \tag{65}$$

$$(-1)^{n-k+i}\frac{\partial^i G(\alpha,\beta,x,t)}{\partial x^i} > 0, \ 0 \leq i \leq \alpha_1, \ x \in (a,b). \tag{66}$$

As $n - i - n(\alpha,i) = m(\alpha,i)$ for $i \notin \alpha$ and $n - i - m(\beta,i) = n(\beta,i)$ for $i \notin \beta$, from (64) and (65) one readily gets (21) and (22), respectively. □

Remark 2. *Inequalities (21) and (22) are results new with respect to Theorems 3.3 and 3.4 of [1]. Likewise, (30) is also new with respect to Theorem 2.1 of [1].*

Next, we will assess the dependence of $G(x,t)$ and some of its partial derivatives with regards to the extremes a and b.

Lemma 2. *Fixed $t \in [a,b]$, $H(x,t)$ is the solution of the problem*

$$LH = 0, \ x \in (a,b);$$

$$\frac{\partial^{\alpha_j} H(a,t)}{\partial x^{\alpha_j}} = 0, \ \alpha_j \in \alpha; \ \frac{\partial^{\beta_j} H(b,t)}{\partial x^{\beta_j}} = -\frac{\partial^{\beta_j+1} G(\alpha,\beta,b,t)}{\partial x^{\beta_j+1}}, \ \beta_j \in \beta. \tag{67}$$

Likewise, $I(x,t)$ is the solution of the problem

$$LI = 0, \ x \in (a,b);$$

$$\frac{\partial^{\alpha_j} I(a,t)}{\partial x^{\alpha_j}} = -\frac{\partial^{\alpha_j+1} G(\alpha,\beta,a,t)}{\partial x^{\alpha_j+1}}, \ \alpha_j \in \alpha; \ \frac{\partial^{\beta_j} I(b,t)}{\partial x^{\beta_j}} = 0, \ \beta_j \in \beta. \tag{68}$$

Proof. The proof of (67) follows the same steps as that of [13], Lemma 3.3 with $x_1 = a$ and $k = 1$ and will not be repeated. The proof of (68) is also similar. □

Theorem 4. *Let us assume that (α,β) are admissible boundary conditions. If $\alpha_1 = 0$ and either*

$$\beta_{n-k} < n-1$$

or
$$\beta_{n-k} = n-1 \quad \text{and} \quad (-1)^{n(\beta,j)} a_j(b) \leq 0, \ 0 \leq j \leq n-1, \ j \notin \beta, \tag{69}$$

with at least one $l \notin \beta$ such that $0 \leq l \leq n-1$ and

$$(-1)^{n(\beta,l)} a_l(b) < 0, \tag{70}$$

then

$$(-1)^{n(\beta,j)} \frac{\partial^j H(x,t)}{\partial x^j} < 0, \ (x,t) \in (a,b) \times (a,b), \ \beta_1 \leq j \leq \beta_B, \tag{71}$$

and

$$(-1)^{n-k} \frac{\partial^j H(x,t)}{\partial x^j} > 0, \ (x,t) \in (a,b) \times (a,b), \ 0 \leq j < \beta_1. \tag{72}$$

If $\alpha_1 > 0$ and either

$$\beta_{n-k} < n-1$$

or

$$\beta_{n-k} = n-1 \quad \text{and} \quad (-1)^{n(\beta,j)} a_j(b) \leq 0, \ 0 \leq j \leq n-1, \ j \notin \beta, \tag{73}$$

then

$$(-1)^{n-k-j} \frac{\partial^j H(x,t)}{\partial x^j} > 0, \ (x,t) \in (a,b) \times (a,b), \ 0 \leq j \leq \beta_B. \tag{74}$$

Proof. Let us suppose that $\alpha_1 = 0$. Fixed $t \in [a,b]$, from Lemma 2 we know that $H(x,t) = \sum_{i=1}^{n-k} h_{\beta_i}(x,t)$, where $h_{\beta_i}(x,t)$ is the solution of

$$Lh_{\beta_i} = 0, \ x \in (a,b); \quad \frac{\partial^{\alpha_j} h_{\beta_i}(a,t)}{\partial x^{\alpha_j}} = 0, \ \alpha_j \in \alpha;$$

$$\frac{\partial^{\beta_j} h_{\beta_i}(b,t)}{\partial x^{\beta_j}} = 0, \ \beta_j \in \beta \setminus \beta_i; \quad \frac{\partial^{\beta_i} h_{\beta_i}(b,t)}{\partial x^{\beta_i}} = -\frac{\partial^{\beta_i+1} G(b,t)}{\partial x^{\beta_i+1}}. \tag{75}$$

Note that if $\beta_i + 1 \in \beta$ then $h_{\beta_i}(x,t) \equiv 0$ due to the disfocality of L on $[a,b]$. That implies that we only need to take into account those β_i such that $\beta_i + 1 \notin \beta$.

If $\beta_{n-k} < n-1$ then $\beta_i < n-1$ for $1 \leq i \leq n-k$ and we can apply (22) and (75) to obtain

$$(-1)^{n(\beta,\beta_i+1)} \frac{\partial^{\beta_i} h_{\beta_i}(b,t)}{\partial x^{\beta_i}} < 0, \tag{76}$$

which combined with the properties properties 2 (as commented at the end of the Introduction the homogeneous boundary conditions in (75) are at least $(\beta_B - 1)$-alternate), 5 and 6 of Lemma 1, and the fact that $n(\beta, \beta_i + 1) = n(\beta, \beta_i)$ when $\beta_i + 1 \notin \beta$, yields

$$(-1)^{n(\beta,j)} \frac{\partial^j h_{\beta_i}(x,t)}{\partial x^j} < 0, \ (x,t) \in (a,b) \times (a,b), \ j \in \beta, \ j \leq \beta_B, \tag{77}$$

and

$$(-1)^{n(\beta,j)} \frac{\partial^j h_{\beta_i}(x,t)}{\partial x^j} > 0, \ (x,t) \in (a,b) \times (a,b), \ j \notin \beta, \ j \leq \beta_B. \tag{78}$$

As $\frac{\partial^j h_{\beta_i}(b,t)}{\partial x^j} = 0$ for $\beta_1 \leq j \leq \beta_B$ and $\frac{\partial^j h_{\beta_i}(b,t)}{\partial x^j} \neq 0$ for $0 \leq j < \beta_1$, from (77) and (78), the facts that $\beta_B \leq \beta_i$ and $n(\beta,j) = n-k$ for $j < \beta_1$, and the decomposition of $H(x,t)$ in $h_{\beta_i}(x,t)$, one gets to (71) and (72).

On the contrary, if $\beta_{n-k} = n-1$ then (77) and (78) hold for all h_{β_i} but for $h_{\beta_{n-k}}$, since in that case the sign of $\dfrac{\partial^{\beta_{n-k}} h_{\beta_{n-k}}(b,t)}{\partial x^{\beta_{n-k}}}$ is the opposite of that of $\dfrac{\partial^n G(b,t)}{\partial x^n}$, which Theorem 3 does not yield. In that case we need to revert to the definition of L. Thus, from (1) and (4) one has

$$\frac{\partial^n G(b,t)}{\partial x^n} = -\sum_{l=0}^{n-1} a_l(b) \frac{\partial^l G(b,t)}{\partial x^l} = -\sum_{l=0, l \notin \beta}^{n-1} a_l(b) \frac{\partial^l G(b,t)}{\partial x^l}. \tag{79}$$

From (22), (69), (70), (75) and (79) one gets to $\dfrac{\partial^n G(b,t)}{\partial x^n} > 0$ and $\dfrac{\partial^{n-1} h_{\beta_{n-k}}(b,t)}{\partial x^{n-1}} < 0$. Applying properties Properties 2, 5 and 6 of Lemma 1 one obtains again (77) and (78), and taking into account the decomposition of $H(x,t)$ in $h_{\beta_i}(x,t)$ one finally gets (71) and (72).

The proof of (74) in the case $\alpha_1 > 0$ can be done following the same reasoning. □

Remark 3. Condition (70) can be removed if $\beta \neq \{k, k+1, \ldots, n-1\}$. Such a condition is needed in the case $\beta = \{k, k+1, \ldots, n-1\}$ to grant $\dfrac{\partial^{n-1} h_{\beta_{n-k}}(b,t)}{\partial x^{n-1}} < 0$, since $\dfrac{\partial^{n-1} h_{\beta_{n-k}}(b,t)}{\partial x^{n-1}} = 0$ implies $H(x,t) = h_{\beta_{n-k}}(x,t) \equiv 0$ by the disfocality of L on $[a,b]$. However, if $\beta \neq \{k, k+1, \ldots, n-1\}$) then there are other non-trivial $h_{\beta_i}(x,t)$ which guarantee the non-triviality of $H(x,t)$.

Corollary 1. Let $b_1 < b_2$. Under the conditions of Theorem 4, if $\alpha_1 = 0$ then

$$(-1)^{n(\beta,j)} \frac{\partial^j G_{a,b_2}(x,t)}{\partial x^j} < (-1)^{n(\beta,j)} \frac{\partial^j G_{a,b_1}(x,t)}{\partial x^j}, \quad (x,t) \in (a,b) \times (a,b), \; \beta_1 \leq j \leq \beta_B, \tag{80}$$

and

$$(-1)^{n-k} \frac{\partial^j G_{a,b_2}(x,t)}{\partial x^j} > (-1)^{n-k} \frac{\partial^j G_{a,b_1}(x,t)}{\partial x^j} > 0, \quad (x,t) \in (a,b) \times (a,b), \; 0 \leq j \leq \beta_1. \tag{81}$$

If $\alpha_1 > 0$ then

$$(-1)^{n-k-j} \frac{\partial^j G_{a,b_2}(x,t)}{\partial x^j} > (-1)^{n-k-j} \frac{\partial^j G_{a,b_1}(x,t)}{\partial x^j} > 0, \quad (x,t) \in (a,b) \times (a,b), \; 0 \leq j \leq \beta_B. \tag{82}$$

Proof. The proof is immediate from Theorem 4. □

Theorem 5. Let us assume that (α, β) are admissible boundary conditions.
If $\alpha_1 = 0$ and either

$$\alpha_k < n-1$$

or

$$\alpha_k = n-1 \text{ and } (-1)^{m(\alpha,j)} a_j(a) \leq 0, \; 0 \leq j \leq n-1, \; j \notin \alpha, \tag{83}$$

then

$$(-1)^{n-k} \frac{\partial^j I(x,t)}{\partial x^j} < 0, \quad (x,t) \in (a,b) \times (a,b), \; 0 \leq j \leq \alpha_A. \tag{84}$$

If $\alpha_1 > 0$ and either

$$\alpha_k < n-1$$

or

$$\alpha_k = n-1 \text{ and } (-1)^{m(\alpha,j)} a_j(a) \leq 0, \; 0 \leq j \leq n-1, \; j \notin \alpha, \tag{85}$$

with at least one $l \notin \alpha$ such that $0 \leq l \leq n-1$ and

$$(-1)^{m(\alpha,l)} a_l(a) < 0, \tag{86}$$

then
$$(-1)^{m(\alpha,j)} \frac{\partial^j I(x,t)}{\partial x^j} < 0, \quad (x,t) \in (a,b) \times (a,b), \quad \alpha_1 \leq j \leq \alpha_A. \tag{87}$$

and
$$(-1)^{n-k-j} \frac{\partial^j I(x,t)}{\partial x^j} < 0, \quad (x,t) \in (a,b) \times (a,b), \quad 0 \leq j < \alpha_1. \tag{88}$$

Proof. The proof is similar to that of Theorem 4. □

Remark 4. *As before, condition (86) can be removed if $\alpha \neq \{n-k, n-k+1, \ldots, n-1\}$.*

Corollary 2. *Let $a_2 < a_1$. Under the conditions of Theorem 5, if $\alpha_1 = 0$ then*
$$(-1)^{n-k} \frac{\partial^j G_{a_2,b}(x,t)}{\partial x^j} > (-1)^{n-k} \frac{\partial^j G_{a_1,b}(x,t)}{\partial x^j} > 0, \quad (x,t) \in (a_1,b) \times (a_1,b), \quad 0 \leq j \leq \alpha_A. \tag{89}$$

If $\alpha_1 > 0$ then
$$(-1)^{m(\alpha,j)} \frac{\partial^j G_{a_2,b}(x,t)}{\partial x^j} > (-1)^{m(\alpha,j)} \frac{\partial^j G_{a_1,b}(x,t)}{\partial x^j}, \quad (x,t) \in (a_1,b) \times (a_1,b), \quad \alpha_1 \leq j \leq \alpha_A. \tag{90}$$

and
$$(-1)^{n-k-j} \frac{\partial^j G_{a_2,b}(x,t)}{\partial x^j} > (-1)^{n-k-j} \frac{\partial^j G_{a_1,b}(x,t)}{\partial x^j} > 0, \quad (x,t) \in (a_1,b) \times (a_1,b), \quad 0 \leq j < \alpha_1. \tag{91}$$

Remark 5. *If $\alpha_1 = 0$, it can happen that $\alpha_A \neq \beta_1$ (more concretely $\alpha_A = \beta_1 - 1$). In that case the statement (i) of [1], Theorem 3.4 (see (14)) does not seem to be valid for $l = \beta_1$ and $b_1 = b_2$, unless an approach not based on the sign of I and its derivatives was used by the authors to prove that assertion. The lack of an explicit proof of that theorem complicates any further analysis, but one cannot help having the impression that the statement is incorrect. The same comment is applicable to the statement (ii) of [1], Theorem 3.4 in the case $\alpha_1 > 0$, $a_1 = a_2$ (see (15)), which seems only valid for $l = 0, \ldots, \beta_B$ and not for $l = \alpha_1$ when $\beta_B \neq \alpha_1$.*

2.2. The Case of p-Alternate Boundary Conditions

When the boundary conditions are *p*-alternate, the lack of simultaneous boundary conditions at a and b for any derivative lower than p suggests no need for the immediately higher derivative to change the sign on (a,b), at least as a consequence of Rolle's theorem. The following theorem shows that this is to some extent the case under certain hypotheses.

Theorem 6. *Let us assume that (α, β) are p-alternate admissible boundary conditions.*
If $\alpha_1 = 0$ and either
$$\beta_{n-k} < n-1$$
or
$$\beta_{n-k} = n-1 \quad \text{and} \quad (-1)^{n(\beta,j)} a_j(x) \leq 0, \ x \in [a,b], \ 0 \leq j \leq n-1, \ j \notin \beta, \tag{92}$$

with at least one $l \notin \beta$ such that $0 \leq l \leq n-1$ and
$$(-1)^{n(\beta,l)} a_l(x) < 0, \ x \in [a,b], \tag{93}$$

then
$$(-1)^{n-k} \frac{\partial^j G(\alpha,\beta,x,t)}{\partial x^j} > 0, \quad (x,t) \in (a,b) \times (a,b), \quad 0 \leq j \leq \beta_1, \tag{94}$$

$$(-1)^{n(\beta,j)} \frac{\partial^j G(\alpha,\beta,x,t)}{\partial x^j} < 0, \quad (x,t) \in (a,b) \times (a,b), \quad \beta_1 \leq j \leq \beta_B, \tag{95}$$

and, if $\beta_{n-k}, p > \beta_B$,

$$(-1)^{n(\beta,\beta_B+1)} \frac{\partial^{\beta_B+1} G(\alpha,\beta,x,t)}{\partial x^{\beta_B+1}} > 0, \quad (x,t) \in (a,b) \times (a,b). \tag{96}$$

If $\alpha_1 > 0$ and either

$$\alpha_k < n-1$$

or

$$\alpha_k = n-1 \quad \text{and} \quad (-1)^{m(\alpha,j)} a_j(x) \leq 0, \ x \in [a,b], \ 0 \leq j \leq n-1, \ j \notin \alpha, \tag{97}$$

with at least one $l \notin \alpha$ such that $0 \leq l \leq n-1$ and

$$(-1)^{m(\alpha,l)} a_l(x) < 0, \ x \in [a,b], \tag{98}$$

then

$$(-1)^{n+k-j} \frac{\partial^j G(\alpha,\beta,x,t)}{\partial x^j} > 0, \quad (x,t) \in (a,b) \times (a,b), \ 0 \leq j \leq \alpha_1, \tag{99}$$

$$(-1)^{m(\alpha,j)} \frac{\partial^j G(\alpha,\beta,x,t)}{\partial x^j} > 0, \quad (x,t) \in (a,b) \times (a,b), \ \alpha_1 \leq j \leq \alpha_A, \tag{100}$$

and, if $\alpha_k, p > \alpha_A$,

$$(-1)^{m(\alpha,\alpha_A+1)} \frac{\partial^{\alpha_A+1} G(\alpha,\beta,x,t)}{\partial x^{\alpha_A+1}} > 0, \quad (x,t) \in (a,b) \times (a,b). \tag{101}$$

Proof. Let us tackle the case $\alpha_1 = 0$ first. From Theorem 3, concretely (23), we already know that (94) holds for $0 \leq j \leq \beta_1$ (note that $n(\beta,\beta_1) = n-k-1$).

Next, let us assume that $x > t$. From the definition of H one has

$$\frac{\partial^j G_{a,b}(\alpha,\beta,x,t)}{\partial x^j} = \frac{\partial^j G_{a,x}(\alpha,\beta,x,t)}{\partial x^j} + \int_x^b \frac{\partial}{\partial s} \frac{\partial^j G_{a,s}(\alpha,\beta,x,t)}{\partial x^j} ds$$

$$= \frac{\partial^j G_{a,x}(\alpha,\beta,x,t)}{\partial x^j} + \int_x^b \frac{\partial^j H_{a,s}(\alpha,\beta,x,t)}{\partial x^j} ds, \ (x,t) \in (a,b) \times (a,b). \tag{102}$$

$G_{a,x}(\alpha,\beta,x,t)$ is the Green function of the problem (4) when $b = x$, so it satisfies the boundary conditions related to β at x, that is

$$\frac{\partial^j G_{a,x}(\alpha,\beta,x,t)}{\partial x^j} = 0, \ t \in (a,x), \ j \in \beta. \tag{103}$$

On the other hand, from the hypotheses and Theorem 4 it follows that

$$(-1)^{n(\beta,j)} \frac{\partial^j H_{a,s}(\alpha,\beta,x,t)}{\partial x^j} < 0, \ (x,t) \in (a,s) \times (a,s), \ t < x \leq s \leq b, \ \beta_1 \leq j \leq \beta_B. \tag{104}$$

From (102), (103) and (104) one finally gets (95) for $x > t$ and $\beta_1 \leq j \leq \beta_B$.
Let us focus now on the case $x \leq t$. As before one has

$$\frac{\partial^j G_{a,b}(\alpha,\beta,x,t)}{\partial x^j} = \frac{\partial^j G_{a,t}(\alpha,\beta,x,t)}{\partial x^j} + \int_t^b \frac{\partial^j H_{a,s}(\alpha,\beta,x,t)}{\partial x^j} ds, \ (x,t) \in (a,b) \times (a,b). \tag{105}$$

$G_{a,t}(\alpha,\beta,x,t)$ is the Green function of the problem (4) when $b = t$, so it satisfies the boundary conditions related to β at t, that is

$$\frac{\partial^j G_{a,t}(\alpha,\beta,t,t)}{\partial x^j} = 0, \ t \in (a,b), \ j \in \beta. \tag{106}$$

If $n-1 \notin \beta$, $G_{a,t}(\alpha,\beta,x,t)$ is n-times continuously differentiable in (a,t), satisfies $LG_{a,t}(\alpha,\beta,x,t) = 0$ for $x \in (a,t)$ and n homogeneous boundary conditions at a and b. Since L is disfocal on $[a,b]$, it is also disfocal on $[a,t]$ and therefore $G_{a,t}(\alpha,\beta,x,t) \equiv 0$ for $x \in [a,t]$. From here, (104) and (105) one gets (95). On the contrary, if $n-1 \in \beta$, from the properties of the Green function (see [8], Chapter 3, page 105, property (ii))) it is straightforward to show that $G_{a,t}(\alpha,\beta,x,t)$ is n-times continuously differentiable on (a,t), satisfies $LG_{a,t}(\alpha,\beta,x,t) = 0$ for $x \in (a,t)$, $n-1$ homogeneous boundary conditions at a and b and the boundary condition

$$\lim_{x \to t^-} \frac{\partial^{n-1} G_{a,t}(\alpha,\beta,x,t)}{\partial x^{n-1}} = -1, \ t \in (a,b). \tag{107}$$

As noted in the Introduction, $p \geq \beta_B - 1$. We can apply Properties 2, 5 and 6 of Lemma 1 to (107), as well as the definition of $n(\beta,j)$, to yield

$$(-1)^{n(\beta,j)} \frac{\partial^j G_{a,t}(\alpha,\beta,x,t)}{\partial x^j} < 0, \ x \in (a,t), \ t \in (a,b), \ j \in \beta, \ j \leq p+1, \tag{108}$$

and

$$(-1)^{n(\beta,j)} \frac{\partial^j G_{a,t}(\alpha,\beta,x,t)}{\partial x^j} > 0, \ x \in (a,t), \ t \in (a,b), \ j \notin \beta, \ j \leq p+1. \tag{109}$$

From (104), (105) and (108) one gets (95) for the case $x \leq t$.

To address (96), let us note that if both β_{n-k}, $p > \beta_B$ then $\beta_B \notin \alpha$, $\beta_B + 1 \in \alpha$ and $\beta_B + 1 \notin \beta$ due to the definition of β_B and the p-alternate property of the boundary conditions (α,β). In that case we can define the boundary conditions $(\alpha,\check{\beta})$ by adding $\beta_B + 1$ and removing β_{n-k} to/from β, that is $\check{\beta} = \{\beta\setminus\beta_{n-k}\} \cup (\beta_B + 1)$. Then, fixed $t \in [a,b]$, the function $g_4(x) = G(\alpha,\check{\beta},x,t) - G(\alpha,\beta,x,t)$ is n times continuously differentiable on $[a,b]$ and satisfies

$$Lg_4 = 0, \ x \in (a,b);$$

$$g_4^{(\alpha_j)}(a) = 0, \ \alpha_j \in \alpha; \quad g_4^{(\beta_j)}(b) = 0, \ \beta_j \in \beta\setminus\beta_{n-k};$$

$$g_4^{(\beta_B+1)}(b) = -\frac{\partial^{\beta_B+1} G(\alpha,\beta,b,t)}{\partial x^{\beta_B+1}}. \tag{110}$$

From (22) and (110) it follows that

$$(-1)^{n(\beta,\beta_B+1)} g_4^{(\beta_B+1)}(b) < 0. \tag{111}$$

Applying property 2 of Lemma 1 to (111) (note that $p \geq \beta_B + 1$) one has

$$(-1)^{n(\beta,\beta_B+1)} g_4^{(\beta_B+1)}(x) < 0, \ x \in (a,b). \tag{112}$$

Likewise, applying (95) to $G(\alpha,\check{\beta},x,t)$ (note that $\check{\beta}_{n-k} < n-1$) one has

$$(-1)^{n(\check{\beta},\beta_B+1)} \frac{\partial^{\beta_B+1} G(\alpha,\check{\beta},x,t)}{\partial x^{\beta_B+1}} < 0, \ x \in (a,b), \tag{113}$$

which is also

$$(-1)^{n(\beta,\beta_B+1)} \frac{\partial^{\beta_B+1} G(\alpha,\check{\beta},x,t)}{\partial x^{\beta_B+1}} > 0, \ x \in (a,b). \tag{114}$$

Combining (112) and (114) one finally gets to

$$(-1)^{n(\beta,\beta_B+1)} \frac{\partial^{\beta_B+1} G(\alpha,\beta,x,t)}{\partial x^{\beta_B+1}}$$

$$= (-1)^{n(\beta,\beta_B+1)} \frac{\partial^{\beta_B+1} G(\alpha, \check{\beta}, x, t)}{\partial x^{\beta_B+1}} - (-1)^{n(\beta,\beta_B+1)} g_4^{(\beta_B+1)}(x) > 0, \; x \in (a, b), \quad (115)$$

which is (96).

The proof of (99)–(101) can be done using the same auxiliar Green function $G'(\beta, \alpha, x, t)$ of (57), applying (63) to (94)–(96) and taking into account that $n(\alpha, j) + m(\alpha, j) = n - j - 1$ when $j \in \alpha$. □

2.3. The Strongly Admissible Case

Last, but not least, we will prove a result on the strongly admissible case, extending the order of the partial derivatives of $G(x, t)$ for which the sign is constant in (a, b) up to the $(n-1)$-th order.

Theorem 7. *Let us assume that (α, β) are strongly admissible boundary conditions and that*

$$(-1)^{m(\alpha,j)} a_j(x) \leq 0, \; x \in [a, b], \; 0 \leq j \leq n - 1. \quad (116)$$

If $n - 1 \in \alpha$ let us assume that there exists at least one $l_\alpha \notin \alpha$ such that

$$(-1)^{m(\alpha,l_\alpha)} a_{l_\alpha}(a) < 0. \quad (117)$$

If $n - 1 \in \beta$ let us assume that there exists at least one $l_\beta \notin \beta$ such that

$$(-1)^{m(\alpha,l_\beta)} a_{l_\beta}(b) < 0. \quad (118)$$

If either of the following two conditions holds

1. $\alpha_1 = 0$ and either $\{\beta_B + 1, \ldots n - 1\} \subset \alpha$ or $\{\beta_B + 2, \ldots n - 1\} \subset \beta$,
2. $\alpha_1 > 0$ and either $\{\alpha_A + 2, \ldots n - 1\} \subset \alpha$ or $\{\alpha_A + 1, \ldots n - 1\} \subset \beta$,

then

$$(-1)^{m(\alpha,j)} \frac{\partial^j G(x, t)}{\partial x^j} > 0, \; (x, t) \in (a, b) \times (a, b), \; 0 \leq j \leq n - 1. \quad (119)$$

Proof. The key of this theorem is to prove that, fixed $t \in [a, b]$, $\frac{\partial^n G(x,t)}{\partial x^n} \geq 0$ for $x \in (a, b)$. This, added to the property of the Green functions (see [8], Chapter 3, page 105) that states that

$$\lim_{x \to t^+} \frac{\partial^{n-1} G(x, t)}{\partial x^{n-1}} = 1 + \lim_{x \to t^-} \frac{\partial^{n-1} G(x, t)}{\partial x^{n-1}}, \quad (120)$$

and the presence of one homogeneous boundary condition in $\frac{\partial^{n-1} G(x,t)}{\partial x^{n-1}}$ at either a or b, guarantees that $\frac{\partial^{n-1} G(x,t)}{\partial x^{n-1}}$ does not change sign on $x \in (a, b)$. The same absence of change of the sign of the partial derivatives of $G(x, t)$ of lower orders follows immediately from this fact and the strong admissibility of the homogeneous boundary conditions.

To prove the non-negative sign of $\frac{\partial^n G(x,t)}{\partial x^n}$ on (a, b) for fixed $t \in [a, b]$, let us focus first on its value at the extremes a and b. Thus, from the definition of L one has

$$\frac{\partial^n G(a, t)}{\partial x^n} = -\sum_{l=0}^{n-1} a_l(a) \frac{\partial^l G(a, t)}{\partial x^l} = -\sum_{l=0, l \notin \alpha}^{n-1} a_l(a) \frac{\partial^l G(a, t)}{\partial x^l}, \quad (121)$$

and

$$\frac{\partial^n G(b, t)}{\partial x^n} = -\sum_{l=0}^{n-1} a_l(b) \frac{\partial^l G(b, t)}{\partial x^l} = -\sum_{l=0, l \notin \beta}^{n-1} a_l(b) \frac{\partial^l G(b, t)}{\partial x^l}. \quad (122)$$

From Theorem 3 and the hypotheses (116), (117), it is straightforward to show that $\frac{\partial^n G(a,t)}{\partial x^n} > 0$ if $n - 1 \in \alpha$ and $\frac{\partial^n G(a,t)}{\partial x^n} \geq 0$ else. As for $\frac{\partial^n G(b,t)}{\partial x^n}$, if $l \notin \beta$, then $l \in \alpha$ and the strong admissibility forces that $m(\alpha,l) = n(\beta,l)$. From here, Theorem 3 and the hypotheses (116), (118), again, one gets that $\frac{\partial^n G(b,t)}{\partial x^n} > 0$ if $n-1 \in \beta$ and $\frac{\partial^n G(b,t)}{\partial x^n} \geq 0$ otherwise.

Next, let us do a similar comparison for the partial derivatives of lower order. If $n-1 \in \alpha$, from Taylor's theorem there must be a $\delta > 0$ such that

$$\frac{\partial^{n-1} G(x,t)}{\partial x^{n-1}} > 0, \quad x \in (a, a+\delta). \tag{123}$$

Applying Taylor's theorem recursively and taking into account (21) one proves that there exists a $\delta_1 > 0$ such that

$$(-1)^{m(\alpha,i)} \frac{\partial^i G(\alpha,\beta,x,t)}{\partial x^i} > 0, \quad x \in (a, a+\delta_1), \ 0 \leq i \leq n-1. \tag{124}$$

As for b, (22) already gives

$$(-1)^{n(\beta,n-1)} \frac{\partial^{n-1} G(b,t)}{\partial x^{n-1}} = \frac{\partial^{n-1} G(b,t)}{\partial x^{n-1}} > 0. \tag{125}$$

Applying again Taylor's theorem recursively and taking into account (22) one has that there must be a $\delta_2 > 0$ such that

$$(-1)^{n(\beta,i)} \frac{\partial^i G(\alpha,\beta,x,t)}{\partial x^i} > 0, \quad x \in (b-\delta_2, b], \ i \notin \beta, \ 0 \leq i \leq n-1, \tag{126}$$

and

$$(-1)^{n(\beta,i)} \frac{\partial^i G(\alpha,\beta,x,t)}{\partial x^i} < 0, \quad x \in (b-\delta_2, b), \ i \in \beta, \ 0 \leq i \leq n-1. \tag{127}$$

From (123) and (125) it is clear that $\frac{\partial^{n-1} G(x,t)}{\partial x^{n-1}}$ has the same (positive, in this case) sign on $x \in (a, a+\delta_1) \cup (b-\delta_2, b]$. We can prove by induction that this same sign property is valid for all partial derivatives of lower order, namely, that the signs given by (124), (126) and (127) are the same for each partial derivative. Thus, let us suppose that the sign of the partial derivative of order $l+1$ is the same in the neighborhoods of a and b, and is given by (124). If $l \in \beta$, then by Taylor's theorem, the sign of the derivative of order l must be the opposite of the sign of the derivative of order $l+1$ in the neighborhood of b. Likewise, $m(\alpha,l) = m(\alpha,l+1)+1$, so from (124) the sign of the derivative of order l must also be the opposite of the sign of the derivative of order $l+1$ in the neighborhood of a. Therefore, the sign of the partial derivatives of order l must coincide at the proximity of a and b. Likewise, if $l \in \alpha$ then by Taylor's theorem the sign of the derivative of order l must be the same as the sign of the derivative of order $l+1$ in the neighborhood of a, whereas the sign of the derivative of order l at b is given by $(-1)^{n(\beta,l)}$. If $l+1 \notin \beta$ then from (126) and since $n(\beta,l) = n(\beta,l+1)$ the sign of the derivative of order $l+1$ at b must also coincide with that of the derivative of order l at b. If $l+1 \in \beta$ then $n(\beta,l) = n(\beta,l+1)+1$, so from (127) the sign of the derivative of order $l+1$ at b must also coincide with that of the derivative of order l at b. That means, again, that the signs of the partial derivatives of $G(x,t)$ of order l must also coincide at the neighborhoods of a and b.

A similar reasoning can be done for the case $n-1 \in \beta$, leading to the same conclusions.

Once we have that the signs of the partial derivatives of $G(x,t)$ on the vicinity of a and b are the same, regardless of the order, and knowing already from Theorem 6 (note that the strongly admissible conditions are $(n-1)$-alternate) that the sign of $\frac{\partial^i G(x,t)}{\partial x^i}$ is constant on (a,b) for $0 \leq i \leq \beta_B$ (case $\alpha_1 = 0, \beta_{n-k} = \beta_B$), $0 \leq i \leq \beta_B + 1$ (case $\alpha_1 = 0, \beta_{n-k} > \beta_B$), $0 \leq i \leq \alpha_A$ (case $\alpha_1 > 0, \alpha_k = \alpha_A$) or $0 \leq i \leq \alpha_A + 1$ (case $\alpha_1 > 0, \alpha_k > \alpha_A$), and determined by (124) in all cases (it is straightforward to check), it remains to prove that the sign of $\frac{\partial^i G(x,t)}{\partial x^i}$ is constant on (a,b) for the rest of values of i up to $n-1$. We will do it by reduction to the absurd. Thus, let us suppose that there is an order l for which

$\frac{\partial^l G(x,t)}{\partial x^l}$ changes sign on (a,b). Since the sign at the vicinity of the extremes is the same, there must be at least an even number of sign changes on (a,b). Let us call $x_{1,l}$ the minimum of these points and $x_{2,l}$ the maximum of these points. Clearly the sign of $\frac{\partial^l G(x,t)}{\partial x^l}$ must be the same for $x \in (a, x_{1,l})$ and $x \in (x_{2,l}, b)$, and be given by (124).

Let us assume that $\{l, \ldots, n-1\} \subset \alpha$. Then by Rolle's Theorem we can obtain a sequence of zeroes $x_{1,j}, j = l, \ldots, n-1$, such that $x_{1,l} > x_{1,l+1} > \ldots > x_{1,n-2} > a$, for which the sign of $\frac{\partial^j G(x,t)}{\partial x^j}$ is constant on $(a, x_{1,j})$, and again given by (124). Since $\frac{\partial^{n-1} G(x,t)}{\partial x^{n-1}}$ has a discontinuity at $x = t$, there must be a smallest point $x_{1,n-1} < x_{1,n-2}$ where there is a change of sign of $\frac{\partial^{n-1} G(x,t)}{\partial x^{n-1}}$ from positive (see (124)) to negative, but from (120) it is clear that such a point cannot be $x_{1,n-1} = t$, so it must be a zero of $\frac{\partial^{n-1} G(x,t)}{\partial x^{n-1}}$. From the mean value theorem there must exist an $x^* \in (a, x_{1,n-1})$ such that $\frac{\partial^n G(x^*,t)}{\partial x^n} < 0$. However, the above reasoning implies that the sign of all partial derivatives of orders from l to $n-1$ is given by (124) for $x \in (a, x_{1,n-1})$, and from (116), that also means that the sign of $\frac{\partial^n G(x,t)}{\partial x^n}$ must be non-negative for all $x \in (a, x_{1,n-1})$, which is a contradiction.

A similar argument can be used if $\{l, \ldots, n-1\} \subset \beta$ and if $\alpha_1 > 0$, which completes the proof. □

Remark 6. *If $a_l(a) = 0$ for all $j \notin \alpha$, then the hypothesis (117) of the Theorem 7 can be replaced by any combination of $a_l(x)$ that grants $\frac{\partial^n G(x,t)}{\partial x^n} > 0$ for $x \in (a, a + \delta)$. Likewise, if $a_l(b) = 0$ for all $j \notin \beta$, then the hypothesis (118) of the Theorem 7 can be replaced by any combination of $a_l(x)$ that grants $\frac{\partial^n G(x,t)}{\partial x^n} > 0$ for $x \in (b - \delta, b)$.*

Remark 7. *One cannot help wondering if, with the right combinations of signs of $a_l(x)$ in $[a,b]$, it is possible to guarantee the conservation of sign of each partial derivative of G with respect to x in $[a,b]$ regardless of how α_j and β_j alternate in the case of strongly admissible conditions (that is, without imposing Conditions 1 and 2 in Theorem 7). Even though that assertion looks quite plausible, its proof has been elusive to the authors so far.*

3. Discussion

The results presented in this paper provide information about the sign and dependence on the extremes a and b of the Green function of the problem (4) and its derivatives when the two-point boundary conditions are admissible, property which encompasses many types of boundary conditions usually covered in the literature (for instance, conjugate or focal boundary conditions). By doing so, this paper extends (and to a small degree corrects, as discussed in the Remark 5) the results of Eloe and Ridenhour in [1], a fine piece of Green function theory that is considered a reference in the subject. The paper goes beyond to address the p-alternate and strongly admissible cases, for which results on the signs of higher derivatives on the interval are provided. Thus, whilst both [1] and the Section 2.1 yield sign results only for derivatives up to $\max(\alpha_1, \beta_1)$-th order, in the case of p-alternate they are supplied for derivatives up to $\alpha_A + 1$ (if $\alpha_1 > 0$) and $\beta_B + 1$ (if $\alpha_1 = 0$) orders, and in the case of strongly admissible conditions, for derivatives up to $(n-1)$-th order. As stated in the Introduction, this is relevant since the maximum value of the integer μ of the problem (6) which allows a cone-based approach is limited by the order of the highest derivative of $G(x,t)$ with constant sign, so that finding results for higher derivatives of $G(x,t)$ permits increasing the applicability of the cone theory to such problems.

One question that is left open is whether it is possible to find conditions on the sign of the coefficients of L which grant a constant sign of every derivative of $G(x,t)$ on (a,b) up to the $(n-1)$-th order, for any strongly admissible boundary conditions. We hypothesize an affirmative response, but a proper proof is still pending.

To conclude, other areas that can benefit from an extension of these sign findings are those of boundary conditions mixing different derivatives or those with integral conditions. The determination of the sign of the Green function of fractional boundary value problems is also a topic that has raised interest recently, as part of more sophisticated mechanisms to find solutions of other related non-linear fractional boundary value problems (see for instance [23–26]). However, there is a lot to do in this area, since most of these cases require the explicit calculation of the associated Green function, and this calculation is only possible in the simplest ones. A more generic approach that provided signs without having to solve fractional differential equations, similar to that presented here, would, therefore, be very welcome.

Author Contributions: Conceptualization, P.A.B.; methodology, P.A.B. and L.J.; investigation, P.A.B.; validation, P.A.B. and L.J.; writing—original draft preparation, P.A.B.; writing—review and editing, L.J.; visualization, P.A.B. and L.J.; supervision, P.A.B.; project administration, L.J.; funding acquisition, L.J. All authors have read and agreed to the published version of the manuscript.

Funding: This work has been supported by the Spanish Ministerio de Economía, Industria y Competitividad (MINECO), the Agencia Estatal de Investigación (AEI) and Fondo Europeo de Desarrollo Regional (FEDER UE) grant MTM2017-89664-P.

Conflicts of Interest: The authors declare no conflict of interest.

References

1. Eloe, P.W.; Ridenhour, J. Sign properties of Green's functions for a family of two-point boundary value problems. *Proc. Am. Math. Soc.* **1994**, *120*, 443–452.
2. Butler, G.; Erbe, L. Integral comparison theorems and extremal points for linear differential equations. *J. Diff. Equ.* **1983**, *47*, 214–226. [CrossRef]
3. Peterson, A. Green's functions for focal type boundary value problems. *Rocky Mountain J. Math.* **1979**, *9*, 721–732. [CrossRef]
4. Peterson, A. Focal Green's functions for fourth-order differential equations. *J. Math. Anal. Appl.* **1980**, *75*, 602–610. [CrossRef]
5. Elias, U. Green's functions for a nondisconjugate differential operator. *J. Diff. Equ.* **1980**, *37*, 319–350. [CrossRef]
6. Peterson, A.; Ridenhour, J. Comparison theorems for Green's functions for focal boundary value problems. In *World Scientific Series in Applicable Analysis*; Recent Trends in Differential Equations; Agarwal, R.P., Ed.; World Scientific Publishing Co. Pte. Ltd.: Singapore, 1992; Volume 1, pp. 493–506.
7. Nehari, Z. Disconjugate linear differential operators. *Trans. Am. Math. Soc.* **1967**, *129*, 500–516. [CrossRef]
8. Coppel, W. *Disconjugacy*; Springer: Berlin, Germany, 1971.
9. Krein, M.G.; Rutman, M.A. *Linear Operators Leaving a Cone Invariant in a Banach Space*; American Mathematical Society Translation Series 1; Cañada, A., Drábek, P., Fonda, A., Eds.; American Mathematical Society: Providence, RI, USA, 1962; Volume 10, pp. 199–325.
10. Krasnosel'skii, M.A. *Positive Solutions of Operator Equations*; P. Noordhoff Ltd.: Groningen, The Netherlands, 1964.
11. Keener, M.S.; Travis, C.C. Positive cones and focal points for a class of nth order differential equations. *Trans. Am. Math. Soc.* **1978**, *237*, 331–351. [CrossRef]
12. Schmitt, K.; Smith, H.L. Positive solutions and conjugate points for systems of differential equations. *Nonlinear Anal. Theory Methods Appl.* **1978**, *2*, 93–105. [CrossRef]
13. Eloe, P.W.; Hankerson, D.; Henderson, J. Positive solutions and conjugate points for multipoint boundary value problems. *J. Diff. Equ.* **1992**, *95*, 20–32. [CrossRef]
14. Eloe, P.W.; Henderson, J. Focal point characterizations and comparisons for right focal differential operators. *J. Math. Anal. Appl.* **1994**, *181*, 22–34. [CrossRef]
15. Almenar, P.; Jódar, L. Solvability of N-th order boundary value problems. *Int. J. Diff. Equ.* **2015**, *2015*, 1–19. [CrossRef]
16. Almenar, P.; Jódar, L. Improving results on solvability of a class of n-th order linear boundary value problems. *Int. J. Diff. Equ.* **2016**, *2016*, 1–10. [CrossRef]
17. Almenar, P.; Jódar, L. Solvability of a class of n-th order linear focal problems. *Math. Modell. Anal.* **2017**, *22*, 528–547. [CrossRef]
18. Sun, Y.; Sun, Q.; Zhang, X. Existence and nonexistence of positive solutions for a higher-order three-point boundary value problem. *Abstr. Appl. Anal.* **2014**, *2014*, 1–7. [CrossRef]

19. Hao, X.; Liu, L.; Wu, Y. Iterative solution to singular nth-order nonlocal boundary value problems. *Boundary Val. Prob.* **2015**, *2015*, 1–10. [CrossRef]
20. Eloe, P.W.; Neugebauer, J.T. Avery Fixed Point Theorem applied to Hammerstein integral equations. *Electr. J. Diff. Equ.* **2019**, *2019*, 1–20.
21. Webb, J.R.L. New fixed point index results and nonlinear boundary value problems. *Bull. Lond. Math. Soc.* **2017**, *49*, 534–547. [CrossRef]
22. Greguš, M. *Third Order Linear Differential Equations*; Mathematics and its Applications; Springer: Groningen, The Netherlands, 1987; Volume 22.
23. Jiang, D.; Yuan, C. The positive properties of the Green function for Dirichlet-type boundary value problems of nonlinear fractional differential equations and its application. *Nonlinear Anal. Theory Methods Appl.* **2010**, *72*, 710–719. [CrossRef]
24. Wang, Y.; Liu, L. Positive properties of the Green function for two-term fractional differential equations and its application. *J. Nonlinear Sci. Appl.* **2017**, *10*, 2094–2102. [CrossRef]
25. Zhang, L.L.; Tian, H. Existence and uniqueness of positive solutions for a class of nonlinear fractional differential equations. *Adv. Diff. Equ.* **2017**, *2017*, 1–19. [CrossRef]
26. Wang, Y. The Green's function of a class of two-term fractional differential equation boundary value problem and its applications. *Adv. Diff. Equ.* **2020**, *2020*, 1–20. [CrossRef]

© 2020 by the authors. Licensee MDPI, Basel, Switzerland. This article is an open access article distributed under the terms and conditions of the Creative Commons Attribution (CC BY) license (http://creativecommons.org/licenses/by/4.0/).

Article

Powers of the Stochastic Gompertz and Lognormal Diffusion Processes, Statistical Inference and Simulation

Eva María Ramos-Ábalos [1,*,†], Ramón Gutiérrez-Sánchez [1,†] and Ahmed Nafidi [2,†]

1. Department of Statistics and Operational Research, Faculty of Science, University of Granada, Avda. Fuentenueva, S/N, 18071 Granada, Spain; ramongs@ugr.es
2. Department of Mathematics and Informatics, LAMSAD, National School of Applied Sciences Berrechid, University of Hassan 1, Avenue de l'université, BP 280 Berrechid, Morocco; ahmed.nafidi@uhp.ac.ma
* Correspondence: ramosa@ugr.es; Tel.: +34-958-240-493
† These authors contributed equally to this work.

Received: 5 March 2020; Accepted: 13 April 2020; Published: 15 April 2020

Abstract: In this paper, we study a new family of Gompertz processes, defined by the power of the homogeneous Gompertz diffusion process, which we term the powers of the stochastic Gompertz diffusion process. First, we show that this homogenous Gompertz diffusion process is stable, by power transformation, and determine the probabilistic characteristics of the process, i.e., its analytic expression, the transition probability density function and the trend functions. We then study the statistical inference in this process. The parameters present in the model are studied by using the maximum likelihood estimation method, based on discrete sampling, thus obtaining the expression of the likelihood estimators and their ergodic properties. We then obtain the power process of the stochastic lognormal diffusion as the limit of the Gompertz process being studied and go on to obtain all the probabilistic characteristics and the statistical inference. Finally, the proposed model is applied to simulated data.

Keywords: powers of stochastic Gompertz diffusion models; powers of stochastic lognormal diffusion models; estimation in diffusion process; stationary distribution and ergodicity; trend function; application to simulated data

1. Introduction

Stochastic processes are used to model stochastic phenomena in various fields of science, engineering, economics and finance. An important category among these processes is that of Stochastic Diffusion Processes (SDP), which have received considerable attention recently, due on the one hand to their diverse applications in stochastic modelling, and on the other, to their value in addressing probabilistic statistical problems, especially those involving statistical inference. In consequence, these processes have been widely studied, and much research has been undertaken to resolve these issues of statistical inference, with particular respect to the estimation of parameters; see, among others, Bibby and Sorensen [1], Prakasa Rao [2], Chang and Cheng [3], Beskos et al. [4], Stramer and Yan [5], Shoji and Ozaki [6], Durham and Gallant [7] and Fan [8], without forgetting the works of Yenkie and Diwekar [9] and Kloeden et al. [10] and the important bibliography cited in these works.

There has been much recent interest in applying SDP, and many researchers are working on the construction of stochastic processes in order to model phenomena of interest. These processes are used in areas such as the stochastic economy, new technologies, interest rates, courses of action, insurance, finance in general, cell growth, radiotherapy, chemotherapy, emissions from energy consumption and the emissions of CO_2 and greenhouse gases. Research results have been applied to various processes, both in the homogeneous and in the non-homogeneous cases and many particular SDP have been proposed, such as Katsamaki and Skiadas [11] in the case of the exponential model, Skiadas and Giovanis [12] in the case of the Bass model, Giovanis and Skiadas [13] in the case of the logistic model, Gutiérrez et al. [14] in the case of the Rayleigh model and Román-Román et al. [15] in the case of the lognormal with exogenous factors.

Among the above-mentioned processes is the Stochastic Gompertz Diffusion Process (SGDP), which was first proposed by Ricciardi [16], who defined it in the homogeneous case by means of stochastic differential equations, for use in studies of population growth. It was subsequently used by Dennis and Patil [17] in ecology modelling. With respect to the Kolmogorov equations, it was defined by Nafidi [18], in a general way and for both the univariate and the multivariate cases.

In various papers, Gutiérrez et al. [19–21], Ferrante et al. [22], Román-Román et al. [23] and Giorno and Nobile [24], have highlighted the importance of this process, and many subsequent extensions have appeared, especially regarding the non-homogeneous case with exogenous factors (external variables) that affect the drift coefficient. In general, these extensions take one of the following two forms:

With external information (when no functional form is available): the exogenous factors are completely determined by the observed data (monthly, annual, etc.) and to obtain their functional forms interpolation methods, among others, can be used. This methodology has been applied by Gutiérrez et al. [25,26], Rupsys et al. [27] and Badurally Adam et al. [28]. In all these papers it is assumed that the coefficient drift is a linear combination of exogenous factors, obtained by linear interpolation.

Without external information: in this case there are no observed data for the exogenous factors, but they are functions of time and of certain parameters. For example, the case in which the deceleration factor is affected by exogenous factors was developed by Gutiérrez et al. [29]. Ferrante et al. [30] studied the Gompertz process in which exogenous factors are obtained as the sum of two exponential functions and Albano and Giorno [31] did so considering logarithmic exogenous factor.

The lognormal SDP and the SGDP, in turn, have been extended to the multivariate case with delay, by Frank [32], and to the bivariate case without delay by Gutiérrez et al. [33], and an application has been devised to model the emissions of CO_2 in Spain [34]. Other recent papers that have addressed questions related to SGDP include Hu [35] and Zou et al. [36].

In the present study, we define and examine a new extension of the Gompertz and lognormal diffusion processes, based on the homogeneous version of these processes, i.e., their power. Thus, we obtain two families of homogeneous diffusion processes. Firstly, we show that Gompertzian and lognormal diffusions are stable by power transformation. Them we define the proposed model as the solution to a stochastic differential equation. From this, we obtain: the explicit expression of the process, the Probability Transition Density Function (PTDF), the moments of different orders and, in particular, the conditioned and unconditioned trends of the process; the ergodicity of the process and its stationary distribution and the process parameters, estimated by maximum likelihood,with discrete sampling, determining the asymptotic properties of the likelihood estimators and the approximated confidence interval of the parameters.

In addition, we obtain the probabilistic and statistical characteristics of the lognormal process power, as a particular case of the process being studied, when the deceleration factor tends toward zero. Finally, the process and the methodology presented are applied to simulated data obtained from the explicit expression of the solution to the characteristic state equation for the process.

2. The Model and Its Basic Probabilistic Characteristics

2.1. An Overview of the Homogeneous Gompertz Stochastic Diffusion Process

Let $\{X(t); t \in [t_0, T]; t_0 \geq 0\}$ be a stochastic process taking values on $(0, \infty)$, $X(t)$ is a Gompertz diffusion process with parameters α, β and σ and which is denoted by $\text{Gomp}(\alpha; \beta; \sigma)$ if $X(t)$ satisfies Ito's Stochastic Differential Equation (SDE) as follows (see [16,18,20,37]):

$$dX(t) = [\alpha X(t) - \beta X(t) \log X(t)]\, dt + \sigma X(t) dw_t \quad ; \quad P(X(t_0) = X_{t_0}) = 1 \tag{1}$$

In the literature, the constant α ($\in \mathbb{R}$) is the intrinsic growth rate; the β ($\in \mathbb{R}$) constant is the deceleration factor, the $\sigma > 0$ constant is the diffusion coefficient, $X_{t_0} > 0$ is a fixed real number and w_t denotes the one-dimensional standard Wiener process.

The analytical expression of the unique solution to Equation (1) is given by (see, for example, [21,37])

$$X(t) = \exp\left\{ e^{-\beta(t-t_0)} \log X_{t_0} + \frac{\alpha - \sigma^2/2}{\beta}\left(1 - e^{-\beta(t-t_0)}\right) + \sigma \int_{t_0}^{t} e^{-\beta(t-\tau)} dw(\tau) \right\} \tag{2}$$

From this, we deduce that the process $X(t)$ is distributed as the following one-dimensional lognormal distribution:

$$\Lambda_1\left(e^{-\beta(t-t_0)} \log X_{t_0} + \frac{(\alpha - \sigma^2/2)}{\beta}\left(1 - e^{-\beta(t-t_0)}\right); \frac{\sigma^2}{2\beta}\left(1 - e^{-2\beta(t-t_0)}\right)\right)$$

It has been shown (see [21]), that for $\beta > 0$, $X(t)$ is ergodic and that the stationary distribution has a lognormal distribution. Hence, we have:

$$X(\infty) \sim \Lambda_1\left(\frac{\alpha - \sigma^2/2}{\beta} \; ; \; \frac{\sigma^2}{2\beta}\right) \tag{3}$$

2.2. The Proposed Model

Let $\{X(t); t \in [t_0, T]; t_0 \geq 0\}$ be a $\text{Gomp}(\alpha; \beta; \sigma)$. Then, the γ-power of the Stochastic Gompertz Diffusion Process (γ-PSGDP) $X(t)$ is defined by

$$x_\gamma(t) = X^\gamma(t); \qquad \gamma \in \mathbb{R}^* \tag{4}$$

The process $\{x_\gamma(t); t \in [t_0, T]; t_0 \geq 0\}$ is also a diffusion process with values in $(0, \infty)$ and has the drift and diffusion coefficients are shown below.

By applying Ito's formula to the transform given in Equation (4), we have

$$\begin{aligned} dx_\gamma(t) &= \gamma X^{\gamma-1}(t)\left[\alpha X(t) - \beta X(t) \log X(t)\right] dt + \gamma \sigma X^\gamma(t) dW_t + \gamma(\gamma - 1)\frac{\sigma^2}{2} X^\gamma(t) dt \\ &= [\alpha \gamma X^\gamma(t) - \beta \gamma X^\gamma(t) \log X(t)] dt + \gamma \sigma X^\gamma(t) dW_t \end{aligned}$$

Then, after some algebraic rearrangement, we obtain

$$dx_\gamma(t) = [ax_\gamma(t) - \beta x_\gamma(t) \log x_\gamma(t)] dt + cx_\gamma(t) dw(t)$$

This shows that the process $x_\gamma(t)$ is also a $\text{Gomp}(a; \beta; c)$ process, where:

$a = \gamma\alpha + \gamma(\gamma-1)\frac{\sigma^2}{2}$ and $c = \gamma\sigma$ and the drift and diffusion coefficients are given respectively by:

$$A_1(x) = \left(\gamma\alpha + \frac{\gamma(\gamma-1)\sigma^2}{2}\right)x - \beta x \log(x)$$

$$A_2(x) = \gamma^2\sigma^2 x^2$$

The model proposed in this paper belongs to the family of processes γ-PSGDP $\{x_\gamma(t); t \in [t_0, T]; t_0 \geq 0\}$ defined by the following SDE:

$$dx_\gamma(t) = A_1(x_\gamma(t))dt + \sqrt{A_2(x_\gamma(t))}dw(t) \quad ; \quad P(x_\gamma(t_0) = x_{t_0}) = 1$$

2.3. Probabilistic Characteristics of the γ-PSGDP

Under the initial condition given, the unique solution of the SDE Equation (5) can be obtained using the relations expressed by Equations (2) and (4), from which we have

$$x_\gamma(t) = \exp\left\{e^{-\beta(t-t_0)}\log x_{t_0} + \frac{\gamma(\alpha - \sigma^2/2)}{\beta}\left(1 - e^{-\beta(t-t_0)}\right) + \gamma\sigma \int_{t_0}^t e^{-\beta(t-\tau)}dw(\tau)\right\} \quad (5)$$

We then deduce that $x_\gamma(t)$ is distributed as a one dimensional lognormal distribution $\Lambda_1(\mu(s, t, x_{t_0}), \gamma^2\sigma^2\lambda^2(t_0, t))$, where $\mu(s, t, x_{t_0})$ and $\lambda^2(t_0, t)$ are given by

$$\mu(s, t, x_{t_0}) = e^{-\beta(t-t_0)}\log x_{t_0} + \frac{\gamma(\alpha - \sigma^2/2)}{\beta}\left(1 - e^{-\beta(t-t_0)}\right)$$

$$\lambda^2(t_0, t) = \frac{1}{2\beta}\left(1 - e^{-2\beta(t-t_0)}\right)$$

From the homogeneity of the process, we know that $x_\gamma(t) \mid x_\gamma(s) = x_s$ has the lognormal distribution $\Lambda_1(\mu(s, t, x_s), \sigma^2\lambda^2(s, t))$, and then the PTDF of the process is

$$f(y, t \mid x, s) = \frac{1}{y}\left[2\pi\gamma^2\sigma^2\lambda^2(s, t)\right]^{-1/2}\exp\left(-\frac{[\log(y) - \mu(s, t, x)]^2}{2\gamma^2\sigma^2\lambda^2(s, t)}\right)$$

The rth conditional moment of the process is given by

$$E\left(x_\gamma^r(t) \mid x_\gamma(s) = x_s\right) = \exp\left\{r\mu(s, t, x_s) + \frac{r^2\gamma^2\sigma^2}{2}\lambda^2(s, t)\right\}$$

from which the Conditional Trend Function (CTF) gives

$$E(x_\gamma(t) \mid x_\gamma(s) = x_s) = \exp\left\{e^{-\beta(t-s)}\log x_s + \frac{\gamma(\alpha - \sigma^2/2)}{\beta}\left(1 - e^{-\beta(t-s)}\right)\right.$$
$$\left. + \frac{\gamma^2\sigma^2}{4\beta}\left(1 - e^{-2\beta(t-s)}\right)\right\} \quad (6)$$

Assuming the initial condition $P(x_\gamma(t_0) = x_{t_0}) = 1$, the Trend Function (TF) of the process is

$$E(x_\gamma(t)) = \exp\left\{e^{-\beta(t-t_0)}\log(x_{t_0}) + \frac{\gamma(\alpha - \sigma^2/2)}{\beta}\left(1 - e^{-\beta(t-t_0)}\right)\right.$$
$$\left. + \frac{\gamma^2\sigma^2}{4\beta}\left(1 - e^{-2\beta(t-t_0)}\right)\right\} \quad (7)$$

From Equation (3), we deduce that for $\beta > 0$, the stationary distribution of the process is also a lognormal distribution and thus we have:

$$x_\gamma(\infty) \sim \Lambda_1\left(\frac{\gamma(\alpha - \sigma^2/2)}{\beta}; \frac{\gamma^2\sigma^2}{2\beta}\right) \qquad (8)$$

Therefore, the asymptotic trend function of the process (for $\beta > 0$) is given by

$$E[x_\gamma(\infty)] = \exp\left(\frac{\gamma(\alpha - \sigma^2/2)}{\beta} + \frac{\gamma^2\sigma^2}{4\beta}\right)$$

The limit of the trend function in Equation (7) (when t tends to ∞) coincides with this asymptotic trend function.

3. Statistical Inference on the Model

3.1. Likelihood Parameter Estimation

In the present study, with discrete sampling, we estimate the parameters α, σ^2 and β of the model by applying Maximum Likelihood (ML) methodology, following the same scheme as in Gutiérrez et al. [21]. To do so, we consider a discrete sampling of the process $x_\gamma(t_1) = x_1, x_\gamma(t_2) = x_2, \ldots, x_\gamma(t_n) = x_n$ for times $t_1, t_2, \ldots, t_n$ and assume, moreover, that the length of the time intervals $[t_{i-1}, t_i]$ ($i = 2, \ldots, n$) is equal to constant h i.e., $t_i - t_{i-1} = h$ and an initial distribution $P[x_\gamma(t_1) = x_1] = 1$. Then the associated likelihood function can be obtained by the following expression:

$$\mathbb{L}(x_1, \ldots, x_n, \alpha, \beta, \sigma^2) = \prod_{j=2}^{n} f(x_j, t_j \mid x_{j-1}, t_{j-1})$$

The variable change can be used to work with a known probability function and to calculate the maximum probability estimators in a simpler way, considering the following transformation: $v_1 = x_1, v_{i,\beta} = \lambda_\beta^{-1}(\log(x_i) - e^{-\beta h}\log(x_{i-1}))$, for $i = 2, \ldots, n$ and denoting $\mathbf{V}_\beta = (v_{2,\beta}, \ldots, v_{n,\beta})'$. Thus, in terms of $\mathbf{V}_\beta$, the likelihood function is expressed as follows:

$$\mathbb{L}_{\mathbf{V}_\beta}(a_\gamma, \beta, c_\gamma^2) = \left[2\pi c_\gamma^2 \lambda_\beta^2\right]^{-(n-1)/2} \exp\left(-\frac{1}{2c_\gamma^2}(\mathbf{V}_\beta - \nu_\beta a_\gamma \mathbf{U})'(\mathbf{V}_\beta - \nu_\beta a_\gamma \mathbf{U})\right)$$

where $a_\gamma = \gamma\left(\alpha - \frac{\sigma^2}{2}\right)$, $c_\gamma = \gamma\sigma$, $\nu_\beta = \lambda_\beta^{-1}(1 - e^{-\beta h})/\beta$, $\lambda_\beta^2 = \frac{1}{2\beta}(1 - e^{-2h\beta})$ and $\mathbf{U} = (1, \ldots, 1)'$ is a vector of the order $(n-1)$.

By differentiating the log-likelihood function with respect to a_γ and c_γ^2, we obtain the following equations:

$$\mathbf{U}'\mathbf{V}_\beta = \hat{a}_\gamma \nu_\beta \mathbf{U}'\mathbf{U}$$
$$(n-1)\hat{c}_\gamma^2 = (\mathbf{V}_\beta - \hat{a}_\gamma \nu_\beta \mathbf{U})'(\mathbf{V}_\beta - \hat{a}_\gamma \nu_\beta \mathbf{U})$$

The third likelihood equation is obtained by differentiating the log-likelihood function with respect to β and by using the effect that $\mathbf{V}_\beta = \lambda_\beta^{-1}(J_x - e^{-\beta h}I_x)$ with $J_x = (\log(x_2), \ldots, \log(x_n))'$ and $I_x = (\log(x_1), \ldots, \log(x_{n-1}))'$. After various operations, we have

$$I_x'(\mathbf{V}_\beta - \hat{a}_\gamma \nu_\beta \mathbf{U}) = 0$$

Taking into account that $\mathbf{U'U} = n - 1$ and after algebraic rearrangement (not shown), the ML estimators of $\mathbf{a}_\gamma$ and c_γ^2 are

$$(n-1)\hat{\mathbf{a}}_\gamma = \mathbf{V}_\beta^{-1}\mathbf{U'V}_\text{fi} \qquad (9)$$

$$(n-1)\hat{c}_\gamma^2 = \mathbf{V}_\beta'\mathbf{H_U V}_\beta \qquad (10)$$

The ML estimator of β is given by

$$\hat{\beta} = \frac{1}{h}\log\left(\frac{\mathbf{I}_x'\mathbf{H_U I}_x}{\mathbf{I}_x'\mathbf{H_U J}_x}\right) \qquad (11)$$

where $\mathbf{H_U} = \mathbf{I}_{n-1} - \frac{1}{n-1}\mathbf{UU'}$ is idempotent and a symmetric matrix and $\mathbf{I}_{n-1}$ denotes the identity matrix.

3.2. Asymptotic Properties of the Parameter Drift Estimators

Let X be a random variable with a distribution function given by Equation (8); then $\log(X)$ is distributed as a normal distribution $N_1\left(\frac{\gamma(\alpha-\sigma^2/2)}{\beta}; \frac{\gamma^2\sigma^2}{2\beta}\right)$. If $\beta > 0$, the process under consideration has ergodic properties, and for $\theta^* = (a_\gamma, \beta) \in (a_{\gamma,1}, a_{\gamma,2}) \times (\beta_1, \beta_2)$, with $\beta_1 > 0$, we have

$$\mathcal{L}_\theta\left(\sqrt{T}(\hat{\theta} - \theta)\right) \to \mathcal{N}_2\left(0, \mathbb{I}^{-1}(\theta)\right) \quad ; \quad \text{when} \quad T \to \infty \qquad (12)$$

$\mathbb{I}(\theta)$ is the information matrix and is given by $\mathbb{I}(\theta) = \mathbb{E}_\theta\left(\frac{\dot{A}_1(X)\dot{A}_1^*(X)}{A_2(X)}\right)$

where $\dot{A}_1(x)$ is the following vector: $\dot{A}_1(x) = \left(\frac{\partial A_1(x)}{\partial \alpha}; \frac{\partial A_1(x)}{\partial \beta}\right)^*$

Then, we have

$$\mathbb{I}(\theta) = \frac{1}{\gamma^2\sigma^2}\mathbb{E}_\theta\begin{pmatrix} \gamma^2 & -\gamma\log(X) \\ -\gamma\log(X) & \log^2(X) \end{pmatrix} = \frac{1}{\sigma^2}\begin{pmatrix} 1 & -\frac{\alpha-\sigma^2/2}{\beta} \\ -\frac{\alpha-\sigma^2/2}{\beta} & \frac{\sigma^2}{2\beta} + \frac{(\alpha-\sigma^2/2)^2}{\beta^2} \end{pmatrix}$$

and the inverse is

$$\mathbb{I}^{-1}(\theta) = \begin{pmatrix} \sigma^2 + \frac{2}{\beta}(\alpha - \frac{\sigma^2}{2})^2 & 2\alpha - \sigma^2 \\ 2\alpha - \sigma^2 & 2\beta \end{pmatrix} \qquad (13)$$

An approximated, asymptotic confidence region of θ and an approximated, asymptotic marginal confidence interval of α and β can be obtained from Equations (12) and (13). The above-mentioned region is given, for a large T, by

$$P\left[T(\theta - \hat{\theta})^* \hat{\mathbb{I}}(\theta)(\theta - \hat{\theta}) \leq \chi_{2,\xi}^2\right] = 1 - \xi$$

obtaining $\hat{\mathbb{I}}(\theta)$ by replacing the parameters by their estimators and where $\chi_{2,\xi}^2$ represents the upper 100ξ per cent points of the chi squared distribution with two degrees of freedom.

The $\xi\%$ confidence (marginal) intervals for parameters α and β are given, for a large T, by

$$P\left(\alpha \in \left[\hat{\alpha} \pm \frac{1}{\gamma}\lambda_\xi \left(\frac{\hat{\beta}\hat{\sigma}^2 + 2(\hat{\alpha} - \hat{\sigma}^2/2)^2}{\hat{\beta}T}\right)^{1/2}\right]\right) = 1 - \xi \qquad (14)$$

$$P\left(\beta \in \left[\hat{\beta} \pm \lambda_\xi (2\hat{\beta}/T)^{1/2}\right]\right) = 1 - \xi \qquad (15)$$

where λ_ξ represents the 100ξ per cent points of the normal standard distribution.

Note that in Equations (14) and (15) we have assumed that σ is known with a value $\sigma = \hat{\sigma}$.

4. Powers of the Lognormal Diffusion Process

The Stochastic Lognormal Diffusion Process (SLDP) is known to be a particular case of the Gompertz diffusion process when the deceleration factor $\beta = 0$ (see, for example [21]). Then, the power of the SLDP can be obtained from that of the SGDP by tending β to zero.

Then, if the SLDP $Y(t)$ is given by the following SDE:

$$dY(t) = \alpha Y(t)dt + \sigma Y(t)dw_t$$

The resulting γ-PSLDP ($y_\gamma(t) = Y^\gamma(t)$) is governed by the following SDE:

$$dy_\gamma(t) = \left(\gamma\alpha + \frac{\gamma(\gamma-1))\sigma^2}{2}\right)y_\gamma dt + \gamma\sigma y_\gamma dw(t) \qquad (16)$$

The same approach can be used to derive all the probabilistic properties and statistics for the γ-PSLDP process, taking $\beta = 0$ on the perspective equations established for the properties of γ-PSGDP in the previous sections, except as regards the symptotic properties of the drift parameter estimators (we already know that there is no asymptotic distribution in the case of the SLDP). For the latter case, we can obtain the exact distributions of the estimators, together with the confidence intervals for the process parameters (see [21]).

4.1. Estimated Trend Functions

In the same way as in Gutiérrez et al. [21], by Zehna's theorem [38], the Estimated Conditional Trend (ECT) and the Estimated Trend (ET) functions can be obtained from Equations (6) and (7) by replacing the parameters by their estimators. Furthermore, we can obtain an approximated and asymptotic confidence interval of the ETF and ECTF by means of the approximated and asymptotic confidence interval of the parameters given by Equations (14) and (15).

5. Simulation and Application

The trajectory of the model can be obtained by simulating the exact solution of SDE Equation (4) obtained in Equation (5). From this explicit solution, the simulated trajectories of the process are obtained from the following discretising time interval $[t_0, T]$: $t_i = t_0 + ih$, for $i = 1, \ldots, N$ (N is an integer and h is the discretization step), taking into account that the random variable in the latter expression $\sigma(w_t) - w(t_1)$ is distributed as a one-dimensional normal distribution $\mathcal{N}(0, \sigma^2(t - t_1))$ ([39]).

Table 1 shows the simulated data and the ETF for different powers, considering $h = 1$, $N = 30$, and the initial value $x_1 = 0.99$. We estimate the parameters by maximum likelihood, reserving the values observed for the time $t = 30$ for comparison with the corresponding prediction by the model. The results are shown in Table 2.

Table 1. Simulated data and estimated trend function.

Time	$x_1(t)$	ETF-x_1	$x_{1.5}(t)$	ETF-$x_{1.5}$	$x_2(t)$	ETF-x_2
1	0.99	0.99	0.99	0.99	0.99	0.99
2	2.1831	2.1832	3.2364	3.2369	4.7957	4.7960
3	3.5272	3.5271	6.6380	6.6385	12.4861	12.4876
4	4.7180	4.7181	10.2628	10.2613	22.3149	22.3122
5	5.6288	5.6286	13.3648	13.3620	31.7343	31.7261
6	6.2651	6.2645	15.6878	15.6818	39.2796	39.2767
7	6.6848	6.6846	17.2845	17.2802	44.7154	44.7063
8	6.9539	6.9531	18.3316	18.3276	48.3607	48.3586
9	7.1220	7.1211	18.9998	18.9933	50.7075	50.7176
10	7.2251	7.2250	19.4136	19.4087	52.1922	52.2041
11	7.2894	7.2887	19.6703	19.6649	53.1189	53.1268
12	7.3285	7.3277	19.8262	19.8219	53.7088	53.6943
13	7.3520	7.3514	19.9247	19.9177	54.0539	54.0414
14	7.3663	7.3658	19.9776	19.9761	54.2598	54.2531
15	7.3742	7.3746	20.0117	20.0115	54.3836	54.3818
16	7.3792	7.3799	20.0323	20.0330	54.4489	54.4601
17	7.3820	7.3831	20.0497	20.0461	54.4903	54.5076
18	7.3841	7.3851	20.0598	20.0540	54.5492	54.5364
19	7.3849	7.3863	20.0641	20.0588	54.5629	54.5539
20	7.3862	7.3870	20.0648	20.0617	54.5623	54.5645
21	7.3875	7.3874	20.0633	20.0635	54.5783	54.5710
22	7.3877	7.3877	20.0654	20.0645	54.5922	54.5749
23	7.3885	7.3879	20.0662	20.0652	54.5997	54.5773
24	7.3882	7.3880	20.0587	20.0656	54.6148	54.5787
25	7.3881	7.3880	20.0626	20.0658	54.6020	54.5796
26	7.3883	7.3881	20.0638	20.0660	54.5914	54.5801
27	7.3890	7.3881	20.0599	20.0661	54.6196	54.5804
28	7.3878	7.3881	20.0549	20.0661	54.6297	54.5806
29	7.3873	7.3881	20.0507	20.0661	54.6110	54.5807
Prediction						
30	7.3872	7.3881	20.0473	20.0662	54.6221	54.5808

Table 2. Starting values used in the simulation and estimation of the parameters.

	σ	α	β
Starting Values	0.0001	1	0.5

γ	$\hat{\sigma}$	$\hat{\alpha}$	$\hat{\beta}$
1	0.0000852	0.999952	0.500008
1.5	0.0001498	1.00043	0.500377
2	0.0001606	1.00003	0.500052

Figure 1 shows the fit and the prediction obtained for $x_\gamma(t)$ using the ETF ($\gamma = 1$ $\gamma = 1.5$ and $\gamma = 2$) (see Table 1).

Figure 2 shows 10 simulated trajectories for $x_\gamma(t)$ ($\gamma = 1$ $\gamma = 1.5$ and $\gamma = 2$), taking as the values for α, β and σ those obtained by maximum likelihood estimation (see Table 2). For each trajectory, 2901 data are generated by considering $h = 0.01$, and initial value $x_1 = 0.99$.

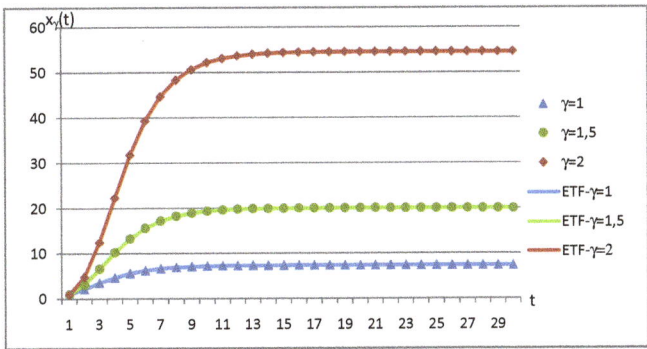

Figure 1. Fit and prediction based on ETF.

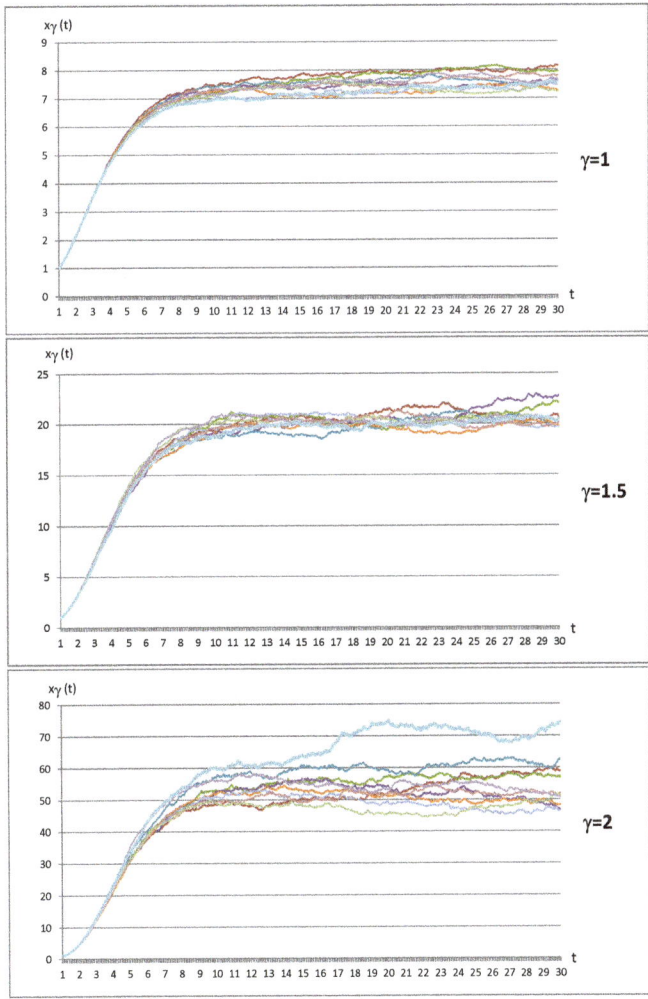

Figure 2. Fit and prediction based on ETF.

Figure 3 shows a trajectory whose values are the average of those obtained in the simulation of 100 trajectories, with the ETF. The values used in the simulation and the results obtained by estimating the parameters are shown in Table 3.

Table 3. Starting values used in the simulation and estimation of the parameters.

	σ	α	β
Starting values	0.0001	1	0.5
γ	$\hat{\sigma}$	$\hat{\alpha}$	$\hat{\beta}$
1.5	0.0000106801	1.00006	0.50003

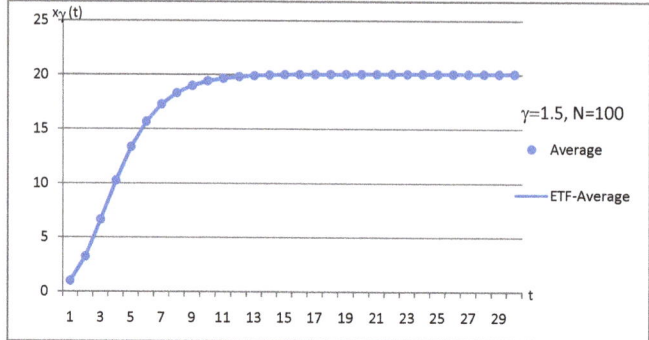

Figure 3. Fit and prediction based on ETF.

The variation of the mean and standard error of the estimators is studied, taking into account how N and h change. The results are shown in Table 4.

20 process paths are simulated with N observations each. The parameters are estimated using the equations (ref Eq11), (ref Eq12) and (ref Eq13), obtaining a vector of 20 components corresponding to the different estimators. For these, the sample mean is calculated and the Standard Error (SE).

The next step is to study the evolution of the mean and the standard error of the estimators with respect to the variation in the number N and in h. The results of this study are shown in Table 4.

The true parameter values considered in this simulation are $\alpha = 1$, $\beta = 0.5$, $\sigma = 0.0001$ and the start point is $x_1 = 0.99$, and $t_1 = 0$ and $\gamma = 1.5$.

The calculations have been made using the Mathematica program, in which a program has been implemented.

Table 4. Mean and standard error of the estimators.

h	N	Mean ($\hat{\sigma}$)	SE ($\hat{\sigma}$)	Mean ($\hat{\alpha}$)	SE ($\hat{\alpha}$)	Mean ($\hat{\beta}$)	SE ($\hat{\beta}$)
0.05	100	0.025108	0.114736	1.000132	0.000503	0.500144	0.000439
0.05	500	0.000112	0.000005	0.999637	0.000839	0.499770	0.000809
0.05	1000	0.000116	0.000005	1.000181	0.000953	0.500090	0.000915
0.1	100	0.000106	0.000008	1.000027	0.000350	0.500007	0.000262
0.1	500	0.000123	0.000010	1.000020	0.000654	0.500044	0.000647
0.1	1000	0.000141	0.000016	0.999081	0.000736	0.499156	0.000672
0.5	100	0.000143	0.000030	1.000046	0.000253	0.500002	0.000274
0.5	500	0.000329	0.000069	0.999171	0.000616	0.499202	0.000570
0.5	1000	0.000491	0.000141	0.998581	0.000730	0.498779	0.000584
1	100	0.000230	0.000074	0.999610	0.000381	0.499638	0.000359
1	500	0.000584	0.000217	0.999034	0.000597	0.499092	0.000541
1	1000	0.000908	0.000318	0.997592	0.001211	0.498045	0.000923

6. Conclusions

This article presents a study of the Gamma Power Stochastic Gompertz Diffusion Process (γ-PSGDP), including all its probabilistic properties and the corresponding statistical inference. As a particular case in the limit comparison test, we also study the Gamma Power Stochastic Lognormal Diffusion Process (γ-PSLDP).

A simulation study was conducted, analysing different process trajectories.

In the future, it will be possible to apply these models to fit real data and to obtain goodness of fit results between the processes and the data. We will also study the possibility of defining all these processes in their non-homogeneous form, by introducing exogenous factors, and considering the use of numerical methods to obtain the estimates.

Author Contributions: All the authors have collaborated equally in the realization of this work, both in theoretical and applied developments. Similarly in the writing and review of it. All authors have read and agreed to the published version of the manuscript.

Funding: This research has been funded by "Programa Operativo FEDER de Andalucía 2014-2020 A-FQM-228-UGR18".

Conflicts of Interest: The authors declare no conflict of interest.

Abbreviations

The following abbreviations are used in this manuscript:

SDP	Stochastic Diffusion Processes
SGDP	Stochastic Gompertz Diffusion Process
PTDF	Probability Transition Density Function
SDE	Stochastic Differential Equation
γ-PSGDP	γ-Power of the Stochastic Gompertz Diffusion Process
γ-PSLDP	γ- Power of the Stochastic Lognormal Diffusion Process
CTF	Conditional Trend Function
TF	Trend Function (TF)
ML	Maximum Likelihood
SLDP	Stochastic Lognormal Diffusion Process
ECT	Estimated Conditional Trend
ET	Estimated Trend
SE	Standard Error

References

1. Bibby, B.M.; Sorensen, M. Martingale estimation functions for discretely observed diffusion processes. *Bernoulli* **1995**, *1*, 17–39. [CrossRef]
2. Prakasa Rao, B.S.L. *Statistical Inference for Diffusion Type Process*; Ed. Arnold: London, UK; Oxford University Press: New York, NY, USA, 1999.
3. Chang, J.; Chen, S.X. On the approximate maximum likelihood estimation for diffusion processes. *Ann. Stat.* **2011**, *39*, 2820–2851. [CrossRef]
4. Beskos, A.; Papaspiliopoulos, O.; Roberts, G.O.; Fearnhead, P. Exact and computationally eficient likelihood-based estimation for discretely observed diffusion processes (with discussion). *J. R. Stat. Soc. Ser. B (Stat. Methodol.)* **2006**, *68*, 333–382. [CrossRef]
5. Stramer, O.; Yan, J. On the simulated likelihood of discretely observed diffusion processes and comparison to closed-form approximation. *J. Comput. Graph. Stat.* **2007**, *16*, 672–691. [CrossRef]
6. Shoji, I.; Ozaki, T. Comparative study of estimation methods for continuous time stochastic processes. *J. Time Ser. Anal.* **1997**, *18*, 485–506. [CrossRef]
7. Durham, G.B.; Gallant, A.R. Numerical techniques for maximum likelihood estimation of the continuous-times diffusion processes. *J. Bus. Econ. Stat.* **2002**, *20*, 297–316. [CrossRef]

8. Fan, J. A selective overview of nonparametric methods in financial econometrics. *Stat. Sci.* **2005**, *20*, 317–337. [CrossRef]
9. Yenkie, K.M.; Urmila, D. The "No Sampling Parameter Estimation (NSPE)" algorithm for stochastic differential equations. *Chem. Eng. Res. Des.* **2018**, *129*, 376–383. [CrossRef]
10. Kloeden, P.E.; Platen, E.; Schurz, H. *Numerical Solution of SDE through Computer Experiments*; Springer: Heidelberg/Berlin, Germany, 1994.
11. Katsamaki, A.; Skiadas, C.H. Analytic solution and estimation of parameters on a stochastic exponential model for a technological diffusion process application. *Appl. Stoch. Model. Data Anal.* **1995**, *11*, 59–75. [CrossRef]
12. Skiadas, C.H.; Giovani, A.N. A stochastic Bass innovation diffusion model for studying the growth of electricity consumption in Greece. *Appl. Stoch. Model. Data Anal.* **1997**, *13*, 85–101. [CrossRef]
13. Giovanis, A.N.; Skiadas, C.H. A stochastic logistic innovation diffusion model studying the electricity consumption in Greece and the United States. *Technol. Forecast. Soc. Chang.* **1999**, *61*, 253–264. [CrossRef]
14. Gutiérrez, R.; Gutiérrez-Sánchez, R.; Nafidi, A. The stochastic Rayleigh diffusion model: Statistical inference and computational aspects. Applications to modelling of real cases. *Appl. Math. Comput.* **2006**, *175*, 628–644. [CrossRef]
15. Román-Román, P.; Serrano-Pérez, J.J.; Torres-Ruiz, F. Some notes about inference for the lognormal diffusion process with exogenous factors. *Mathematics* **2018**, *6*, 85. [CrossRef]
16. Ricciardi, L. Diffusion processes and related topics in biology. In *Lecture Notes in Biomathematics*; Springer: Berlin, Germany, 1977.
17. Dennis, B.; Patil, G.B. Application in Ecology. In *Lognormal Distributions: Theory and Applications*; Crow, E.L., Shimizu, K., Eds.; Marcel Dekker: New York, NY, USA, 1988; pp. 310–330.
18. Nafidi, A. Lognormal Diffusion Process with Exogenous Factors, Extensions from the Gompertz Diffusion Process. Ph.D. Thesis, Granada University, Granada, Spain, 1997. (In Spanish)
19. Gutiérrez, R.; Nafidi, A.; Gutiérrez-Sánchez, R. Inference in the stochastic Gompertz diffusion model with continuous sampling. In *Monografías del Seminario García de Galdeano*; Torrens, J.J., Madaune-Tort, M., Trujillo, D., López de Silanes, M.C., Palacios, M., Sanz, G., Eds.; Prensas de la Universidad de Zaragoza: Zaragoza, Spain, 2004; Volume 31, pp. 247–253.
20. Gutiérrez, R.; Nafidi, A.; Gutiérrez-Sánchez, R. Forecasting total natural-gas consumption in Spain by using the stochastic Gompertz innovation diffusion model. *Appl. Energy* **2005**, *80*, 115–124. [CrossRef]
21. Gutiérrez, R.; Gutiérrez-Sánchez, R. ; Nafidi, A. Modelling and forecasting vehicle stocks using the trends of stochastic Gompertz diffusion models: The case of Spain. *Appl. Stoch. Models Bus. Ind.* **2009**, *25*, 385–405. [CrossRef]
22. Ferrante, L.; Bompade, S.; Possati, L.; Leone, L. Parameter estimation in a Gompertzian stochastic-model for tumor growth. *Biometrics* **2000**, *56*, 1076–1081. [CrossRef]
23. Román-Román, P.; Serrano-Pérez, J.J.; Torres-Ruiz, F. A Note on estimation of multi-sigmoidal Gompertz functions with random noise. *Mathematics* **2019**, *7*, 541. [CrossRef]
24. Giorno, V.; Nobile, A.G. Restricted Gompertz-type diffusion processes with periodic regulation functions. *Mathematics* **2019**, *7*, 555. [CrossRef]
25. Gutiérrez, R.; Nafidi, A.; Gutiérrez-Sánchez. R.; Román P. ; Torres, F. Inference in Gompertz type non homogeneous stochastic systems by means of discrete sampling. *Cybern. Syst.* **2005**, *36*, 203–216.
26. Gutiérrez, R.; Gutiérrez-Sánchez, R.; Nafidi, A. Electricity consumption in Morocco: Stochastic Gompertz diffusion analysis with exogenous factors. *Appl. Energy* **2006**, *83*, 1139–1151. [CrossRef]
27. Rupsys, P.; Bartkevicius, E.; Petrauskas, E. A univariate stochastic Gompertz model for tree diameter modeling. *Trends Appl. Sci. Res.* **2011**, *6*, 134–153. [CrossRef]
28. Badurally Adam, N.R.; Elahee, M.K.; Dauhoo, M.Z. Forecasting of peak electricity demand in Mauritius using non-homogeneous Gompertz diffusion process. *Energy* **2011**, *36*, 6763–6769. [CrossRef]
29. Gutiérrez, R.; Gutiérrez-Sánchez, R.; Nafidi, A. A generalization of the Gompertz diffusion model: Statistical inference and application. In *Monografías del Seminario García de Galdeano*; Madaune-Tort, M., Trujillo, D., López de Silanes, M.C., Palacios, M., Sanz, G., Torrens, J.J., Eds.; Prensas de la Universidad de Zaragoza: Zaragoza, Spain, 2006; Volume 33, pp. 273–280.

30. Ferrante, L.; Bompade, S.; Possati; L.; Leone, L.; Montanari, M.P. A stochastic formulation of the Gompertzian growth model for in vitro bactericidad kinetics: Parameter estimation and extinction probability. *Biom. J.* **2005**, *47*, 309–318. [CrossRef]
31. Albano, G.; Giorno, V. A stochastic model in tumor growth. *J. Theor. Biol.* **2006**, *242*, 329–336. [CrossRef]
32. Frank, T.D. Multivariate Markov processes for stochastic systems with delays: Application to the stochastic Gompertz model with delay. *Phys. Rev. E* **2002**, *66*, 011914. [CrossRef]
33. Gutiérrez, R.; Gutiérrez-Sánchez, R.; Nafidi, A. A bivariate stochastic Gompertz diffusion model: Statistical aspects and application to the joint modeling of the Gross Domestic Product and CO2 emissions in Spain. *Environmetrics* **2008**, *19*, 643–658. [CrossRef]
34. Gutiérrez, R.; Gutiérrez-Sánchez, R.; Nafidi, A. Trend analysis using nonhomogeneous stochastic diffusion processes, emission of CO_2; Kyoto protocol in Spain. *Stoch. Environ. Res. Risk Assess.* **2008**, *22*, 55–66.
35. Hu, G. Invariant distribution of stochastic Gompertz equation under regime switching. *Math. Comput. Simul.* **2014**, *97*, 192–206. [CrossRef]
36. Zou, W.; Li, W.; Wang, K. Ergodic method on optimal harvesting for a stochastic Gompertz-type diffusion process. *Appl. Math. Lett.* **2013**, *26*, 170–174. [CrossRef]
37. Gutiérrez, R.; Gutiérrez-Sánchez, R.; Nafidi, A.; Ramos-Ábalos, E. A τ-power stochastic gamma diffusion process: Computational statistical inference and simulation aspects. A real example. *Appl. Math. Comput.* **2012**, *219*, 1576–1588.
38. Zehna, P.W. Invariance of maximum likelihood estimators. *Ann. Math. Stat.* **1966**, *37*, 744, 1966. [CrossRef]
39. Gutiérrez-Sánchez, R.; Nafidi, A.; Pascual, A.; Ramos-Ábalos, E. Three parameter gamma-type growth curve, using a stochastic gamma diffusion model: Computational statistical aspects and simulation. *Math. Comput. Simul.* **2011**, *82*, 234–243. [CrossRef]

© 2020 by the authors. Licensee MDPI, Basel, Switzerland. This article is an open access article distributed under the terms and conditions of the Creative Commons Attribution (CC BY) license (http://creativecommons.org/licenses/by/4.0/).

Article

Asymptotic Behavior of Solutions of the Third Order Nonlinear Mixed Type Neutral Differential Equations

Osama Moaaz [1,†], Dimplekumar Chalishajar [2,*,†] and Omar Bazighifan [3,4,†]

1. Department of Mathematics, Faculty of Science, Mansoura University, Mansoura 35516, Egypt; o_moaaz@mans.edu.eg
2. Department of Applied Mathematics, Virginia Military Institute (VMI) 435 Mallory Hall, Lexington, VA 24450, USA
3. Department of Mathematics, Faculty of Science, Hadhramout University, Hadhramout 50512, Yemen; o.bazighifan@gmail.com
4. Department of Mathematics, Faculty of Education, Seiyun University, Hadhramout 50512, Yemen
* Correspondence: chalishajardn@vmi.edu
† These authors contributed equally to this work.

Received: 29 January 2020; Accepted: 20 March 2020; Published: 1 April 2020

Abstract: The objective of our paper is to study asymptotic properties of the class of third order neutral differential equations with advanced and delayed arguments. Our results supplement and improve some known results obtained in the literature. An illustrative example is provided.

Keywords: oscillation; third order; mixed neutral differential equations

1. Introduction

Equations with neutral terms are of particular significance, as they arise in many applications including systems of control, electrodynamics, mixing liquids, neutron transportation, networks and population models; see [1].

Asymptotic properties of solutions of second/third order differential equations have been subject to intensive research in the literature. This problem for differential equations with respective delays has received a great deal of attention in the last years; see for examples, [2–21].

This paper deals with the oscillation and asymptotic behavior of solutions of the class of third-order, nonlinear, mixed-type, neutral differential equations

$$\left(r\left(t\right)\left(z''\left(t\right)\right)^{\alpha}\right)' + q_1\left(t\right) f_1\left(x\left(\sigma_1\left(t\right)\right)\right) + q_2\left(t\right) f_2\left(x\left(\sigma_2\left(t\right)\right)\right) = 0, \tag{1}$$

where

$$z\left(t\right) = x\left(t\right) + p_1\left(t\right) x\left(\tau_1\left(t\right)\right) + p_2\left(t\right) x\left(\tau_2\left(t\right)\right)$$

and we will assume the following assumptions hold:

(M_1) $r \in C\left(\left[t_0, \infty\right), \left(0, \infty\right)\right)$, $\int_{t_0}^{\infty} r^{-1/\alpha}\left(s\right) ds = \infty$ and α is a ratio of odd positive integers;
(M_2) $p_i \in C\left(\left[t_0, \infty\right), \left[0, c_i\right]\right)$ where c_i are constants for $i = 1, 2$ and $c_1 + c_2 < 1$;
(M_3) $\tau_i, \sigma_i \in C\left(\left[t_0, \infty\right), \mathbb{R}\right)$, $\tau_1\left(t\right) < t$, $\sigma_1\left(t\right) < t$, $\tau_2\left(t\right) > t$, $\sigma_2\left(t\right) > t$, $\sigma_i\left(\tau_i\left(t\right)\right) = \tau_i\left(\sigma_i\left(t\right)\right)$ and $\lim_{t\to\infty} \tau_i\left(t\right) = \lim_{t\to\infty} \sigma_i\left(t\right) = \infty$ for $i = 1, 2$;
(M_4) $q_i \in C\left(\left[t_0, \infty\right), \left(0, \infty\right)\right)$ for $i = 1, 2$;

(M5) $f_1, f_2 \in C(\mathbb{R}, \mathbb{R})$, $f_1(x)/x^\beta \geq k_1 > 0$ and $f_2(x)/x^\gamma \geq k_2$ for $x \neq 0$ where β and γ are ratios of odd positive integers.

By a solution of Equation (1), we mean a non-trivial real function $x \in C([t_x, \infty))$, $t_x \geq t_0$, with $z(t)$, $z'(t)$ and $r_1(t)(z''(t))^\alpha$ being continuously differentiable for all $t \in [t_x, \infty)$, and satisfying (1) on $[t_x, \infty)$. A solution of Equation (1) is called oscillatory if it has arbitrary large zeros; otherwise it is called nonoscillatory. Equation (1) is said to be oscillatory if all its solutions are oscillatory.

Han et al. in [22] studied the asymptotic properties of the solutions of equation

$$\left(r(t)\left(z''(t)\right)\right)' + q_1(t) x (\sigma_1(t)) + q_2(t) x (\sigma_2(t)) = 0, \tag{2}$$

where $z(t) = x(t) + p_1(t) x(\tau_1(t)) + p_2(t) x(\tau_2(t))$.

Baculíková and Džurina [5] studied the oscillation of the third-order equation

$$\left(r(t)\left(x'(t)\right)^\alpha\right)'' + q(t) f(x(\tau(t))) + p(t) h(x(\sigma(t))) = 0,$$

where $\tau(t) \leq t$ and $\sigma(t) \geq t$.

Thandapani and Rama [23] established some oscillation theorems for equation

$$\left(r(t)\left(z''(t)\right)\right)' + q_1(t) x^\alpha (\sigma_1(t)) + q_2(t) x^\beta (\sigma_2(t)) = 0,$$

where $z(t) = x(t) + p_1(t) x(\tau_1(t)) + p_2(t) x(\tau_2(t))$, and the authors used the Recati technique.

The aim of this paper is to discuss the asymptotic behavior of solutions of a class of third-order, nonlinear, mixed-type, neutral differential equations. We established sufficient conditions to ensure that the solution of Equation (1) is oscillatory or tended to zero. The results of this study basically generalize and improve the previous results. An illustrative example is provided.

2. Auxiliary Lemmas

In order to prove our results, we shall need the next auxiliary lemmas.

Lemma 1. *Assume that* $f(y) = Uy - Vy^{\frac{\eta+1}{\eta}}$, *where* U *and* V *are constants,* $V > 0$ *and* η *is a quotient of odd positive integers. Then* f *imposes its maximum value on* $\mathbb{R}$ *at* $y^* = \left(\frac{U\eta}{V(\eta+1)}\right)^\eta$ *and*

$$\max_{y \in \mathbb{R}} f = f(y^*) = \frac{\eta^\eta}{(\eta+1)^{\eta+1}} U^{\eta+1} V^{-\eta}.$$

Lemma 2 ([24]). *Assume that* $A \geq 0$ *and* $B \geq 0$. *If* $\delta > 1$, *then*

$$(A+B)^\delta \leq 2^{\delta-1}\left(A^\delta + B^\delta\right)$$

Moreover, if $0 < \delta < 1$, *then* $(A+B)^\delta \leq \left(A^\delta + B^\delta\right)$.

Lemma 3 ([17]). *If the function* y *satisfies* $y^{(i)} > 0$, $i = 0, 1, ..., n$, *and* $y^{(n+1)} < 0$, *then*

$$\frac{y(t)}{t^n/n!} \geq \frac{y'(t)}{t^{n-1}/(n-1)!}.$$

Lemma 4 ([23]). *Assume that $u(t) > 0$, $u'(t) > 0$, $u''(t) > 0$ and $u'''(t) < 0$ on (T, ∞). Then,*

$$\frac{u(t)}{u'(t)} \geq \frac{t-T}{2} \geq \frac{\mu t}{2}$$

for $t \geq T$ and some $\mu \in (0,1)$.

Lemma 5. *Let x be a positive solution of Equation (1). Then z has only one of the following two properties eventually:*
(i) $z(t) > 0$, $z'(t) > 0$ and $z''(t) > 0$;
(ii) $z(t) > 0$, $z'(t) < 0$ and $z''(t) > 0$.

Proof. The proof is similar to that of Lemma 2.1 of [10] and hence the details are omitted. □

Lemma 6. *Let x be a positive solution of Equation (1), and z has the property (ii). If $\beta = \gamma$ and*

$$\int_{t_0}^{\infty} \int_v^{\infty} \left(\frac{1}{r(u)} \int_u^{\infty} (k_1 q_1(s) + k_2 q_2(s))ds \right)^{1/\alpha} du\, dv = \infty, \tag{3}$$

then the solution x of Equation (1) converges to zero as $t \to \infty$.

Proof. Let x be a positive solution of Equation (1). Since z satisfies the property (ii), we get $\lim_{t \to \infty} z(t) = \delta \geq 0$. Next, we will prove that $\delta = 0$. Suppose that $\delta > 0$, then we have for all $\varepsilon > 0$ and t enough large $\delta < z(t) < \delta + \varepsilon$. By choosing $\varepsilon < \frac{1 - c_1 - c_2}{c_1 + c_2} \delta$, we obtain

$$\begin{aligned}
x(t) &= z(t) - p_1(t) x(\tau_1(t)) - p_2(t) x(\tau_2(t)) \\
&> \delta - (c_1 + c_2) z(\tau_1(t)) \\
&> \delta - (c_1 + c_2)(\delta + \varepsilon) \\
&> L(\delta + \varepsilon) > L z(t),
\end{aligned}$$

where $L = \frac{\delta - (c_1 + c_2)(\delta + \varepsilon)}{\delta + \varepsilon} > 0$. Thus, from (1) and (M5), we have

$$\begin{aligned}
0 &\geq \left(r(t) (z''(t))^\alpha \right)' + k_1 q_1(t) x^\beta(\sigma_1(t)) + k_2 q_2(t) x^\beta(\sigma_2(t)) \\
&\geq \left(r(t) (z''(t))^\alpha \right)' + L^\beta (k_1 q_1(t) + k_2 q_2(t)) z^\beta(\sigma_2(t)),
\end{aligned}$$

and so,

$$\left(r(t) (z''(t))^\alpha \right)' \leq -L^\beta \delta^\beta (k_1 q_1(t) + k_2 q_2(t)).$$

By integrating this inequality two times from t to ∞, we get

$$-z'(t) > L^{\beta/\alpha} \delta^{\beta/\alpha} \int_t^{\infty} \left(\frac{1}{r(u)} \int_u^{\infty} (k_1 q_1(s) + k_2 q_2(s))ds \right)^{1/\alpha} du.$$

Integrating the last inequality from t_1 to ∞, we have

$$z(t_1) > L^{\beta/\alpha} \delta^{\beta/\alpha} \int_{t_1}^{\infty} \int_v^{\infty} \left(\frac{1}{r(u)} \int_u^{\infty} (k_1 q_1(s) + k_2 q_2(s))ds \right)^{1/\alpha} du\, dv.$$

Thus, we are led to a contradiction with (3). Then, $\lim_{t \to \infty} z(t) = 0$; moreover, the fact that $x(t) \leq z(t)$ implies $\lim_{t \to \infty} x(t) = 0$. □

3. Main Results

In this section, we will establish new oscillation criteria for solutions of the Equation (1). For the sake of convenience, we insert the next notation:

$$R_u(t) := \int_u^t \frac{1}{r^{1/\alpha}(s)} ds,$$

$$R_u^*(t) := \min_{t \geq t_0} \{R_u(t), R_u(\tau_1(t))\}$$

and

$$q_i^*(t) := \min_{t \geq t_0} \{q_i(t), q_i(\tau_1(t)), q_i(\tau_2(t))\}, \ i = 1, 2.$$

Theorem 1. *Assume that (M_1)–(M_5) and (3) hold. Let $\beta = \gamma \geq \alpha$, $\sigma_1(t) \leq \tau_1(t)$ and $\sigma_1'(t) > 0$. If there exists a positive function $\rho \in C^1([t_0, \infty))$ such that*

$$\limsup_{t \to \infty} \int_{t_0}^t \left(\Theta_1(s) - \left(1 + c_1^\beta + \frac{c_2^\beta}{2^{\beta-1}}\right) \frac{1}{(\alpha+1)^{\alpha+1}} \frac{(\rho_+'(s))^{\alpha+1} r(\sigma_1(s))}{(\rho(s)\sigma_1'(s))^\alpha} \right) ds = \infty, \quad (4)$$

where $\rho_+'(s) = \max\{\rho'(s), 0\}$ and

$$\Theta_1(t) = \frac{\mu^\alpha v^{\beta-\alpha}}{2^{2\beta+\alpha-2}} \rho(t) \sigma_1^\beta(t) (k_1 q_1^*(t) + k_2 q_2^*(t)),$$

then every solution of equation (1) either oscillates or tends to zero as $t \to \infty$.

Proof. Let x be non-oscillatory solution of Equation (1). Without loss of generality, we assume that $x(t) > 0$; then there exists a $t_1 \geq t_0$ such that $x(t) > 0$, $x(\tau_i(t)) > 0$ and $x(\sigma_i(t)) > 0$ for $t \geq t_1$ and $i = 1, 2$. From Lemma 5, we have that z has the property (i) or the property (ii). From Lemma 6, if $z(t)$ has the property (ii), then we obtain $\lim_{t \to \infty} x(t) = 0$. Next, let z have the property (i). Using (1) and (M_5), we obtain

$$\left(r(t)(z''(t))^\alpha\right)' + k_1 q_1(t) x^\beta(\sigma_1(t)) + k_2 q_2(t) x^\beta(\sigma_2(t)) \leq 0.$$

Thus, we get

$$\begin{aligned} 0 \geq & \left(r(t)(z''(t))^\alpha\right)' + k_1 q_1(t) x^\beta(\sigma_1(t)) + k_2 q_2(t) x^\beta(\sigma_2(t)) \\ & + c_1^\beta [\left(r(\tau_1(t))(z''(\tau_1(t)))^\alpha\right)' + k_1 q_1(\tau_1(t)) x^\beta(\sigma_1(\tau_1(t))) \\ & + k_2 q_2(\tau_1(t)) x^\beta(\sigma_2(\tau_1(t)))] + \frac{c_2^\beta}{2^{\beta-1}} [\left(r(\tau_2(t))(z''(\tau_2(t)))^\alpha\right)' \\ & + k_1 q_1(\tau_2(t)) x^\beta(\sigma_1(\tau_2(t))) + k_2 q_2(\tau_2(t)) x^\beta(\sigma_2(\tau_2(t)))]. \end{aligned}$$

That is

$$\left(r(t)\left(z''(t)\right)^{\alpha}\right)' + c_1^{\beta}\left(r(\tau_1(t))\left(z''(\tau_1(t))\right)^{\alpha}\right)' + \frac{c_2^{\beta}}{2^{\beta-1}}\left(r(\tau_2(t))\left(z''(\tau_2(t))\right)^{\alpha}\right)'$$
$$+ k_1 q_1^*(t)\left(x^{\beta}(\sigma_1(t)) + c_1^{\beta}x^{\beta}(\sigma_1(\tau_1(t))) + \frac{c_2^{\beta}}{2^{\beta-1}}x^{\beta}(\sigma_1(\tau_2(t)))\right)$$
$$+ k_2 q_2^*(t)\left(x^{\beta}(\sigma_2(t)) + c_1^{\beta}x^{\beta}(\sigma_2(\tau_1(t))) + \frac{c_2^{\beta}}{2^{\beta-1}}x^{\beta}(\sigma_2(\tau_2(t)))\right) \leq 0. \quad (5)$$

From Lemma 2, we obtain

$$z^{\beta}(t) \leq (x(t) + c_1(t) x(\tau_1(t)) + c_2(t) x(\tau_2(t)))^{\beta}$$
$$\leq 4^{\beta-1}\left(x^{\beta}(t) + c_1^{\beta}x^{\beta}(\tau_1(t)) + \frac{c_2^{\beta}}{2^{\beta-1}}x^{\beta}(\tau_2(t))\right), \quad (6)$$

which with (5) gives

$$\left(r(t)\left(z''(t)\right)^{\alpha}\right)' + c_1^{\beta}\left(r(\tau_1(t))\left(z''(\tau_1(t))\right)^{\alpha}\right)' + \frac{c_2^{\beta}}{2^{\beta-1}}\left(r(\tau_2(t))\left(z''(\tau_2(t))\right)^{\alpha}\right)'$$
$$+ \frac{k_1}{4^{\beta-1}} q_1^*(t) z^{\beta}(\sigma_1(t)) + \frac{k_2}{4^{\beta-1}} q_2^*(t) z^{\beta}(\sigma_2(t)) \leq 0.$$

This implies that

$$\left(r(t)\left(z''(t)\right)^{\alpha}\right)' + c_1^{\beta}\left(r(\tau_1(t))\left(z''(\tau_1(t))\right)^{\alpha}\right)' + \frac{c_2^{\beta}}{2^{\beta-1}}\left(r(\tau_2(t))\left(z''(\tau_2(t))\right)^{\alpha}\right)'$$
$$+ \frac{1}{4^{\beta-1}}\left(k_1 q_1^*(t) + k_2 q_2^*(t)\right) z^{\beta}(\sigma_1(t)) \leq 0. \quad (7)$$

Now, we define

$$\omega_1(t) = \rho(t) \frac{r(t)\left(z''(t)\right)^{\alpha}}{\left(z'(\sigma_1(t))\right)^{\alpha}}.$$

Then $\omega_1(t) > 0$. By differentiating, we get

$$\omega_1'(t) = \frac{\rho'(t)}{\rho(t)} \omega_1(t) + \rho(t) \frac{\left(r(t)\left(z''(t)\right)^{\alpha}\right)'}{\left(z'(\sigma_1(t))\right)^{\alpha}} - \alpha \rho(t) \frac{r(t)\left(z''(t)\right)^{\alpha}}{\left(z'(\sigma_1(t))\right)^{\alpha+1}} z''(\sigma_1(t)) \sigma_1'(t).$$

Since $\left(r(t)\left(z''(t)\right)^{\alpha}\right)' < 0$ and $\sigma_1(t) < t$, we obtain

$$r(t)\left(z''(t)\right)^{\alpha} \leq r(\sigma_1(t))\left(z''(\sigma_1(t))\right)^{\alpha},$$

and hence

$$\omega_1'(t) \leq \frac{\rho_+'(t)}{\rho(t)} \omega_1(t) - \alpha \frac{\sigma_1'(t)}{\rho^{1/\alpha}(t) r^{1/\alpha}(\sigma_1(t))} \omega_1^{\frac{\alpha+1}{\alpha}}(t) + \rho(t) \frac{\left(r(t)\left(z''(t)\right)^{\alpha}\right)'}{\left(z'(\sigma_1(t))\right)^{\alpha}}.$$

Using Lemma 1 with

$$\eta = \alpha, \quad U = \frac{\rho'_+(t)}{\rho(t)}, \quad V = \alpha \frac{\sigma'_1(t)}{\rho^{1/\alpha}(t) r^{1/\alpha}(\sigma_1(t))} \quad \text{and} \quad y = w_1,$$

we obtain

$$w'_1(t) \leq \rho(t) \frac{\left(r(t)(z''(t))^\alpha\right)'}{(z'(\sigma_1(t)))^\alpha} + \frac{1}{(\alpha+1)^{\alpha+1}} \frac{(\rho'_+(t))^{\alpha+1} r(\sigma_1(t))}{(\rho(t) \sigma'_1(t))^\alpha}. \tag{8}$$

Further, we define the function

$$w_2(t) = \rho(t) \frac{r(\tau_1(t))(z''(\tau_1(t)))^\alpha}{(z'(\sigma_1(t)))^\alpha}.$$

Then $w_2(t) > 0$. By differentiating w_2 and using $\sigma_1(t) \leq \tau_1(t)$, we find

$$w'_2(t) \leq \frac{\rho'(t)}{\rho(t)} w_2(t) + \rho(t) \frac{\left(r(\tau_1(t))(z''(\tau_1(t)))^\alpha\right)'}{(z'(\sigma_1(t)))^\alpha} - \alpha \frac{\sigma'_1(t)}{\rho^{1/\alpha}(t) r^{1/\alpha}(\sigma_1(t))} w_2^{\frac{\alpha+1}{\alpha}}(t).$$

Using Lemma 1, we obtain

$$w'_2(t) \leq \rho(t) \frac{\left(r(\tau_1(t))(z''(\tau_1(t)))^\alpha\right)'}{(z'(\sigma_1(t)))^\alpha} + \frac{1}{(\alpha+1)^{\alpha+1}} \frac{(\rho'_+(t))^{\alpha+1} r(\sigma_1(t))}{(\rho(t) \sigma'_1(t))^\alpha}. \tag{9}$$

Next, we define another function

$$w_3(t) = \rho(t) \frac{r(\tau_2(t))(z''(\tau_2(t)))^\alpha}{(z'(\sigma_1(t)))^\alpha}.$$

Thus $w_3(t) > 0$. By differentiating, and similar to (9) we have

$$w'_3(t) \leq \rho(t) \frac{\left(r(\tau_2(t))(z''(\tau_2(t)))^\alpha\right)'}{(z'(\sigma_1(t)))^\alpha} + \frac{1}{(\alpha+1)^{\alpha+1}} \frac{(\rho'_+(t))^{\alpha+1} r(\sigma_1(t))}{(\rho(t) \sigma'_1(t))^\alpha}. \tag{10}$$

From (8)–(10), we get

$$w'_1(t) + c_1^\beta w'_2(t) + \frac{c_2^\beta}{2^{\beta-1}} w'_3(t) \leq \frac{\rho(t)}{(z'(\sigma_1(t)))^\alpha} \left(\left(r(t)(z''(t))^\alpha\right)' + \right.$$

$$\left. + c_1^\beta \left(r(\tau_1(t))(z''(\tau_1(t)))^\alpha\right)' + \frac{c_2^\beta}{2^{\beta-1}} \left(r(\tau_2(t))(z''(\tau_2(t)))^\alpha\right)' \right)$$

$$+ \left(1 + c_1^\beta + \frac{c_2^\beta}{2^{\beta-1}}\right) \frac{1}{(\alpha+1)^{\alpha+1}} \frac{(\rho'_+(t))^{\alpha+1} r(\sigma_1(t))}{(\rho(t) \sigma'_1(t))^\alpha},$$

which with (7) gives

$$w'_1(t) + c_1^\beta w'_2(t) + \frac{c_2^\beta}{2^{\beta-1}} w'_3(t) \leq -\frac{\rho(t)}{4^{\beta-1}} \left(k_1 q_1^*(t) + k_2 q_2^*(t)\right) \frac{z^\beta(\sigma_1(t))}{(z'(\sigma_1(t)))^\alpha}$$

$$+ \left(1 + c_1^\beta + \frac{c_2^\beta}{2^{\beta-1}}\right) \frac{1}{(\alpha+1)^{\alpha+1}} \frac{(\rho'_+(t))^{\alpha+1} r(\sigma_1(t))}{(\rho(t) \sigma'_1(t))^\alpha}. \tag{11}$$

Using Lemma 4, we have, for some $\mu \in (0,1)$,

$$\frac{z(\sigma_1(t))}{z'(\sigma_1(t))} \geq \frac{\mu}{2}\sigma_1(t).$$

From property (i), we get

$$z(t) = z(t_1) + \int_{t_1}^{t} z'(s)\,ds$$
$$\geq (t - t_1) z'(t_1) \geq \frac{v}{2} t, \qquad (12)$$

for some $v > 0$ and for t enough large. Therefore, for some $\mu \in (0,1)$ and $v > 0$, we find

$$\frac{z^{\beta}(\sigma_1(t))}{(z'(\sigma_1(t)))^{\alpha}} \geq \frac{\mu^{\alpha} v^{\beta-\alpha}}{2^{\alpha}} \sigma_1^{\beta}(t).$$

Combining the last inequality with (11), we obtain

$$w_1'(t) + c_1^{\beta} w_2'(t) + \frac{c_2^{\beta}}{2^{\beta-1}} w_3'(t) \leq -\Theta(t)$$
$$+ \left(1 + c_1^{\beta} + \frac{c_2^{\beta}}{2^{\beta-1}}\right) \frac{1}{(\alpha+1)^{\alpha+1}} \frac{(\rho'_+(t))^{\alpha+1} r(\sigma_1(t))}{(\rho(t) \sigma_1'(t))^{\alpha}}.$$

Integrating the above inequality from t_1 to t, we have

$$\int_{t_1}^{t} \left(\Theta(s) - \left(1 + c_1^{\beta} + \frac{c_2^{\beta}}{2^{\beta-1}}\right) \frac{1}{(\alpha+1)^{\alpha+1}} \frac{(\rho'_+(s))^{\alpha+1} r(\sigma_1(s))}{(\rho(s) \sigma_1'(s))^{\alpha}}\right) ds$$
$$\leq w_1(t_1) + c_1^{\beta} w_2(t_1) + \frac{c_2^{\beta}}{2^{\beta-1}} w_3(t_1).$$

Taking the superior limit as $t \to \infty$, we get a contradiction with (4). The proof is complete. □

Remark 1. *In the Theorem 1, if $\sigma_1(t) \geq \tau_1(t)$ and $\tau_1'(t) > 0$, then the assumption (4) is replaced by*

$$\limsup_{t \to \infty} \int_{t_0}^{t} \left(\Theta_1(s) - \left(1 + c_1^{\beta} + \frac{c_2^{\beta}}{2^{\beta-1}}\right) \frac{1}{(\alpha+1)^{\alpha+1}} \frac{(\rho'_+(s))^{\alpha+1} r(\tau_1(s))}{(\rho(s) \tau_1'(s))^{\alpha}}\right) ds = \infty.$$

Theorem 2. *Assume that (M_1)–(M_5) and (3) hold. Let $\beta = \gamma \geq \alpha$ and $r'(t) > 0$. If there exists a positive function $\rho \in C^1([t_0, \infty))$ such that*

$$\limsup_{t \to \infty} \int_{t_0}^{t} \left(\Theta_2(s) - \left(1 + c_1^{\beta} + \frac{c_2^{\beta}}{2^{\beta-1}}\right) \frac{1}{(\alpha+1)^{\alpha+1}} \frac{(\rho'_+(t))^{\alpha+1}}{\left(\rho(t) R_{t_0}^*(t)\right)^{\alpha}}\right) ds = \infty, \qquad (13)$$

where

$$\Theta_2(t) = \frac{v^{\beta-\alpha}}{2^{3\beta-\alpha-2}t^{2\alpha}} \rho(t) \sigma_1^{\beta+\alpha}(t) (k_1 q_1^*(t) + k_2 q_2^*(t)),$$

then every solution of Equation (1) either oscillates or tends to zero as $t \to \infty$.

Proof. Proceeding as in the proof of Theorem 1, we have that (7) holds. Since $(r(t)(z''(t))^\alpha)' < 0$, we obtain

$$z'(t) = z'(t_1) + \int_{t_1}^t \frac{[r(s)(z''(s))^\alpha]^{1/\alpha}}{r^{1/\alpha}(s)} ds$$

$$\geq \left[r(t)(z''(t))^\alpha\right]^{1/\alpha} R_{t_1}(t). \tag{14}$$

Now, we define

$$\omega_1(t) = \rho(t) \frac{r(t)(z''(t))^\alpha}{z^\alpha(t)}.$$

Then $\omega_1(t) > 0$. By differentiating ω_1 and using (14), we get

$$\omega_1'(t) \leq \frac{\rho_+'(t)}{\rho(t)} \omega_1(t) - \alpha \frac{R_{t_1}(t)}{\rho^{1/\alpha}(t)} \omega_1^{\frac{\alpha+1}{\alpha}}(t) + \rho(t) \frac{(r(t)(z''(t))^\alpha)'}{z^\alpha(t)}.$$

Using Lemma 1 with $\eta = \alpha$, $U = \frac{\rho_+'(t)}{\rho(t)}$, $V = \alpha \frac{R_{t_1}(t)}{\rho^{1/\alpha}(t)}$ and $y = \omega_1$, we obtain

$$\omega_1'(t) \leq \rho(t) \frac{(r(t)(z''(t))^\alpha)'}{z^\alpha(t)} + \frac{1}{(\alpha+1)^{\alpha+1}} \frac{(\rho_+'(t))^{\alpha+1}}{(\rho(t) R_{t_1}(t))^\alpha}. \tag{15}$$

Next, we define a function

$$\omega_2(t) = \rho(t) \frac{r(\tau_1(t))(z''(\tau_1(t)))^\alpha}{z^\alpha(t)}. \tag{16}$$

Then $\omega_2(t) > 0$. Since $z''(t) > 0$ and $\tau_1(t) < t$, we obtain $z'(t) > z'(\tau_1(t))$. Hence, from (14), we find

$$z'(t) > \left[r(\tau_1(t))(z''(\tau_1(t)))^\alpha\right]^{1/\alpha} R_{t_1}(\tau_1(t)). \tag{17}$$

for $t \geq t_2 \geq t_1$. By differentiating (16) and using (17), we get

$$\omega_2'(t) \leq \frac{\rho'(t)}{\rho(t)} \omega_2(t) - \alpha \frac{R_{t_1}(\tau_1(t))}{\rho^{1/\alpha}(t)} \omega_2^{\frac{\alpha+1}{\alpha}}(t) + \rho(t) \frac{(r(\tau_1(t))(z''(\tau_1(t)))^\alpha)'}{z^\alpha(t)}.$$

By using Lemma 1, we obtain

$$\omega_2'(t) \leq \rho(t) \frac{(r(\tau_1(t))(z''(\tau_1(t)))^\alpha)'}{z^\alpha(t)} + \frac{1}{(\alpha+1)^{\alpha+1}} \frac{(\rho_+'(t))^{\alpha+1}}{(\rho(t) R_{t_1}(\tau_1(t)))^\alpha}. \tag{18}$$

Additionally, we define another function

$$\omega_3(t) = \rho(t) \frac{r(\tau_2(t))(z''(\tau_2(t)))^\alpha}{z^\alpha(t)}. \tag{19}$$

Thus $w_3(t) > 0$. Using $(r(t)(z''(t))^\alpha)' < 0$, $\tau_2(t) > t$ and (14), we note that

$$z'(t) > \left[r(\tau_2(t))(z''(\tau_2(t)))^\alpha\right]^{1/\alpha} R_{t_1}(t). \tag{20}$$

By differentiating (19) and using (20) and Lemma 1, we get

$$w_3'(t) \leq \rho(t) \frac{(r(\tau_2(t))(z''(\tau_2(t)))^\alpha)'}{z^\alpha(t)} + \frac{1}{(\alpha+1)^{\alpha+1}} \frac{(\rho_+'(t))^{\alpha+1}}{(\rho(t) R_{t_1}(t))^\alpha}. \tag{21}$$

From (7), (15), (18) and (21), we find

$$w_1'(t) + c_1^\beta w_2'(t) + \frac{c_2^\beta}{2^{\beta-1}} w_3'(t) \leq -\frac{\rho(t)}{4^{\beta-1}}(k_1 q_1^*(t) + k_2 q_2^*(t)) \frac{z^\beta(\sigma_1(t))}{z^\alpha(t)}$$

$$+ \left(1 + c_1^\beta + \frac{c_2^\beta}{2^{\beta-1}}\right) \frac{1}{(\alpha+1)^{\alpha+1}} \frac{(\rho_+'(t))^{\alpha+1}}{\left(\rho(t) R_{t_1}^*(t)\right)^\alpha}. \tag{22}$$

Using (12) and Lemma 6, we have

$$\frac{z^\beta(\sigma_1(t))}{z^\alpha(t)} \geq \frac{v^{\beta-\alpha}}{2^{\beta-\alpha} t^{2\alpha}} \sigma_1^{\beta+\alpha}(t).$$

As in the proof of Theorem 1, we are led to a contradiction with (13). This completes the proof. □

In the following Theorems, we are concerned with the oscillation of solutions of Equation (1) when $\alpha = 1$ and $r(t) = 1$.

Theorem 3. *Assume that (M_1)-(M_5) and (3) hold. Let $0 < \beta < 1 < \gamma$ and τ_i^{-1} exists for $i = 1, 2$. If the inequalities*

$$y'''(t) + \left(\frac{k_1}{\lambda_1}\right)^{\lambda_1} \left(\frac{k_2}{4^{\gamma-1} \lambda_2}\right)^{\lambda_2} \frac{(q_1^*(t))^{\lambda_1} (q_2^*(t))^{\lambda_2}}{\left(1 + c_1^\beta + c_2^\beta\right)} y\left(\tau_i^{-1}(\sigma_j(t))\right) \leq 0, \tag{23}$$

where $i, j = 1, 2, i \neq j$, $\lambda_1 = \frac{\gamma-1}{\gamma-\beta}$ and $\lambda_2 = \frac{1-\beta}{\gamma-\beta}$, have oscillatory solutions, then every solution of Equation (1) is oscillatory.

Proof. Let x non-oscillatory solution of Equation (1). Without loss of generality we assume that $x > 0$; then, there exists a $t_1 \geq t_0$ such that $x(t) > 0$, $x(\tau_i(t)) > 0$ and $x(\sigma_i(t)) > 0$ for $t \geq t_1$ and $i = 1, 2$. By Lemma 6, we get that $z(t) > 0$, $z''(t) > 0$ and $z'''(t) < 0$. Now, we define a function

$$y(t) = z(t) + c_1^\beta z(\tau_1(t)) + c_2^\beta z(\tau_2(t)). \tag{24}$$

Thus $y(t) > 0$ and $y''(t) > 0$. From (1) and (M_5), we obtain

$$z'''(t) \leq -k_1 q_1(t) x^\beta(\sigma_1(t)) - k_2 q_2(t) x^\gamma(\sigma_2(t)). \tag{25}$$

Combining (24) with (25), we get

$$\begin{aligned}y'''(t) &= z'''(t) + c_1^\beta z'''(\tau_1(t)) + c_2^\beta z'''(\tau_2(t)) \\ &\leq -k_1 q_1(t) x^\beta(\sigma_1(t)) - k_2 q_2(t) x^\gamma(\sigma_2(t)) \\ &\quad - c_1^\beta \left(-k_1 q_1(t) x^\beta(\sigma_1(\tau_1(t))) - k_2 q_2(t) x^\gamma(\sigma_2(\tau_1(t)))\right) \\ &\quad - c_2^\beta \left(-k_1 q_1(t) x^\beta(\sigma_1(\tau_2(t))) - k_2 q_2(t) x^\gamma(\sigma_2(\tau_2(t)))\right),\end{aligned}$$

and so,

$$\begin{aligned}y'''(t) &\leq -k_1 q_1^*(t) \left(x^\beta(\sigma_1(t)) + c_1^\beta x^\beta(\sigma_1(\tau_1(t))) + c_2^\beta x^\beta(\sigma_1(\tau_2(t)))\right) \\ &\quad - k_2 q_2^*(t) \left(x^\gamma(\sigma_2(t)) + c_1^\beta x^\gamma(\sigma_2(\tau_1(t))) + c_2^\beta x^\gamma(\sigma_2(\tau_2(t)))\right).\end{aligned}$$

By Lemma 2, since $c_1 + c_2 < 1$ and $\beta < 1 < \gamma$, we obtain

$$y'''(t) + k_1 q_1^*(t) z^\beta(\sigma_1(t)) + k_2 q_2^*(t) \left(x^\gamma(\sigma_2(t)) + c_1^\gamma x^\gamma(\sigma_2(\tau_1(t))) + \frac{c_2^\gamma}{2^{\gamma-1}} x^\gamma(\sigma_2(\tau_2(t)))\right) \leq 0.$$

This implies

$$y'''(t) + k_1 q_1^*(t) z^\beta(\sigma_1(t)) + \frac{k_2}{4^{\gamma-1}} q_2^*(t) z^\gamma(\sigma_2(t)) \leq 0. \tag{26}$$

Using Lemma 6, we have two cases for $z'(t)$. If $z'(t) > 0$, we find

$$y'''(t) + k_1 q_1^*(t) z^\beta(\sigma_1(t)) + \frac{k_2}{4^{\gamma-1}} q_2^*(t) z^\gamma(\sigma_1(t)) \leq 0. \tag{27}$$

Using arithmetic-geometric mean inequality with $u_1 = \frac{k_1}{\lambda_1} q_1^*(t) z^\beta(\sigma_1(t))$ and $u_2 = \frac{k_2}{4^{\gamma-1}\lambda_2} q_2^*(t) z^\gamma(\sigma_1(t))$, we get

$$\begin{aligned}\lambda_1 u_1 + \lambda_2 u_2 &\geq u_1^{\lambda_1} u_2^{\lambda_2} \\ &= \left(\frac{k_1}{\lambda_1}\right)^{\lambda_1} \left(\frac{k_2}{4^{\gamma-1}\lambda_2}\right)^{\lambda_2} (q_1^*(t))^{\lambda_1} (q_2^*(t))^{\lambda_2} z(\sigma_1(t)).\end{aligned} \tag{28}$$

Since $\tau_1(t) < t < \tau_2(t)$, we note that

$$y(t) \leq \left(1 + c_1^\beta + c_2^\beta\right) z(\tau_2(t)).$$

Hence, from (28), (27) becomes

$$y'''(t) + \left(\frac{k_1}{\lambda_1}\right)^{\lambda_1} \left(\frac{k_2}{4^{\gamma-1}\lambda_2}\right)^{\lambda_2} \frac{(q_1^*(t))^{\lambda_1} (q_2^*(t))^{\lambda_2}}{\left(1 + c_1^\beta + c_2^\beta\right)} y\left(\tau_2^{-1}(\sigma_1(t))\right) \leq 0. \tag{29}$$

Then, the condition (23) implies (29) has oscillatory solution, which contradicts $y(t) > 0$.

Let $z'(t) < 0$. As in the previous case, we get

$$y'''(t) + \left(\frac{k_1}{\lambda_1}\right)^{\lambda_1} \left(\frac{k_2}{4\gamma - 1 \lambda_2}\right)^{\lambda_2} \frac{(q_1^*(t))^{\lambda_1} (q_2^*(t))^{\lambda_2}}{\left(1 + c_1^\beta + c_2^\beta\right)} y\left(\tau_1^{-1}(\sigma_2(t))\right) \leq 0. \tag{30}$$

Hence, the condition (23) implies (30) has oscillatory solution, which contradicts $y(t) > 0$. This contradiction completes the proof. □

Remark 2. *There are numerous results concerning the oscillation of the equation*

$$y'''(t) + q(t) y(\sigma(t)) = 0,$$

(see [2,18,20,21]), which include Hille and Nehari types, Philos type, etc.

Assume that

$$\tau_i(t) = t + (-1)^i \tilde{\tau}_i, \quad \sigma_i(t) = t - (-1)^i \tilde{\sigma}_i, \tag{31}$$

where $\tilde{\tau}_i, \tilde{\sigma}_i$ are positive constants for $i = 1, 2$. It is well known (see [9]) that the differential inequalities (29) and (30) are oscillatory if

$$\liminf_{t \to \infty} \int_{t-(\tilde{\tau}_2 + \tilde{\sigma}_1)/3}^{t} (\tilde{\tau}_2 + \tilde{\sigma}_1)^2 (q_1^*(t))^{\lambda_1} (q_2^*(t))^{\lambda_2} > \frac{9}{2e} \left(\frac{\lambda_1}{k_1}\right)^{\lambda_1} \left(\frac{4\gamma - 1 \lambda_2}{k_2}\right)^{\lambda_2} \tag{32}$$

and

$$\liminf_{t \to \infty} \int_t^{t + \tilde{\tau}_1 + \tilde{\sigma}_2} (s - t)^2 (q_1^*(t))^{\lambda_1} (q_2^*(t))^{\lambda_2} > 2 \left(\frac{\lambda_1}{k_1}\right)^{\lambda_1} \left(\frac{4\gamma - 1 \lambda_2}{k_2}\right)^{\lambda_2}, \tag{33}$$

respectively. Hence, we conclude the following theorem:

Theorem 4. *Assume that $0 < \beta < 1 < \gamma$ and (31) hold. If (32) and (33) hold, then every solution of Equation (1) is oscillatory.*

Remark 3. *In the case where $\alpha = 1$, $r(t) = 1$ and $p_i(t) = 0$, Equation (1) becomes*

$$x'''(t) + q_1(t) f_1(x(\sigma_1(t))) + q_2(t) f_2(x(\sigma_2(t))) = 0. \tag{34}$$

Baculikova and Dzurina [5] proved that every nonoscillatory solution x of (34) satisfies $x' < 0$. Thus, Theorems 3 and 4 improve the results in [5].

Remark 4. *A manner similar to the Theorem 3, we can study the oscillation of solutions of Equation (1) when $0 < \gamma < 1 < \beta$.*

Remark 5. *If $\alpha = 1$, $f_1(x) = x^\beta$, $f_2(x) = x^\gamma$, $\tau_1(t) = t - \tilde{\tau}_1$, $\sigma_1(t) = t - \tilde{\sigma}_1$, $\tau_2(t) = t + \tilde{\tau}_2$, $\sigma_2(t) = t + \tilde{\sigma}_2$ and $\tilde{\tau}_i, \tilde{\sigma}_i$ are positive constants, then Theorem 1 extends Theorem 2.5 and 2.7 in [23].*

Remark 6. *The results of Theorem 3 can be extended to the third-order differential equation*

$$((z(t))^\alpha)''' + q_1(t) f_1(x(\sigma_1(t))) + q_2(t) f_2(x(\sigma_2(t))) = 0;$$

the details are left to the reader.

Example 1. Consider the equation

$$\left(x + \frac{1}{3}x\left(\frac{1}{3}t\right) + \frac{1}{3}x(2t)\right)''' + \frac{q_0}{t^3}x\left(\frac{1}{2}t\right) + \frac{q_1}{t^3}x(2t) = 0, \tag{35}$$

where $q_0 > 0$. We note that $\alpha = \beta = \gamma = 1$, $r(t) = 1$, $p_1(t) = p_2(t) = 1/3$, $\tau_1(t) = 1/3t$, $\sigma_1(t) = 1/2t$, $\tau_2(t) = \sigma_2(t) = 2/t$ and $q^*(t) = q_0/t^3$. Hence, it is easy to see that

$$\int_{t_0}^{\infty} \frac{1}{r^{1/\alpha}(s)} ds = \infty.$$

Now, if we set $\rho(s) := t$ and $k_1 = k_2 = 1$, then we have

$$\Theta_1(t) = \frac{q_0}{2s}.$$

Thus, we find

$$\limsup_{t \to \infty} \int_{t_0}^{t} \left(\Theta_1(s) - \left(1 + c_1^\beta + \frac{c_2^\beta}{2^{\beta-1}}\right) \frac{1}{(\alpha+1)^{\alpha+1}} \frac{(\rho'_+(s))^{\alpha+1} r(\sigma_1(s))}{(\rho(s)\sigma'_1(s))^{\alpha}}\right) ds$$

$$= \limsup_{t \to \infty} \int_{t_0}^{t} \left(\frac{q_0}{2s} - \frac{5}{6s}\right) ds.$$

Thus, the conditions become

$$q_0 > 1.66.$$

Thus, by using Theorem 1, Equation (35) is either oscillatory if $q_0 > 1.66$ or tends to zero as $t \to \infty$.

Author Contributions: O.M. and O.B.: Writing original draft, and writing review and editing. D.C.: Formal analysis, writing review and editing, funding and supervision. All authors have read and agreed to the published version of the manuscript.

Funding: The authors received no direct funding for this work.

Acknowledgments: The authors thank the reviewers for for their useful comments, which led to the improvement of the content of the paper.

Conflicts of Interest: There are no competing interests for the authors.

References

1. Hale, J.K. *Theory of Functional Differential Equations*; Springer: New York, NY, USA, 1977.
2. Agarwal, R.P.; Bohner, M.; Li, T.; Zhang, C. Oscillation of third-order nonlinear delay differential equations. *Taiwanese J. Math.* **2013**, *17*, 545–558. [CrossRef]
3. Baculikova, B.; Dzurina, J. On the asymptotic behavior of a class of third order nonlinear neutral differential equations. *Cent. Eur. J. Math.* **2010**, *8*, 1091–1103. [CrossRef]
4. Baculikova, B.; Dzurina, J. Oscillation of third-order neutral differential equations. *Math. Comput. Modell.* **2010**, *52*, 215–226. [CrossRef]
5. Baculikova, B.; Dzurina, J. Some Properties of Third-Order Differential Equations with Mixed Arguments. *J. Math.* **2013**, *2013*, 528279. [CrossRef]

6. Bazighifan, O.; Cesarano, C. Some New Oscillation Criteria for Second-Order Neutral Differential Equations with Delayed Arguments. *Mathematics* **2019**, *7*, 619. [CrossRef]
7. Bazighifan, O.; Elabbasy, E.M.; Moaaz, O. Oscillation of higher-order differential equations with distributed delay. *J. Inequal. Appl.* **2019**, *55*, 1–9. [CrossRef]
8. Chatzarakis, G.E.; Elabbasy, E.M.; Bazighifan, O. An oscillation criterion in 4th-order neutral differential equations with a continuously distributed delay. *Adv. Differ. Equ.* **2019**, *336*, 1–9.
9. Das, P. Oscillation criteria for odd order neutral equations. *J. Math. Anal. Appl.* **1994**, *188*, 245–257. [CrossRef]
10. Elabbasy, E.M.; Hassan, T.S.; Moaaz, O. Oscillation behavior of second-order nonlinear neutral differential equations with deviating arguments. *Opusc. Math.* **2012**, *32*, 719–730. [CrossRef]
11. Elabbasy, E.M.; Moaaz, O. On the asymptotic behavior of third-order nonlinear functional differential equations. *Serdica Math. J.* **2016**, *42*, 157–174.
12. Elabbasy, E.M.; Barsoum, M.Y.; Moaaz, O. Boundedness and oscillation of third order neutral differential equations with deviating arguments. *J. Appl. Math. Phys.* **2015**, *3*, 1367–1375. [CrossRef]
13. Elabbasy, E.M.; Moaaz, O. On the oscillation of third order neutral differential equations. *Asian J. Math. Appl.* **2016**, *2016*, 0274.
14. Elabbasy, E.M.; Moaaz, O.; Almehabresh, E.S. Oscillation Properties of Third Order Neutral Delay Differential Equations. *Appl. Math.* **2016**, *7*, 1780–1788. [CrossRef]
15. Bazighifan, O.; Cesarano, C. A Philos-Type Oscillation Criteria for Fourth-Order Neutral Differential Equations. *Symmetry* **2020**, *12*, 379. [CrossRef]
16. Elabbasy, E.M.; Moaaz, O. Oscillation Criteria for third order nonlinear neutral differential equations with deviating arguments. *Int. J. Sci. Res.* **2016**, *5*, 1.
17. Kiguradze, I.T.; Chanturia, T.A. *Asymptotic Properties of Solutions of Nonautonomous Ordinary Differential Equations*; Kluwer Academic Publishers: Drodrcht, The Netherlands, 1993.
18. Ladas, G.; Sficas, Y.G.; Stavroulakis, I.P. Necessary and sufficent conditions for oscillation of higher order delay differential equations. *Trans. Am. Math. Soc.* **1984**, *285*, 81–90. [CrossRef]
19. Moaaz, O.; Awrejcewicz, J.; Bazighifan, O. A New Approach in the Study of Oscillation Criteria of Even-Order Neutral Differential Equations. *Mathematics* **2020**, *8*, 197. [CrossRef]
20. Saker, S.H.; Dzurina, J. On the oscillation of certain class of third-order nonlinear delay differential equations. *Math. Bohemica* **2010**, *135*, 225–237.
21. Saker, S.H. Oscillation criteria of Hille and Nehari types for third-order delay differential equations. *Comm. Appl. Anal.* **2007**, *11*, 451–468.
22. Han, Z.; Li, T.; Zhang, C.; Sun, S. Oscillatory Behavior of Solutions of Certain Third-Order Mixed Neutral Functional Differential Equations. *Bull. Malays. Math. Sci. Soc.* **2012**, *35*, 611–620.
23. Thandapani, E.; Rama, R. Oscillatory behavior of solutions of certain third order mixed neutral differential equations. *Tamkang J. Math.* **2013**, *44*, 99–112. [CrossRef]
24. Thandapani, E.; Li, T. On the oscillation of third-order quasi-linear neutral functional differential equations. *Arch. Math.* **2011**, *47*, 181–199.

© 2020 by the authors. Licensee MDPI, Basel, Switzerland. This article is an open access article distributed under the terms and conditions of the Creative Commons Attribution (CC BY) license (http://creativecommons.org/licenses/by/4.0/).

Article

Differential Equations Associated with Two Variable Degenerate Hermite Polynomials

Kyung-Won Hwang [1] and Cheon Seoung Ryoo [2,*]

[1] Department of Mathematics, Dong-A University, Busan 604-714, Korea; khwang@dau.ac.kr
[2] Department of Mathematics, Hannam University, Daejeon 34430, Korea
* Correspondence: ryoocs@hnu.kr

Received: 14 January 2020; Accepted: 5 February 2020; Published: 10 February 2020

Abstract: In this paper, we introduce the two variable degenerate Hermite polynomials and obtain some new symmetric identities for two variable degenerate Hermite polynomials. In order to give explicit identities for two variable degenerate Hermite polynomials, differential equations arising from the generating functions of degenerate Hermite polynomials are studied. Finally, we investigate the structure and symmetry of the zeros of the two variable degenerate Hermite equations.

Keywords: differential equations; symmetric identities; degenerate Hermite polynomials; complex zeros

MSC: 05A19; 11B83; 34A30; 65L99

1. Introduction

The classical Hermite numbers H_n and Hermite polynomials $H_n(x)$ are usually defined by the generating functions:

$$e^{2xt-t^2} = \sum_{n=0}^{\infty} H_n(x) \frac{t^n}{n!}$$

and

$$e^{-t^2} = \sum_{n=0}^{\infty} H_n \frac{t^n}{n!}.$$

Clearly, $H_n = H_n(0)$.

It can be seen that these numbers and polynomials play an important role in various areas of mathematics and physics, including numerical theory, combinations, special functions, and differential equations. Many interesting properties about them have been explored (see [1–6]). For example, in mathematics and physics, the Hermite polynomials are a classical orthogonal polynomial sequence. In probability, they appears as the Edgeworth series; in combinatorics, they arise in the umbral calculus as an example of an Appell sequence; in numerical analysis, they play a role in Gaussian quadrature; and in physics, they give rise to the eigenstates of the quantum harmonic oscillator. The polynomials $H_n(x)$ satisfy the Hermite differential equation:

$$\frac{d^2H(x)}{dx^2} - 2x\frac{dH(x)}{dx} + 2nH(x) = 0, n = 0, 1, 2, \ldots.$$

Thus, the Hermite polynomials $H_n(x)$ satisfy the second-order linear differential equation:

$$v'' - 2xv' + 2nv = 0.$$

The special polynomials of two variables provided a new means of analysis for the solution of large classes of partial differential equations often encountered in physical problems. Most of the special polynomials of mathematical physics and their generalization have been proposed by physical problems. As another application of the Hermite differential equation for $H_n(x,y)$, we recall that the two variable Hermite polynomials $H_n(x,y)$ defined by the generating function (see [2]):

$$\sum_{n=0}^{\infty} H_n(x,y)\frac{t^n}{n!} = e^{xt+yt^2} \tag{1}$$

are the solution of heat equation:

$$\frac{\partial}{\partial y}H_n(x,y) = \frac{\partial^2}{\partial x^2}H_n(x,y), \quad H_n(x,0) = x^n. \tag{2}$$

Observe that

$$H_n(2x,-1) = H_n(x).$$

Motivated by their potential and importance for applications in certain problems in probability, combinatorics, number theory, differential equations, numerical analysis, and other fields of mathematics and physics, several kinds of some special polynomials and numbers were recently studied by many authors (see [1–7]). Many mathematicians have studied the area of the degenerate Bernoulli polynomials, degenerate Euler polynomials, and degenerate tangent polynomials (see [8–11]). One of the important aspect of the study of any degenerate polynomials is to find their definition. Recently, Haroon and Khan [12] proposed the degenerate Hermite-Bernoulli polynomials, which are formulated in terms of p-adic invariant integral on $\mathbb{Z}_p$:

$$\sum_{n=0}^{\infty} {}_HB_n(x,y|\lambda)\frac{t^n}{n!} = \frac{\log(1+\lambda t)^{\frac{1}{\lambda}}}{(1+\lambda t)^{\frac{1}{\lambda}}-1}(1+\lambda t)^{\frac{x}{\lambda}}(1+\lambda t^2)^{\frac{y}{\lambda}}.$$

Mathematicians have studied the differential equations arising from the generating functions of special polynomials (see [13–18]). Based on the results so far, in the present work, a new class of degenerate Hermite polynomials is constructed. We can derive the differential equations generated from the generating function of two variable degenerate Hermite polynomials. By using the coefficients of this differential equation, we have explicit identities for the two variable degenerate Hermite polynomials.

We remember that the classical Stirling numbers of the first kind $S_1(n,k)$ and the second kind $S_2(n,k)$ are defined by the relations (see [8–12]):

$$(x)_n = \sum_{k=0}^{n} S_1(n,k)x^k \text{ and } x^n = \sum_{k=0}^{n} S_2(n,k)(x)_k,$$

respectively. Here, $(x)_n = x(x-1)\cdots(x-n+1)$ denotes the falling factorial polynomial of order n. We also have:

$$\sum_{n=m}^{\infty} S_2(n,m)\frac{t^n}{n!} = \frac{(e^t-1)^m}{m!} \text{ and } \sum_{n=m}^{\infty} S_1(n,m)\frac{t^n}{n!} = \frac{(\log(1+t))^m}{m!}. \tag{3}$$

The rest of the paper is organized as follows. In Section 2, we introduce the two variable degenerate Hermite polynomials and obtain the basic properties of these polynomials. In Section 3, we give some symmetric identities for two variable degenerate Hermite polynomials. In Section 4, we derive the differential equations generated from the generating function of two variable degenerate Hermite polynomials. Using the coefficients of this differential equation, we have explicit identities for the two variable degenerate Hermite polynomials. In Section 5, we investigate the zeros of the two variable degenerate Hermite equations by using a computer. Further, we observe the pattern of scattering phenomenon for the zeros of two variable degenerate Hermite equations. Our paper ends with Section 6, where the conclusions and future directions of this work are presented.

2. Basic Properties for the Two Variable Degenerate Hermite Polynomials

In this section, a new class of the two variable degenerate Hermite polynomials is considered. Further, some properties of these polynomials are also obtained.

We define the two variable degenerate Hermite polynomials $\mathcal{H}_n(x,y,\lambda)$ by means of the generating function:

$$\sum_{n=0}^{\infty} \mathcal{H}_n(x,y,\lambda) \frac{t^n}{n!} = (1+\lambda)^{\frac{xt}{\lambda}} (1+\lambda)^{\frac{yt^2}{\lambda}}. \tag{4}$$

Since $(1+\lambda)^{\frac{1}{\lambda}} \to e^t$ as $\lambda \to 0$, it is evident that Equation (4) reduces to Equation (1). Observe that Khan's degenerate Hermite-Bernoulli polynomials $_H B_n(x,y|\lambda)$ vary from the two variable degenerate Hermite polynomials $\mathcal{H}_n(x,y,\lambda)$. Another application of the differential equation for $\mathcal{H}_n(x,y,\lambda)$ is as follows: Note that:

$$F(t,x,y,\lambda) = (1+\lambda)^{\frac{xt}{\lambda}} (1+\lambda)^{\frac{yt^2}{\lambda}}$$

satisfies

$$\frac{\log(1+\lambda)}{\lambda} \frac{\partial F(t,x,y,\lambda)}{\partial y} - \frac{\partial^2 F(t,x,y,\lambda)}{\partial x^2} = 0.$$

Substitute the series in Equation (4) for $F(t,x,y,\lambda)$ to get:

$$\frac{\partial}{\partial y} \mathcal{H}_n(x,y,\lambda) - \frac{\lambda}{\log(1+\lambda)} \frac{\partial^2}{\partial x^2} \mathcal{H}_n(x,y,\lambda) = 0.$$

Thus, the two variable degenerate Hermite polynomials $\mathcal{H}_n(x,y,\lambda)$ in the generating function (4) are the solution of equation:

$$\frac{\partial}{\partial y} \mathcal{H}_n(x,y,\lambda) = \frac{\lambda}{\log(1+\lambda)} \frac{\partial^2}{\partial x^2} \mathcal{H}_n(x,y,\lambda), \quad \mathcal{H}_n(x,0,\lambda) = \left(\frac{\lambda}{\log(1+\lambda)}\right)^n x^n. \tag{5}$$

Since $\frac{\lambda}{\log(1+\lambda)} \to 1$ as λ approaches zero, it is apparent that Equation (5) descends to Equation (2).

By Equation (3), we also need the binomial theorem: for a variable x,

$$(1+\lambda)^{xt/\lambda} = \sum_{m=0}^{\infty} \left(\frac{tx}{\lambda}\right)_m \frac{\lambda^m}{m!}$$

$$= \sum_{m=0}^{\infty} \left(\sum_{l=0}^{m} S_1(m,l) \left(\frac{tx}{\lambda}\right)^l \frac{\lambda^m}{m!}\right) \tag{6}$$

$$= \sum_{l=0}^{\infty} \left(\sum_{m=l}^{\infty} S_1(m,l) x^l \lambda^{m-l} \frac{l!}{m!}\right) \frac{t^l}{l!}.$$

The generating function (4) is useful for deriving several properties of the two variable degenerate Hermite polynomials $\mathcal{H}_n(x,y,\lambda)$. For example, we have the following expression for these polynomials:

Theorem 1. *For any positive integer n, we have:*

$$H_n(x,y) = n! \sum_{k=0}^{[\frac{n}{2}]} \frac{y^k x^{n-2k}}{k!(n-2k)!},$$

where $[\ ]$ denotes taking the integer part.

Proof. By Equations (4) and (6), we have:

$$\sum_{n=0}^{\infty} H_n(x,y,\lambda) \frac{t^n}{n!} = (1+\lambda)^{\frac{xt}{\lambda}} (1+\lambda)^{\frac{yt^2}{\lambda}}$$

$$= \sum_{k=0}^{\infty} \left(\frac{y \log(1+\lambda)}{\lambda}\right)^k \frac{t^{2k}}{k!} \sum_{l=0}^{\infty} \left(\frac{x \log(1+\lambda)}{\lambda}\right)^l \frac{t^l}{l!}$$

$$= \sum_{n=0}^{\infty} \left(\sum_{k=0}^{[\frac{n}{2}]} \left(\frac{\log(1+\lambda)}{\lambda}\right)^{n-k} y^k x^{n-2k} \frac{n!}{k!(n-2k)!} \right) \frac{t^n}{n!}.$$

On comparing the coefficients of $\frac{t^n}{n!}$, the expected result of Theorem 1 is achieved. □

When $\lambda \to 0$, then we have:

$$H_n(x,y) = n! \sum_{k=0}^{[\frac{n}{2}]} \frac{y^k x^{n-2k}}{k!(n-2k)!}.$$

The following basic properties of the two variable degenerate Hermite polynomials $\mathcal{H}_n(x,y,\lambda)$ are derived from (4). We, therefore, chose to omit the details involved.

Theorem 2. *For any positive integer n, we have:*

(1) $\mathcal{H}_n(x,0,\lambda) = \sum_{m=n}^{\infty} S_1(m,n) x^n \lambda^{m-n} \frac{n!}{m!}.$

(2) $\mathcal{H}_n(x,y,\lambda) = n! \sum_{k=0}^{[\frac{n}{2}]} \left(\sum_{m=k}^{\infty} \frac{S_1(m,k) y^k \lambda^{m-k}}{m!} \sum_{m=n-2k}^{\infty} \frac{S_1(m,n-2k) x^{n-2k} \lambda^{m-n+2k}}{m!} \right).$

(3) $\mathcal{H}_n(x_1+x_2,y,\lambda) = \sum_{l=0}^{n} \binom{n}{l} \mathcal{H}_l(x_1,y,\lambda) \left(\frac{\log(1+\lambda)}{\lambda}\right)^{n-l} x_2^{n-l}.$

(4) $\mathcal{H}_n(x_1+x_2,y,\lambda) = \sum_{l=0}^{n} \binom{n}{l} \mathcal{H}_{n-l}(x_1,y,\lambda) \sum_{m=l}^{\infty} S_1(m,l) x_2^l \lambda^{m-l} \frac{l!}{m!}.$

(5) $\mathcal{H}_n(x,y_1+y_2,\lambda) = \sum_{k=0}^{[\frac{n}{2}]} \mathcal{H}_k(x,y_1,\lambda) \left(\frac{\log(1+\lambda)}{\lambda}\right)^{n-2k} y_2^{n-2k} \frac{n!}{k!(n-2k)!}.$

(6) $\mathcal{H}_n(x,y_1+y_2,\lambda) = n! \sum_{k=0}^{[\frac{n}{2}]} \sum_{m=k}^{\infty} \frac{S_1(m,k) y_2^k \lambda^{m-k}}{m!(n-2k)!} \mathcal{H}_{n-2k}(x,y_1,\lambda).$

(7) $\mathcal{H}_n(x_1+x_2,y_1+y_2,\lambda) = \sum_{l=0}^{n} \binom{n}{l} \mathcal{H}_l(x_1,y_1,\lambda) \mathcal{H}_{n-l}(x_2,y_2,\lambda).$

3. Symmetric Identities for the Two Variable Degenerate Hermite Polynomials

In this section, we give some new symmetric identities for the two variable degenerate Hermite polynomials. We also get some explicit formulas and properties for the two variable degenerate Hermite polynomials.

Let $a, b > 0$ and $a \neq b$. We start with:

$$\mathcal{F}(t, \lambda) = (1+\lambda)^{\frac{abxt}{\lambda}} (1+\lambda)^{\frac{a^2 b^2 y t^2}{\lambda}}.$$

Then, the expression for $\mathcal{F}(t, \lambda)$ is symmetric in a and b:

$$\mathcal{F}(t, \lambda) = \sum_{m=0}^{\infty} \mathcal{H}_m(ax, a^2 y, \lambda) \frac{(bt)^m}{m!} = \sum_{m=0}^{\infty} b^m \mathcal{H}_m(ax, a^2 y, \lambda) \frac{t^m}{m!}.$$

On similar lines, we can obtain that:

$$\mathcal{F}(t, \lambda) = \sum_{m=0}^{\infty} \mathcal{H}_m(bx, b^2 y, \lambda) \frac{(at)^m}{m!} = \sum_{m=0}^{\infty} a^m \mathcal{H}_m(bx, b^2 y, \lambda) \frac{t^m}{m!}.$$

Comparing the coefficients of $\frac{t^m}{m!}$ in the last two equations, we have the following theorem.

Theorem 3. *Let $a, b > 0$ and $a \neq b$. The following identity holds true:*

$$a^m \mathcal{H}_m(bx, b^2 y, \lambda) = b^m \mathcal{H}_m(ax, a^2 y, \lambda).$$

Again, we now use:

$$\mathcal{G}(t, \lambda) = \frac{abt(1+\lambda)^{\frac{abxt}{\lambda}} (1+\lambda)^{\frac{a^2 b^2 y t^2}{\lambda}} \left((1+\lambda)^{\frac{abt}{\lambda}} - 1\right)}{\left((1+\lambda)^{\frac{at}{\lambda}} - 1\right)\left((1+\lambda)^{\frac{bt}{\lambda}} - 1\right)}.$$

For $\lambda \in \mathbb{C}$, we define the degenerate Bernoulli polynomials given by the generating function:

$$\sum_{n=0}^{\infty} \mathcal{B}_n(x, \lambda) \frac{t^n}{n!} = \frac{t}{(1+\lambda)^{\frac{t}{\lambda}} - 1} (1+\lambda)^{\frac{xt}{\lambda}}.$$

When $x = 0$ and $\mathcal{B}_n(\lambda) = \mathcal{B}_n(0, \lambda)$ are called the degenerate Bernoulli numbers. The first few follow immediately from this generating function,

$$\mathcal{B}_0(\lambda) = \frac{\lambda}{\log(1+\lambda)},$$

$$\mathcal{B}_1(\lambda) = -\frac{1}{2} + x,$$

$$\mathcal{B}_2(\lambda) = \frac{\log(1+\lambda)}{6\lambda} - \frac{x \log(1+\lambda)}{\lambda} + \frac{x^2 \log(1+\lambda)}{\lambda},$$

$$\mathcal{B}_3(\lambda) = \frac{x \log(1+\lambda)^2}{2\lambda^2} - \frac{3x^2 \log(1+\lambda)^2}{2\lambda^2} + \frac{x^3 \log(1+\lambda)^2}{\lambda^2},$$

$$\mathcal{B}_4(\lambda) = -\frac{\log(1+\lambda)^3}{30\lambda^3} + \frac{x^2 \log(1+\lambda)^3}{\lambda^3} - \frac{2x^3 \log(1+\lambda)^3}{\lambda^3} + \frac{x^4 \log(1+\lambda)^3}{\lambda^3}.$$

For each integer $k \geq 0$, $\mathcal{T}_k(n) = 0^k + 1^k + 2^k + \cdots + (n-1)^k$ is called the sum of integers. A generalized falling factorial sum $\sigma_k(n, \lambda)$ can be defined by the generation function:

$$\sum_{k=0}^{\infty} \sigma_k(n, \lambda) \frac{t^k}{k!} = \frac{(1+\lambda)^{\frac{(n+1)t}{\lambda}} - 1}{(1+\lambda)^{\frac{t}{\lambda}} - 1}.$$

Note that $\lim_{\lambda \to 0} \sigma_k(n, \lambda) = \mathcal{T}_k(n)$. From $\mathcal{G}(t, \lambda)$, we get the following result:

$$\mathcal{G}(t, \lambda) = \frac{abt(1+\lambda)^{\frac{abxt}{\lambda}}(1+\lambda)^{\frac{a^2b^2yt^2}{\lambda}}\left((1+\lambda)^{\frac{abt}{\lambda}} - 1\right)}{\left((1+\lambda)^{\frac{at}{\lambda}} - 1\right)\left((1+\lambda)^{\frac{bt}{\lambda}} - 1\right)}$$

$$= \frac{abt}{\left((1+\lambda)^{\frac{at}{\lambda}} - 1\right)} (1+\lambda)^{\frac{abxt}{\lambda}} (1+\lambda)^{\frac{a^2b^2yt^2}{\lambda}} \frac{\left((1+\lambda)^{\frac{abt}{\lambda}} - 1\right)}{\left((1+\lambda)^{\frac{bt}{\lambda}} - 1\right)}$$

$$= b \sum_{n=0}^{\infty} \mathcal{B}_n(\lambda) \frac{(at)^n}{n!} \sum_{n=0}^{\infty} \mathcal{H}_n(bx, b^2y, \lambda) \frac{(at)^n}{n!} \sum_{n=0}^{\infty} \sigma_k(a-1, \lambda) \frac{(bt)^n}{n!}$$

$$= \sum_{n=0}^{\infty} \left(\sum_{i=0}^{n} \sum_{m=0}^{i} \binom{n}{i}\binom{i}{m} a^i b^{n+1-i} \mathcal{B}_m(\lambda) \mathcal{H}_{i-m}(bx, b^2y, \lambda) \sigma_{n-i}(a-1, \lambda) \right) \frac{t^n}{n!}.$$

In a similar fashion, we have:

$$\mathcal{G}(t, \lambda) = \frac{abt}{\left((1+\lambda)^{\frac{bt}{\lambda}} - 1\right)} (1+\lambda)^{\frac{abxt}{\lambda}} (1+\lambda)^{\frac{a^2b^2yt^2}{\lambda}} \frac{\left((1+\lambda)^{\frac{abt}{\lambda}} - 1\right)}{\left((1+\lambda)^{\frac{at}{\lambda}} - 1\right)}$$

$$= a \sum_{n=0}^{\infty} \mathcal{B}_n(\lambda) \frac{(bt)^n}{n!} \sum_{n=0}^{\infty} \mathcal{H}_n(bx, b^2y, \lambda) \frac{(bt)^n}{n!} \sum_{n=0}^{\infty} \sigma_k(a-1, \lambda) \frac{(at)^n}{n!}$$

$$= \sum_{n=0}^{\infty} \left(\sum_{i=0}^{n} \sum_{m=0}^{i} \binom{n}{i}\binom{i}{m} b^i a^{n+1-i} \mathcal{B}_m(\lambda) \mathcal{H}_{i-m}(ax, a^2y, \lambda) \sigma_{n-i}(b-1, \lambda) \right) \frac{t^n}{n!}.$$

By comparing the coefficients of $\frac{t^m}{m!}$ on the right-hand sides of the last two equations, we have the theorem below.

Theorem 4. Let $a, b > 0$ and $a \neq b$. The the following identity holds true:

$$\sum_{i=0}^{n} \sum_{m=0}^{i} \binom{n}{i}\binom{i}{m} a^i b^{n+1-i} \mathcal{B}_m(\lambda) \mathcal{H}_{i-m}(bx, b^2y, \lambda) \sigma_{n-i}(a-1, \lambda)$$

$$= \sum_{i=0}^{n} \sum_{m=0}^{i} \binom{n}{i}\binom{i}{m} b^i a^{n+1-i} \mathcal{B}_m(\lambda) \mathcal{H}_{i-m}(ax, a^2y, \lambda) \sigma_{n-i}(b-1, \lambda).$$

Making the N-times derivative for Equation (4) with respect to t, we have:

$$\left(\frac{\partial}{\partial t}\right)^N F(t,x,y,\lambda) = \left(\frac{\partial}{\partial t}\right)^N (1+\lambda)^{\frac{xt}{\lambda}} (1+\lambda)^{\frac{yt^2}{\lambda}} = \sum_{m=0}^{\infty} \mathcal{H}_{m+N}(x,y,\lambda)\frac{t^m}{m!}. \quad (7)$$

By multiplying the exponential series $e^{xt} = \sum_{m=0}^{\infty} x^m \frac{t^m}{m!}$ in both sides of Equation (7) and the Cauchy product, we get:

$$e^{-n\left(\frac{\log(1+\lambda)}{\lambda}\right)t}\left(\frac{\partial}{\partial t}\right)^N F(t,x,y,\lambda)$$

$$= \left(\sum_{m=0}^{\infty}(-n)^m\left(\frac{\log(1+\lambda)}{\lambda}\right)^m \frac{t^m}{m!}\right)\left(\sum_{m=0}^{\infty} \mathcal{H}_{m+N}(x,y,\lambda)\frac{t^m}{m!}\right) \quad (8)$$

$$= \sum_{m=0}^{\infty}\left(\sum_{k=0}^{m}\binom{m}{k}(-n)^{m-k}\left(\frac{\log(1+\lambda)}{\lambda}\right)^{m-k}\mathcal{H}_{N+k}(x,y,\lambda)\right)\frac{t^m}{m!}.$$

By Equation (8) and the Leibniz rule, we have:

$$e^{-n\left(\frac{\log(1+\lambda)}{\lambda}\right)t}\left(\frac{\partial}{\partial t}\right)^N F(t,x,y,\lambda)$$

$$= \sum_{k=0}^{N}\binom{N}{k}n^{N-k}\left(\frac{\log(1+\lambda)}{\lambda}\right)^{N-k}\left(\frac{\partial}{\partial t}\right)^k \left(e^{-n\left(\frac{\log(1+\lambda)}{\lambda}\right)t} F(t,x,y,\lambda)\right) \quad (9)$$

$$= \sum_{m=0}^{\infty}\left(\sum_{k=0}^{N}\binom{N}{k}n^{N-k}\left(\frac{\log(1+\lambda)}{\lambda}\right)^{N-k}\mathcal{H}_{m+k}(x-n,y,\lambda)\right)\frac{t^m}{m!}.$$

Hence, by Equations (8) and (9), comparing the coefficients of $\frac{t^m}{m!}$ gives the below theorem.

Theorem 5. *Let m, n, N be nonnegative integers. Then,*

$$\sum_{k=0}^{m}\binom{m}{k}(-n)^{m-k}\left(\frac{\log(1+\lambda)}{\lambda}\right)^{m-k}\mathcal{H}_{N+k}(x,y,\lambda)$$
$$= \sum_{k=0}^{N}\binom{N}{k}n^{N-k}\left(\frac{\log(1+\lambda)}{\lambda}\right)^{N-k}\mathcal{H}_{m+k}(x-n,y,\lambda). \quad (10)$$

If we take $m = 0$ in (10), then we have the following:

Theorem 6. *For $N = 0, 1, 2, \ldots$, we have:*

$$\mathcal{H}_N(x,y,\lambda) = \sum_{k=0}^{N}\binom{N}{k}n^{N-k}\left(\frac{\log(1+\lambda)}{\lambda}\right)^{N-k}\mathcal{H}_k(x-n,y,\lambda).$$

4. Differential Equations Associated with Two Variable Degenerate Hermite Polynomials

In this section, we study the differential equations with coefficients $a_i(N, x, y, \lambda)$ arising from the generating functions of the two variable degenerate Hermite polynomials:

$$\left(\frac{\partial}{\partial t}\right)^N F(t, x, y, \lambda) - a_0(N, x, y, \lambda) F(t, x, y, \lambda) - \cdots - a_N(N, x, y, \lambda) t^N F(t, x, y, \lambda) = 0.$$

By using the coefficients of this differential equation, we can derive explicit identities for the two variable degenerate Hermite polynomials $\mathcal{H}_n(x, y, \lambda)$. Recall that:

$$F = F(t, x, y, \lambda) = (1+\lambda)^{\frac{xt}{\lambda}} (1+\lambda)^{\frac{yt^2}{\lambda}} = \sum_{n=0}^{\infty} \mathcal{H}_n(x, y, \lambda) \frac{t^n}{n!}, \quad \lambda, x, t \in \mathbb{C}. \tag{11}$$

Then, by Equation (11), we have:

$$\begin{aligned} F^{(1)} &= \frac{\partial}{\partial t} F(t, x, y, \lambda) \\ &= \frac{\partial}{\partial t} \left((1+\lambda)^{\frac{xt}{\lambda}} (1+\lambda)^{\frac{yt^2}{\lambda}} \right) \\ &= \left(\frac{(x+2yt) \log(1+\lambda)}{\lambda} \right) (1+\lambda)^{\frac{xt}{\lambda}} (1+\lambda)^{\frac{yt^2}{\lambda}} \\ &= \left(\frac{(x+2yt) \log(1+\lambda)}{\lambda} \right) F(t, x, y, \lambda), \end{aligned} \tag{12}$$

$$\begin{aligned} F^{(2)} &= \frac{\partial}{\partial t} F^{(1)}(t, x, y, \lambda) \\ &= \left(\frac{2y \log(1+\lambda)}{\lambda} \right) F(t, x, y, \lambda) + \left(\frac{(x+2yt) \log(1+\lambda)}{\lambda} \right) F^{(1)}(t, x, y, \lambda) \\ &= \left(\left(\frac{\log(1+\lambda)}{\lambda} \right) 2y + \left(\frac{\log(1+\lambda)}{\lambda} \right)^2 x^2 \right) F(t, x, y, \lambda) \\ &\quad + \left(\left(\frac{\log(1+\lambda)}{\lambda} \right)^2 4xy \right) t F(t, x, y, \lambda) \\ &\quad + \left(\left(\frac{\log(1+\lambda)}{\lambda} \right)^2 (2y)^2 \right) t^2 F(t, x, y, \lambda). \end{aligned} \tag{13}$$

Continuing this process as shown in Equation (13), we can guess that:

$$F^{(N)} = \left(\frac{\partial}{\partial t}\right)^N F(t, x, y, \lambda) = \sum_{i=0}^{N} a_i(N, x, y, \lambda) t^i F(t, x, y, \lambda), (N = 0, 1, 2, \ldots). \tag{14}$$

Differentiating Equation (14) with respect to t, we have:

$$\begin{aligned}
F^{(N+1)} = \frac{\partial F^{(N)}}{\partial t} &= \sum_{i=0}^{N} a_i(N,x,y,\lambda) i t^{i-1} F(t,x,y,\lambda) \\
&\quad + \sum_{i=0}^{N} a_i(N,x,y,\lambda) t^i F^{(1)}(t,x,y,\lambda) \\
&= \sum_{i=0}^{N} i a_i(N,x,y,\lambda) t^{i-1} F(t,x,y,\lambda) \\
&\quad + \sum_{i=0}^{N} \frac{x \log(1+\lambda)}{\lambda} a_i(N,x,y,\lambda) t^i F(t,x,y,\lambda) \\
&\quad + \sum_{i=0}^{N} \frac{2y \log(1+\lambda)}{\lambda} a_i(N,x,y,\lambda) t^{i+1} F(t,x,y,\lambda) \\
&= \sum_{i=0}^{N-1} (i+1) a_{i+1}(N,x,y,\lambda) t^i F(t,x,y,\lambda) \\
&\quad + \sum_{i=0}^{N} \frac{x \log(1+\lambda)}{\lambda} a_i(N,x,y,\lambda) t^i F(t,x,y,\lambda) \\
&\quad + \sum_{i=1}^{N+1} \frac{2y \log(1+\lambda)}{\lambda} a_{i-1}(N,x,y,\lambda) t^i F(t,x,y,\lambda).
\end{aligned} \quad (15)$$

Now, replacing N by $N+1$ in Equation (14), we find:

$$F^{(N+1)} = \sum_{i=0}^{N+1} a_i(N+1,x,y,\lambda) t^i F(t,x,y,\lambda). \quad (16)$$

Comparing the coefficients on both sides of Equations (15) and (16), we obtain:

$$\begin{aligned}
a_0(N+1,x,y,\lambda) &= a_1(N,x,y,\lambda) + \frac{x \log(1+\lambda)}{\lambda} a_0(N,x,y,\lambda), \\
a_N(N+1,x,y,\lambda) &= \frac{x \log(1+\lambda)}{\lambda} a_N(N,x,y,\lambda) + \frac{2y \log(1+\lambda)}{\lambda} a_{N-1}(N,x,y,\lambda), \\
a_{N+1}(N+1,x,y,\lambda) &= \frac{2y \log(1+\lambda)}{\lambda} a_N(N,x,y,\lambda),
\end{aligned} \quad (17)$$

and:

$$\begin{aligned}
a_i(N+1,x,y,\lambda) &= (i+1) a_{i+1}(N,x,y,\lambda) + \frac{x \log(1+\lambda)}{\lambda} a_i(N,x,y,\lambda) \\
&\quad + \frac{2y \log(1+\lambda)}{\lambda} a_{i-1}(N,x,y,\lambda), (1 \leq i \leq N-1).
\end{aligned} \quad (18)$$

In addition, by Equation (14), we have:

$$F(t,x,y,\lambda) = F^{(0)}(t,x,y,\lambda) = a_0(0,x,y,\lambda) F(t,x,y,\lambda). \quad (19)$$

By Equation (19), we get:

$$a_0(0,x,y,\lambda) = 1. \quad (20)$$

It is not difficult to show that:

$$\frac{x\log(1+\lambda)}{\lambda}F(t,x,y,\lambda) + \frac{2y\log(1+\lambda)}{\lambda}tF(t,x,y,\lambda)$$
$$= F^{(1)}(t,x,y,\lambda)$$
$$= \sum_{i=0}^{1} a_i(1,x,y,\lambda)t^i F(t,x,y,\lambda) \quad (21)$$
$$= a_0(1,x,y,\lambda)F(t,x,y,\lambda) + a_1(1,x,y,\lambda)tF(t,x,y,\lambda).$$

Thus, by Equations (12) and (21), we also get:

$$a_0(1,x,y,\lambda) = \frac{x\log(1+\lambda)}{\lambda}, \quad a_1(1,x,y,\lambda) = \frac{2y\log(1+\lambda)}{\lambda}. \quad (22)$$

From Equation (17), we note that:

$$a_0(N+1,x,y,\lambda) = a_1(N,x,y,\lambda) + \frac{x\log(1+\lambda)}{\lambda}a_0(N,x,y,\lambda),$$
$$a_0(N,x,y,\lambda) = a_1(N-1,x,y,\lambda) + \frac{x\log(1+\lambda)}{\lambda}a_0(N-1,x,y,\lambda),$$
$$\cdots \quad (23)$$
$$a_0(N+1,x,y,\lambda) = \sum_{i=0}^{N}\left(\frac{x\log(1+\lambda)}{\lambda}\right)^i a_1(N-i,x,y,\lambda)$$
$$+ \left(\frac{\log(1+\lambda)}{\lambda}\right)^{N+1} x^{N+1},$$

$$a_N(N+1,x,y,\lambda) = \frac{x\log(1+\lambda)}{\lambda}a_N(N,x,y,\lambda) + \frac{2y\log(1+\lambda)}{\lambda}a_{N-1}(N,x,y,\lambda),$$
$$a_{N-1}(N,x,y,\lambda) = \frac{x\log(1+\lambda)}{\lambda}a_{N-1}(N-1,x,y,\lambda)$$
$$+ \frac{2y\log(1+\lambda)}{\lambda}a_{N-2}(N-1,x,y,\lambda),\ldots \quad (24)$$
$$a_N(N+1,x,y,\lambda) = (N+1)x(2y)^N\left(\frac{\log(1+\lambda)}{\lambda}\right)^{N+1},$$

and:

$$a_{N+1}(N+1,x,y,\lambda) = \frac{2y\log(1+\lambda)}{\lambda}a_N(N,x,y,\lambda),$$
$$a_N(N,x,y,\lambda) = \frac{2y\log(1+\lambda)}{\lambda}a_{N-1}(N-1,x,y,\lambda),\ldots \quad (25)$$
$$a_{N+1}(N+1,x,y,\lambda) = \left(\frac{\log(1+\lambda)}{\lambda}\right)^{N+1}(2y)^{N+1}.$$

For $i=1$ in Equation (18), we have:

$$a_1(N+1,x,y,\lambda) = 2\sum_{k=0}^{N}\left(\frac{x\log(1+\lambda)}{\lambda}\right)^k a_2(N-k,x,y,\lambda)$$
$$+ \frac{2y\log(1+\lambda)}{\lambda}\sum_{k=0}^{N}\left(\frac{x\log(1+\lambda)}{\lambda}\right)^k a_0(N-k,x,y,\lambda), \quad (26)$$

Continuing this process, we can deduce that, for $1 \leq i \leq N-1$,

$$a_i(N+1,x,y,\lambda) = (i+1)\sum_{k=0}^{N}\left(\frac{x\log(1+\lambda)}{\lambda}\right)^k a_{i+1}(N-k,x,y,\lambda)$$
$$+\frac{2y\log(1+\lambda)}{\lambda}\sum_{k=0}^{N}\left(\frac{x\log(1+\lambda)}{\lambda}\right)^k a_{i-1}(N-k,x,y,\lambda).$$
(27)

Note that, here, the matrix $a_i(j,x,y,\lambda)_{0\leq i,j\leq N+1}$ is given by:

$$\begin{pmatrix} 1 & \frac{x\log(1+\lambda)}{\lambda} & \frac{2y\log(1+\lambda)}{\lambda}+\left(\frac{x\log(1+\lambda)}{\lambda}\right)^2 & \cdots & \cdot \\ 0 & \frac{2y\log(1+\lambda)}{\lambda} & \left(\frac{\log(1+\lambda)}{\lambda}\right)^2 4xy & \cdots & \cdot \\ 0 & 0 & \left(\frac{2y\log(1+\lambda)}{\lambda}\right)^2 & \cdots & \cdot \\ \vdots & \vdots & \vdots & \ddots & \cdot \\ 0 & 0 & 0 & \cdots & \left(\frac{2y\log(1+\lambda)}{\lambda}\right)^{N+1} \end{pmatrix}$$

Therefore, by Equations (20)–(27), we obtain the following theorem.

Theorem 7. *For $N = 0,1,2,\ldots$, the differential equation:*

$$\left(\frac{\partial}{\partial t}\right)^N F(t,x,y,\lambda) - \left(\sum_{i=0}^{N} a_i(N,x,y)t^i\right) F(t,x,y,\lambda) = 0$$

has a solution

$$F = F(t,x,y,\lambda) = (1+\lambda)^{\frac{xt}{\lambda}}(1+\lambda)^{\frac{yt^2}{\lambda}},$$

where:

$$a_0(N+1,x,y,\lambda) = \sum_{i=0}^{N}\left(\frac{x\log(1+\lambda)}{\lambda}\right)^i a_1(N-i,x,y,\lambda)$$
$$+\left(\frac{\log(1+\lambda)}{\lambda}\right)^{N+1} x^{N+1},$$

$$a_N(N+1,x,y,\lambda) = (N+1)x(2y)^N\left(\frac{\log(1+\lambda)}{\lambda}\right)^{N+1},$$

$$a_{N+1}(N+1,x,y,\lambda) = \left(\frac{\log(1+\lambda)}{\lambda}\right)^{N+1}(2y)^{N+1},$$

$$a_i(N+1,x,y,\lambda) = (i+1)\sum_{k=0}^{N}\left(\frac{x\log(1+\lambda)}{\lambda}\right)^k a_{i+1}(N-k,x,y,\lambda)$$
$$+\frac{2y\log(1+\lambda)}{\lambda}\sum_{k=0}^{N}\left(\frac{x\log(1+\lambda)}{\lambda}\right)^k a_{i-1}(N-k,x,y,\lambda), (1\leq i\leq N-2).$$

Here is a plot of the surface for this solution.

In Figure 1a, we chose $-2 \le x \le 2, -2 \le t \le 2, \lambda = 1/2$, and $y = 0.1$. In Figure 1b, we chose $-2 \le y \le 2, -1 \le t \le 1, \lambda = 1/2$, and $x = 0.1$.

Making the N-times derivative for Equation (4) with respect to t, we have:

$$\left(\frac{\partial}{\partial t}\right)^N F(t,x,y,\lambda) = \sum_{m=0}^{\infty} \mathcal{H}_{m+N}(x,y,\lambda)\frac{t^m}{m!}. \tag{28}$$

By Equation (28) and Theorem 7, we have:

$$a_0(N,x,y,\lambda)F(t,x,y,\lambda) + a_1(N,x,y,\lambda)tF(t,x,y,\lambda) + \cdots + a_N(N,x,y,\lambda)t^N F(t,x,y,\lambda)$$
$$= \sum_{m=0}^{\infty} \mathcal{H}_{m+N}(x,y,\lambda)\frac{t^m}{m!}.$$

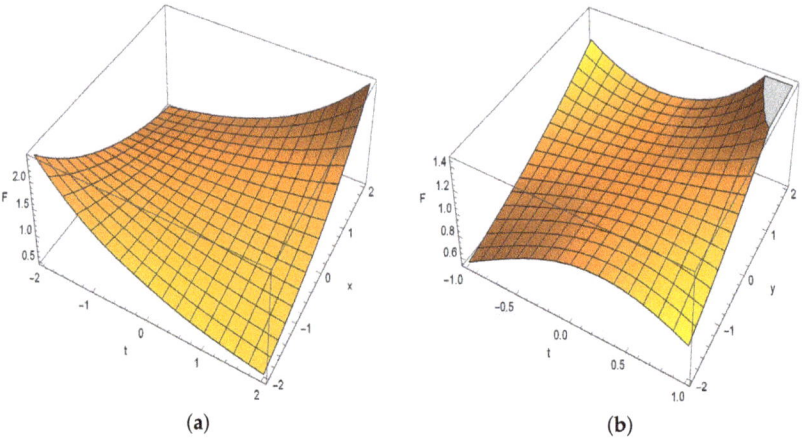

Figure 1. The surface for the solution of $F(t,x,y,\lambda) = 0$. (a) $-2 \le x \le 2, -2 \le t \le 2, \lambda = 1/2$, and $y = 0.1$; (b) $-2 \le y \le 2, -1 \le t \le 1, \lambda = 1/2$, and $x = 0.1$

Hence, we have the following theorem.

Theorem 8. For $N = 0, 1, 2, \ldots$, we get:

$$\mathcal{H}_{m+N}(x,y,\lambda) = \sum_{i=0}^{m} \frac{\mathcal{H}_{m-i}(x)a_i(N,x,y,\lambda)m!}{(m-i)!}. \tag{29}$$

If we take $m = 0$ in Equation (29), then we have the corollary below.

Corollary 1. For $N = 0, 1, 2, \ldots$, we have:

$$\mathcal{H}_N(x,y,\lambda) = a_0(N,x,y,\lambda),$$

where:

$$a_0(0,x,y,\lambda) = 1,$$

$$a_0(N+1,x,y,\lambda) = \sum_{i=0}^{N}\left(\frac{x\log(1+\lambda)}{\lambda}\right)^i a_1(N-i,x,y,\lambda)$$
$$+ \left(\frac{\log(1+\lambda)}{\lambda}\right)^{N+1} x^{N+1}.$$

The first few of them are:

$$\mathcal{H}_0(x,y,\lambda) = 1,$$

$$\mathcal{H}_1(x,y,\lambda) = \frac{x\log(1+\lambda)}{\lambda},$$

$$\mathcal{H}_2(x,y,\lambda) = \frac{2y\log(1+\lambda)}{\lambda} + \frac{x^2(\log(1+\lambda))^2}{\lambda^2},$$

$$\mathcal{H}_3(x,y,\lambda) = \frac{6xy(\log(1+\lambda))^2}{\lambda^2} + \frac{x^3(\log(1+\lambda))^3}{\lambda^3},$$

$$\mathcal{H}_4(x,y,\lambda) = \frac{12y^2(\log(1+\lambda))^2}{\lambda^2} + \frac{12x^2y(\log(1+\lambda))^3}{\lambda^3} + \frac{x^4(\log(1+\lambda))^4}{\lambda^4},$$

$$\mathcal{H}_5(x,y,\lambda) = \frac{60xy^2(\log(1+\lambda))^3}{\lambda^3} + \frac{20x^3y(\log(1+\lambda))^4}{\lambda^4} + \frac{x^5(\log(1+\lambda))^5}{\lambda^5}.$$

5. Zeros of the Two Variable Degenerate Hermite Polynomials

This section shows the benefits of supporting theoretical prediction through numerical experiments and finding a new interesting pattern of the zeros of the two variable degenerate Hermite equations $\mathcal{H}_n(x,y,\lambda) = 0$. By using a computer, the two variable degenerate Hermite polynomials $\mathcal{H}_n(x,y,\lambda)$ can be determined explicitly. We investigated the zeros of the two variable degenerate Hermite equations $\mathcal{H}_n(x,y,\lambda) = 0$. The zeros of the $\mathcal{H}_n(x,y,\lambda) = 0$ for $n = 30, y = 2, -2, 2 + i, -2 - i, \lambda = 1/2$, and $x \in \mathbb{C}$ are displayed in Figure 2.

In Figure 2a, we chose $n = 30$ and $y = 2$. In Figure 2b, we chose $n = 30$ and $y = -2$. In Figure 2c, we chose $n = 30$ and $y = 2 + i$. In Figure 2d, we chose $n = 30$ and $y = -2 - i$.

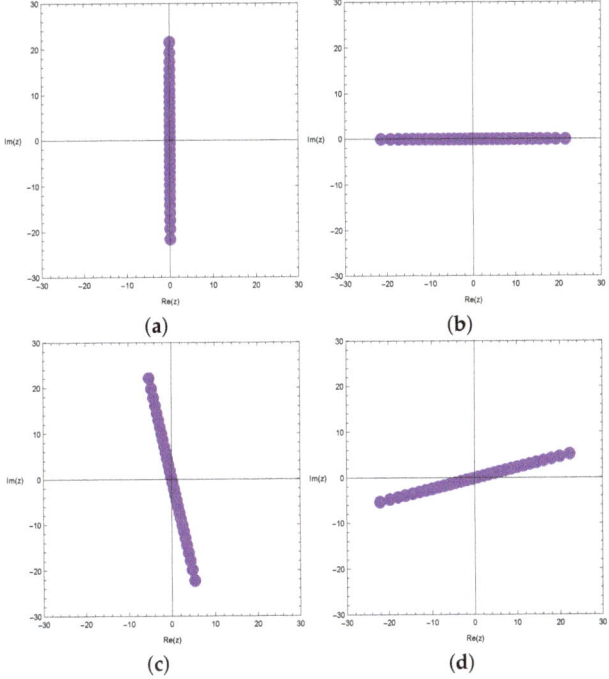

Figure 2. Zeros of $\mathcal{H}_n(x,y,\lambda) = 0$. (a) $n = 30$ and $y = 2$; (b) $n = 30$ and $y = -2$; (c) $n = 30$ and $y = 2 + i$; (d) $n = 30$ and $y = -2 - i$.

Stacks of zeros of the two variable degenerate Hermite equations $\mathcal{H}_n(x,y,\lambda) = 0$ for $1 \leq n \leq 30, \lambda = 1/2$ from a 3D structure are presented in Figure 3.

In Figure 3a, we chose $y = 2$. In Figure 3b, we chose $y = -2$. In Figure 3c, we chose $y = 2 + i$. In Figure 3d, we chose $y = -2 - i$.

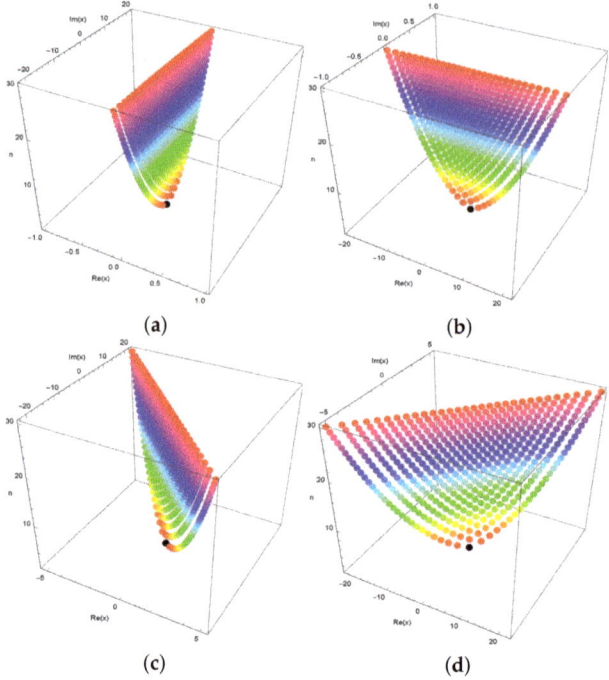

Figure 3. Stacks of zeros of $\mathcal{H}_n(x,y,\lambda) = 0, 1 \leq n \leq 30$. (a) $y = 2$; (b) $y = -2$; (c) $y = 2 + i$; (d) $y = -2 - i$.

Our numerical results for approximate solutions of real zeros of the two variable degenerate Hermite equations $\mathcal{H}_n(x,y,\lambda) = 0$ are displayed (Tables 1 and 2).

Table 1. Numbers of real and complex zeros of $\mathcal{H}_n(x,y,\lambda) = 0$.

	$y = 2, \lambda = 1/2$		$y = -2, \lambda = 1/2$.	
Degree n	Real Zeros	Complex Zeros	Real Zeros	Complex Zeros
1	1	0	1	0
2	0	2	2	0
3	1	2	3	0
4	0	4	4	0
5	1	4	5	0
6	0	6	6	0
7	1	6	7	0
8	0	8	8	0
9	1	8	9	0
10	0	10	10	0

We observed a remarkable regular structure of the complex roots of the two variable degenerate Hermite equations $\mathcal{H}_n(x,y,\lambda) = 0$ and also hoped to verify the same kind of regular structure of the complex roots of the two variable degenerate Hermite equations $\mathcal{H}_n(x,y,\lambda) = 0$ (Table 1).

The plot of the real zeros of the two variable degenerate Hermite equations $\mathcal{H}_n(x,y,\lambda) = 0$ for the $1 \leq n \leq 30, \lambda = 1/2$ structure is presented in Figure 4.

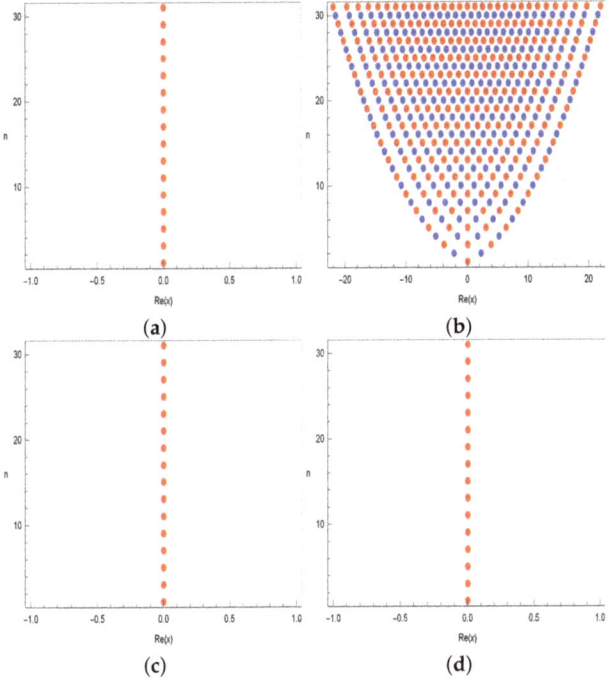

Figure 4. Real zeros of $\mathcal{H}_n(x,y,\lambda) = 0$ for $1 \leq n \leq 40$. (a)$y = 2$; (b) $y = -2$; (c) $y = 2 + i$; (d) $y = -2 - i$.

In Figure 4a, we chose $y = 2$. In Figure 4b, we chose $y = -2$. In Figure 4c, we chose $y = 2 + i$. In Figure 4d, we chose $y = -2 - i$.

Next, we calculated an approximate solution satisfying $\mathcal{H}_n(x,y,\lambda) = 0, x \in \mathbb{C}$. The results are given in Table 2. In Table 2, we chose $y = -2$ and $\lambda = 1/2$.

Table 2. Approximate solutions of $\mathcal{H}_n(x,y,\lambda) = 0, x \in \mathbb{R}$.

Degree n	x
1	0
2	−2.2209, 2.2209
3	−3.8468, 0, 3.8468
4	−5.1846, −1.6479, 1.6479, 5.1846
5	−6.3452, −3.0108, 0, 3.0108, 6.3452
6	−7.3830, −4.1958, −1.3697, 1.3697, 4.1958, 7.3830
7	−8.3295, −5.2564, −2.5639, 0, 2.5639, 5.2564, 8.3295

6. Conclusions and Future Directions

In this article, we defined the two variable degenerate Hermite polynomials and obtained some new symmetric identities for two variable degenerate Hermite polynomials. We constructed differential equations arising from the generating function of the two variable degenerate Hermite polynomials $\mathcal{H}_n(x,y,\lambda)$. We also investigated the symmetry of the zeros of the two variable degenerate Hermite equations $\mathcal{H}_n(x,y,\lambda) = 0$ for various variables x and y. As a result, we found that the distribution of the zeros of two variable degenerate Hermite equations $\mathcal{H}_n(x,y,\lambda) = 0$ is a very regular pattern. Therefore, we made the following series of conjectures with numerical experiments:

Let us use the following notations. $R_{\mathcal{H}_n(x,y,\lambda)}$ denotes the number of real zeros of $\mathcal{H}_n(x,y,\lambda) = 0$ lying on the real plane $Im(x) = 0$, and $C_{\mathcal{H}_n(x,y,\lambda)}$ denotes the number of complex zeros of $\mathcal{H}_n(x,y,\lambda) = 0$. Since n is the degree of the polynomial $\mathcal{H}_n(x,y,\lambda)$, we have $R_{\mathcal{H}_n(x,y,\lambda)} = n - C_{\mathcal{H}_n(x,y,\lambda)}$.

We can see a good regular pattern of the complex roots of the two variable degenerate Hermite equations $\mathcal{H}_n(x,y,\lambda) = 0$ for y and λ. Therefore, the following conjecture is possible.

Conjecture 1. *Let n be an odd positive integer. For $y > 0$ or $y \in \mathbb{C} \setminus \{y \mid y < 0\}$, prove or disprove that:*

$$R_{\mathcal{H}_n(x,y,\lambda)} = 1, \quad C_{\mathcal{H}_n(x,y,\lambda)} = 2\left[\frac{n}{2}\right], \quad \mathcal{H}_n(0,y,\lambda) = 0,$$

where $\mathbb{C}$ is the set of complex numbers.

Conjecture 2. *For $y < 0$, prove or disprove that:*

$$R_{\mathcal{H}_n(x,y,\lambda)} = n, \quad C_{\mathcal{H}_n(x,y,\lambda)} = 0.$$

As a result of investigating more y and λ variables, it is still unknown whether the Conjectures 1 and 2 are true or false for all variables y and λ.

We observed that solutions of the two variable degenerate Hermite equations $\mathcal{H}_n(x,y,\lambda) = 0$ have no $Re(x) = a$ reflection symmetry for $a \in \mathbb{R}$. It is expected that solutions of the two variable degenerate Hermite equations $\mathcal{H}_n(x,y,\lambda) = 0$ have $Re(x) = 0$ reflection symmetry (see Figures 2–4).

Conjecture 3. *Prove that $\mathcal{H}_n(x,y,\lambda), x \in \mathbb{C}, y > 0$, has $Im(x) = 0$ reflection symmetry analytic complex functions. Prove that $\mathcal{H}_n(x,y,\lambda), x \in \mathbb{C}, y < 0$, has $Re(x) = 0$ reflection symmetry analytic complex functions.*

Finally, we considered the more general problems. How many zeros does $\mathcal{H}_n(x,y,\lambda)$ have? We were not able to decide if $\mathcal{H}_n(x,y,\lambda) = 0$ had n distinct solutions. We would like to know the number of complex zeros $C_{\mathcal{H}_n(x,y,\lambda)}$ of $\mathcal{H}_n(x,y,\lambda) = 0$.

Conjecture 4. *Prove or disprove that $\mathcal{H}_n(x,y,\lambda) = 0$ has n distinct solutions.*

As a result of investigating more n variables, it is still unknown whether the conjecture is true or false for all variables n (see Tables 1 and 2).

We expect that research in these directions will make a new approach using the numerical method related to the research of the two variable degenerate Hermite equations $\mathcal{H}_n(x,y,\lambda) = 0$, which appear in applied mathematics and mathematical physics.

Author Contributions: We are all equally contributed to write this paper. All authors have read and agreed to the published version of the manuscript.

Funding: This work was supported by the Dong-A university research fund.

Conflicts of Interest: The authors declare no conflict of interest.

References

1. Andrews, L.C. *Special Functions for Engineers and Mathematicians*; Macmillan. Co.: New York, NY, USA, 1985.
2. Appell, P.; Hermitt Kampé de Fériet, J. *Fonctions Hypergéométriques et Hypersphériques: Polynomes d Hermite*; Gauthier-Villars: Paris, France, 1926.
3. Erdelyi, A.; Magnus, W.; Oberhettinger, F.; Tricomi, F.G. *Higher Transcendental Functions*; Krieger: New York, NY, USA, 1981; Volume 3.
4. Andrews, G.E.; Askey, R.; Roy, R. *Special Functions*; Cambridge University Press: Cambridge, UK, 1999.
5. Arfken, G. *Mathematical Methods for Physicists*, 3rd ed.; Academic Press: Orlando, FL, USA, 1985.
6. Roman, S. *The Umbral Calculus, Pure and Applied Mathematics*; Academic Press, Inc.: New York, NY, USA; Harcourt Brace Jovanovich Publishes: San Diego, CA, USA, 1984; Volume 111.
7. Ozden, H.; Simsek, Y. A new extension of q-Euler numbers and polynomials related to their interpolation functions. *Appl. Math. Lett.* **2008**, *21*, 934–938. [CrossRef]
8. Carlitz, L. Degenerate Stiling, Bernoulli and Eulerian numbers. *Utilitas Math.* **1979**, *15*, 51–88.
9. Young, P.T. Degenerate Bernoulli polynomials, generalized factorial sums, and their applications. *J. Number Theorey* **2008**, *128*, 738–758. [CrossRef]
10. Cenkci, M.; Howard, F.T. Notes on degenerate numbers. *Discrete Math.* **2007**, *307*, 2359–2375. [CrossRef]
11. Ryoo, C.S. Notes on degenerate tangent polynomials. *Glob. J. Pure Appl. Math.* **2015**, *11*, 3631–3637.
12. Haroon, H.; Khan, W.A. Degenerate Bernoulli numbers and polynomials associated with degenerate Hermite polynomials. *Commun. Korean Math. Soc.* **2018**, *33*, 651–669.
13. Kim, T.; Kim, D.S. Identities involving degenerate Euler numbers and polynomials arising from non-linear differential equations. *J. Nonlinear Sci. Appl.* **2016**, *9*, 2086–2098. [CrossRef]
14. Kim, T.; Kim, D.S.; Kwon, H.I.; Ryoo, C.S. Differential equations associated with Mahler and Sheffer-Mahler polynomials. *Nonlinear Funct. Anal. Appl.* **2019**, *24*, 453–462.
15. Ryoo, C.S. A numerical investigation on the structure of the zeros of the degenerate Euler-tangent mixed-type polynomials. *J. Nonlinear Sci. Appl.* **2017**, *10*, 4474–4484 [CrossRef]
16. Ryoo, C.S. Differential equations associated with tangent numbers. *J. Appl. Math. Inform.* **2016**, *34*, 487–494. [CrossRef]
17. Ryoo, C.S. Some identities involving Hermitt Kampé de Fériet polynomials arising from differential equations and location of their zeros. *Mathematics* **2019**, *7*, 23. [CrossRef]
18. Ryoo, C.S.; Agarwal, R.P.; Kang, J.Y. Differential equations associated with Bell-Carlitz polynomials and their zeros. *Neural Parallel Sci. Comput.* **2016**, *24*, 453–462.

© 2020 by the authors. Licensee MDPI, Basel, Switzerland. This article is an open access article distributed under the terms and conditions of the Creative Commons Attribution (CC BY) license (http://creativecommons.org/licenses/by/4.0/).

Article

Influence of Single- and Multi-Wall Carbon Nanotubes on Magnetohydrodynamic Stagnation Point Nanofluid Flow over Variable Thicker Surface with Concave and Convex Effects

Anum Shafiq [1], Ilyas Khan [2,*], Ghulam Rasool [3], El-Sayed M. Sherif [4,5] and Asiful H. Sheikh [4]

1 School of Mathematics and Statistics, Nanjing University of Information Science and Technology, Nanjing 210044, China; anumshafiq@ymail.com
2 Faculty of Mathematics and Statistics, Ton Duc Thang University, Ho Chi Minh City 72915, Vietnam
3 School of Mathematical Sciences, Yuquan Campus, Zhejiang University, Hangzhou 310027, China; grasool@zju.edu.cn
4 Center of Excellence for Research in Engineering Materials (CEREM), King Saud University, P.O. Box, 800, Al-Riyadh 11421, Saudi Arabia; esherif@ksu.edu.sa (E.-S.M.S.); aseikh@ksu.edu.sa (A.H.S.)
5 Electrochemistry and Corrosion Laboratory, Department of Physical Chemistry, National Research Centre, El-Behoth St. 33, Dokki, Cairo 12622, Egypt
* Correspondence: ilyaskhan@tdtu.edu.vn

Received: 8 December 2019; Accepted: 2 January 2020; Published: 8 January 2020

Abstract: This paper reports a theoretical study on the magnetohydrodynamic flow and heat exchange of carbon nanotubes (CNTs)-based nanoliquid over a variable thicker surface. Two types of carbon nanotubes (CNTs) are accounted for saturation in base fluid. Particularly, the single-walled and multi-walled carbon nanotubes, best known as SWCNTs and MWCNTs, are used. Kerosene oil is taken as the base fluid for the suspension of nanoparticles. The model involves the impact of the thermal radiation and induced magnetic field. However, a tiny Reynolds number is assumed to ignore the magnetic induction. The system of nonlinear equations is obtained by reasonably adjusted transformations. The analytic solution is obtained by utilizing a notable procedure called optimal homotopy analysis technique (O-HAM). The impact of prominent parameters, such as the magnetic field parameter, Brownian diffusion, Thermophoresis, and others, on the dimensionless velocity field and thermal distribution is reported graphically. A comprehensive discussion is given after each graph that summarizes the influence of the respective parameters on the flow profiles. The behavior of the friction coefficient and the rate of heat transfer (Nusselt number) at the surface ($y = 0$) are given at the end of the text in tabular form. Some existing solutions of the specific cases have been checked as the special case of the solution acquired here. The results indicate that MWCNTs cause enhancement in the velocity field compared with SWCNTs when there is an increment in nanoparticle volume fraction. Furthermore, the temperature profile rises with an increment in radiation estimator for both SWCNT and MWCNT and, finally, the heat transfer rate lessens for increments in the magnetic parameter for both types of nanotubes.

Keywords: kerosene oil-based fluid; stagnation point; carbon nanotubes; variable thicker surface; thermal radiation

---

1. Introduction

The idea of nanofluid was first introduced by Choi [1] in 1995. In his pioneering study, Choi named nanofluids as one of the most essential type of fluids for an enhanced heat transfer rate. Nanofluids are formed by suspending nanoparticles of interested metals in the base fluid. To date, different types of nanoparticles as well as the base fluids are used in the literature. Some of them are magnetic nanoparticles, polymeric nanoparticles, carbon nanotubes, liposomes, quantum dots, metallic nanoparticles, dendrimers, polymeric nanoparticles, and many others. The base fluids are always water, oil, and/or ethylene glycol. The metallic nanoparticles are alumina, carbides, copper, metal oxides, and nitrides, whereas non-metallic nanoparticles are graphite and the well-known carbon nanotubes. Researchers have used different combinations of nanoparticles and base fluids, however, nobody gave a final decision about which combination of nanoparticles and base fluid can give a better enhancement in the heat transfer rate (see, for example, the works of [2–18]).

In addition to the above discussion, each type of nanoparticle and base fluid has its unique importance. In this work, carbon nanotubes (CNTs) were used as nanoparticles suspended in Kerosene oil, chosen as base fluid. CNTs are elongated, tubular structure, and 1–2 nm in diameter (see, for example, the works of [19–23]). However, the best CNTs are those arranged in the form of hexagonal network of carbon atoms rolled up to form a steam-less hollow cylinder (Choi and Zhang [24]). CNTs were first discovered by a Japanese physicist Sumio Iijma in 1991 for multiple wall nanotubes (Sumio [25]). However, it took less than two years before single wall nanotubes were discovered. Several researchers these days are taking keen interest in studying CNTs owing to their unique nanostructures, high thermal conductivity, and exceptional mechanical strength and corrosion resistance. These novel characteristics of CNTs make them useful in industry such as solar cell, nanotube transistors, lithium ion batteries, chemical sensors, and so on. The theoretical and experimental researchers usually use two types of CNTs, namely single-walled carbon nanotubes (SWNTs) and multi-walled carbon nanotubes (MWNTs). Choi [26] found anomalous thermal conductivity enhancement in nanotubes' suspensions. Hone [27] studied, in details, the thermal properties of CNTs. Kamali and Binesh [28] numerically investigated heat transfer enhancement in non-Newtonian nanofluids using CNTs. Prajapati et al. [29], Kumaresan and Velraj [30], and Wang et al. [31] also provided some interesting studies on CNTs. Khan et al. [32] examined heat transfer using CNTs and fluid flow with Navier slip boundary conditions. Noreen et al. [33] used CNTs and analyzed thermal and velocity slips on Magnetohydrodynamics (MHD) peristaltic flow in an asymmetric channel. Noreen and Khan [34] studied heat transfer using individual MWCNTs owing to the metachronal beating of cilia. Ebaid and Sharif [35] suspended CNTs in a base fluid and studied the effect of a magnetic field on fluid motion and enhanced the rate of heat transfer of nanofluids using CNTs (also see the works of [36–39]). Zhang et al. [40] examined the effects of surface modification on the thermal conductivity and stability of the suspension formulated using the CNTs. Aman et al. [41] suspended CNTs in four different types of molecular liquid and studied heat transfer enhancement in the free convection flow of Maxwell nanofluids. Zhang experimentally investigated the heat transfer of CNTs membranes. Soleimani et al. [42] studied the impact of carbon nanotubes-based nanofluid on oil recovery efficiency using core flooding. Details about the MWCNTs were provided by Taheriang et al. [43] for enhanced thermophysical properties of nanofluids in a critical review. Wang et al. [44], in their review paper, studied mechanisms and applications of CNTs in terahertz devices. Pop et a. [45] studied the well-known stagnation point fluid flow over a stretching sheet involving the heat transfer factor due to radiation. Sharma and Singh [46] investigated stagnation point flow past a linearly stretching sheet with additional effects of variable thermal conductivity, heat source/sink, and MHD. Some other important studies with some interesting experimental findings are given in the works of [47–60] and cross references cited therein.

The basic objective of this examination is to report an MHD flow of nanofluid with Kerosene oil as base fluid and CNTs as nanoparticles over a variable thicker surface. Such formulation is not found in the literature so far. Two types of CNTs (SWCNTs and MWCNTs) are chosen. The problem is first arranged in suitable nonlinear differential equations using reasonable transformations. Analytic solution is obtained by utilizing a notable procedure called the optimal homotopy analysis technique (O-HAM). Several plots are generated to discuss the physical behavior of embedded parameters on the dimensionless velocity field and thermal distribution. The results for skin-friction (wall-drag) coefficient and rate of heat transfer (Nusselt factor) are computed in tabular data form. The present results are successfully reduced to the published results in the literature when compared.

2. Mathematical Formulation

The present communication reports a theoretical study on the magnetohydrodynamic flow and heat exchange of carbon nanotubes (CNTs)-based nanoliquid over a variable thicker surface. Kerosene oil is taken as the base fluid for the suspension of nanoparticles. Two types of carbon nanotubes (CNTs) are accounted for saturation in base fluid. Flow phenomenon is investigated in the presence of applied magnetic field. SWCNTs and MWCNTs are utilized as nanomaterials and kerosine oil is as base liquid. The impact of radiation and viscous dissipation are considered in the heat analysis. The thickness of the surface mentioned by $y = B(x+b)^{\frac{1-m}{2}}$ is variable. The ambient temperature is taken to be constant. The physical model can be seen in Figure 1.

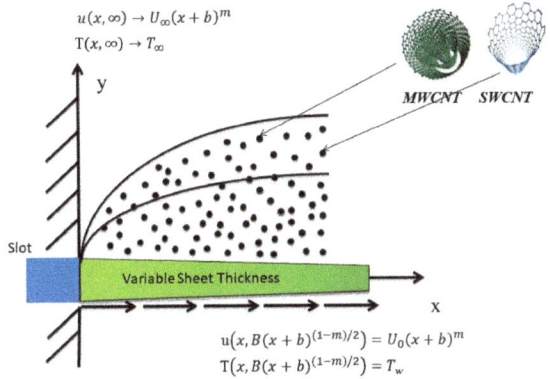

Figure 1. Physical flow model. MWCNT, multi-walled carbon nanotube; SWCNT, single-walled carbon nanotube.

The boundary layer equations for the aforementioned problem (see, for example, the works of [49–52]) are as follows:

$$\frac{\partial u}{\partial x} + \frac{\partial v}{\partial y} = 0, \qquad (1)$$

$$u\frac{\partial u}{\partial x} + v\frac{\partial u}{\partial y} = U_e \frac{dU_e}{dx} + \nu_{nf}\frac{\partial^2 u}{\partial y^2} - \frac{\sigma B_0^2}{\rho_{nf}}(U - U_e), \qquad (2)$$

$$u\frac{\partial T}{\partial x} + v\frac{\partial T}{\partial y} = \alpha_{nf}\frac{\partial^2 T}{\partial y^2} - \frac{1}{\rho C_p}\frac{\partial q_r}{\partial y} + \frac{\nu_{nf}}{\rho C_p}\left(\frac{\partial u}{\partial y}\right)^2, \qquad (3)$$

where velocity components are (u, v) along the x- and y-axes, respectively; T denotes the temperature; (ν_{nf}, α_{nf}) denotes the effective kinematic viscosity and thermal diffusivity of nanoliquid, respectively; (U_w, U_e) are defined as the stretching surface and free stream velocity, respectively; q_r defines the radiative heat flux; and C_p defines the specific heat. The fruitful characteristics of nanoliquids may be defined using the properties of base liquid and carbon nanotubes and the solid volume fraction of CNTs in the base liquids (see, for example, the works of [49–52]) as follows:

$$\mu_{nf} = \frac{\mu_f}{(1-\varphi)^{2.5}}, \; \nu_{nf} = \frac{\mu_{nf}}{\rho_{nf}}, \; \rho_{nf} = (1-\varphi)\,\rho_f + \varphi\rho_s(C_p)_{CNT},$$

$$\alpha_{nf} = \frac{k_{nf}}{\rho_{nf}(C_p)_{nf}}, \; \frac{k_{nf}}{k_f} = \frac{(1-\varphi)+2\varphi\frac{k_{CNT}}{k_{CNT}-k_f}\ln\frac{k_{CNT}+k_f}{2k_f}}{(1-\varphi)+2\varphi\frac{k_f}{k_{CNT}-k_f}\ln\frac{k_{CNT}+k_f}{2k_f}}, \qquad (4)$$

where viscosity of nanoliquid is defined by μ_{nf}; nanoparticle volume fraction is defined by φ; and density of liquid and CNTs are defined by ρ_f and ρ_{CNT}, respectively. Base liquid's thermal conductivity is defined by k_f; thermal conductivity of nanoliquids is defined by k_{nf}; the specific heat of nanoliquids, base liquid, and carbon nanotubes are defined by $(C_p)_{nf}$, $(C_p)_f$, and $(C_p)_{CNTs}$, respectively; while thermal conductivity of CNTs is defined by k_{CNT}, with boundary conditions as follows:

$$\text{at } y = B(x+b)^{\frac{1-m}{2}}, \ U_w(x) = U = U_0(x+b)^m, \ T = T_w, \ v = 0, \\ \text{as } y \to \infty, \ u \to U_\infty(x+b)^m = U_{e(x)}, \ T \to T_\infty. \tag{5}$$

Using Rosseland approximation, we get the accompanying articulation

$$q_r = \frac{-4\,\sigma^*}{3\,K^*}\frac{\partial T^4}{\partial y} = -\frac{16\sigma^*}{3K^*}T^3\frac{\partial T}{\partial y}, \tag{6}$$

where the mean absorption coefficient is defined by k^*, the Stefan–Boltzmann constant is defined by σ^*, and T^4 defined using Taylor series expansion about T_∞ is

$$T^4 = 4T_\infty^3 T - 3T_\infty^4.$$

The state of surface firmly relies on m. It should be noted that, for $m = 1$, the surface is flat; the thickness of the wall rises for $m < 1$ and the surface shape becomes of the outer convex type. The wall thickness decreases for $m > 1$ and, consequently, the surface shape becomes of the inner concave type; m is accountable for the motion type, that is, for $m = 0$, the motion is linear with constant velocity. Motion deceleration and accelerated are defined by $m < 1$ and $m > 1$. Employing the transformations

$$\xi = y\sqrt{\frac{U_0(x+b)^{m-1}}{2v_f}(m+1)}, \ \psi = \sqrt{\frac{1}{m+1}2v_f U_0(x+b)^{m+1}}F(\xi), \ \Theta(\xi) = \frac{T-T_\infty}{T_w-T_\infty}, \\ u = U_0(x+b)^m F'(\xi), \ v = -\sqrt{\frac{m+1}{2}v_f U_0(x+b)^{m-1}}\left[F + \xi F'\frac{m-1}{m+1}\right], \tag{7}$$

the incompressibility condition is consequently fulfilled and Equations (2), (3), and (5) are lessened to the following:

$$\left(\frac{1}{(1-\varphi)^{2.5}\left(1-\varphi+\varphi\frac{\rho_{CNT}}{\rho_f}\right)}\right)F''' + FF'' - \frac{2m}{m+1}(F')^2 + \frac{2m}{m+1}A^2 - M^2(F'-A) = 0, \tag{8}$$

$$\left(\frac{k_{nf}/k_f}{\left(1-\varphi+\varphi\frac{(\rho c_p)_{CNT}}{(\rho c_p)_f}\right)}\right)\left(1+\frac{4}{3k_{nf}/k_f}R_d\right)\Theta'' + \Pr F\Theta' + \Pr Ec(F'')^2 = 0, \tag{9}$$

$$F'(\alpha) = 1, \ F(\alpha) = \alpha\frac{1-m}{1+m}, \ \Theta(\alpha) = 0, \ \text{at } \alpha = B\sqrt{\frac{m+1}{2}\frac{U_0}{v_f}}, \\ F'(\infty) \to A, \ \Theta(\infty) \to 1 \ \alpha \to \infty, \tag{10}$$

where $\alpha = B\sqrt{\frac{m+1}{2}\frac{U_0}{v_f}}$ is the wall thickness parameter. Putting

$$F(\xi) = f(\eta) = f(\xi - \alpha).$$

The final equations in one variable form are given here:

$$\left(\frac{1}{(1-\varphi)^{2.5}(1-\varphi+\varphi\frac{\rho_{CNT}}{\rho_f})}\right)f''' + ff'' - \frac{2m}{m+1}(f')^2 + \frac{2m}{m+1}A^2 - M^2(f'(\eta) - A) = 0, \quad (11)$$

$$\left(\frac{k_{nf}/k_f}{\left(1-\varphi+\varphi\frac{(\rho c_p)_{CNT}}{(\rho c_p)_f}\right)}\right)\left(1 + \frac{4}{3k_{nf}/k_f}R_d\right)\theta'' + \Pr f\theta' + \Pr Ec(f'')^2 = 0, \quad (12)$$

$$\begin{aligned}&f'(0) = 1,\ f(0) = \alpha\frac{1-m}{1+m},\ \theta(0) = 0,\ \text{at } \eta = 0,\\ &f'(\infty) \to A,\ \theta(\infty) \to 1\ \ \eta \to \infty,\end{aligned} \quad (13)$$

where A is defined as ratio parameter, M is defined as magnetic parameter, R_d is defined as radiation estimator, $\Pr$ is defined as Prandtl number, and Ec is defined as the well-known Eckert number. Mathematically,

$$A = \frac{U_\infty}{U_0},\ M = \sqrt{\frac{\sigma B_0^2}{\rho c_p}},\ R_d = \frac{4\sigma^* T^3}{k_f K^*},\ \Pr = \frac{\mu c_p}{k},\ Ec = \frac{U_w^2}{c_p(T_w - T_\infty)}.$$

The expression of friction coefficient (wall-drag) and local Nusselt (the heat transfer) number are

$$C_f = \frac{\tau_w}{\rho f U_w^2},\ Nu_x = \frac{(x+b)q_w}{k_f(T_\infty - T_w)},$$

$$\tau_w = \mu_{nf}\left(\frac{\partial u}{\partial y}\right)_{y=B(x+b)^{\frac{1-m}{2}}},\ q_w = -\kappa_{nf}\left(\frac{\partial T}{\partial y}\right)_{y=B(x+b)^{\frac{1-m}{2}}}.$$

The dimensionless forms of the above parameters are

$$C_f Re_x^{\frac{1}{2}} = \frac{1}{(1-\varphi)^{2.5}}\sqrt{\frac{1}{2}(m+1)}f'''(0),\ Nu_x Re^{\frac{-1}{2}} = -\frac{k_{nf}}{k_f}\sqrt{\frac{1}{2}(m+1)}\theta'(0),$$

where $Re_x = \frac{U_w(x+b)}{\nu_f}$ denotes Reynolds number.

3. Mathematical Analysis

3.1. OHAM (BVPh 2.0)

The governing problems are explained using BVPh 2.0, via the homotopy analysis method (HAM)-based Mathematica package. The BVPh 2.0 is simple to utilize. It simply needs to compose the required governing problems. For each governing equation, we select the proper auxiliary linear operators and accurate initial guess for every undetermined function. Expression of linear operators and initial guesses are

$$f_0(\eta) = A\eta + (1-A)(1-\exp(-\eta)) - \alpha\frac{m-1}{m+1}, \quad (14)$$

$$\theta_0(\eta) = 1 - \exp(-\eta), \quad (15)$$

$$L_f(f) = \frac{d^3 f}{d\eta^3} - \frac{df}{d\eta},\ L_\theta(\theta) = \frac{d^2\theta}{d\eta^2} - \theta, \quad (16)$$

with

$$L_f[E_1 + E_2\exp(\eta) + E_3\exp(-\eta)] = 0, \quad (17)$$

$$L_\theta[E_4\exp(\eta) + E_5\exp(-\eta)] = 0, \quad (18)$$

above $E_i (i = 1, 2, \ldots, 5)$ shows arbitrary constants.

3.2. Optimal Convergence Analysis

The estimations of convergence control parameters (h_f, h_θ) in kerosene oil nanoliquids for both type of nanotubes, that is, SWCNTs and MWCNTs, are calculated through the boundary value problem solver package BVPh 2.0. We now continue to attain the solution of governing equations via Boundary value problem solver package BVPh 2.0. These governing equations hold two unknown convergence estimates (h_f, h_θ). Optimal estimates of these parameters are calculated by the total minimum error. It ought to be seen that the convergence estimates assume an essential part in the frame of the homotopy analysis method (HAM), and HAM differs from other analytical techniques. To enormously diminish the Central Processing Unit (CPU) time, the average residual error at the kth-order of estimate is characterized by

$$\varepsilon_\theta^f(h_f, h_\theta) = \frac{1}{N+1} \sum_{j=0}^{N} \left[\sum_{i=0}^{N} (f_i)_{\eta = j\pi} \right]^2, \qquad (19)$$

and

$$\varepsilon_\theta^f(h_f, h_\theta) = \frac{1}{N+1} \sum_{j=0}^{N} \left[\sum_{i=0}^{k} (f_i)_{\eta = j\pi}, \sum_{i=0}^{k} (\theta_i)_{\eta = j\pi} \right]^2. \qquad (20)$$

The optimal estimates of h_f and h_θ for single-walled (SWCNT) kerosene oil are $h_f = -0.338076$, $h_\theta = -0.112645$ and those for multi-walled (MWCNT) kerosene oil are $h_f = -0.373325$, $h_\theta = -0.122976$. Figures 2 and 3 are drawn to see the relative total residual errors for SWCNT and MWCNT kerosene oil, respectively.

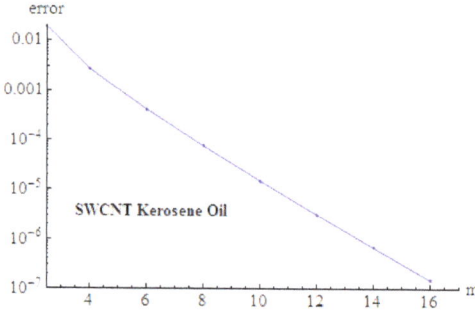

Figure 2. Total error vs. order of approximations.

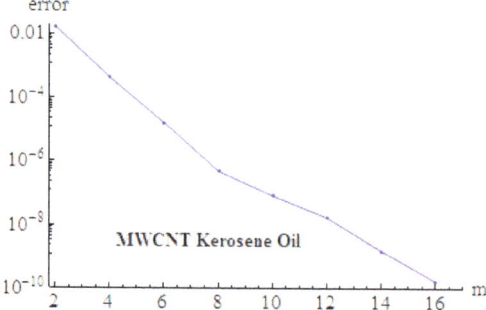

Figure 3. Total error vs. order of approximations.

4. Results and Discussion

The flow and heat transfer of CNTs (SWCNTs and MWCNTs) with kerosene oil as a base liquid are investigated. The governing set of nonlinear differential equations is numerically solved. The impact of A on $f'(\eta)$ is plotted in Figure 4 for kerosene oil for SWCNTs and MWCNTs. It is noticed that the velocity field rises for the increment in A for both $A > 1$ and $A < 1$. On the other side, for $A > 1$ and $A < 1$, the related thickness of boundary layer has a reverse trend, but for $A = 1$, no boundary layer is found. This means that the surface and ambient velocities are the same. The velocity field is dominant for single-walled tubes as compared with multi-walled tubes. Figure 5 illustrates the impact of ϕ on $f'(\eta)$. It is observed that velocity distribution is the mounting function for the increment in ϕ for SWCNTs and MWCNTs. The increment in nanomaterial volume fraction leads to rise in the convective flow. It is likewise noticed that $f'(\eta)$ enhances for kerosene oil nanoliquid for MWCNT as compared with SWCNT. The significance of magnetic parameter on the velocity field is outlined in Figure 6. The velocity $f'(\eta)$ and related boundary layer decrease for larger estimates of the magnetic estimator. The increment in M demonstrates the rise in resistive power (Lorentz force) and, therefore, the velocity of the liquid reduces. It is additionally noticed that velocity distribution is dominated for MWCNT as compared with SWCNT kerosene oil. Figure 7 is drawn for the behavior of m on $f'(\eta)$. It is analyzed that, for the higher power index, the velocity profile shows reduction for SWCNTs and MWCNTs using base liquid kerosene oil.

An analysis of Ec on $\theta(\eta)$ is portrayed in Figure 8. It is worth mentioning that $\theta(\eta)$ becomes higher for the increment in Ec for both SWCNT and MWCNT. The increment in Eckert number leads to larger drag forces between the fluid materials. Consequently, more heat is induced and the temperature distribution increases. The influence of ϕ on $\theta(\eta)$ is drawn in Figure 9. Here, temperature distribution reduces with the increase in ϕ. Additionally, the increment in the nanomaterial volume fraction causes the improvement of the convective heat phenomenon from heated liquid along the cold surface, and consequently, temperature reduces. It is additionally noted that the temperature distribution is dominant for MWCNT as compared with SWCNT. Figure 10 outlines the significance of the temperature in light of an adjustment in the estimations of the radiation parameter R_d for MWCNT and SWCNT. Obviously, the temperature distribution and related boundary layer thickness improve for higher values of the radiation estimator R_d. It is obvious that the surface heat flux increments under the effect of thermal radiation. Consequently, temperature enhances inside the boundary layer region.

Table 1 displays the thermophysical characteristics of the base liquid kerosene oil with carbon nanotubes. Table 2 is set up for the square average residual errors of governing problems at various orders of approximations. It is noticed that the square average residual error diminishes as the request of estimation rises for SWCNTs and MWCNTs using kerosene oil. Table 3 is drawn for the numerical values of the friction estimator for various estimates of different related parameters. It is examined that the friction estimator is larger for higher estimates of m, M, ϕ, and α, while it decreases for larger A for both SWCNTs and MWCNTs. Table 4 is set up for the numerical estimates of local Nusselt number for different values of different appropriate parameters. It is examined that the local Nusselt parameter rises for larger A, R_d, Ec, and ϕ, while it lessens for increments in α and M for both single-walled and multi-walled carbon nanotubes. Table 5 exhibits the relative investigation of the skin fraction coefficient with the past work of Pop et al. [45] and Sharma and Singh [46] in limiting cases. It is established that all the outcomes have a decent understanding.

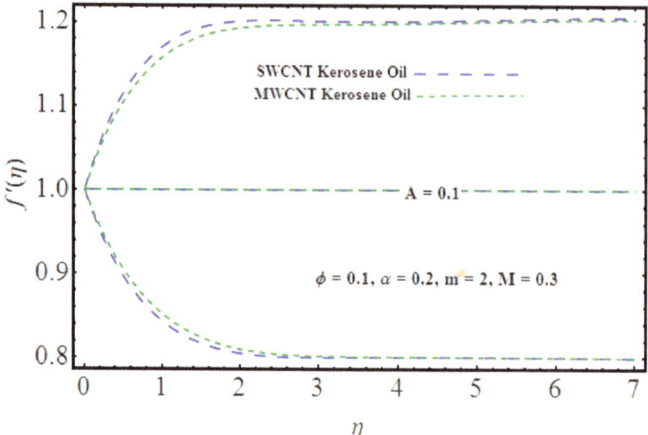

Figure 4. Impact of A on velocity $f'(\eta)$.

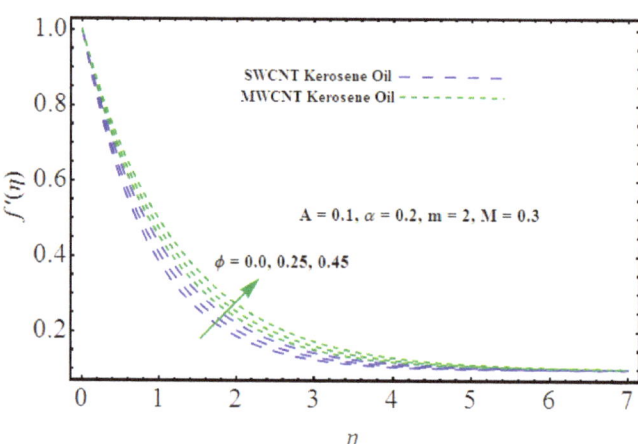

Figure 5. Impact of ϕ on velocity $f'(\eta)$.

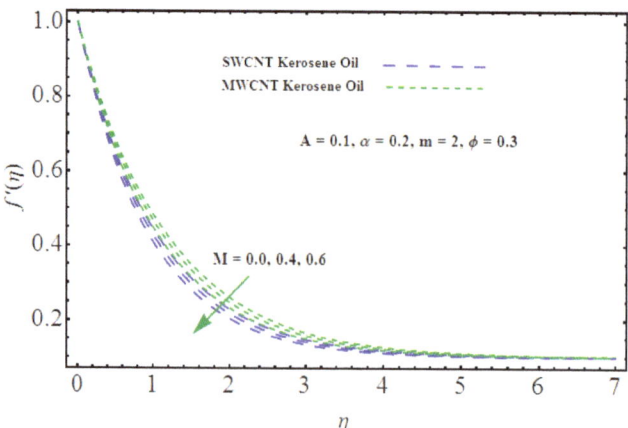

Figure 6. Impact of M on velocity $f'(\eta)$.

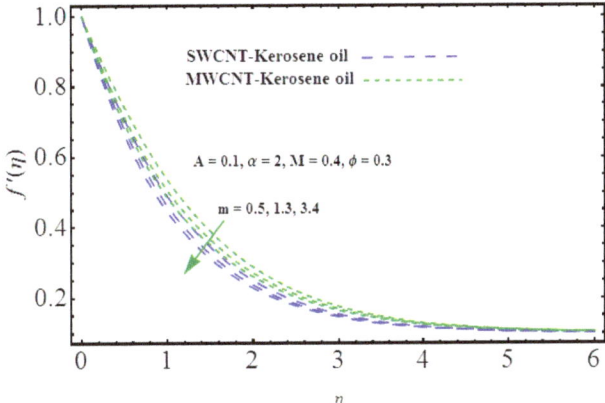

Figure 7. Impact of m on velocity $f'(\eta)$.

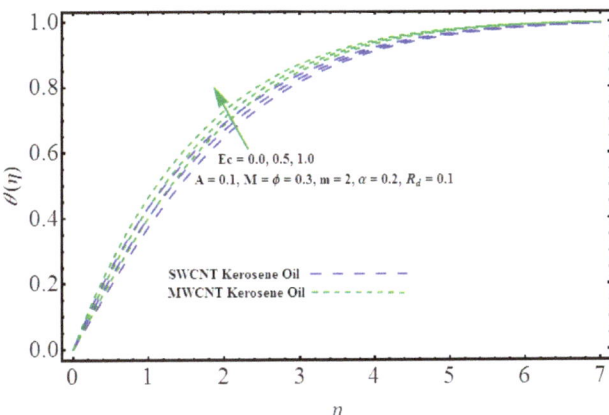

Figure 8. Impact of Ec on velocity $\theta(\eta)$.

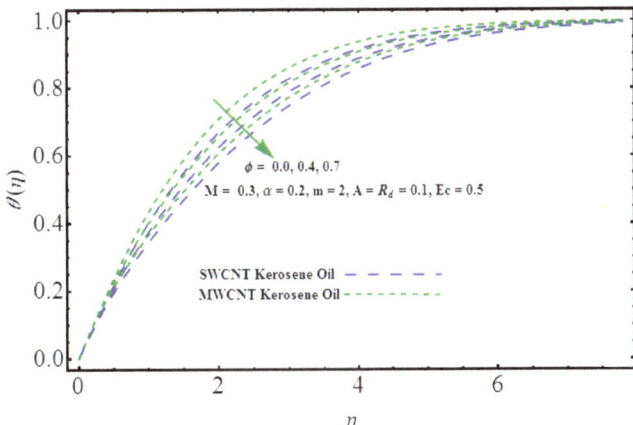

Figure 9. Impact of ϕ on velocity $\theta(\eta)$.

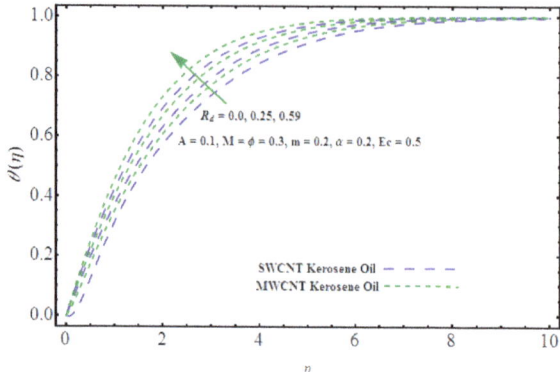

Figure 10. Impact of Ec on velocity $\theta(\eta)$.

Table 1. Data of thermophysical properties for the given particles and fluid. MWCNT, multi-walled carbon nanotube; SWCNT, single-walled carbon nanotube.

Properties	Base Fluid	Particles	
	Kerosene Oil	SWCNT	MWCNT
ρ	783	2600	1600
c_p	2090	425	796
k	0.145	6600	3000

Table 2. Mean square residual errors.

	SWCNT		MWCNT	
k	ε_k^f	ε_k^θ	ε_k^f	ε_k^θ
2	3.42849×10^{-6}	0.0372506	8.92410×10^{-4}	0.0171806
4	4.19794×10^{-7}	2.72977×10^{-2}	1.12544×10^{-6}	4.46668×10^{-3}
8	8.05829×10^{-8}	4.16494×10^{-3}	3.82436×10^{-9}	4.90227×10^{-7}
12	8.02125×10^{-10}	3.18276×10^{-6}	1.07810×10^{-10}	1.75135×10^{-8}
14	7.59501×10^{-10}	1.08974×10^{-7}	1.08022×10^{-11}	1.57636×10^{-9}
16	2.32740×10^{-11}	1.51108×10^{-8}	8.61210×10^{-13}	1.70647×10^{-10}

Table 3. Skin friction (wall drag) data.

α	A	ϕ	M	m	$-C_f Re_x^{\frac{1}{2}}$	
					SWCNT	MWCNT
0.0	0.1	0.3	0.3	2	2.55298	2.75297
0.2					2.69267	2.69167
0.4					2.63276	2.63177
0.2	0.0	0.3	0.3	2	2.78805	2.78805
	0.1				2.69106	2.69106
	0.2				2.54457	2.54457
0.2	0.1	0.0	0.3	2	3.65298	3.75297
		0.2			3.69267	3.69167
		0.4			3.62276	3.62377
0.2	0.1	0.3	0.0	2	2.61848	2.60848
			0.2		2.65576	2.64576
			0.4		2.77469	2.75469
0.2	0.1	0.3	0.3	0.0	3.51365	3.52365
				0.2	3.55576	3.53476
				0.4	3.57869	3.57469

Table 4. Heat transfer (Nusselt number) data.

α	A	ϕ	M	R_d	Ec	$-Re_x^{-1/2}Nu_x$ SWCNT	$-Re_x^{-1/2}Nu_x$ MWCNT
0.0	0.1	0.3	0.3	0.1	0.5	5.84027	5.94427
0.2						5.64158	5.65078
0.4						5.32405	5.36445
0.2	0.0	0.3	0.3	0.1	0.5	5.72660	5.72560
	0.1					5.78191	5.77091
	0.2					5.82148	5.81148
0.2	0.1	0.0	0.3	0.1	0.5	3.45876	3.45876
		0.2				3.55896	3.55896
		0.4				3.78451	3.78451
0.2	0.1	0.3	0.0	0.1	0.5	5.65429	5.65469
			0.2			5.64380	5.64280
			0.4			5.62461	5.61561
0.2	0.1	0.3	0.3	0.0	0.5	6.25429	6.25129
				0.2		7.34380	7.34281
				0.4		7.62461	7.63461
0.2	0.1	0.3	0.3	0.1	0.0	4.15419	4.15229
					0.2	4.34280	4.35282
					0.4	4.62263	4.62462

Table 5. Comparison with previous literature.

A	Pop et al. [45]	Sharma and Singh [46]	Present Results
0.1	−0.9694	−0.969386	−0.96937
0.2	−0.9181	−0.9181069	−0.91813
0.5	−0.6673	−0.667263	−0.66723
0.7			−0.43345
0.8			−0.29921
0.9			−0.15457
1.0			0.00000

5. Conclusions

The present communication reports a theoretical study on the magnetohydrodynamic flow and the heat exchange of carbon nanotube (CNT)-based nanoliquid over a variable thicker surface. Kerosene oil is taken as the base fluid for the suspension of nanoparticles. Two types of carbon nanotubes (CNTs) are accounted for saturation in base fluid, particularly the single-walled and multi-walled carbon nanotubes, best known as SWCNTs and MWCNTs. The system of nonlinear equations is gained by a reasonable transformation. Analytic solution is obtained by utilizing a notable procedure called the optimal homotopic analysis technique. The key points are given below:

- MWCNTs causes enhancement in velocity field as compared with SWCNTs when there is an increment in the nanoparticle volume fraction ϕ.
- Higher values given to the magnetic number reduce the flow velocity and are dominant for MWCNTs as compared with SWCNTs.
- Temperature profile rises with an increment in radiation estimator for both SWCNT and MWCNT.
- Augmented values of Eckert number enhance the thermal distribution, but lesser for SWCNT as compared with MWCNT.
- Friction coefficient rises for increments in m, M, ϕ, and α for both type of nanotubes.
- Heat transfer rate lessens for increments in α and M for both SWCNT and MWCNT.

Author Contributions: Conceptualization, A.S. and G.R.; methodology, A.S.; software, A.S., G.R., I.K.; validation, A.S., I.K. and G.R.; formal analysis, A.S., I.K. and G.R., E.-S.M.S. and A.H.S.; investigation, A.S. and G.R.; resources, I.K., E.-S.M.S. and A.H.S.; data curation, A.S.; writing—original draft preparation, A.S., I.K. and G.R.; writing—review and editing, A.S., I.K. and G.R., E.-S.M.S. and A.H.S.; visualization, A.S. and G.R.; supervision, A.S.; project administration, A.S.; funding acquisition, I.K., E.-S.M.S. and A.H.S. All authors have read and agreed to the published version of the manuscript.

Funding: This research was funded by Researchers Supporting Project number (RSP-2019/33), King Saud University, Riyadh, Saudi Arabia.

Acknowledgments: Researchers Supporting Project number (RSP-2019/33), King Saud University, Riyadh, Saudi Arabia. The first author is supported by the Talented Young Scientist Program of Ministry of Science and Technology of China (Pakistan-19-007).

Conflicts of Interest: The authors declare no conflict of interests.

Nomenclature

The following abbreviations have been used in this text:

Name/Title	Description	Unit
x, y	Cartesian (horizontal and vertical) coordinates	m
u, v	Velocity (horizontal and vertical) components	$\frac{m}{s}$
ν_{nf}	Kinematic viscosity of the nanofluid	$\frac{m^2}{s}$
μ_{nf}	Dynamic viscosity of the nanofluid	Pa·s
B	Magnetic field strength	$(\Omega m)^{-1}$
ρ_{fl}	Density of the base fluid	Kg·m^{-3}
ρ_{nf}	Density of the nanofluid	
k	Thermal conductivity	W·m^{-1}·K^{-1}
α	Thermal diffiusivity	m^2·s^{-1}
T, T_w, T_∞	Temperature distributions	K
U_w	Stretching velocity	m·s^{-1}
C, C_w, C_∞	Concentration distributions	
D_B	Brownian diffusion	
D_T	Thermophoresis	
Nu	Nusselt number	
Sh	Sherwood number	
C_f	Drag force coefficient	
N_b	Brownian diffusion parameter	
N_t	Thermophoresis parameter	
b	Integer	
SWCNTs	Single-walled carbon nanotubes	
MWCNTs	Multi-walled carbon nanotubes	
CNTs	Carbon nanotubes	
MHD	Magnetohydrodynamic	

References

1. Choi, S. *Enhancing Thermal Conductivity of Fluids with Nanoparticles*; ASME Publications-Fed: San Francisco, CA, USA, 1995; pp. 99–106.
2. Buongiorno, J. Convective transport in nanofluids. *J. Heat Transf.* **2006**, *128*, 240–250. [CrossRef]
3. Buongiorno, J.; Venerus, D.C.; Prabhat, N.; McKrell, T.; Townsend, J.; Christianson, R.; Tolmachev, Y.V.; Keblinski, P.; Hu, L.W.; Alvarado, J.L.; et al. A benchmark study on the thermal conductivity of nanofluids. *J. Appl. Phys.* **2009**, *106*, 094312. [CrossRef]
4. Rasool, G.; Zhang, T.; Shafiq, A. Second grade nanofluidic flow past a convectively heated vertical Riga plate. *Phys. Scr.* **2019**, *94*, 125212. [CrossRef]
5. Rasool, G.; Zhang, T. Characteristics of chemical reaction and convective boundary conditions in Powell-Eyring nanofluid flow along a radiative Riga plate. *Heliyon* **2019**, *5*, e01479. [CrossRef] [PubMed]
6. Rasool, G.; Shafiq, A.; Khalique, C.M.; Zhang, T. Magnetohydrodynamic Darcy Forchheimer nanofluid flow over nonlinear stretching sheet. *Phys. Scr.* **2019**, *94*, 105221. [CrossRef]

7. Rasool, G.; Zhang, T.; Shafiq, A.; Durur, H. Influence of chemical reaction on Marangoni convective flow of nanoliquid in the presence of Lorentz forces and thermal radiation: A numerical investigation. *J. Adv. Nanotechnol.* **2019**, *1*, 32–49. [CrossRef]
8. Rasool, G.; Zhang, T.; Shafiq, A. Marangoni effect in second grade forced convective flow of water based nanofluid. *J. Adv. Nanotechnol.* **2019**, *1*, 50–61. [CrossRef]
9. Lund, L.A.; Omar, Z.; Khan, I.; Dero, S. Multiple solutions of $Cu - C_6H_9NaO_7$ and $Ag - C_6H_9NaO_7$ nanofluids flow over nonlinear shrinking surface. *J. Cent. South Univ.* **2019**, *26*, 1283–1293. [CrossRef]
10. Lund, L.A.; Omar, Z.; Khan, I.; Raza, J.; Bakouri, M.; Tlili, I. Stability Analysis of Darcy-Forchheimer Flow of Casson Type Nanofluid Over an Exponential Sheet: Investigation of Critical Points. *Symmetry* **2019**, *11*, 412. [CrossRef]
11. Rasool, G.; Zhang, T.; Chamkha, A.J.; Shafiq, A.; Tlili, I.; Shahzadi, G. Entropy Generation and Consequences of Binary Chemical Reaction on MHD Darcy–Forchheimer Williamson Nanofluid Flow Over Non-Linearly Stretching Surface. *Entropy* **2020**, *22*, 18. [CrossRef]
12. Lund, L.A.; Omar, Z.; Khan, I. Steady incompressible magnetohydrodynamics Casson boundary layer flow past a permeable vertical and exponentially shrinking sheet: A stability analysis. *Heat Transf. Asian Res.* **2019**. [CrossRef]
13. Rasool, G.; Zhang, T. Darcy-Forchheimer nanofluidic flow manifested with Cattaneo-Christov theory of heat and mass flux over non-linearly stretching surface. *PLoS ONE* **2019**, *14*, e0221302. [CrossRef] [PubMed]
14. Shafiq, A.; Khan, I.; Rasool, G.; Seikh, A.H.; Sherif, E.S.M. Significance of Double Stratification in Stagnation Point Flow of Third-Grade Fluid towards a Radiative Stretching Cylinder. *Mathematics* **2019**, *7*, 1103. [CrossRef]
15. Shafiq, A.; Zari, I.; Rasool, G.; Tlili, I.; Khan, T.S. On the MHD Casson Axisymmetric Marangoni Forced Convective Flow of Nanofluids. *Mathematics* **2019**, *7*, 87. [CrossRef]
16. Rasool, G.; Shafiq, A.; Tlili, I. Marangoni convective nano-fluid flow over an electromagnetic actuator in the presence of first order chemical reaction. *Heat Transf. Asian Res.* **2019**. [CrossRef]
17. Rasool, G.; Shafiq, A.; Durur, H. Darcy-Forchheimer relation in Magnetohydrodynamic Jeffrey nanofluid flow over stretching surface. In *Discrete and Continuous Dynamical Systems—Series S*; American Institute of Mathematical Sciences: San Jose, CA, USA, 2019.
18. Rasool, G.; Shafiq, A.; Khalique, C.M. Marangoni forced convective Casson type nanofluid flow in the presence of Lorentz force generated by Riga plate. In *Discrete and Continuous Dynamical Systems—Series S*; American Institute of Mathematical Sciences: San Jose, CA, USA, 2019.
19. Kim, Y.J.; Shin, T.S.; Choi, H.D.; Kwon, J.H.; Chung, Y.C.; Yoon, H.G. Electrical conductivity of chemically modified multiwalled carbon nanotube/epoxy composites. *Carbon* **2005**, *43*, 23–30. [CrossRef]
20. Xue, Q. Model for thermal conductivity of carbon nanotube based composites. *Phys. B Condens. Matter* **2005**, *368*, 302–307. [CrossRef]
21. Liu, M.S.; Lin, M.C.C.; I-Te, H.; Wang, C.C. Enhancement of thermal conductivity with carbon nanotube for nanofluids. *Int. Commun. Heat Mass Transf.* **2005**, *32*, 1202–1210. [CrossRef]
22. Ding, Y.; Alias, H.; Wen, D.; Williams, R.A. Heat transfer of aqueous suspensions of carbon nanotubes (CNT nanofluids). *Int. J. Heat Mass Transf.* **2006**, *49*, 240–250. [CrossRef]
23. Ma, X.; Su, F.; Chen, J.; Bai, T.; Han, Z. Enhancement of bubble absorption process using a CNTs-ammonia binary nanofluid. *Int. Commun. Heat Mass. Transf.* **2009**, *36*, 657–660. [CrossRef]
24. Choi, J.; Zhang, Y. Properties and Applications of Single-, Double- and Multi-Walled Carbon Nanotubes. In *Aldrich Materials Science*; Sigma-Aldrich Co. LLC: Steinheim, Germany, 1995.
25. Sumio, I. Helical microtubules of graphitic carbon. *Nature* **1991**, *354*, 354–561.
26. Choi, S.U.S.; Zhang, Z.G.; Yu, W.; Lockwood, F.E.; Grulke, E.A. Anomalous thermal conductivity enhancement in nanotube suspensions. *Appl. Phys. Lett.* **2001**, *79*, 2252. [CrossRef]
27. Hone, J. Carbon nanotubes: Thermal properties. In *Dekker Encyclopedia of Nanoscience and Nanotechnology*; CRC Press: New York, NY, USA, 2004; pp. 603–610.
28. Kamali, R.; Binesh, A. Numerical investigation of heat transfer enhancement using carbon nanotube-based non-Newtonian nanofluids. *Int. Commun. Heat Mass Transf.* **2010**, *37*, 1153–1157. [CrossRef]
29. Prajapati, V.; Sharma, P.K.; Banik, A. Carbon nanotubes and its applications. *Int. J. Pharm. Sci. Res.* **2011**, *2*, 1099–1107.

30. Kumaresan, V.; Velraj, R.; Das, S.K. Convective heat transfer characteristics of secondary refrigerant based CNT nanofluids in a tubular heat exchanger. *Int. J. Refrig.* **2012**, *35*, 2287–2296. [CrossRef]
31. Wang, J.; Zhu, J.; Zhang, X.; Chen, Y. Heat transfer and pressure drop of nanofluids containing carbon nanotubes in laminar flows. *Exp. Therm. Fluid Sci.* **2013**, *44*, 716–721. [CrossRef]
32. Khan, W.A.; Khan, Z.H.; Rahi, M. Fluid flow and heat transfer of carbon nanotubes along a flat plate with Navier slip boundary. *Appl. Nanosci.* **2013**. [CrossRef]
33. Noreen, S.A.; Nadeem, S.; Khan, Z.H. Thermal and velocity slip effects on the MHD peristaltic flow with carbon nanotubes in an asymmetric channel: Application of radiation therapy. *Appl. Nanosci.* **2013**. [CrossRef]
34. Noreen, S.A.; Khan, Z.H. Heat transfer study of an individual multiwalled carbon nanotube due to metachronal beating of cilia. *Int. Commun. Heat Mass Transf.* **2014**, *59*, 114–119.
35. Ebaid, A.; al Sharif, M. Application of Laplace transform for the exact effect of a magnetic field on heat transfer of carbon-nanotubes suspended nanofluids. *Zeitschrift für Naturforschung A* **2015**, *70*, 471–475. [CrossRef]
36. Haq, R.U.; Nadeem, S.; Khan, Z.H.; Noor, N.F.M. Convective heat transfer in MHD slip flow over a stretching surface in the presence of carbon nanotubes. *Phys. B Condens. Matter* **2015**, *457*, 40–47. [CrossRef]
37. Kandasamy, R.; Mohamad, R.; Ismoen, M. Impact of chemical reaction on Cu, Al2O3, and SWCNTs-nanofluid flow under slip conditions. *Eng. Sci. Technol. Int. J.* **2016**, *19*, 700–709. [CrossRef]
38. Kandasamy, R.; Muhaimin, I.; Mohammad, R. Single walled carbon nanotubes on MHD unsteady flow over a porous wedge with thermal radiation with variable stream conditions. *Alex. Eng. J.* **2016**, *55*, 275–285. [CrossRef]
39. Mohammad, R.; Kandasamy, R. Nanoparticle shapes on electric and magnetic force in water, ethylene glycol and engine oil based Cu, Al and SWCNTs. *J. Mol. Liq.* **2017**, *237*, 54–64. [CrossRef]
40. Zhang, P.; Hong, W.; Wu, J.F.; Liu, G.Z.; Xiao, J.; Chen, Z.B.; Cheng, H.B. Effects of surface modification on the suspension stability and thermal conductivity of carbon nanotubes nanofluids. *Energy Procedia* **2015**, *69*, 699–705. [CrossRef]
41. Aman, S.; Khan, I.; Ismail, Z.; Salleh, M.Z.; Al-Mdall, Q.M. Heat transfer enhancement in free convection flow of CNTs Maxwell nanofluids with four different types of molecular liquid. *Sci. Rep.* **2017**, *7*, 2445. [CrossRef]
42. Soleimani, H.; Baig, M.K.; Yahya, N.; Khodapanah, L.K.; Sabet, M.; Demiral, B.M.R.; Burda, M. Impact of carbon nanotubes based nanofluid on oil recovery efficiency using core flooding. *Results Phys.* **2018**, *9*, 39–48. [CrossRef]
43. Taheriang, H.; Alvarado, J.L.; Languri, E.M. Enhanced thermophysical properties of multiwalled carbon nanotubes based nanofluids. Part 1: Critical review. *Renew. Sustain. Energy Rev.* **2018**, *82*, 4326–4336. [CrossRef]
44. Wang, R.; Xie, L.; Hameed, S.; Wang, C.; Ying, Y. Mechanisms and applications of carbon nanotubes in terahertz devices: A review. *Carbon* **2018**, *132*, 42–58. [CrossRef]
45. Pop, S.; Grosan, T.; Pop, I. Radiation effects on the flow near the stagnation point of a stretching sheet. *Technol. Mech.* **2004**, *25*, 100–106.
46. Sharma, P.; Singh, G. Effects of variable thermal conductivity and heat source/sink on MHD flow near a stagnation point on a linearly stretching sheet. *J. Appl. Fluid Mech.* **2009**, *2*, 13–21.
47. Haq, R.U.; Khan, Z.H.; Khan, W.A. Thermophysical effects of carbon nanotubes on MHD flow over a stretching surface. *Phys. E Low-Dimens. Syst. Nanostruct.* **2014**, *63*, 215–222. [CrossRef]
48. Haq, R.U.; Khan, Z.H.; Khan, W.A.; Shah, I.A. Viscous Dissipation Effects in Water Driven Carbon Nanotubes along a Stream Wise and Cross Flow Direction. *Int. J. Chem. React. Eng.* **2017**, *15*. [CrossRef]
49. Hussain, S.T.; Haq, R.U.; Khan, Z.H.; Nadeem, S. Water drivenflow of carbon nanotubes in a rotating channel. *J. Mol. Liq.* **2016**, *214*, 136–144. [CrossRef]
50. Haq, R.U.; Rashid, I.; Khan, Z.H. Effects of aligned magneticfield and CNTs in two different base fluids over a moving slip surface. *J. Mol. Liq.* **2017**, *243*, 682–688. [CrossRef]
51. Haq, R.U.; Kazmi, S.N.; Mekkaoui, T. Thermal management of water based SWCNTs enclosed in a partiallyheated trapezoidal cavity via FEM. *Int. J. Heat Mass Transf.* **2017**, *112*, 972–982. [CrossRef]
52. Haq, R.U.; Soomro, F.A.; Hammouch, Z.; Rehman, S.U. Heat exchange within the partially heated C-shape cavity filled with thewater based SWCNTs. *Int. J. Heat Mass Transf.* **2018**, *127*, 506–514. [CrossRef]

53. Karimipour, A.; Bagherzadeh, S.A.; Taghipour, A.; Abdollahi, A.; Safaei, M.R. A novel nonlinear regression model of SVR as a substitute for ANN to predict conductivity of MWCNT-CuO/water hybrid nanofluid based on empirical data. *Phys. A Stat. Mech. Appl.* **2019**, *521*, 89–97. [CrossRef]
54. Karimipour, A.; D'Orazio, A.; Goodarzi, M. Develop the lattice Boltzmann method to simulate the slip velocity and temperature domain of buoyancy forces of FMWCNT nanoparticles in water through a micro flow imposed to the specified heat flux. *Phys. A Stat. Mech. Appl.* **2018**, *509*, 729–745. [CrossRef]
55. Safaei, M.R.; Togun, H.; Vafai, K.; Kazi, S.N.; Badarudin, A. Investigation of Heat Transfer Enhancement in a Forward-Facing Contracting Channel Using FMWCNT Nanofluids. *Numer. Heat Transf. Part A Appl.* **2014**, *66*, 1321–1340. [CrossRef]
56. Jalali, E.; Akbari, O.A.; Sarafraz, M.M.; Abbas, T.; Safaei, M.R. Heat Transfer of Oil/MWCNT Nanofluid Jet Injection Inside a Rectangular Microchannel. *Symmetry* **2019**, *11*, 757. [CrossRef]
57. Aghaei, A.; Sheikhzadeh, G.A.; Goodarzi, M.; Hasani, H.; Damirchi, H.; Afrand, M. Effect of horizontal and vertical elliptic baffles inside an enclosure on the mixed convection of a MWCNTs-water nanofluid and its entropy generation. *Eur. Phys. J. Plus* **2018**, *133*, 486. [CrossRef]
58. Esfe, M.H.; Emami, R.; Amiri, M.K. Experimental investigation of effective parameters on MWCNT–TiO2/SAE50 hybrid nanofluid viscosity. *J. Therm. Anal. Calorim.* **2019**. [CrossRef]
59. Bagherzadeh, S.A.; D'Orazio, A.; Karimipour, A.; Goodarzi, M.; Bach, Q.V. A novel sensitivity analysis model of EANN for F-MWCNTs–Fe3O4/EG nanofluid thermal conductivity: Outputs predicted analytically instead of numerically to more accuracy and less costs. *Phys. A Stat. Mech. Appl.* **2019**, *521*, 406–415. [CrossRef]
60. Ghasemi, A.; Hassani, M.; Goodarzi, M.; Afrand, M.; Manafi, S. Appraising influence of COOH-MWCNTs on thermal conductivity of antifreeze using curve fitting and neural network. *Phys. A Stat. Mech. Appl.* **2019**, *514*, 36–45. [CrossRef]

© 2020 by the authors. Licensee MDPI, Basel, Switzerland. This article is an open access article distributed under the terms and conditions of the Creative Commons Attribution (CC BY) license (http://creativecommons.org/licenses/by/4.0/).

Article

Numerical Solutions for Multi-Term Fractional Order Differential Equations with Fractional Taylor Operational Matrix of Fractional Integration

İbrahim Avcı * and Nazim I. Mahmudov

Department of Mathematics, Eastern Mediterranean University, Famagusta, TR 99628, Northern Cyprus, via Mersin-10, Turkey; nazim.mahmudov@emu.edu.tr
* Correspondence: ibrahim.avci@emu.edu.tr

Received: 15 December 2019; Accepted: 31 December 2019; Published: 7 January 2020

Abstract: In this article, we propose a numerical method based on the fractional Taylor vector for solving multi-term fractional differential equations. The main idea of this method is to reduce the given problems to a set of algebraic equations by utilizing the fractional Taylor operational matrix of fractional integration. This system of equations can be solved efficiently. Some numerical examples are given to demonstrate the accuracy and applicability. The results show that the presented method is efficient and applicable.

Keywords: fractional differential equations; numerical solutions; Riemann-Liouville fractional integral; Caputo fractional derivative; fractional Taylor vector

MSC: 26A33, 34A08

---

1. Introduction

Fractional calculus is an emerging field of mathematics, which is a generalisation of differentiation and integration to non-integer orders. The history of fractional calculus is almost as long as the history of classical calculus, beginning with some speculations of Leibniz (1695, 1697) and Euler (1730). However, fractional calculus and fractional differential equations (FDEs) are increasingly becoming popular in recent years. The progressively developing history of this old and yet novel topic can be found in [1–5]. In fact, fractional calculus provides the mathematical modeling of some important phenomena like social and natural in a more powerful way than the classical calculus. During the last few decades, many applications were reported in many branches of science and engineering such as chaotic systems [6,7], fluid mechanics [8], viscoelasticity [9], optimal control problems [10,11], chemical kinetics [12,13], electrochemistry [14], biology [15], physics [16], bioengineering [17], finance [18], social sciences [19], economics [20,21], optics [22], chemical reactions [23], rheology [24], and so on. Due to the importance of FDEs, the solutions of them are attracting widespread interest. On the other hand, analytical solutions are not always possible for solving them. Therefore, numerical techniques becomes more important for solving such equations.

There are various numerical methods have been developed for solving FDEs in literature such as predictor-corrector method [25], Laplace transforms [26], Taylor collocation method [27], variational iteration method and homotopy perturbation method [8] (Chapter 6), Adomian decomposition method [28], Tau method [29], inverse Laplace transform [30], Haar wavelet collocation method [31], generalized block pulse operational matrix [32], shifted Legendre-tau method [33], fractional multi-step differential transformed method [34], q-homotopy analysis transform method [35], conformable Laplace transform [36], fractional B-splines collocation method [37], finite difference method [38], homotopy analysis method [39] and so on.

Multi-term fractional differential equations are one of the most important type of FDEs, which is a system of mixed fractional and ordinary differential equations and involving more than one fractional differential operators. Nowadays, they are widely appearing for modelling of many important processes, especially for multirate systems. Their numerical solution is then a strong subject that deserves high attention. In this paper, motivated by the results reported in [40,41] for solving a smaller class of problems where the highest order of derivative is an integer and involving at most one noninteger order derivative, we go further and establish a method for numerical solutions for higher order and arbitrary multi-term fractional differential equations which have a general form

$$D^{\alpha}y(t) = f\left(t, y(t), D^{\beta_0}y(t), D^{\beta_1}y(t), ..., D^{\beta_k}y(t)\right), \; t \in [0, R] \quad (1)$$

where D^{α} representing the Caputo fractional derivative of order $\alpha > 0$ and we assume that $0 < \beta_0 < \beta_1 < ... < \beta_k < \alpha$, $y^{(p)} = Y_p$, $p = 0, 1, ...n$ where $n - 1 < \alpha < n$.

Multi-term fractional order differential equations also have useful properties and they can describe complex multi-rate physical processes in a various way and can be applied in many fields, see e.g., [2,4,26,42]. Basset [43] and Bagley–Torvik [44] equations can be given as important examples for smaller class of multi-term fractional differential equations. Existence, uniqueness and stability of solution for multi-term fractional differential equations are discussed in [45–49]. Because of difficulty of finding the exact solutions for such equations, many new numerical techniques have been developed to investigate the numerical solutions such as Adams method [50], Haar wavelet method [51], differential transform method [52], Adams–Bashforth–Moulton method [53], collocation method based on shifted Chebyshev polynomials of the first kind [54], Boubaker polynomials method [55], matrix Mittag–Leffler functions [56], differential transform method [57] and so on.

Our main purpose is to present an effective, reliable method to approximate initial value problem for the Equation (1). In order to reach this aim, we rewrite and focus the general type of Caputo multi-term fractional differential equation given in Equation (1) in the following linear form

$$D^{\alpha}y(t) = \sum_{i=0}^{k} u_i D^{\beta_i}y(t) + u_{k+1}y(t) + f(t), \; 0 \leq t \leq R, \quad (2)$$

subject to the

$$y^{(p)}(0) = Y_p, \; p = 0, 1, ..., n-1 \text{ where } n - 1 < \alpha < n$$
$$u_i \; (i = 0, 1, ..., k) \text{ are known coefficients and} \quad (3)$$
$$0 < \beta_0 < \beta_1 < ... < \beta_k < \alpha$$

Here, we also state that the highest order α need not to be an integer. This equation is important in applications due to the fact it can treat the problems with fractional force, therefore it is suitable for being treated within fractional operators of Caputo type.

In this work, a numerical approach based on fractional Taylor vector is proposed to solve the initial value problem of general type of multi-term fractional differential equations which is given in Equations (2) and (3). The core idea of this method is to employ the operational matrix of fractional integration based on fractional Taylor vector to given problem and reduce it to a set of algebraic equations which can be efficiently solved.

The structure of the manuscript is organized as follows. In Section 2, we briefly introduce some preliminary ideas of fractional calculus and necessary definitions. In Section 3, an operational matrix of fractional integration based on fractional taylor vector is derived. In Section 4, we present the numerical algorithm to solve the given equation and a pseudo-code for matlab is also provided in Algorithm 1. In Section 5, the presented method is applied to six examples to demonstrate the efficiency. A final conclusion is presented in the last section.

2. Preliminary Knowledge

In this section, we recall some fundamental definitions and preliminary facts of fractional calculus.

2.1. The Fractional Integral and Derivative

Definition 1. *The Riemann–Liouville fractional integral to order α of an integrable function $y(t)$ is defined to be*

$$I^\alpha y(t) = \begin{cases} \dfrac{1}{\Gamma(\alpha)} \int_0^t (t-s)^{\alpha-1} y(q)\, ds, & \alpha > 0 \\ y(t), & \alpha = 0 \end{cases} \tag{4}$$

When applied to a power function, it yields the following result:

$$I^\alpha (t)^c = \frac{\Gamma(c+1)}{\Gamma(c+\alpha+1)} (t)^{c+\alpha}, \ \alpha \geq 0, \ c > -1 \tag{5}$$

The operator has a semigroup property, namely

$$I^\alpha I^\beta y(t) = I^\beta I^\alpha y(t), \ \alpha, \beta > 0$$

and it is linear, namely

$$I^\alpha (A_1 y_1(t) + A_2 y_2(t)) = A_1 I^\alpha y_1(t) + A_2 I^\alpha y_2(t)$$

for any two functions y_1, y_2 and constants A_1, A_2.

Definition 2. *The fractional derivative of $y(t)$ of the order α in the Caputo sense is given as*

$$D^\alpha y(t) = I^{j-\alpha} \left(\frac{d^j}{dt^j} y(t) \right), \ j-1 < \alpha \leq j, \ j \in \mathbb{N} \tag{6}$$

2.2. Some Properties

1. The Riemann-Liouville fractional integral and Caputo fractional derivative do not usually commute with each other. The following Newton–Leibniz identity gives an important relation between them:

$$I^\alpha (D^\alpha y(t)) = y(t) - \sum_{i=0}^{j-1} y^{(i)}(0) \frac{t^i}{i!} \tag{7}$$

2. The Caputo fractional derivative also has the following substitution identity. If we write $y_1(q) = y(qR)$ and $q = t/R$, then

$$D^\alpha y(t) = \frac{1}{R^\alpha} D^\alpha y_1(q) \tag{8}$$

where $j-1 < \alpha \leq j, \ j \in \mathbb{N}$

3. Operational Matrix of Fractional Integration for Fractional Taylor Vector

3.1. Fractional Taylor Basis Vector

We shall make use of the fractional Taylor vector,

$$T_{m\delta}(t) = \left[1, t^\delta, t^{2\delta}, \ldots, t^{m\delta} \right] \tag{9}$$

for $m \in \mathbb{N}$ and $\delta > 0$ in the work of this paper.

3.2. Approximation of Function

Suppose that $T_{m\delta}(t) \subset H$, where H is the space of all square integrable functions on the interval $[0,1]$. For any $y \in H$, since $S = span\left\{1, t^\delta, t^{2\delta}, ..., t^{m\delta}\right\}$ is a finite dimensional vector space in H, then, y has a unique best approximation $y_* \in S$, so that

$$\forall \hat{y} \in S, \ \|y - y_*\| \leq \|y - \hat{y}\|$$

Therefore, the function y is approximated by fractional Taylor vector as following

$$y \simeq y_* = \sum_{i=0}^{m} c_i t^{i\delta} = C^T T_{m\delta}(t) \tag{10}$$

where $T_{m\delta}(t)$ denote the fractional Taylor vector and

$$C^T = [c_0, c_1, c_2, ..., c_m] \tag{11}$$

are the unique coefficients.

3.3. Fractional Taylor Operational Matrix of Integration

By using the property of Riemann-Liouville fractional integral given in Equations (5) and (9), we get

$$I^\alpha(T_{m\delta}(t)) = \left[\frac{1}{\Gamma(\alpha+1)}t^\alpha, \frac{\Gamma(\delta+1)}{\Gamma(\delta+\alpha+1)}t^{\delta+\alpha}, \frac{\Gamma(2\delta+1)}{\Gamma(2\delta+\alpha+1)}t^{2\delta+\alpha}, ..., \frac{\Gamma(m\delta+1)}{\Gamma(m\delta+\alpha+1)}t^{m\delta+\alpha}\right]$$
$$= t^\alpha M_\alpha T_{m\delta}(t) \tag{12}$$

where

$$M_\alpha = diag\left[\frac{1}{\Gamma(\alpha+1)}, \frac{\Gamma(\delta+1)}{\Gamma(\delta+\alpha+1)}, \frac{\Gamma(2\delta+1)}{\Gamma(2\delta+\alpha+1)}, ..., \frac{\Gamma(m\delta+1)}{\Gamma(m\delta+\alpha+1)}\right]$$

denotes the operational matrix of integration.

If we define G_α as

$$G_\alpha = \left[\frac{1}{\Gamma(\alpha+1)}, \frac{\Gamma(\delta+1)}{\Gamma(\delta+\alpha+1)}, \frac{\Gamma(2\delta+1)}{\Gamma(2\delta+\alpha+1)}, ..., \frac{\Gamma(m\delta+1)}{\Gamma(m\delta+\alpha+1)}\right]$$

then, we can rewrite the Equation (10) as

$$I^\alpha(T_{m\delta}(t)) = t^\alpha G_\alpha * T_{m\delta}(t) \tag{13}$$

where $*$ denotes the operation of multiplying matrices term by term.

4. The Numerical Algorithm

In this section, to solve the given multi-term fractional differential equation in Equations (2) and (3), we employ the fractional Taylor method. The algorithm of method is given below.

Firstly, by using the transformation $q = t/R$, we replace the variable $t \in [0, R]$ with $q \in [0, 1]$. Now, by using Equation (8) in Equation (2), we get

$$\frac{1}{R^\alpha} D^\alpha y_1(q) = \sum_{i=0}^{k} \frac{1}{R^{\beta_i}} u_i D^{\beta_i} y_1(q) + u_{k+1} y_1(q) + f_1(q), \ 0 \leq s \leq 1 \tag{14}$$

where $f_1(q) = f(qR)$ and $y_1(q) = y(qR)$. Similar to Equation (10) we approximate the $y_1(q)$ as

$$y_1(q) = \sum_{i=0}^{m} c_i q^{i\delta} = C^T T_{m\delta}(q) \tag{15}$$

such that $T_{m\delta}(q) = [1, q^{\delta}, q^{2\delta}, ..., q^{m\delta}]^T$ is the fractional Taylor vector and the unique coefficients C^T is given in Equation (11).

Next, applying the Riemann–Liouville fractional integral on both side of (14), we get

$$\frac{1}{R^{\alpha}} \left[y_1(q) - \sum_{j=0}^{n-1} y_1^{(j)}(0^+) \frac{t^j}{j!} \right] = \sum_{i=0}^{k} \frac{1}{R^{\beta_i}} u_i I^{\alpha-\beta_i} \left[y_1(q) - \sum_{j=0}^{n_i-1} y_1^{(j)}(0^+) \frac{t^j}{j!} \right]$$

$$+ u_{k+1} I^{\alpha} y_1(q) + I^{\alpha} f_1(q) \tag{16}$$

where $y^{(p)}(0) = V_p$, $p = 0, 1, ..., n-1$ where $n_i - 1 < \beta_i < n_i$.

Hence, by substituting initial conditions (3), we get

$$\frac{1}{R^{\alpha}} [y_1(q)] = \sum_{i=0}^{k} \frac{1}{R^{\beta_i}} u_i I^{\alpha-\beta_i} [y_1(q)] + u_{k+1} I^{\alpha} y_1(q) + h_1(q) \tag{17}$$

such that $h_1(q) = I^{\alpha} f_1(q) + \frac{1}{R^{\alpha}} \left(\sum_{j=0}^{n-1} V_j \frac{t^j}{j!} \right) + \sum_{i=0}^{k} \frac{1}{R^{\beta_i}} u_i I^{\alpha-\beta_i} \left(\sum_{j=0}^{n_i-1} V_j \frac{t^j}{j!} \right)$.

Now, by using the Equation (12), we approximate the fractional order integrals in Equation (17) and we have

$$\frac{1}{R^{\alpha}} \left[C^T T_{m\delta}(q) \right] = \sum_{i=0}^{k} \frac{1}{R^{\beta_i}} u_i C^T q^{\alpha-\beta_i} \left(G_{\alpha-\beta_i} * T_{m\delta}(q) \right)$$

$$+ u_{k+1} q^{\alpha} C^T \left(G_{\alpha} * T_{m\delta}(q) \right) + h_1(q) \tag{18}$$

Finally, by taking the collocation points $q_j = j/m$ ($j = 0, 1, ..., m$) in Equation (18), we get $m+1$ linear algebraic equations. This linear system can be solved for the unknown vector C^T. Consequently, $y_1(q)$ can be approximated by Equation (15).

MATLAB Implementation of Method

The pseudocode given in Algorithm 1 below allows us to use proposed method in MATLAB for obtain a numerical solution of given problem [58].

Algorithm 1: Fractional Taylor Method
───
$[A, b] = fractionalTaylor(alpha, beta, Uk, func, t0, R, y0, m, delta)$
% Input

% $alpha$ is the highest order of fractional derivative of given equation
% $beta$ is the order of fractional derivatives other than alpha. $beta$ must be a vector with decending ordered values
% Uk is the vector of coefficients
% $func$ is defining the right hand side of given problem
% $t0$ and R denotes the left and right endpoints
% $y0$ is the initial conditions
% m denotes the number of steps
% $delta$ is a real number greater than zero. We usually take $delta = 1$ or $delta =$ fractional part of $alpha$

% Output

% A is an $(m+1) \times (m+1)$ matrix
% b is an $(m+1) \times 1$ matrix

% using $fractionalTaylor.m$, where command $fractionalTaylor.m$ is defined by the Equation (18), gives us the linear system $AC = B$ which is $(m+1)$
% algebraic equations with unknown coefficients C^T
% Next step is to use matlab function $linsolve(A,b)$ to solve obtained algebraic equation for unknown coefficient vector C^T with dimension $(m+1)$.
$C = linsolve(A, b)$
% Output

% C is an $(m+1) \times 1$ matrix which is the solution of linear system $AC = B$

% Next step is substituting obtained coefficients to $approxSoln()$ as input, where the command $approxSoln()$ defined by Equation (15), we get the approximate solution of given problem
$[s, y] = approxSoln(C)$
% Input

% C is the vector of coefficients obtained in previous step.

% Output

% s is the nodes on $[t0, R]$ in which the approximate solution calculated
% y is the numerical solution evaluated in the points of s.

───

5. Illustrative Examples

To illustrate the applicability and effectiveness of the presented method, we give six examples in this section. In each example, we apply the fractional Taylor operational matrix method which is presented in previous section and the approximate results compared with analytical solutions. Obtained results indicate that the proposed technique is very effective for multi-term fractional differential equations. In order to solve the numerical computations, MATLAB version R2015a has been used.

For choosing δ, we usually take either $\delta = 1$ or $\delta = \alpha - \lfloor \alpha \rfloor$, the fractional part of α.

5.1. Example 1

Consider the following form of multi-order fractional differential equation [59]

$$D^\alpha y(t) = u_0 D^{\beta_0} y(t) + u_1 D^{\beta_1} y(t) + u_2 D^{\beta_2} y(t) + u_3 D^{\beta_3} y(t) + f(t), \quad 0 \leq t \leq R, \quad (19)$$
$$y(0) = V_0, \quad y'(0) = V_1$$

We let $\alpha = 2, V_0 = V_1 = 0, R = 1$, the coefficients $u_0 = u_2 = -1, u_1 = 2, u_3 = 0$ and $\beta_0 = 0, \beta_1 = 1, \beta_2 = \frac{1}{2}$ and the function $f(t)$ is

$$f(t) = t^7 + \frac{2048}{429\sqrt{\pi}} t^{6.5} - 14t^6 + 42t^5 - t^2 - \frac{8}{3\sqrt{\pi}} t^{1.5} + 4t - 2.$$

where the exact solution is $y(t) = t^7 - t^2$.

We apply the given procedure which is implemented in previous section for solving the Equation (19) step by step.

Firstly, change variable $t \in [0, R]$ to $q \in [0, 1]$ by using $q = t/R$.

Now, we use the Equation (8) and get

$$\frac{1}{R^\alpha} D^\alpha y_1(q) = \frac{u_0}{R^{\beta_0}} D^{\beta_0} y_1(q) + \frac{u_1}{R^{\beta_1}} D^{\beta_1} y_1(q) + \frac{u_2}{R^{\beta_2}} D^{\beta_2} y_1(q) + \frac{u_3}{R^{\beta_3}} D^{\beta_3} y_1(q) + f_1(q) \quad (20)$$

where $0 \leq q \leq 1$.

Next, using Equation (7) we get

$$\frac{1}{R^\alpha}(y_1(q) - y_1(0) - qy_1'(0)) = \frac{u_0}{R^{\beta_0}} I^{\alpha-\beta_0}(y_1(q) - y_1(0) - qy_1'(0))$$
$$+ \frac{u_1}{R^{\beta_1}} I^{\alpha-\beta_1}(y_1(q) - y_1(0) - qy_1'(0))$$
$$+ \frac{u_2}{R^{\beta_2}} I^{\alpha-\beta_2}(y_1(q) - y_1(0) - qy_1'(0))$$
$$+ \frac{u_3}{R^{\beta_3}} I^{\alpha-\beta_3}(y_1(q) - y_1(0) - qy_1'(0))$$
$$+ I^\alpha f_1(q). \quad (21)$$

Now, using Equation (21) and substituting initial conditions $y(0) = V_0, y'(0) = V_1$ into equation

$$\frac{1}{R^\alpha}(C^T T_{m\delta}(q) - V_0 - RqV_1) = \frac{u_0}{R^{\beta_0}} I^{\alpha-\beta_0}(C^T T_{m\delta}(q) - V_0 - RqV_1)$$
$$+ \frac{u_1}{R^{\beta_1}} I^{\alpha-\beta_1}(C^T T_{m\delta}(q) - V_0 - RqV_1)$$
$$+ \frac{u_2}{R^{\beta_2}} I^{\alpha-\beta_2}(C^T T_{m\delta}(q) - V_0 - RqV_1)$$
$$+ \frac{u_3}{R^{\beta_3}} I^{\alpha-\beta_3}(C^T T_{m\delta}(q) - V_0 - RqV_1)$$
$$+ I^\alpha f_1(q). \quad (22)$$

From Equation (12), we have

$$\frac{1}{R^\alpha}(C^T T_{m\delta}(q) - V_0 - RqV_1)$$

$$= \frac{u_0}{R^{\beta_0}} q^{\alpha-\beta_0} C^T(G_{\alpha-\beta_0} * T_{m\delta}(q)) - \frac{u_0 q^{\alpha-\beta_0}}{R^{\beta_0}\Gamma(\alpha-\beta_0+1)} V_0 - \frac{u_0 q^{\alpha-\beta_0+1}}{R^{\beta_0}\Gamma(\alpha-\beta_0+2)} V_1$$

$$+ \frac{u_1}{R^{\beta_1}} q^{\alpha-\beta_1} C^T(G_{\alpha-\beta_1} * T_{m\delta}(q)) - \frac{u_1 q^{\alpha-\beta_1}}{R^{\beta_1}\Gamma(\alpha-\beta_1+1)} V_0 - \frac{u_1 q^{\alpha-\beta_1+1}}{R^{\beta_1}\Gamma(\alpha-\beta_1+2)} V_1$$

$$+ \frac{u_2}{R^{\beta_2}} q^{\alpha-\beta_2} C^T(G_{\alpha-\beta_2} * T_{m\delta}(q)) - \frac{u_2 q^{\alpha-\beta_2}}{R^{\beta_2}\Gamma(\alpha-\beta_2+1)} V_0 - \frac{u_2 q^{\alpha-\beta_2+1}}{R^{\beta_2}\Gamma(\alpha-\beta_2+2)} V_1$$

$$+ \frac{u_3}{R^{\beta_3}} q^{\alpha-\beta_3} C^T(G_{\alpha-\beta_3} * T_{m\delta}(q)) - \frac{u_3 q^{\alpha-\beta_3}}{R^{\beta_3}\Gamma(\alpha-\beta_3+1)} V_0 - \frac{u_3 s^{\alpha-\beta_3+1}}{R^{\beta_3}\Gamma(\alpha-\beta_3+2)} V_1$$

$$+ I^\alpha f_1(q). \tag{23}$$

Now, taking $R = 1$ in Equation (23) and putting the given values for V_0, V_1, u_i, β_i where $i = 0, 1, 2, 3$ into this equation, we get

$$C^T T_{m\delta} = 2q^1 C^T(G_1 * T_{m\delta}(q)) - q^{3/2} C^T(G_{3/2} * T_{m\delta}(q)) - q^2 C^T(G_1 * T_{m\delta}(q)) + I^2 f_1(q) \tag{24}$$

Finally, taking the collocation points $q_j = j/m$ ($j = 0, 1, ..., m$) generates a linear algebraic system of dimension $m + 1$ with unknown vector C^T. In order to solve this system by using presented method and comparing the results, we choose $\delta = 1$ and different values of m.

To show the efficiency, we compared the numerical results with the method given in [59].

Table 1, compares the obtained results for absolute error with $m = 4, 6, 7$. We observe from Table 1 that, the absolute errors for presented method are smaller and the numerical solution is more accurate for the same size of m.

Table 1. The comparison absolute errors of the present scheme and method given in [59] with $m = 4, 6, 7$.

t	Present Method $m = 4$	Method in [59] $m = 4$	Present Method $m = 6$	Method in [59] $m = 6$	Present Method $m = 7$	Method in [59] $m = 7$
0.2	0.0116	0.0844	6.81430698097618 × 10⁻⁷	0.0044	1.040834086 × 10⁻¹⁶	2.81025203108243 × 10⁻¹⁵
0.4	0.0032	0.3501	1.01100805164899 × 10⁻⁴	0.0079	2.498001805 × 10⁻¹⁶	6.63358257213531 × 10⁻¹⁵
0.6	0.0108	0.6734	1.2907314422994 × 10⁻⁵	0.0143	1.665334537 × 10⁻¹⁶	3.27515792264421 × 10⁻¹⁵
0.8	0.0037	1.0234	1.16246682382747 × 10⁻⁴	0.0214	3.330669074 × 10⁻¹⁶	4.25770529943748 × 10⁻¹⁴
1.0	0.0026	1.6700	1.11299947542775 × 10⁻⁵	0.0280	1.110223025 × 10⁻¹⁶	2.43819897540083 × 10⁻¹³

In Figures 1–3, we present the graphical representation of comparison between exact solution and the numerical solutions obtained by proposed method and the method of [59] for the problem (19) with $m = 4, 6, 7$ respectively. From these results, we can conclude that $m = 4$ and $m = 6$ give larger absolute error, while $m = 7$ gives smaller absolute error (10^{-16}) and more precise numerical solution. These comparisons also shows that the results obtained by proposed method is closer to the exact solution than the results of [59].

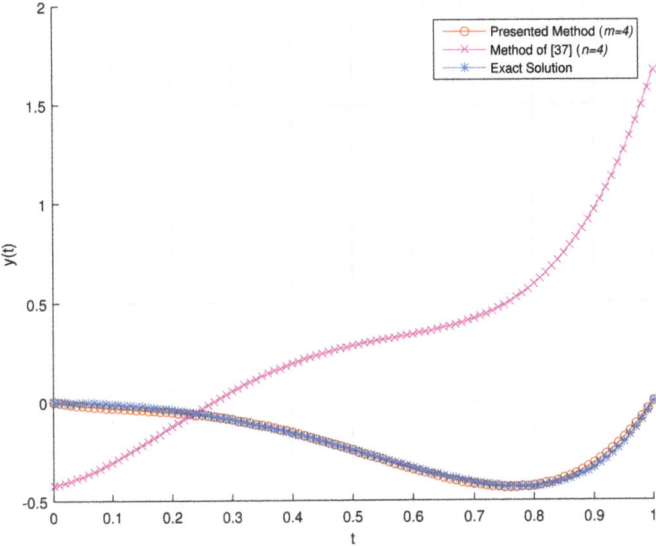

Figure 1. The comparison between exact solution and the numerical solutions obtained by proposed method and the method of [59] with $m, n = 4$.

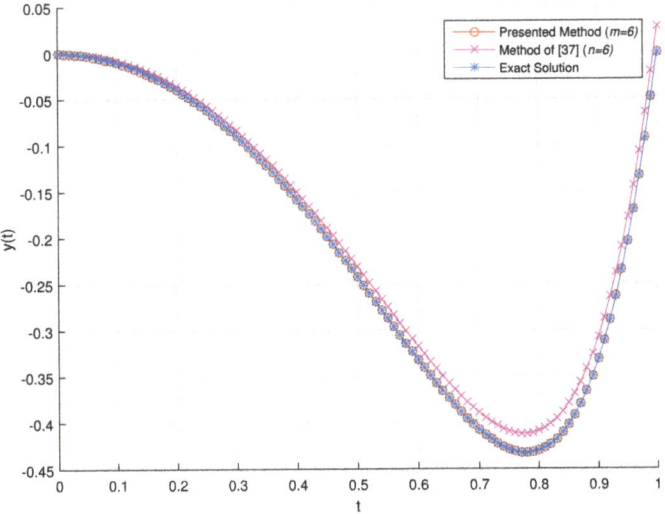

Figure 2. The comparison between exact solution and the numerical solutions obtained by proposed method and the method of [59] with $m, n = 6$.

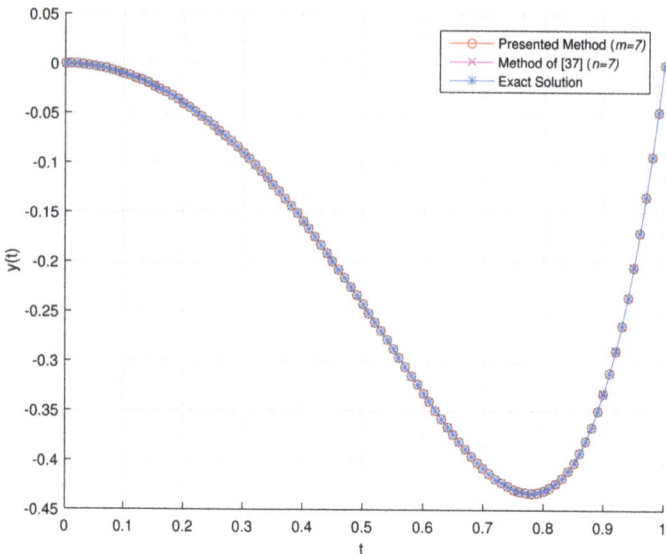

Figure 3. The comparison between exact solution and the numerical solutions obtained by proposed method and the method of [59] with $m, n = 7$.

In Figure 4, we show the graphical representation of absolute errors obtained by using proposed method and the method of [59] with $m, n = 6$.

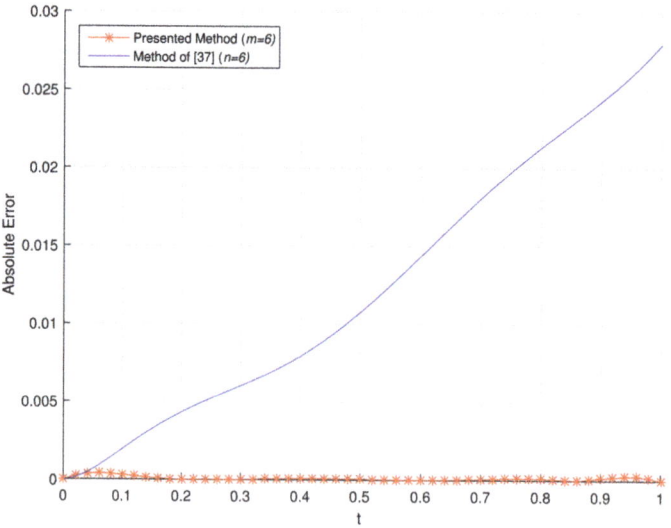

Figure 4. The behaviour of absolute errors obtained by using proposed method and the method of [59] with $m, n = 6$.

From Figure 4, we can conclude that the absolute error obtained by our method is remaining smaller and stable while the absolute error of other method is increasing in the interval $[0, 1]$.

In Figures 5 and 6, we give the graphical representation of absolute errors obtained by using proposed method with $m = 4, 7$ respectively.

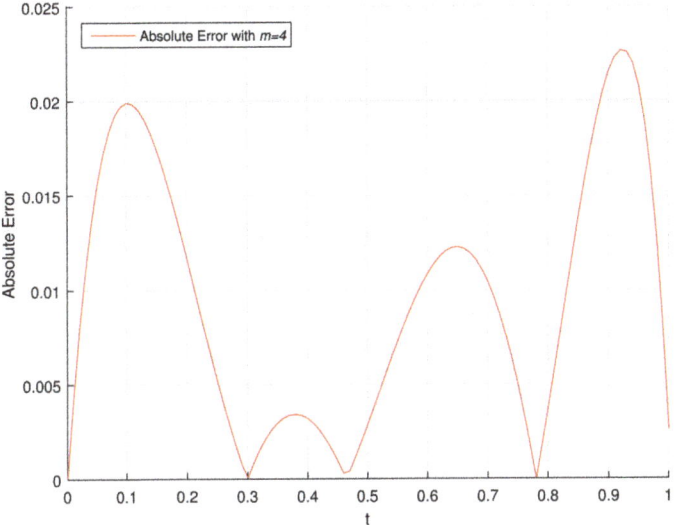

Figure 5. The absolute error with $m = 4$.

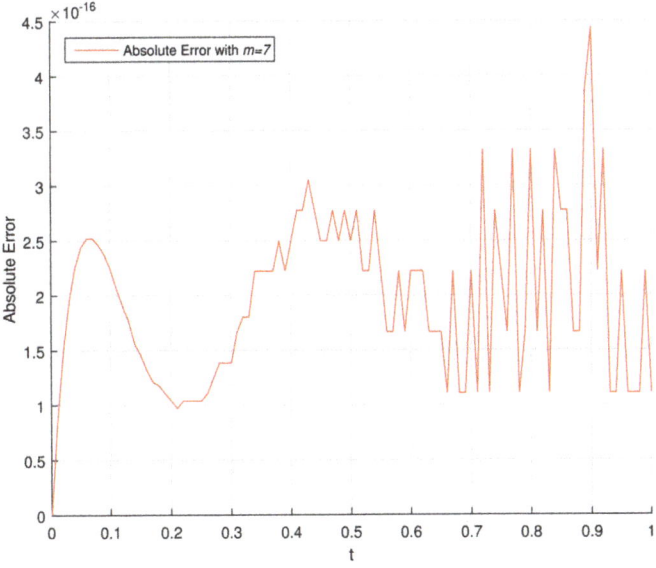

Figure 6. The absolute error with $m = 7$.

A pseudo-code for MATLAB implementation of Example 1 is given in Algorithm 2 below :

Algorithm 2: Fractional Taylor Method

alpha = 2;
beta = [1, 1/2, 0];
Uk = [2, −1, −1];
func =@(t) t^7 + 2048/(429 ∗ sqrt(pi)) ∗ $t^{6.5}$ − 14 ∗ t^6 + 42 ∗ t^5 − t^2 − ...
  8/(3 ∗ sqrt(pi)) ∗ $t^{1.5}$ + 4 ∗ t − 2;
t0 = 0; R = 1;
y0 = [0; 0];
m = 4;
delta = 1;
[A, b] = fractionalTaylor(alpha, beta, Uk, func, t0, R, y0, m, delta)
C = linsolve(A, b)
[s, y] = approxSoln(C)

5.2. Example 2

In this example, we consider the Equation (19) with $\alpha = 2$, $V_0 = V_1 = 0$, the coefficients $u_0 = u_2 = -1$, $u_1 = 0$, $u_3 = 2$ and $\beta_0 = 0$, $\beta_2 = \frac{2}{3}$, $\beta_3 = \frac{5}{3}$ and the function is

$$f(t) = t^3 + 6t - \frac{12}{\Gamma(\frac{7}{3})}t^{\frac{4}{3}} + \frac{6}{\Gamma(\frac{10}{3})}t^{\frac{7}{3}}.$$

The exact solution of this equation is $y(t) = t^3$ [59].

Applying the same procedure to given problem as presented in Example 1, we get the following equation

$$C^T T_{m\delta} = 2q^{1/3}C^T(G_{1/3} * T_{m\delta}(q)) - q^{4/3}C^T(G_{4/3} * T_{m\delta}(q)) - q^2 C^T(G_2 * T_{m\delta}(q)) + I^2 f_1(q) \quad (25)$$

As we stated in previous example, collocating this equation at the nodes $q_j = j/m$ ($j = 0, 1, ..., m$) generates a system of algebraic equations. In this example, to solve this sysem for C^T, we choose $\delta = 1, 1.5$ and different values of m.

Table 2 shows the results for obtained absolute errors by using presented method with $m = 2, 3$. From these results, we can see that, there is satisfactory agreement between the exact solution and numerical solutions. The absolute error is achieved about 10^{-15}. We also note that, the proposed method gives better results for $m = 2$ by taking $\delta = 1.5$.

Table 2. The absolute errors with m = 2, 3.

t	$\delta = 1, m = 2$	$\delta = 1.5, m = 2$	$\delta = 1, m = 3$
0	0	0	0
0.1	0.010209105	1.3×10^{-17}	7.42×10^{-17}
0.2	0.008778787	4.68×10^{-17}	1.232×10^{-16}
0.3	0.001709047	1.11×10^{-16}	1.769×10^{-16}
0.4	0.005000117	2.082×10^{-16}	2.637×10^{-16}
0.5	0.005348703	3.608×10^{-16}	4.163×10^{-16}
0.6	0.006663287	5.829×10^{-16}	6.661×10^{-16}
0.7	0.037035855	8.882×10^{-16}	9.992×10^{-16}
0.8	0.091769001	1.2212×10^{-15}	1.5543×10^{-15}
0.9	0.176862723	1.6653×10^{-15}	1.9984×10^{-15}
1.0	0.2983170221	2.2204×10^{-15}	2.8866×10^{-15}

In Figure 7a, we show the graphical representation of obtained numerical solution and the exact solution of the given problem. Figure 7b presents the obtained absolute error by using proposed method with $m = 3$.

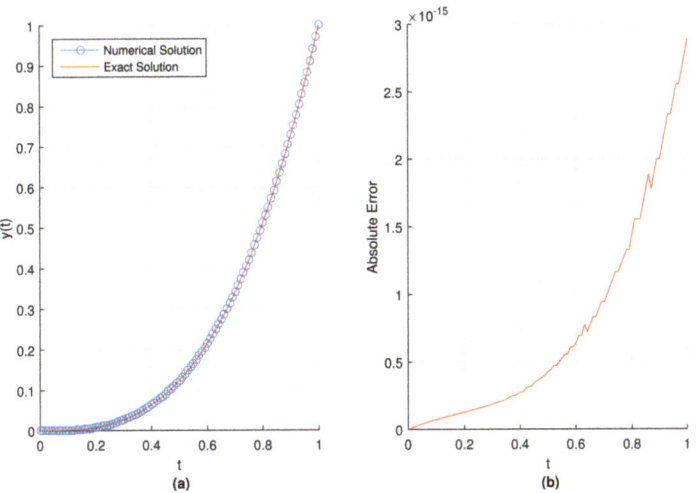

Figure 7. (a) The numerical and the exact solutions with $m = 3$. (b) The absolute error with $m = 3$.

5.3. Example 3

Consider the multi-term fractional order initial value problem [54]

$$D^{(2.2)}y(t) + 1.3D^{(1.5)}y(t) + 2.6y(t) = \sin(2t), \qquad (26)$$

with initial conditions

$$y(0) = y\prime(0) = y''(0) = 0,$$

where the equation have the series solution given by [52]

$$y_s(t) = \frac{28561}{3600000}t^6 + \frac{2}{\Gamma(4.2)}t^{3.2} - \frac{13}{5\Gamma(4.9)}t^{3.9} + \frac{169}{50\Gamma(5.6)}t^{4.6}$$
$$- \frac{8}{\Gamma(6.2)}t^{5.2} - \frac{2197}{500\Gamma(6.3)}t^{5.3} - \frac{26}{5\Gamma(6.4)}t^{5.4} + \frac{52}{5\Gamma(6.9)}t^{5.9}. \qquad (27)$$

In order to solve this problem, we choose $\delta = 1$, and $m = 10$.

We give the comparison of series solution and the numerical solution obtained by presented method in Table 3. Table 4 compares the obtained absolute errors by using presented method with the results of [54]. From this compared results, it can be seen that the approximate solution is very close to series solution for a small number of m for the given method.

From the compared results of Table 4, we can conclude that the proposed method has better approach to series solution with a smaller m.

Table 3. Comparison of numerical solution with series solution for Example 3.

t	Series Solution [52]	Present Method $m = 10$
0.0	0	0
0.1	0.000147766	0.000147731
0.2	0.001274983	0.001275552
0.3	0.00439917	0.00440567
0.4	0.010405758	0.010441315
0.5	0.019962077	0.020094648
0.6	0.033452511	0.033841301
0.7	0.050923716	0.051890573
0.8	0.0720381	0.074169634
0.9	0.096035415	0.100321388

Table 4. Comparison of absolute errors for Example 3.

t	Present Method $m = 10$	Method in [54] $m = 20$
0.0	0	0
0.1	3.47449×10^{-8}	5.2560×10^{-7}
0.2	5.69366×10^{-7}	1.7150×10^{-6}
0.3	6.49968×10^{-6}	8.2260×10^{-6}
0.4	3.55576×10^{-5}	3.7820×10^{-5}
0.5	0.000132571	0.0001353
0.6	0.00038879	0.000392
0.7	0.000966858	0.0009704
0.8	0.002131534	0.002135
0.9	0.004285973	0.00429

The graphical representation of comparison between series solution and numerical solutions obtained by presented method and the method of [54] in the interval $[0,1]$ is illustrated in Figure 8.

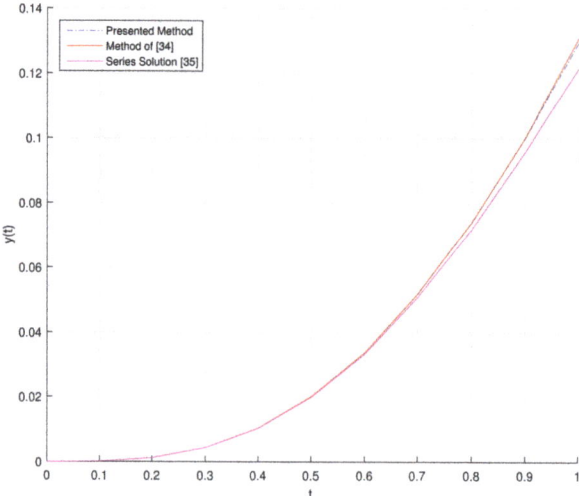

Figure 8. The comparison between series solution and numerical solutions obtained by proposed method and the method of [54] with $m = 10$.

In Figure 9, we show present graphical representation of absolute errors obtained by using proposed method and the method of [54] with $m = 10$.

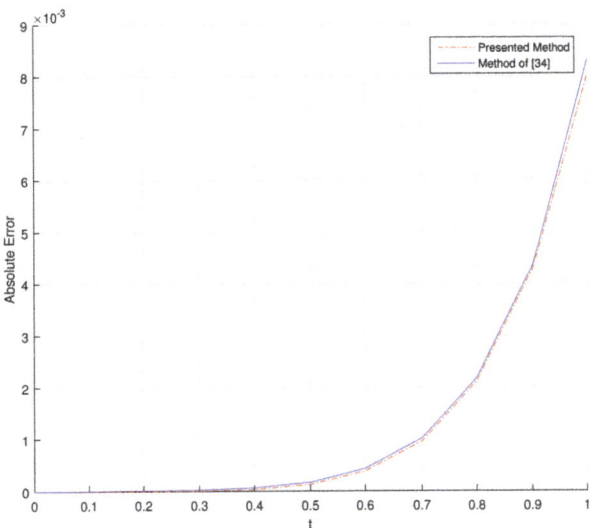

Figure 9. The behaviour of absolute errors obtained by using proposed method and the method of [54].

In Figure 10, we show the graphical representation for series solution and the numerical results of presented method for the interval $[0, 10]$. The results plotted in Figure 10 are in a very good and satisfactory agreement with the series solution given in [52] and the results of [60].

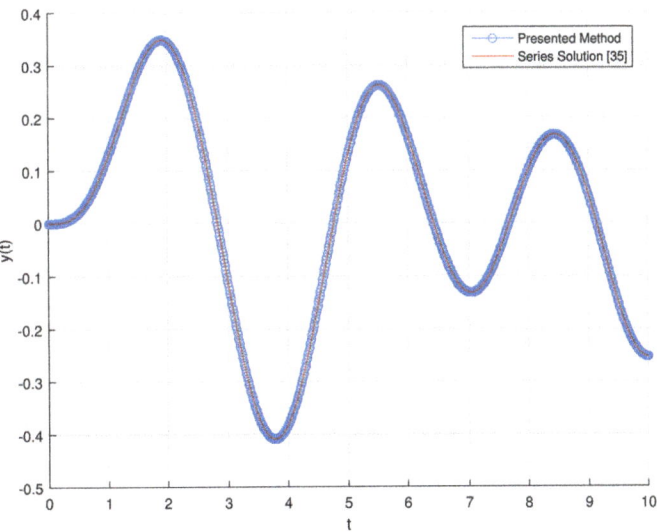

Figure 10. The behaviour of series solution and the numerical solution obtained by proposed method for the interval $[0, 10]$.

5.4. Example 4

Motivated by [50], we consider the following form of fractional differential equation,

$$D^{\alpha}y(t) + y(t) = \begin{cases} \dfrac{2}{\Gamma(3-\alpha)}t^{2-\alpha} + t^2 - t, & \alpha > 1 \\ \dfrac{2}{\Gamma(3-\alpha)}t^{2-\alpha} - \dfrac{1}{\Gamma(2-\alpha)}t^{1-\alpha} + t^2 - t, & \alpha \leq 1 \end{cases} \quad (28)$$

with initial conditions

$$y(0) = 0, \quad y\prime(0) = -1$$

whose exact solution is $y(t) = t^2 - t$.

In order to apply the presented method to Equation (28) and compare the results with methods of [54,61,62], we solve this problem with $\alpha = 0.3, 0.5, 0.7, 1.25, 1.5, 1.85$, and different values for δ and m. The obtained results are presented as below.

In Table 5, we list the results of obtained absolute errors for $\alpha = 0.3, 0.5, 0.7$ by use of presented method. Also, the results for $\alpha = 1.25, 1.5, 1.85$ are given in Table 6.

Table 5. The absolute errors with $m = 3$ and $\alpha < 1$ for Example 4.

t	$\alpha = 0.3$	$\alpha = 0.5$	$\alpha = 0.7$
0	0	0	0
0.1	4.16×10^{-17}	8.33×10^{-17}	1.94×10^{-16}
0.2	8.33×10^{-17}	5.55×10^{-17}	2.78×10^{-16}
0.3	1.11×10^{-16}	2.78×10^{-17}	2.50×10^{-16}
0.4	1.67×10^{-16}	1.39×10^{-16}	2.50×10^{-16}
0.5	1.67×10^{-16}	1.11×10^{-16}	1.67×10^{-16}
0.6	1.67×10^{-16}	5.55×10^{-17}	2.78×10^{-17}
0.7	1.67×10^{-16}	8.33×10^{-17}	8.33×10^{-17}
0.8	3.05×10^{-16}	5.55×10^{-17}	1.11×10^{-16}
0.9	2.08×10^{-16}	1.25×10^{-16}	1.39×10^{-16}
1.0	1.91×10^{-16}	1.26×10^{-16}	8.91×10^{-17}

Table 6. The absolute errors with $m = 3$ and $\alpha > 1$ for Example 4.

t	$\alpha = 1.25$	$\alpha = 1.5$	$\alpha = 1.85$
0.0	0	0	0
0.1	1.39×10^{-17}	2.78×10^{-17}	1.25×10^{-16}
0.2	5.55×10^{-17}	5.55×10^{-17}	1.94×10^{-16}
0.3	5.55×10^{-17}	5.55×10^{-17}	2.22×10^{-16}
0.4	5.55×10^{-17}	2.78×10^{-17}	2.50×10^{-16}
0.5	1.11×10^{-16}	0	2.22×10^{-16}
0.6	1.67×10^{-16}	5.55×10^{-17}	1.67×10^{-16}
0.7	1.94×10^{-16}	5.55×10^{-17}	5.55×10^{-17}
0.8	3.05×10^{-16}	1.39×10^{-16}	5.55×10^{-17}
0.9	1.11×10^{-16}	8.33×10^{-17}	1.39×10^{-17}
1.0	8.21×10^{-17}	1.97×10^{-16}	1.06×10^{-16}

In Figure 11a,b, we present the graphical representation of obtained results for numerical and exact solution of the given problem and absolute error for $\alpha = 1.5$ in the interval $[0, 1]$.

In Figure 12, we plot the graphical representation for behavior of the obtained numerical solution by use of the presented method and the exact solution of the given problem for $\alpha = 1.5$ in the interval $[0, 15]$.

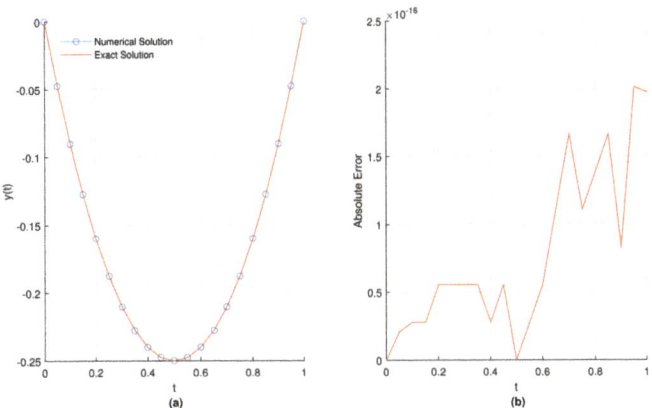

Figure 11. (a) The numerical and exact solutions for $\alpha = 1.5$. (b) The absolute error for $\alpha = 1.5$.

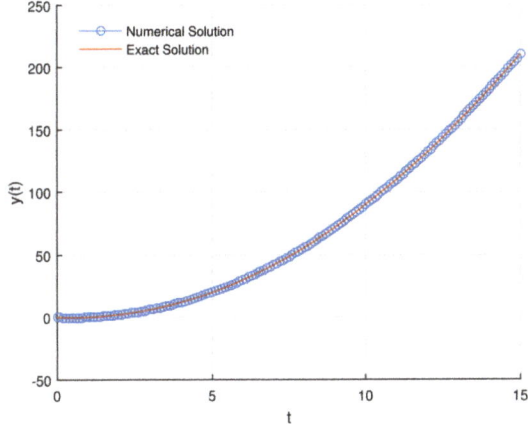

Figure 12. The behaviour of the obtained numerical and exact solutions with $\alpha = 1.5$ for the interval $t \in [0, 15]$.

Table 7 lists the obtained absolute errors for the given problem (28) at $t = 1, 5, 10, 50$ and $\alpha = 1.5$ by use of presented method and some other methods in literature [54,61,62]. From this compared results, we can say that the numerical solution obtained by use of proposed method is in better agreement with the exact solution and obtained absolute error is smaller.

Table 7. Comparison of absolute errors between proposed method and some other numerical methods in literature at $t = 1, 5, 10, 50$ for $\alpha = 1.5$.

t	Presented Method $\delta = 1/2, m = 4$	Method of [63] $n = 20$	Method of [50] $h = 1/320$	Method of [64] $p = 1, T = 1$
1	7.99361×10^{-14}	9.10×10^{-5}	3.42×10^{-3}	-
5	2.55795×10^{-13}	2.42×10^{-3}	-	-
10	1.42109×10^{-13}	5.50×10^{-3}	-	-
50	3.63798×10^{-12}	3.74×10^{-2}	-	1.2

In Figure 13, the behaviour of absolute error for $\alpha = 1.5$ with $m = 4$ and $\delta = 1/2, 1$ at $t \in [0, 50]$ is presented. From this graph, it can be seen that we get better results by taking $\delta = 1/2$ for this example and the numerical solution is very close to exact solution for a small number of m.

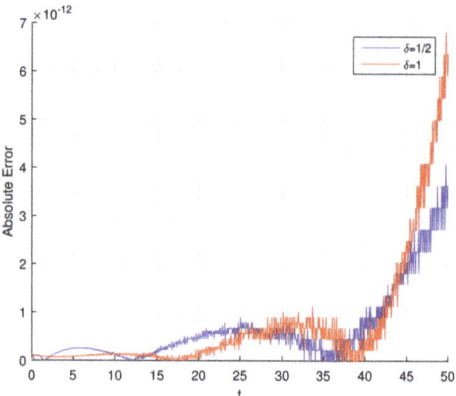

Figure 13. The behaviour of the absolute errors for proposed method where $\alpha = 1.5$, $t \in [0, 50]$ with $m = 4$ and $\delta = 1/2, 1$.

5.5. Example 5

In this example, we consider the following form of linear multi-term fractional differential equation with variable coefficients [65]

$$aD^2y(t) + b(t)D^{\beta_1}y(t) + c(t)Dy(t) + e(t)D^{\beta_2}y(t) + k(t)y(t) = f(t), \quad (29)$$

with,

$$y(0) = 2, \ y\prime(0) = 0$$

where $0 < \beta_2 < 1$, $1 < \beta_1 < 2$ and

$$f(t) = -a - \frac{b(t)}{\Gamma(3-\beta_1)} t^{2-\beta_1} - c(t)t - \frac{e(t)}{\Gamma(3-\beta_2)} t^{2-\beta_2} + k(t)\left(2 - \frac{t^2}{2}\right)$$

whose the exact solution is given by $y(t) = 2 - \frac{t^2}{2}$.

We give the numerical solution for the given problem by proposed method for $a = 1, b(t) = \sqrt{t}, c(t) = t^{\frac{1}{3}}, e(t) = t^{\frac{1}{4}}, k(t) = t^{\frac{1}{5}}, \beta_2 = 0.333, \beta_1 = 1.234$ with $\delta = 1$.

In Table 8, we give the results for maximum errors obtained by use of proposed method and comparison with the results of [65,66]. From this compared results, we can see that the numerical solution obtained by use of proposed method is closer to the exact solution.

Table 8. Maximum errors of Example 5 for $R = 1$ with $m = 3, 4, 5, 6, 10, 20, 40$.

m	Present Method	Method Given in [66]	Method Given in [65]
3	4.44089 × 10⁻¹⁶	4.4409 × 10⁻¹⁶	-
4	6.66134 × 10⁻¹⁶	1.4633 × 10⁻¹³	-
5	4.44089 × 10⁻¹⁶	3.2743 × 10⁻¹²	6.88384 × 10⁻⁵
6	4.44089 × 10⁻¹⁶	1.0725 × 10⁻¹³	-
10	2.22045 × 10⁻¹⁵	-	3.00351 × 10⁻⁶
20	3.47278 × 10⁻¹³	-	1.67837 × 10⁻⁷
40	1.46549 × 10⁻¹³	-	1.02241 × 10⁻⁸

Figure 14 presents the graphical representation for behaviour of numerical and exact solutions with $m = 6$. From this representation, we can see that the numerical solution is in a very good agreement with exact solution.

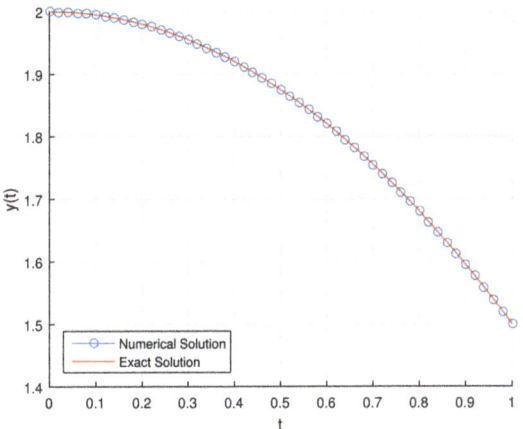

Figure 14. The behaviour of the numerical and exact solutions with $m = 6$.

5.6. Example 6

For the last example, let us consider the below fractional differential equation [63]

$$y\prime(t) + D^{1/2}y(t) - 2y(t) = 0, \ t \in (0, R], \tag{30}$$
$$y(0) = 1$$

which arises, for example, in the study of generalized Basset force occuring when a spherical object sinks in a (relatively dense) incompressible viscous fluid; see [43,67]. By use of Laplace transformation of Caputo derivatives, we get the analytical solution as following

$$y(t) = \frac{2}{3\sqrt{t}} E_{1/2,1/2}(\sqrt{t}) - \frac{1}{6\sqrt{t}} E_{1/2,1/2}(-2\sqrt{t}) - \frac{1}{2\sqrt{\pi t}},$$

where the Mittag–Leffler function $E_{\lambda,\mu}(t)$ with parameters $\lambda, \mu > 0$ is given as

$$E_{\lambda,\mu}(t) = \sum_{k=0}^{\infty} \frac{t^k}{\Gamma(\lambda k + \mu)}.$$

This Mittag–Leffler function and its variations are very significant in fractional calculus and fractional differential equations [68].

In order to solve given problem by use of proposed method and compare the results, we take $t \in (0, 5]$ and use different values of δ and m.

Table 9 lists the exact and obtained numerical solutions by use of presented method and method of [63] for the given problem for $m = 5, 10, 15, 20$. Comparison of this results shows that, even for small values of m, the numerical solution obtained by use of presented method is in a better agreement with exact solution.

Table 9. The resulting values for Example 6, with $R = 5$ in some values of t.

t	Exact	Proposed Method $m = 5$	Method Given in [63] $m = 5$	Proposed Method $m = 10$	Method Given in [63] $m = 10$	Proposed Method $m = 15$	Method Given in [63] $m = 15$	Proposed Method $m = 20$	Method Given in [63] $m = 20$
1	3.42445	3.42415	2.714336	3.425121	3.426525	3.42376044	3.42496	3.424563	3.424807
2	9.69088	9.670891	8.922571	9.692732	9.696794	9.68896761	9.692754	9.691185	9.691706
3	26.6414	26.60757	24.59981	26.64646	26.65929	26.6362145	26.64683	26.64225	26.64381
4	72.6729	72.53849	65.78029	72.68665	72.72038	72.6587861	72.68787	72.6752	72.67936
5	197.77	197.5757	180.1481	197.8077	197.8994	197.731934	197.8112	197.7766	197.7879

In Figures 15a, 16a and 17a, we present the graphical representation of comparison between exact solution and the numerical solutions obtained by using proposed method and the method of [63] with taking $m = 5, 10, 20$ respectively. Also in Figures 15b, 16b and 17b we show the behaviour of absolute errors obtained by proposed method and the method of [63] in the interval $[0, 1]$ with $m = 5, 10, 20$.

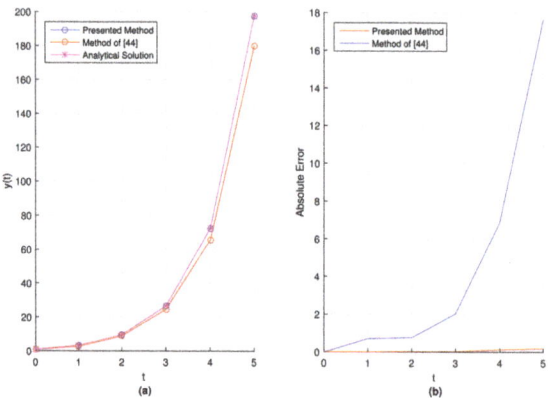

Figure 15. (**a**) The comparison of analytical solution and numerical solutions obtained by the proposed method and the method of [63] with $m = 5$. (**b**) The behaviour of the absolute errors between the exact solution and numerical solutions obtained by our method and the method given in [63] with $m = 5$.

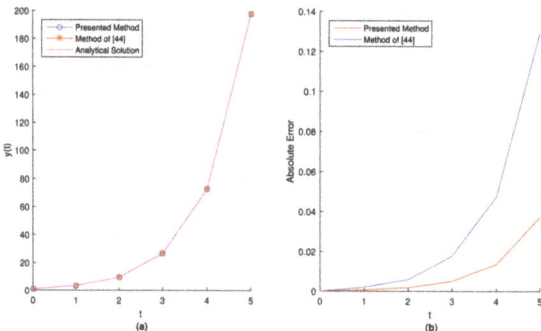

Figure 16. (**a**) The comparison of analytical solution and numerical solutions obtained by the proposed method and the method of [63] with $m = 10$. (**b**) The behaviour of the absolute errors between the exact solution and numerical solutions obtained by our method and the method given in [63] with $m = 10$.

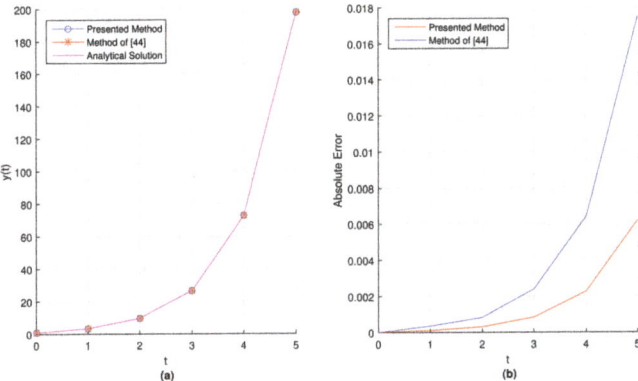

Figure 17. (a) The comparison of analytical solution and numerical solutions obtained by the proposed method and the method of [63] with $m = 20$. (b) The behaviour of the absolute errors between the exact solution and numerical solutions obtained by our method and the method given in [63] with $m = 20$.

From these graphical results represented in Figures 15–17, we can conclude that the absolute error obtained by our method is remaining smaller when compared the absolute error of method given in Reference [63].

6. Conclusions

In this work, an operational matrix based on the fractional Taylor vector is used to numerically solve the multi-term fractional differential equations by reducing them to a set of linear algebraic equations, which simplifies the problem. From comparison of the obtained results with exact solutions and also with results of other methods in the literature, we conclude that the proposed method provides the solution with high accuracy. The findings also show that, even for the small number of steps, we can get satisfactory results by using presented method. All computational results are obtained by using MATLAB.

Author Contributions: Formal analysis, İ.A.; Supervision, N.I.M. All authors contributed equally to this article. All authors have read and agreed to the published version of the manuscript.

Funding: This research received no external funding.

Conflicts of Interest: The authors declare no conflict of interest.

References

1. Oldham, K.B.; Spanier, J. *The Fractional Calculus: Theory and Applications of Differentiation and Integration to Arbitrary Order*; AP: New York, NY, USA, 1974.
2. Miller, K.S.; Ross, B. *An Introduction to the Fractional Calculus and Fractional Differential Equations*; Wiley: New York, NY, USA, 1993.
3. Samko, S.G.; Kilbas, A.A.; Marichev, O.I. *Fractional Integrals and Derivatives: Theory and Applications*; Gordan and Breach: New York, NY, USA, 1993.
4. Diethelm, K. *The Analysis of Fractional Differential Equations*; Springer: Heidelberg, Germany, 2010.
5. Machado, J.T.; Kiryakova, V.; Mainardi, F. Recent history of fractional calculus. *Commun. Nonlinear Sci. Num. Simul.* **2011**, *16*, 1140–1153. [CrossRef]
6. Hajipour, M.; Jajarmi, A.; Baleanu, D. An efficient nonstandard finite difference scheme for a class of fractional chaotic systems. *J. Comput. Nonlinear Dyn.* **2017**, *13*, 021013. [CrossRef]
7. Huang, L.; Bae, Y. Chaotic dynamics of the fractional-love model with an external environment. *Entropy* **2018**, *20*, 53. [CrossRef]

8. Zheng, L.; Zhang, X. *Modeling and Analysis of Modern Fluid Problems. Mathematics in Science and Engineering*; AP: London, UK, 2017.
9. Mainardi, F. *Fractional Calculus and Waves in Linear Viscoelasticity*; ICP: London, UK, 2010.
10. Mahmudov, N.I. Finite-approximate controllability of evolution equations. *Appl. Comput. Math.* **2017**, *16*, 159–167.
11. Baleanu, D.; Jajarmi, A.; Hajipour, M. A new formulation of the fractional optimal control problems involving Mittag—Leffler nonsingular kernel. *J. Optimiz. Theory. App.* **2017**, *175*, 718–737. [CrossRef]
12. Singh, J.; Kumar, D.; Baleanu, D. On the analysis of chemical kinetics system pertaining to a fractional derivative with Mittag-Leffler type kernel. *Chaos Interdiscip. J. Nonlinear Sci.* **2017**, *27*, 103113. [CrossRef]
13. Stoenoiu, C.E.; Bolboacă, S.D.; Jäntschi, L. Model formulation and interpretation–from experiment to theory. *Int. J. Pure Appl. Math.* **2008**, *47*, 9–16.
14. Oldham, K.B. Fractional differential equations in electrochemistry. *Adv. Eng. Softw.* **2010**, *41*, 9–12. [CrossRef]
15. Ertürk, V.S.; Odibat, Z.M.; Momani, S. An approximate solution of a fractional order differential equation model of human T-cell lymphotropic virus I (HTLV-I) infection of CD4+ T-cells. *Comput. Math. Appl.* **2011**, *62*, 996–1002. [CrossRef]
16. Hilfer, R. *Applications of Fractional Calculus in Physics*; World Scientific Press: Singapore, 2000.
17. Magin, R.L. *Fractional Calculus in Bioengineering*; Begell House Publishers: Redding, CA, USA, 2006.
18. Fallahgoul, H.; Focardi, S.; Fabozzi, F. *Fractional Calculus and Fractional Processes With Applications to Financial Economics: Theory and Application*; AP: Cambridge, MA, USA, 2016.
19. Baleanu, D.; Lopes, A.M. (Eds.) *Handbook of Fractional Calculus with Applications, Volume 8: Applications in Engineering, Life and Social Sciences, Part B*; De Gruyter: Berlin, Germany, 2019.
20. Tarasov, V.E. On history of mathematical economics: Application of fractional calculus. *Mathematics* **2019**, *7*, 509. [CrossRef]
21. Ming, H.; Wang, J.; Fečkan, M. The Application of Fractional Calculus in Chinese Economic Growth Models. *Mathematics* **2019**, *7*, 665. [CrossRef]
22. Esen, A.; Sulaiman, T.A.; Bulut, H.; Baskonus, H.M. Optical solitons and other solutions to the conformable space–time fractional Fokas–Lenells equation. *Optik* **2018**, *167*, 150–156 [CrossRef]
23. Zabadal, J.; Vilhena, M.; Livotto, P. Simulation of chemical reactions using fractional derivatives. *Nuovo Cimento. B* **2001**, *116*, 529–545.
24. Yang, F.; Zhu, K.Q. On the definition of fractional derivatives in rheology. *Theor. Appl. Mech. Lett.* **2011**, *1*, 012007. [CrossRef]
25. Diethelm, K.; Ford, N.J.; Freed, A.D. A predictor-corrector approach for the numerical solution of fractional differential equations. *Nonlinear Dyn.* **2002**, *29*, 3–22. [CrossRef]
26. Podlubny, I. *Fractional Differential Equations*; AP: New York, NY, USA, 1999.
27. Çenesiz, Y.; Keskin, Y.; Kurnaz, A. The solution of the Bagley–Torvik equation with the generalized Taylor collocation method. *J. Franklin Inst.* **2010**, *347*, 452–466. [CrossRef]
28. Ray, S.S.; Bera, R.K. Solution of an extraordinary differential equation by Adomian decomposition method. *J. Appl. Math.* **2004**, *2004*, 331–338. [CrossRef]
29. Vanani, S.K.; Aminataei, A. Tau approximate solution of fractional partial differential equations. *Comput. Math. Appl.* **2011**, *62*, 1075–1083. [CrossRef]
30. Rani, D.; Mishra, V.; Cattani, C. Numerical inverse Laplace transform for solving a class of fractional differential equations. *Symmetry* **2019**, *11*, 530. [CrossRef]
31. Khashan, M.M.; Amin, R.; Syam, M.I. A new algorithm for fractional Riccati type differential equations by using Haar wavelet. *Mathematics* **2019**, *7*, 545. [CrossRef]
32. Li, Y.; Sun, N. Numerical solution of fractional differential equations using the generalized block pulse operational matrix. *Comput. Math. Appl.* **2011**, *62*, 1046–1054. [CrossRef]
33. Saadatmandi, A.; Dehghan, M. A tau approach for solution of the space fractional diffusion equation. *Comput. Math. Appl.* **2011**, *62*, 1135–1142. [CrossRef]
34. Abuasad, S.; Yildirim, A.; Hashim, I.; Karim, A.; Ariffin, S.; Gómez-Aguilar, J.F. Fractional multi-step differential transformed method for approximating a fractional stochastic sis epidemic model with imperfect vaccination. *Int. J. Environ. Res. Public Health* **2019**, *16*, 973. [CrossRef] [PubMed]
35. Veeresha, P.; Prakasha, D.G.; Baleanu, D. An efficient numerical technique for the nonlinear fractional Kolmogorov–Petrovskii–Piskunov equation. *Mathematics* **2019**, *7*, 265. [CrossRef]

36. Silva, F.; Moreira, D.; Moret, M. Conformable Laplace transform of fractional differential equations. *Axioms* **2018**, *7*, 55. [CrossRef]
37. Pitolli, F. A fractional B-spline collocation method for the numerical solution of fractional predator-prey models. *Fractal Fract.* **2018**, *2*, 13. [CrossRef]
38. Fazio, R.; Jannelli, A.; Agreste, S. A Finite Difference Method on Non-Uniform Meshes for Time-Fractional Advection–Diffusion Equations with a Source Term. *Appl. Sci.* **2018**, *8*, 960. [CrossRef]
39. Odibat, Z.; Momani, S.; Xu, H. A reliable algorithm of homotopy analysis method for solving nonlinear fractional differential equations. *Appl. Math. Model.* **2010**, *34*, 593–600. [CrossRef]
40. Krishnasamy, V.S.; Razzaghi, M. The numerical solution of the Bagley–Torvik equation with fractional Taylor method. *J. Comput. Nonlin. Dyn.* **2016**, *11*. [CrossRef]
41. Krishnasamy, V.S.; Mashayekhi, S.; Razzaghi, M. Numerical solutions of fractional differential equations by using fractional Taylor basis. *IEEE/CAA J. Autom. Sin.* **2017**, *4*, 98–106.
42. Jiang, H.; Liu, F.; Turner, I.; Burrage, K. Analytical solutions for the multi-term time-fractional diffusion-wave/diffusion equations in a finite domain. *Comput. Math. Appl.* **2012**, *64*, 3377–3388. [CrossRef]
43. Basset, A.B. On the descent of a sphere in a vicous liquid. *Quart. J.* **1910**, *41*, 369–381.
44. Torvik, P.J.; Bagley, R.L. On the Appearance of the fractional derivative in the behavior of real materials. *ASME J. Appl. Mech.* **1984**, *51*, 294–298. [CrossRef]
45. Mahmudov, N.I.; Emin, S.; Bawanah, S. On the Parametrization of Caputo-Type Fractional Differential Equations with Two-Point Nonlinear Boundary Conditions. *Mathematics* **2019**, *7*, 707. [CrossRef]
46. Diethelm, K.; Ford, N.J. Analysis of fractional differential equations. *J. Math. Anal. Appl.* **2002**, *265*, 229–248. [CrossRef]
47. Daftardar-Gejji, V.; Jafari, H. Analysis of a system of nonautonomous fractional differential equations involving Caputo derivatives. *J. Math. Anal. Appl.* **2007**, *328*, 1026–1033. [CrossRef]
48. Aphithana, A.; Ntouyas, S.K.; Tariboon, J. Existence and uniqueness of symmetric solutions for fractional differential equations with multi-order fractional integral conditions. *Bound. Value Probl.* **2015**, *2015*, 68. [CrossRef]
49. Aliev, F.A.; Aliev, N.A.; Safarova, N.A. Transformation of the Mittag-Leffler Function to an Exponential Function and Some of its Applications to Problems with a Fractional Derivative. *Appl. Comput. Math* **2019**, *18*, 316–325.
50. Diethelm, K.; Ford, N.J.; Freed, A.D. Detailed error analysis for a fractional Adams method. *Numer. Algorithms* **2004**, *36*, 31–52. [CrossRef]
51. Lepik, Ü. Solving fractional integral equations by the Haar wavelet method. *Appl. Math. Comput.* **2009**, *214*, 468–478. [CrossRef]
52. Arikoglu, A.; Ozkol, I. Solution of fractional differential equations by using differential transform method. *Chaos Soliton. Fract.* **2007**, *34*, 1473–1481. [CrossRef]
53. Diethelm, K.; Ford, N.J. Multi–order fractional differential equations and their numerical solution. *Appl. Math. Comput.* **2004**, *154*, 621–640. [CrossRef]
54. Saw, V.; Kumar, S. The approximate solution for multi-term the fractional order initial value problem using collocation method based on shifted Chebyshev polynomials of the first kind. In *Information Technology and Applied Mathematics*; Chandra, P., Giri, D., Li, F., Kar, S., Jana, D.K., Eds.; Springer: Singapore, 2019; Volume 699, pp. 53–67.
55. Bolandtalat, A.; Babolian, E.; Jafari, H. Numerical solutions of multi-order fractional differential equations by Boubaker polynomials. *Open Phys.* **2016**, *14*, 226–230. [CrossRef]
56. Popolizio, M. Numerical solution of multiterm fractional differential equations using the matrix Mittag–Leffler functions. *Mathematics* **2018**, *6*, 7. [CrossRef]
57. Rebenda, J. Application of Differential Transform to Multi-Term Fractional Differential Equations with Non-Commensurate Orders. *Symmetry* **2019**, *11*, 1390. [CrossRef]
58. Garrappa, R. Numerical solution of fractional differential equations: A survey and a software tutorial. *Mathematics* **2018**, *6*, 16. [CrossRef]
59. Han, W.; Chen, Y.M.; Liu, D.Y.; Li, X.L.; Boutat, D. Numerical solution for a class of multi-order fractional differential equations with error correction and convergence analysis. *Adv. Differ. Equ. N. Y.* **2018**, *2018*, 253. [CrossRef]

60. El-Mesiry, A.E.M.; El-Sayed, A.M.A.; El-Saka, H.A.A. Numerical methods for multi-term fractional (arbitrary) orders differential equations. *Appl. Math. Comput.* **2005**, *160*, 683–699. [CrossRef]
61. Bhrawy, A.H.; Tharwat, M.M.; Yildirim, A. A new formula for fractional integrals of Chebyshev polynomials: Application for solving multi-term fractional differential equations. *Appl. Math. Model.* **2013**, *37*, 4245–4252. [CrossRef]
62. Deng, W.H.; Li, C. Numerical schemes for fractional ordinary differential equations. *Numer. Model.* **2012**, *16*, 355–374.
63. Esmaeili, S.; Shamsi, M. A pseudo-spectral scheme for the approximate solution of a family of fractional differential equations. *Commun. Nonlinear Sci.* **2011**, *16*, 3646–3654. [CrossRef]
64. Deng, W. Short memory principle and a predictor-corrector approach for fractional differential equations. *J. Comput. Appl. Math.* **2007**, *206*, 174–188. [CrossRef]
65. Maleknejad, K.; Nouri, K.; Torkzadeh, L. Operational matrix of fractional integration based on the shifted second kind Chebyshev polynomials for solving fractional differential equations. *Med. J. Math.* **2016**, *13*, 1377–1390. [CrossRef]
66. Liu, J.; Li, X.; Wu, L. An operational matrix of fractional differentiation of the second kind of Chebyshev polynomial for solving multiterm variable order fractional differential equation. *Math. Probl. Eng.* **2016**, *2016*, 7126080. [CrossRef]
67. Carpinteri, A.; Mainardi, F. (Eds.) Fractional calculus: Some basic problems in continuum and statistical mechanics. In *Fractals and Fractional Calculus in Continuum Mechanics*; Springer Verlag: Wien, Austria; New York, NY, USA, 1997; pp. 291–348.
68. Srivastava, H.M.; Fernandez, A.; Baleanu, D. Some New Fractional-Calculus Connections between Mittag–Leffler Functions. *Mathematics* **2019**, *7*, 485. [CrossRef]

© 2020 by the authors. Licensee MDPI, Basel, Switzerland. This article is an open access article distributed under the terms and conditions of the Creative Commons Attribution (CC BY) license (http://creativecommons.org/licenses/by/4.0/).

Article

Reinterpretation of Multi-Stage Methods for Stiff Systems: A Comprehensive Review on Current Perspectives and Recommendations

Yonghyeon Jeon [1,†], Soyoon Bak [1,†] and Sunyoung Bu [2,*]

1. Department of Mathematics, Kyungpook National University, Daegu 41566, Korea; dydgus1020@naver.com (Y.J.); jiya525@knu.ac.kr (S.B.)
2. Department of Liberal Arts, Hongik university, Sejong 30016, Korea
* Correspondence: syboo@hongik.ac.kr; Tel.: +82-44-860-2121
† These authors contributed equally to this work.

Received: 9 November 2019; Accepted: 26 November 2019; Published: 1 December 2019

Abstract: In this paper, we compare a multi-step method and a multi-stage method for stiff initial value problems. Traditionally, the multi-step method has been preferred than the multi-stage for a stiff problem, to avoid an enormous amount of computational costs required to solve a massive linear system provided by the linearization of a highly stiff system. We investigate the possibility of usage of multi-stage methods for stiff systems by discussing the difference between the two methods in several numerical experiments. Moreover, the advantages of multi-stage methods are heuristically presented even for nonlinear stiff systems through several numerical tests.

Keywords: multi-stage method; multi-step method; Runge–Kutta method; backward difference formula; stiff system

---

1. Introduction

Most time-dependent differential equations are usually solved by multi-stage (one-step) method or multi-step method [1–3]. In general, there seems to be no significant difference in the structure between them when the multi-stage method is applied to get an initial guess for the multi-step method [4]. Nonetheless, a comparison of both methods has attracted quite a lot of interest from the viewpoints of convergence, stability, practical computations, numerical efficiency, etc. [5–11]. Comparisons in this regard do not take into account the impact of advances in computer science and technologies such as artificial intelligence (AI) or parallel computation, etc. Considering the impact, a new perspective to compare the potentials of both methods should be investigated as well as existing comparative studies. First of all, it is well known that the highest order of an A-stable multi-step method is two, so lots of research [12–24] developing higher order methods have focused on either multi-step methods satisfying some less restrictive stability condition or multi-stage methods which combine A-stability with high-order accuracy [2,25–29]. In addition, multi-stage methods such as Runge–Kutta (RK) type methods do not require any additional memory for function values at previous steps since it does not use any previously computed values [30–32]. On the other hand, multi-step methods require additional memory in the sense that they use previously computed function values and have insufficient function values for initial data. Multi-stage methods are comparable with multi-step methods for nonlinear stiff problems and have no restriction to express initial data contrast to the other. There seems not to be such a clear a priori distinction between multi-stage and multi-step methods.

Another interesting point of view to find more efficient methods is quite susceptible to stiffness and nonlinearity of the given problem. For nonlinear stiff problems, a multi-step method is needed to evaluate function values only once at each iteration in a nonlinear solver, whereas multi-stage methods require several function evaluations at each iteration. This disadvantage of the multi-stage method can be ignored by the authors' recent research [33]. The authors showed numerically that one stage of the multi-stage method is equivalent to one step of the multi-step method for simple ordinary differential equation (ODE) systems. However, the multi-step methods such as the backward differentiation formula (BDF) are usually recommended to apply nonlinear stiff problems because the process of solving the nonlinear system of equation is also expensive computationally. In the process of solving nonlinear stiff problems by a multi-stage method, it generally generates a system $M_d \otimes M_s$, where d and s represent the dimension of the given problem and the number of stages used in the multi-stage method, respectively. Here, the notation M_k represents a matrix with the size $k \times k$ and the notation $\otimes$ denotes a Kronecker product. On the other hand, a multi-step method needs to solve only a system of size $d \times d$.

The purpose of this paper is to investigate and compare the properties of the multi-stage and the multi-step methods for d-dimensional stiff problems described by

$$\frac{dy}{dt} = f(t,y) \in \mathbb{R}^d. \tag{1}$$

Most nonlinear stiff problems are solved by multi-step methods rather than multi-stage methods since the multi-stage methods usually transform nonlinear stiff problems into bigger nonlinear systems, as mentioned in the previous paragraph. To solve such nonlinear systems efficiently, one has to consider both nonlinear and linear solvers. The nonlinear systems are usually solved by using an iteration technique such as Newton-like iterations, which incur considerable computation costs. There are various Newton-like iterations. Among them, a simplified Newton iteration is developed in connection with the development of computer process capacity [34–37]. Different nonlinear system solvers generate linear systems correspondingly. It means that the nonlinear system solver should be well-selected to adapt efficient linear solvers such as the eigenvalue decomposition method. Note that efficient linear solvers have also been well-studied [1,2,38,39]. An eigenvalue decomposition combined with simplified Newton iteration can apply to a multi-stage method. The resulting multi-stage method generates the same matrix, regardless of integration or iteration, as an object of decomposition for solving a linear system induced by the simplified Newton iteration. It allows for decomposing the matrix only once throughout the whole process. As a result, applying this combination to multi-stage methods highlights the advantage of multi-stage methods by reducing computational costs to the level of the costs required from multi-step methods without any loss of the original advantages of multi-stage methods, which is the main contention of this paper.

The remaining parts of this paper are as follows. We briefly describe the multi-step and multi-stage methods and simplified Newton iteration in Section 2. To support theoretical analysis, we present preliminary numerical results in Section 3. Finally, in Section 4, all results are summarized and further possibilities are discussed.

2. Preliminary

2.1. Methods

In this subsection, we briefly describe ODE solvers classified by mathematical theory. Numerical methods for ODEs fall naturally into two categories: one is 'multi-stage method' using one starting value at each step and the other is 'multi-step method' or 'multi-value method' based on several values of the solution. We deal with the theories of two methods in terms of convergence and stability. The multi-step method has a critically bad stability property with a higher convergence rate that can not actually be used. Due to these reasons, the third-order RK method (RK3) and the third-order BDF (BDF3) are considered as examples of multi-stage methods and multi-step methods. Note that the higher order multi-step method is also available, but it has very low practical use.

The general form of the multi-step methods [1,3,7,26,38,40,41] is described by

$$y_{n+1} = \sum_{j=0}^{s} a_j y_{n-j} + h \sum_{j=-1}^{s} b_j f(t_{n-j}, y_{n-j}), \quad n \geq s. \tag{2}$$

Here, the coefficients $a_0, \ldots, a_s, b_{-1}, b_0, \ldots, b_s$ are constants. If method (2) use $s + 1$ previous solution values with either $a_s \neq 0$ or $b_s \neq 0$, the method is called an $s + 1$ step method. A BDF method is the most efficient linear multi-step methods among several multi-step methods [40]. It is composed of the coefficients $b_{p-1} = \cdots = b_0 = 0$ and the others chosen such that the method with the convergence order of s convergence order s. Thus, the s-step BDF has s-th convergence order. The BDF3 is given by

$$y_{n+3} - \frac{18}{11} y_{n+2} + \frac{9}{11} y_{n+1} - \frac{2}{11} y_n = \frac{6}{11} h f(t_{n+3}, y_{n+3}). \tag{3}$$

Since implicit A-stable linear multi-step methods have convergence order of at most 2, second-order BDF can be A-stable, but the method can not be A-stable with order more than 3. The stability of BDF3 is almost A-stable [38].

An explicit RK method has been developed by Runge, Heun, and Kutta based on a Euler method [3,40]. Later, an implicit RK was developed for stiff problems based on several quadrature rules. RK methods have the following form:

$$y_{n+1} = y_n + h \sum_{i=1}^{s} b_i k_i,$$

$$k_i = f(t_n + c_i h, y_n + h \sum_{j=1}^{s} a_{ij} k_j), \quad i = 1, \ldots, s, \tag{4}$$

or an equivalent form of Butcher tableau

$$\begin{array}{c|c} c & A \\ \hline & b \end{array}.$$

One can specify a particular RK method by providing the number of stage s and all elements of the Butcher tableau, a_{ij} ($1 \leq i, j \leq s$), b_i and c_i ($i = 1, \ldots, s$). There is a popular implicit RK method for solving the stiff problem, which is called a collocation method. The collocation method is changed depending on the choice of the collocation points. For more details on the collocation method, one can refer to [3,38]. If we select uniform collocation points defined by $c_i = i/3$ ($1 \leq i \leq 3$), we can obtain a third-order collocation method with having the following butcher table:

$$\begin{array}{c|ccc} \frac{1}{3} & \frac{23}{36} & -\frac{4}{9} & \frac{5}{36} \\ \frac{2}{3} & \frac{7}{9} & -\frac{2}{9} & \frac{1}{9} \\ 1 & \frac{3}{4} & 0 & \frac{1}{4} \\ \hline & \frac{3}{4} & 0 & \frac{1}{4} \end{array}. \tag{5}$$

Note that the order of the stage and convergence for the method (5) are both three as shown in the convergence analysis in [3,38]. Furthermore, the stability of (5) demonstrated through Dahlquist's problem is almost L-stable.

2.2. Simplified Newton Iteration and Eigenvalue Decomposition Method

To explicate a simplified Newton iteration proposed by Liniger and Willoughby [10], we consider the following nonlinear system obtained by RK-type methods,

$$z_i = h \sum_{j=1}^{s} a_{ij} f(x_0 + c_j h, y_0 + z_j), \quad i = 1, \ldots, s. \tag{6}$$

Equation (6) is equivalent to a system of equations described by

$$Z = h(A \otimes I_d)) F(Z), \tag{7}$$

where

$$Z = [z_1, \ldots, z_s]^T,$$
$$A = (a_{ij})_{i,j=1}^{s},$$
$$F(Z) = [f(x_0 + c_1 h, y_0 + z_1), \ldots, f(x_0 + c_s h, y_0 + z_s)]^T,$$

and I_d is d-dimensional identity matrix. By applying Newton iteration to the nonlinear system of Equation (7), we can get a linear system of the form

$$(I_{sd} - h(A \otimes I_d)\mathcal{J}) \Delta Z^k = -Z^k + h(A \otimes I_d) F(Z^k),$$
$$Z^{k+1} = Z^k + \Delta Z^k, \tag{8}$$

where $\mathcal{J}$ is a block diagonal matrix that consists of Jacobians $\frac{\partial f}{\partial y}(t_n + c_i h, y_n + z_i)$, $i = 1, \ldots, s$, $Z^k = (z_1^k, \ldots, z_s^k)^T$ is the k-th iterated solution, $\Delta Z^k = (\Delta z_1^k, \ldots \Delta z_s^k)^T$ is the increment, and $F(Z^k)$ denotes for

$$F(Z^k) = (f(x_0 + c_1 h, y_0 + z_1^k), \ldots, f(x_0 + c_s h, y_0 + z_s^k))^T.$$

Usually, one Newton iteration needs several calculations of the Jacobian which requires lots of computational costs. To reduce such costs, all Jacobians $\frac{\partial f}{\partial y}(t_n + c_i h, y_n + z_i)$ are replaced by $\frac{\partial f}{\partial y}(t_n, y_n)$. This process is called 'simplified Newton iteration'. The simplified Newton iteration for (7) leads (9) to the formula

$$(I_{sd} - hA \otimes J) \Delta Z^k = -Z^k + h(A \otimes I_d) F(Z^k),$$
$$Z^{k+1} = Z^k + \Delta Z^k, \tag{9}$$

where $J := \frac{\partial f}{\partial y}(t_n, y_n)$. Each iteration requires s times evaluation of f and the calculations of a $d \cdot s$-dimensional linear system.

Note that, by using the simplified Newton iteration, the matrix $(I - hA \otimes J)$ is the same for all iterations, so the decomposition method for solving the resulting linear system can be needed only once. For the linear system, we consider an eigenvalue decomposition technique in that it decomposes the given $d \cdot s$ dimensional linear system into several s-dimensional linear systems. In the view of computational efficiency, it is more efficient to calculate several small size systems even if it is a complex system, rather than to calculate one big size system. Note that only a simplified Newton iteration (9) enables usage of eigenvalue decompositions that cannot be applicable to traditional Newton iteration (8). The eigenvalue decomposition method for (9) is proposed independently by Butcher [31] and Bickart [30]. The main ideas of the method are eigenvalue decomposition of the matrix $A^{-1} = T \Lambda T^{-1}$ and linear transformation of the vector Z^k. By transforming $W^k = (T^{-1} \otimes I) Z^k$, the iteration (9) becomes equivalent to

$$(h^{-1} \Lambda \otimes I_d - I_3 \otimes J) \Delta W^k = -h^{-1}(\Lambda \otimes I_d) W^k + (T^{-1} \otimes I_d) F\Big((T \otimes I) W^k\Big),$$
$$W^{k+1} = W^k + \Delta W^k. \tag{10}$$

In a general case of the three-stage implicit RK method such as (5), the inverse matrix of A has an eigenvalue decomposition as follows:

$$A^{-1} = T\Lambda T^{-1} = \begin{bmatrix} u_0, & u_1, & -v_1 \end{bmatrix} \begin{bmatrix} \hat{\gamma} & 0 & 0 \\ 0 & \hat{\alpha} & -\hat{\beta} \\ 0 & \hat{\beta} & \hat{\alpha} \end{bmatrix} \begin{bmatrix} u_0, & u_1, & -v_1 \end{bmatrix}^{-1}, \quad (11)$$

where $\hat{\gamma}$ is one real real eigenvalue, $\hat{\alpha} \pm i\hat{\beta}$ are one complex eigenvalue pair and u_0, and $u_1 \pm v_1$ are eigenvectors corresponding to $\hat{\gamma}, \hat{\alpha} \pm i\hat{\beta}$, respectively. Therefore, the matrix in (10) can be rewritten as

$$\begin{bmatrix} \gamma I_d - J & 0 & 0 \\ 0 & \alpha I_d - J & -\beta I_d \\ 0 & \beta I_d & \alpha I_d - J \end{bmatrix} \quad (12)$$

with $\gamma = \hat{\gamma}/h$, $\alpha = \hat{\alpha}/h$, $\beta = \hat{\beta}/h$ so that (10) can be split into two linear systems of dimension d and $2d$, respectively. Moreover, the $2d$-dimensional real valued subsystem can be transformed to the following d-dimensional complex valued system

$$((\alpha + i\beta)I - J)(u + iv) = a + ib. \quad (13)$$

In terms of computational cost, the number of multiplication to solve (13) is approximately $4d^3/3$, since the complex multiplication consists of four real multiplications. Then, the total multiplication number for (12) is about $5d^3/3$, while the number of multiplications for decomposing the untransformed matrix $(I - hA \otimes J)$ in (9) is about $(3d)^3/3$. Thus, we can reduce the number of multiplications to about 80% by calculating (12) instead of directly calculating the inverse of the matrix of $(I - hA \otimes J)$ in (10). Finally, to solve the transformations $Z^k = (T \otimes I)W^k$, it additionally requires a multiplication of $\mathcal{O}(n)$. This difference becomes more apparent as the size of the matrix (or the numbers of stage) increases.

3. Numerical Comparison

In this section, we experiment five commonly used physical examples for comparison of both methods. In Sections 3.1–3.3, the BDF3 method (3) and RK3 (4) with its butcher table (5) are used as an example of multi-step and multi-stage methods, respectively. The initial guess for BDF3 is taken by exact values. Both methods use the traditional Newton iteration for solving nonlinear systems. In Section 3.3, especially, we measure CPU-time to compare the two methods in terms of accuracy and efficiency and simplified Newton iteration is used for a nonlinear solver. In Sections 3.4–3.5, we use RADAU5 and ODE15s representing a multi-stage and a multi-step method, respectively, which numerical codes are well optimized and open-source. Note that RADAU5, one of multi-stage methods, has convergence order 5 and stage order 3 [38] and ODE15s, one of multi-step methods, included MATLAB library, has variable orders from 1 to 5 [42]. Remarkably, RADAU5 has applied the eigenvalue decomposition and simplified Newton iteration. All numerical simulations are executed with the software MATLAB 2010b (Mathworks, Natick, MA, USA) under OS WINDOWS 7 (Microsoft, Redmond, WA, USA). Note that most numerical results in this section are repeatable even if different computational resources are used.

3.1. Simple Linear ODE

As the first example, we consider the Prothero–Robinson problem [29],

$$f(t, y(t)) = v(y(t) - g(t)) + g'(t), \quad t \in (0, 10], \quad y(0) = g(0), \quad (14)$$

which presents a stiffness by varying the parameter ν. The analytic solution of problem is given by $y(t) = g(t)$. To compare the error behaviors of the two methods, we set up the parameter $\nu = -1.0 \times 10^6$ so that the given problem can be highly stiff. Here, the exact solution of this problem is set by $g(t) = \sin(t)$. In Figure 1, we display absolute errors $|y(t_i) - y_i|$ at each integration step in a log scale obtained by the two methods with different time step sizes $h = 2^{-k}$, (a) $k = 1$, (b) $k = 2$ and (c) $k = 3$. One can see that the error of BDF3 (Red) has magnitude (a) 1.0×10^{-7}, (b) 1.0×10^{-8}, and (c) 1.0×10^{-9}. The error of RK3 (Blue) has magnitude (a) 1.0×10^{-9}, (b) 1.0×10^{-10}, and (c) 1.0×10^{-11}. All three graphs in Figure 1 show that RK3 has better accuracy than BDF3. Additionally, to demonstrate the meaning of the stage of multi-stage methods and the step of multi-step methods, we set up a time step size of the multi-step method, BDF3, as $\tilde{h} = h/3$. The result of BDF3 with $\tilde{h} = h/3$ is labeled as BDF3c hereafter. It can be seen that the result from BDF3c (Black) has the same accuracy, compared with RK3. Therefore, it is sufficient to see a comparison of RK3 and BDF3c for further comparison.

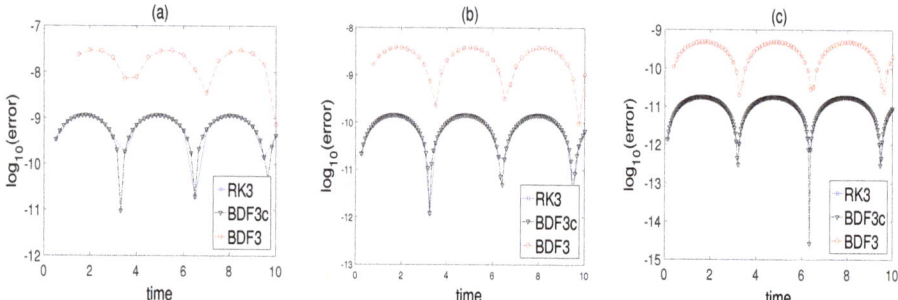

Figure 1. Prothero–Robinson equation: comparing two methods for accuracy by varying step size $h = 2^{-k}$, for (a) $k = 1$, (b) $k = 2$, (c) $k = 3$.

3.2. Nonlinear Stiff ODE System: Multi-Mode Problem

As the second example, we consider a nonlinear ODE system based on the Prothero–Robinson problem. The system is given by

$$f(t, Y(t)) = -\Lambda(Y(t) - g(t) \cdot \mathbf{1}_N)^\delta + g'(t) \cdot \mathbf{1}_N, \quad t \in (0, 10],$$
$$Y(0) = (0, \ldots, 0)^T \in \mathbb{R}^N, \tag{15}$$

where $g(t) = \sin(t)$, $\mathbf{1}_N = (1, \ldots, 1)^T \in \mathbb{R}^N$ and N is the number of dimension. The exact solution is $Y(t) = \sin(t) \cdot \mathbf{1}_N$. The stiffness of (15) can be controlled by the eigenvalues of the matrix Λ, where Λ is diagonal matrix that has elements $\lambda_i = 1.0e + k_i$ $(i = 1, \ldots, N)$, k_i is random integer between 0 and 6. In addition, a linearity of the problem depends on the parameter δ. In this experiment, $\delta = 1$ and $\delta = 5$ are taken for linear and nonlinear cases, respectively. The parameter set $(N, h) = (100, 2^{-3})$ is used for both linear and nonlinear cases.

As similar to the previous subsection, the error behaviors of two methods for both linear and nonlinear cases are observed over time, and the results are plotted in Figure 2. The error is measured as L_∞-norm at each integration step, $||Y(t_i) - Y_i||_\infty$. For the nonlinear case, a traditional Newton iteration is used for a linearization. As mentioned in the previous subsection, BDF3c uses a smaller time step size $\tilde{h} = h/3$ and is compared with RK3 with time step h. Just in case, we mention that BDF3 with time step h is not appropriate to compare RK3 with the same time step size because of the meaning of the stage, explained in the previous subsection. In the linear case, $\delta = 1$, RK3, and BDF3c have similar error behaviors as $1.0 \times 10^{-5.544}$ and $1.0 \times 10^{-5.253}$ at the final time point, respectively. In the nonlinear case, $\delta = 5$, RK3, and BDF3 also have similar error behaviors as $1.0 \times 10^{-5.378}$ and $1.0 \times 10^{-5.123}$ at the final time, respectively.

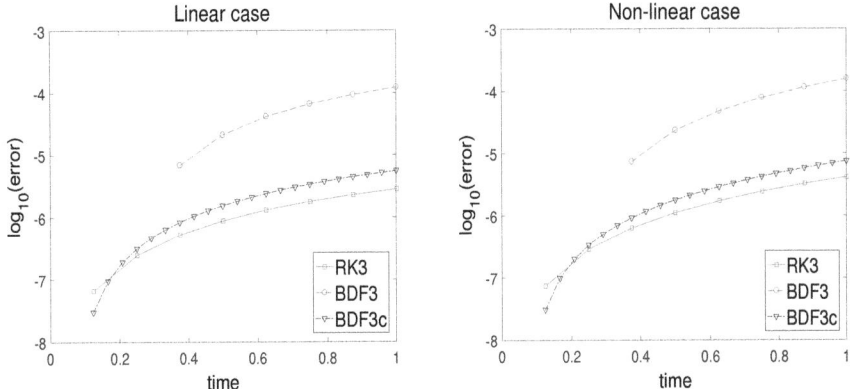

Figure 2. Multi-mode problem: comparing errors of two methods for linear case (**left**) and nonlinear case (**right**).

3.3. Linear PDE—Heat Equation

We consider a linear partial differential equation (PDE), the heat equation generally described by

$$u_t = u_{xx}, \quad (t,x) \in [0,1] \times [0,1] \tag{16}$$

with initial value $u(0,x) = \sin(\pi x) + \frac{1}{2}\sin(3\pi x)$ and boundary conditions $u(t,0) = u(t,1) = 0$. The exact solution is given by $u(x,t) = e^{-\pi^2 t}\sin(\pi x) + \frac{1}{2}e^{-(3\pi)^2 t}\sin(3\pi x)$. This problem is intended to compare two methods for solving big size stiff problems induced from PDE by spatial discretization such as Method of Lines. For the spatial discretization, we use the second-order central difference after evaluating at $x = x_j$ ($x_j = \frac{j}{N}$). Then, the resulting system becomes a N-dimensional system of time dependent ODE. That is, the resulting system can be a big size ODE system depending on the discretization. Note that, to avoid unnecessary computational costs of the multi-stage methods described in the previous sections, we employ the multi-stage methods by combining an efficient linear solver such as an eigenvalue decomposition technique.

To examine the numerical accuracy of two methods for big size stiff systems, we integrate this problem by setting the system size $N = 100$, step size $h = 1/64$ for RK3 and step size $\tilde{h} = 1/192$ for BDF3. For the numerical comparison, we measure L_∞-norm error $Err(t_i) = ||u(x_j, t_i) - u_j^i||_\infty$ in each integration time step where $u_j^i \approx u(x_j, t_i)$. The error behaviors of two methods are plotted in Figure 3, which are measured on a logarithmic scale.

It can be seen that the accuracy of the multi-stage method RK3 with time step h is quite similar to that of the multi-step method BDF3 with time step size $\tilde{h}$. Additionally, to observe of the efficiency for the two methods, CPU-times, and the absolute error are measured at the final time $t = 1$ by varying the resolution of space $N = k \cdot 10^2$ from $k = 1$ to $k = 10$. The results are plotted by absolute error versus CPU-time in Figure 4 with time step sizes $h = 1/100$ and $\tilde{h} = 1/300$.

Figure 4 can be good evidence of the conclusion that RK3 with eigenvalue decomposition technique is more efficient than the BDF3 method. More precisely speaking, BDF3 requires more computational costs to obtain a similar magnitude of accuracy. In addition, RK3 combined with the eigenvalue decomposition technique can obtain higher accuracy for the same cost.

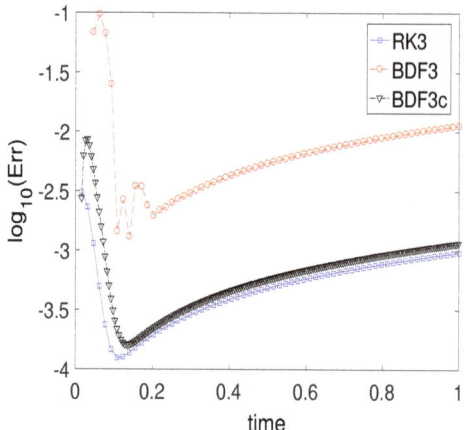

Figure 3. Heat equation: comparing two methods for error behaviors over time.

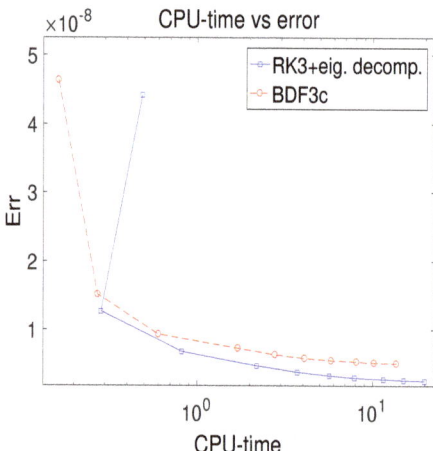

Figure 4. Heat equation: comparing two methods for CPU-time versus error.

3.4. Nonlinear PDE: Medical Akzo Nobel Problem

In this example, we consider one of nonlinear stiff PDE, a reaction-diffusion system with one spatial dimension, described by

$$\begin{cases} u_t = u_{xx} - kuv \\ v_t = -kuv \end{cases} \quad 0 < x < \infty, \quad 0 < t < T, \tag{17}$$

along with the following initial and boundary conditions,

$$u(0, x) = 0, \quad v(0, x) = v_0 \quad \text{for} \quad x > 0,$$

where v_0 is a constant and

$$u(t, 0) = \phi(t) \quad \text{for} \quad 0 < t < T.$$

Semi-discretization of this system yields the nonlinear stiff ODE given by

$$\frac{dy}{dt} = f(t,y), \quad y(0) = g, \quad y \in \mathbb{R}^{2N}, \quad 0 \leq t \leq 20. \qquad (18)$$

The function f is given by

$$f_{2j-1} = \alpha_j \frac{y_{2j+1} - y_{2j-3}}{2\Delta\zeta} + \beta_j \frac{y_{2j-3} - 2y_{2j-1} + y_{2j+1}}{(\Delta\zeta)^2} - ky_{2j-1}y_{2j},$$

$$f_{2j} = -ky_{2j}y_{2j-1},$$

where

$$\alpha_j = \frac{2(j\Delta\zeta - 1)^3}{c^2}, \quad \beta_j = \frac{(j\Delta\zeta - 1)^4}{c^2}, \quad j = 1, \ldots, N$$

with $\Delta\zeta = \frac{1}{N}$, $y_{-1}(t) = \phi(t)$, $y_{2N+1} = y_{2N-1}$ and

$$g = (0, v_0, 0, v_0, \ldots, 0, v_0)^T \in \mathbb{R}^{2N}.$$

The function ϕ is given by

$$\phi(t) = \begin{cases} 2 & \text{for } t \in (0,5], \\ 0 & \text{for } t \in (5, 20]. \end{cases}$$

The parameters k, v_0, and c are set to 100, 1, and 4, respectively. The integer N can be decided by the user. In this experiment, we set N as 200. Since analytic solutions are unavailable, we use reference solutions listed in Table 1 excerpted from [43].

Table 1. Reference solutions for Medical Akzo Nobel problem at the end of the integration interval.

	Reference Solution		Reference Solution
y_{79}	$0.2339942217046434 \times 10^{-3}$	y_{80}	$-0.2339942217046434 \times 10^{-141}$
y_{149}	$0.3595616017506735 \times 10^{-3}$	y_{150}	$0.1649638439865233 \times 10^{-86}$
y_{199}	$0.11737412926802 \times 10^{-3}$	y_{200}	$0.61908071460151 \times 10^{-5}$
y_{239}	$0.68600948191191 \times 10^{-11}$	y_{240}	0.99999973258552

Specifically, results independent of computational resources were measured to compare the efficiency of two methods in this example. The number of times that nonlinear solvers are called (nsolve) and the number of function evaluations (nfeval) are measured by varying relative tolerance (Rtol) and absolute tolerance (Atol) as $(Rtol, Atol) = (10^{-n}, 10^{-n-2})$ $(n = 4, \ldots, 11)$. We also measure an L_∞-norm error at the end time for each tolerance and plot the error in a logarithm scale as a function of nsolve (left) and nfeval (right) in Figure 5. These figures show that RADAU5 generates smaller errors, compared with ODE15s, for paying a similar computational expenses. From a different perspective, RADAU5 requires less computational resources than ODE15s to get similar level of errors. Thus, we can claim that RADAU5 has better performance in terms of computational costs and accuracy than ODE15s.

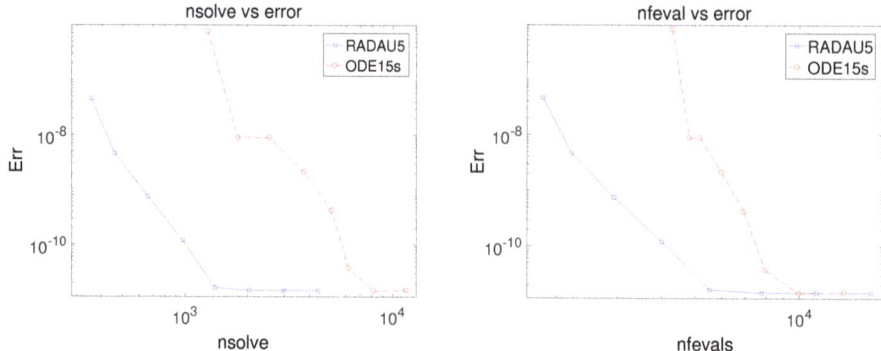

Figure 5. Medical Akzo Nobel problem: nsolve versus error (**left**) and nfeval versus error (**right**).

3.5. Kepler Problem

In this subsection, we consider a two-body Kepler's problem to examine two conservation properties—the Hamiltonian energy and angular momentum—which are indispensable factors in physics. The Kepler's problem describes the Newton's law of gravity revolving around their center of mass placed at the origin in elliptic orbits in the (q_1, q_2)-plan [44]. The equations with unitary masses and gravitational constant are defined by

$$\begin{cases} p_1'(t) = -q_1(q_1^2 + q_2^2)^{(-3/2)}, \\ p_2'(t) = -q_2(q_1^2 + q_2^2)^{(-3/2)}, \\ q_1'(t) = p_1, \\ q_2'(t) = p_2, \end{cases} \qquad (19)$$

with initial conditions $p_1(0) = 0$, $p_2(0) = 2$, $q_1(0) = 0.4$, and $p_1(0) = 0$ on the interval $[0, 100\pi]$. The dynamics are described by Hamiltonian function given by

$$H(p_1, p_2, q_1, q_2) = \frac{1}{2}(p_1^2 + p_2^2) - \frac{1}{\sqrt{q_1^2 + q_2^2}}$$

together with angular momentum L given by

$$L(p_1, p_2, q_1, q_2) = q_1 p_2 - q_2 p_1.$$

The initial Hamiltonian and the initial angular momentum conditions are $H_0 = -0.5$ and $L_0 = 0.8$, respectively.

The conservation properties for the Hamiltonian energy H and angular momentum L are investigated by simulating with the two methods, RADAU5 and ODE15s, with time step size $h = 0.1$ and plot the results in Figure 6. As shown in Figure 6, RADAU5 can conserve both quantities, whereas ODE15s loses the properties as time is going on.

Next, we also consider the movement of comet in planar regulated three-body problem of Sun–Jupiter–Comet. To investigate conservation properties of the two methods, we measure the Hamiltonian energy K and the angular momentum D for the three-body Kepler problem described by

$$x''(t) = \nu \frac{x_S - x}{r_{13}^3} + \mu \frac{x_J - x}{r_{23}^3}, \quad y''(t) = \nu \frac{y_S - y}{r_{13}^3} + \mu \frac{y_J - y}{r_{23}^3}, \qquad (20)$$

where
$$r_{13}^2 = (x_S - x)^2 + (y_S - y)^2, \quad r_{23}^2 = (x_J - x)^2 + (y_J - y)^2,$$
$$x_S = -\mu \cos(t - t_0), \quad y_S = -\mu \sin(t - t_0),$$
$$x_J = \nu \cos(t - t_0), \quad y_J = \nu \sin(t - t_0).$$

The energy and angular momentum of the comet

$$K/2 = \frac{1}{2}(\dot{x}^2 + \dot{y}^2) - \frac{1}{\sqrt{x^2 + y^2}}, \quad D = x\dot{y} - y\dot{x}$$

are constant, when $\mu = 0$ and $\nu = 1$. For this experiment, initial condition is set to

$$x(0) = 5, \quad y(0) = 1, \quad x'(0) = 0, \quad y'(0) = 1,$$

and the initial energy and angular momentum are set to $K_0/2 = 0.3$ and $D_0 = 5$ with parameter step size $h = 1/2\pi$ and $t_0 = 0$.

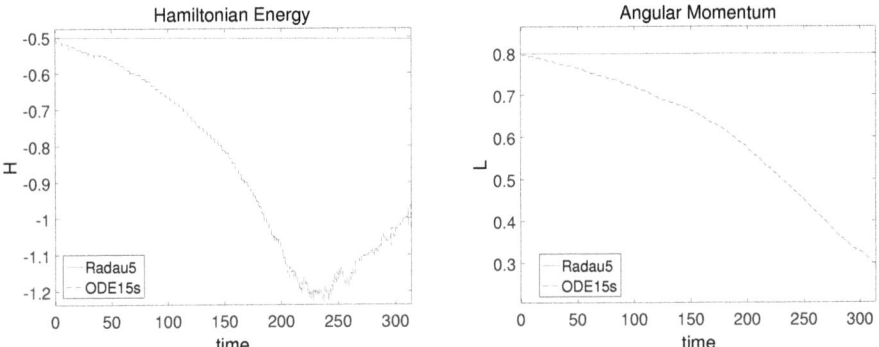

Figure 6. Two-body Kepler problem: comparing two methods in terms of conservation of Hamiltonian and angular momentum.

In Figure 7, one can see that the behaviors of the energy and the momentum over time interval $[0, 100\pi]$. As observed in Figure 7, RADAU5 gives a maximum variation of 8.0356×10^{-5} and 0.0013 for the energy and the momentum, whereas ODE15s presents a variation of 5.9063×10^{-4} and 0.0173 for them. Therefore, one can conclude that RADAU5 has better conservation properties, compared with ODE15s.

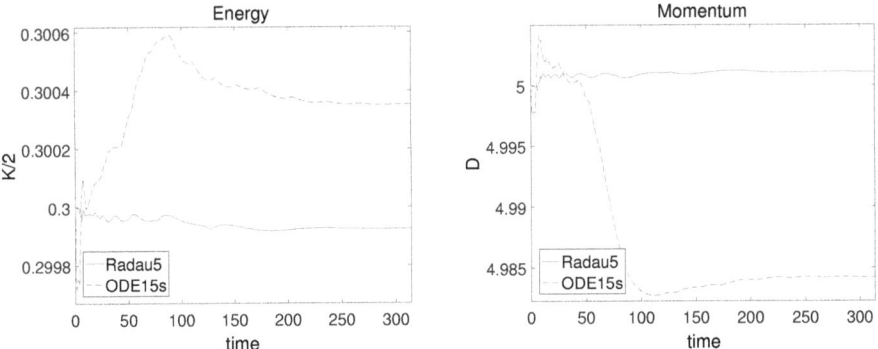

Figure 7. Three-body Kepler problem: comparing two methods in terms of the conservation of total energy and angular momentum.

4. Conclusions and Further Discussion

In this work, we compare multi-stage methods with multi-step methods by investigating the numerical properties of both methods. In a classical approach, nonlinear stiff systems were usually solved by multi-step methods to avoid huge computational complexity induced from linearization of a given nonlinear system. However, the computational costs for the multi-stage method can be reduced sufficiently without loss of stability and conservation, which is possible by using suitable nonlinear and linear solvers such as a Newton-type method and eigenvalue decomposition. It means that the multi-stage method can also be applied to solve nonlinear stiff systems without any damage to computational costs, compared with the multi-step methods. Moreover, it is seen that the multi-stage methods preserve the invariants of the energy and angular momentum in Hamiltonian systems. In addition, it is well-known that a stability property of multi-stage methods is much better than that of multi-step methods.

Overall, one can conclude that the multi-stage method can be a good candidate to solve nonlinear stiff systems. It means that, without any damage to computational costs, multi-stage methods can be applied to long-time simulations and massive physical simulations in fields such as astronomy, meteorology, nuclear fusion, nuclear power, aerospace, machinery, etc.

Author Contributions: Conceptualization, Y.J. and S.B. (Sunyoung Bu); methodology, S.B. (Sunyoung Bu); software, Y.J.; validation, Y.J. and S.B. (Soyoon Bak); formal analysis, Y.J.; investigation, Y.J. and S.B. (Soyoon Bak); resources, Y.J and S.B. (Soyoon Bak); data curation, Y.J.; writing—original draft preparation, S.B. (Sunyoung Bu); writing—review and editing, S.B. (Soyoon Bak); visualization, Y.J.; supervision, S.B. (Sunyoung Bu); project administration, S.B. (Sunyoung Bu); funding acquisition, S.B. (Sunyoung Bu).

Funding: This was supported by the National Research Foundation of Korea (NRF) grant funded by the Korea government (MSIT) (Grant No.: NRF-2019R1H1A2079997) and the R&D programs through NFRI (National Fusion Research Institute) funded by the Ministry of Science and ICT of the Republic of Korea (Grant No. NFRI-EN1841-4). In addition, the corresponding author Sunyoung Bu was partly supported by the National Research Foundation of Korea (NRF) grant funded by the Korea government (MSIT) (Grant No.: NRF-2019R1F1A1058378).

Conflicts of Interest: The authors declare no conflict of interest.

References

1. Atkinson, K.E. Divisions of numerical methods for ordinary differential equations. In *An Introduction to Numerical Analysis*, 2nd ed.; John Wiley & Sons, Inc.: Hoboken, NJ, USA, 1989; pp. 333–462.
2. Gear, C.W. *Numerical Initial Value Problems in Ordinary Differential Equations*; Prentice Hall: Upper Saddle River, NJ, USA, 1971.
3. Hairer, E.; Nørsett, S.P.; Wanner, G. *Solving Ordinary Differential Equations I: Nonstiff Problems*; Springer: Berlin/Heidelberg, Germany; New York, NY, USA, 1996.
4. Kirchgraber, U. Multi-step Method Are Essentially One-step Methods. *Numer. Math.* **1986**, *48*, 85–90. [CrossRef]
5. Alolyan, I.; Simos, T.E. New multiple stages multistep method with best possible phase properties for second order initial/boundary value problems. *J. Math. Chem.* **2019**, *57*, 834–857. [CrossRef]
6. Berg, D.B.; Simos, T.E. Three stages symmetric six-step method with eliminated phase-lag and its derivatives for the solution of the Schrödinger equation. *J. Chem. Phys.* **2017**, *55*, 1213–1235. [CrossRef]
7. Butcher, J.C. *Numerical Analysis of Ordinary Differential Equations: Runge–Kutta and General Linear Methods*; John Wiley & Sons, Inc.: Hoboken, NJ, USA, 1987.
8. Enright, W.; Hull, T.; Lindberg, B. Comparing numerical methods for stiff systems of O.D.E.'s. *BIT Numer. Math.* **1975**, *15*, 10–48. [CrossRef]
9. Fathoni, M.F.; Wuryandari, A.I. Comparison between Euler, Heun, Runge–Kutta and Adams-Bashforth Moulton integration methods in the particle dynamic simulation. In Proceedings of the 2015 4th International Conference on Interactive Digital Media (ICIDM), Bandung, Indonesia, 1–5 December 2015; pp. 1–7.
10. Liniger, W.; Willoughby, R.A. Efficient Integration Methods for Stiff Systems of Ordinary Differential Equations. *SIAM J. Numer. Anal.* **1970**, *7*, 47–66. [CrossRef]
11. Song, X. Parallel Multi-stage and Multi-step Method in ODEs. *J. Comput. Math.* **2000**, *18*, 157–164.

12. Barrio1, M.; Burrage, K.; Burrage, P. Stochastic linear multistep methods for the simulation of chemical kinetics. *J. Chem. Phys.* **2015**, *142*, 064101. [CrossRef]
13. Bu, S. New construction of higher-order local continuous platforms for Error Correction Methods. *J. Appl. Anal. Comput.* **2016**, *6*, 443–462.
14. Cohen, E.B. Analysis of a Class of Multi-stage, Multi-step Runge Kutta Methods. *Comput. Math. Appl.* **1994**, *27*, 103–116. [CrossRef]
15. Guo, L.; Zeng, F.; Turner, I.; Burrage, K.; Karniadakis, G.E. Efficient multistep methods for tempered fractional calculus: Algorithms and Simulations. *SIAM J. Sci. Comput.* **2019**, *41*, A2510–A2535. [CrossRef]
16. Han, T.M.; Han, Y. Solving Implicit Equations Arising from Adams-Moulton Methods. *BIT Numer. Math.* **2002**, *42*, 336–350. [CrossRef]
17. Ghawadri, N.; Senu, N.; Fawzi, F.A.; Ismail, F.; Ibrahim, Z.B. Explicit Integrator of Runge–Kutta Type for Direct Solution of u(4) = f(x, u, ú, ü). *Symmetry* **2019**, *10*, 246. [CrossRef]
18. Kim, S.D.; Kwon, J.; Piao, X.; Kim, P. A Chebyshev Collocation Method for Stiff Initial Value Problems and Its Stability. *Kyungpook Math. J.* **2011**, *51*, 435–456. [CrossRef]
19. Kim, P.; Piao, X.; Jung, W.; Bu, S. A new approach to estimating a numerical solution in the error embedded correction framework. *Adv. Differ. Equ.* **2018**, *68*, 1–21. [CrossRef]
20. Kim, P.; Kim, J.; Jung, W.; Bu, S. An Error Embedded Method Based on Generalized Chebyshev Polynomials. *J. Comput. Phys.* **2016**, *306*, 55–72. [CrossRef]
21. Piao, X.; Bu, S.; Kim, D.; Kim, P. An embedded formula of the Chebyshev collocation method for stiff problems. *J. Comput. Phys.* **2017**, *351*, 376–391. [CrossRef]
22. Xia, K.; Cong, Y.; Sun, G. Symplectic Runge–Kutta methods of high order based on W-transformation. *J. Appl. Anal. Comput.* **2001**, *3*, 1185–1199.
23. Marin, M. Effect of microtemperatures for micropolar thermoelastic bodies. *Struct. Eng. Mech.* **2017**, *61*, 381–387. [CrossRef]
24. Marin, M.; Abd-Alla, A.; Raducanu, D.; Abo-Dahab, S. Structural Continuous Dependence in Micropolar Porous Bodies. *Comput. Mater. Contin.* **2015**, *45*, 107–125.
25. Bak, S. High-order characteristic-tracking strategy for simulation of a nonlinear advection-diffusion equations. *Numer. Methods Partial Differ. Equ.* **2019**, *35*, 1756–1776. [CrossRef]
26. Dahlquist, G. Numerical integration of ordinary differential equations. *Math. Scand.* **1956**, *4*, 33–50. [CrossRef]
27. Pazner, W.; Persson, P. Stage-parallel fully implicit Runge–Kutta solvers for discontinuous Galerkin fluid simulationse. *J. Comput. Phys.* **2017**, *335*, 700–717. [CrossRef]
28. Piao, X.; Kim, P.; Kim, D. One-step L (α)-stable temporal integration for the backward semi-Lagrangian method and its application in guiding center problems. *J. Comput. Phys.* **2018**, *366*, 327–340. [CrossRef]
29. Prothero, A.; Robinson, A. On the Stability and Accuracy of One-step Methods for Solving Stiff Systems of Ordinary Differential Equations. *Math. Comput.* **1974**, *28*, 145–162. [CrossRef]
30. Bickart, T.A. An Efficient Solution Process for Implicit Runge–Kutta Methods. *SIAM J. Numer. Anal.* **1977**, *14*, 1022–1027. [CrossRef]
31. Butcher, J.C. On the Implementation of Implicit Runge–Kutta Methods. *BIT Numer. Math.* **1976**, *16*, 237–240. [CrossRef]
32. Curtiss, C.F.; Hirschfelder, J.O. Integration of Stiff Equations. *Proc. Natl. Acad. Sci. USA* **1952**, *38*, 235–243. [CrossRef]
33. Jeon, Y.; Bu, S.; Bak, S. A comparison of multi-Step and multi-stage methods. *Int. J. Circuits Signal Process.* **2017**, *11*, 250–253.
34. Coper, G.J.; Butcher, J.C. An iteration method for implicit Runge–Kutta methods. *IMA J. Numer. Anal.* **1983**, *3*, 127–140. [CrossRef]
35. Cooper, G.J.; Vignesvaran, R. Some methods for the implementation of implicit Runge–Kutta methods. *J. Comput. Appl. Math.* **1993**, *45*, 213–225. [CrossRef]
36. Frank, R.; Ueberhuber, C. W. Iterated defect correction for the efficient solution of stiff systems of ordinary differential equations. *BIT Numer. Math.* **1977**, *17*, 146–159. [CrossRef]
37. Gonzalez-Pinto, S.; Rojas-Bello, R. Speeding up Newton-type iterations for stiff problems. *J. Comput. Appl. Math.* **2005**, *181*, 266–279. [CrossRef]

38. Hairer, E.; Wanner, G. *Solving Ordinary Differential Equations II: Stiff and Differential Algebraic Problems*; Springer: Berlin/Heidelberg, Germany; New York, NY, USA, 1996.
39. Huang, J.; Jia, J.; Minion, M.L. Accelerating the convergence of spectral deferred correction methods. *J. Comput. Phys.* **2006**, *214*, 633–656. [CrossRef]
40. Süli, E.; Mayers, D.F. *An Introduction to Numerical Analysis*; Cambridge University Press: Cambridge, UK, 2003.
41. Dahlquist, G. A special stability problem for linear multistep methods. *BIT Numer. Math.* **1963**, *3*, 27–43 [CrossRef]
42. Shampine, L.F.; Reichelt, M.W. The MATLAB ODE Suite. *SIAM J. Sci. Comput.* **1997**, *18*, 1–22. [CrossRef]
43. Mazzia, F.; Magherini, C. *Test Set for Initial Value Problem Solvers*; Department of Mathematics, University of Bari: Bari, Italy, 2008.
44. Brugnano, L.; Iavernaro, F.; Trigiante, D. Energy- and Quadratic Invariants–Preserving Integrators Based upon Gauss Collocation Formulae. *SIAM J. Numer. Anal.* **2012**, *50*, 2897–2916. [CrossRef]

© 2019 by the authors. Licensee MDPI, Basel, Switzerland. This article is an open access article distributed under the terms and conditions of the Creative Commons Attribution (CC BY) license (http://creativecommons.org/licenses/by/4.0/).

Article

Second Order Semilinear Volterra-Type Integro-Differential Equations with Non-Instantaneous Impulses

Mouffak Benchohra [1,2], Noreddine Rezoug [1], Bessem Samet [2] and Yong Zhou [3,4,*]

[1] Laboratory of Mathematics, Djillali Liabes University of Sidi Bel-Abbes, P.O. Box 89, Sidi Bel Abbes 22000, Algeria; benchohra@yahoo.com (M.B.); noreddinerezoug@yahoo.fr (N.R.)
[2] Department of Mathematics, College of Science, King Saud University, P.O. Box 2455, Riyadh 11451, Saudi Arabia; bsamet@ksu.edu.sa
[3] Faculty of Information Technology, Macau University of Science and Technology, Macau 999078, China
[4] Faculty of Mathematics and Computational Science, Xiangtan University, Xiangtan 411105, China
* Correspondence: yzhou@xtu.edu.cn

Received: 18 August 2019; Accepted: 12 November 2019; Published: 20 November 2019

Abstract: We consider a non-instantaneous system represented by a second order nonlinear differential equation in a Banach space E. We use the family of linear bounded operators introduced by Kozak, Darbo fixed point method and Kuratowski measure of noncompactness. A new set of sufficient conditions is formulated which guarantees the existence of the solution of the non-instantaneous system. An example is also discussed to illustrate the efficiency of the obtained results.

Keywords: second order differential equations; mild solution; non-instantaneous impulses; Kuratowski measure of noncompactness; Darbo fixed point

1. Introduction

The aim of this paper is to establish a result of the existence of mild solution for a class of the non-autonomous second order nonlinear differential equation with non-instantaneous impulses described in the form

$$\begin{cases} y''(t) = A(t)y(t) + f\left(t, y(t), \int_0^t g(t,s,y(s))ds\right), & t \in (s_i, t_{i+1}], i = 0, \cdots, N, \\ y(t) = \gamma_i(t, y(t_i^-)), & t \in (t_i, s_i], \quad i = 1, \cdots, N, \\ y'(t) = \zeta_i(t, y(t_i^-)), & t \in (t_i, s_i], \quad i = 1, \cdots, N, \\ y(0) = y_0, \ y'(0) = y_1, \end{cases} \quad (1)$$

In this text, E is a reflexive Banach space endowed with a norm $|\cdot|$, $J = [0, a]$, $0 = s_0 < t_1 < s_1 < t_2, \cdots, t_N < s_N < t_{N+1} = a < \infty$. We consider in problem (1) that $y \in C((s_i, t_{i+1}), E)$, $i = 0, 1, \cdots, N$. The functions $\gamma_i(t, y(t_i^-))$ and $\zeta_i(t, y(t_i^-))$ represent noninstantaneous impulses during the intervals $(t_i, s_i]$, $i = 1, \cdots, N$, so impulses at t_i^- have some duration, namely on intervals $(t_i, s_i]$. Further, $A(t) : D(A(t)) \subset E \to E$ is a closed linear operator which generates a evolution system $\{S(t,s)\}_{(t,s) \in D}$ of linear bounded operators, $f : J \times E \times E \to E$, $g \in C(D \times E, E)$, $D = \{(t,s) \in J \times J : s \leq t\}$ and y_0, y_1 are given elements of E.

The theory and application of integrodifferential equations are important subjects in applied mathematics, see, for example [1–8] and recent development of the topic, see the monographs of [9]. In recent times there have been an increasing interest in studying the abstract autonomous second order, see for example [10–14]. Useful for the study of abstract second order equations is the existence of an evolution system $S(t,s)$ for the homogenous equation

$$y''(t) = A(t)y(t), \text{ for } t \geq 0. \tag{2}$$

For this purpose there are many techniques to show the existence of $S(t,s)$ which has been developed by Kozak [15]. In many problems, such as the transverse motion of an extensible beam, the vibration of hinged bars and many other physical phenomena, we deal with the second-order abstract differential equations in the infinite dimensional spaces. On the other hand, recently there exists an extensive literature for the non-autonomous second order see, for example, [16–22].

The dynamics of many evolving processes are subject to abrupt changes such as shocks, harvesting, and natural disaster. These phenomena involve short term perturbations from continuous and smooth dynamics, whose duration is negligible in comparison with the duration of an entire evolution. Particularly, the theory of instantaneous impulsive equations have wide applications in control, mechanics, electrical engineering, biological and medical fields. Recently, Hernandez et al. [23] use first time not instantaneous impulsive condition for semi-linear abstract differential equation of the form

$$\begin{cases} y'(t) = Ay(t) + f(t,y(t)), & t \in (s_i, t_{i+1}], i = 0, \cdots, N, \\ y(t) = g_i(t,y(t)), & t \in (t_i, s_i], i = 1, \cdots, N, \\ y(0) = y_0, \end{cases} \tag{3}$$

and introduced the concepts of mild and classical solution. Wang and Fečkan have changed the conditions $y(t) = g_i(t,y(t))$ in (3) as follows

$$y(t) = g_i(t,y(t_i^+)), \qquad t \in (t_i, s_i], i = 1, \cdots, N.$$

Of course then $y(t_i^+) = g_i(t, y(t_i^-))$, where $y(t_i^+)$ and $y(t_i^-)$ represent respectively the right and left limits of $y(t)$ at $t = t_i$. Motivated by above remark, Wang and Fečkan [24] have shown existence, uniqueness and stability of solutions of such general class of impulsive differential equations. To learn more about this kind of problems, we refer [25–34].

To deal with the above mentioned issues, we investigate necessary and sufficient conditions for the existence of a mild solution of system (1). By virtue of the theory of measure of noncompactness associated with Darbo's and Darbo-Sadovskii's fixed point theorem. This technique was considered by Banas and Goebel [35] and subsequently used in many papers; see, for example, [33,36–39].

A brief outline of this paper is given:. Some preliminaries are presented in Section 2. Section 3, we obtain necessary and sufficient conditions for System (1). An Appropriate example is given to illustrate our results.

2. Basic Definitions and Preliminaries

In this section, we review some basic concepts, notations, and properties needed to establish our main results.

Denote by $C(J,E)$ the space of all continuous E-valued functions on interval J which is a Banach space with the norm

$$\|y\| = \sup_{t \in J} |y(t)|.$$

To treat the impulsive conditions, we define the space of piecewise continuous functions

$$PC(J,E) = \{y : J \to E : y \in C([0,t_1] \cup (t_k, s_k] \cup (s_k, t_{k+1}], E), k = 1, \ldots, N$$
$$\text{and there exist } y(t_k^-), y(t_k^+), y(s_k^-) \text{ and } y(s_k^+) k = 1, \ldots, N \text{ with } y(t_k^-) = y(t_k)$$
$$\text{and } y(s_k^-) = y(s_k)\}.$$

It can be easily proved that $PC(J,E)$ is a Banach space endowed with

$$\|y\|_{PC} = \sup_{t \in J} |y(t)|.$$

For a positive number R, let

$$B_R = \{y \in PC(J, E) : \|y\|_{PC} \leq R\}.$$

be a bounded set in $PC(J, E)$.

$L^r(J, E)$ denotes the space of E-valued Bochner functions on $[0, a]$ with the norm

$$\|y\|_{L^r} = \left(\int_0^a |y(t)|^r dt \right)^{\frac{1}{r}}, \quad r \geq 1.$$

$B(E)$ the Banach space of bounded linear operators from E into E.

First we recall the concept of the evolution operator $S(t, s)$ for problem (2), introduced by Kozak in [15] and recently used by Henríquez, Poblete and Pozo in [20].

Definition 1. *Let $S : D \to B(E)$. The family is said to be an evolution operator generated by the family $\{A(t) : t \in J\}$ if the following conditions are satisfied [15]:*

(e_1) *For each $y \in E$ the function $S(\cdot, \cdot)y : J \times J \to E$ is of class C^1 and*

(i) *for each $t \in J$, $S(t, t) = 0$,*
(ii) *for all $(t, s) \in D$ and for each $y \in E$,*

$$\frac{\partial}{\partial t} S(t,s)y \Big|_{t=s} = y, \quad \frac{\partial}{\partial s} S(t,s)y \Big|_{t=s} = -y.$$

(e_2) *For each $(t, s) \in D$, if $y \in D(A(t))$, then $\frac{\partial}{\partial s} S(t,s)y \in D(A(t))$, the map $(t, s) \mapsto S(t,s)y$ is of class C^2 and*

(i) $\frac{\partial^2}{\partial t^2} S(t,s)y = A(t)S(t,s)y$,
(ii) $\frac{\partial^2}{\partial s^2} S(t,s)y = S(t,s)A(s)y$,
(iii) $\frac{\partial^2}{\partial s \partial t} S(t,s)y \Big|_{t=s} = 0$.

(e_3) *For all $(t, s) \in D$, if $y \in D(A(t))$, then $\frac{\partial}{\partial s} S(t,s)y \in D(A(t))$. Moreover, there exist $\frac{\partial^3}{\partial t^2 \partial s} S(t,s)y$, $\frac{\partial^3}{\partial s^2 \partial t} S(t,s)y$ and*

(i) $\frac{\partial^3}{\partial t^2 \partial s} S(t,s)y = A(t) \frac{\partial}{\partial s} S(t,s)y$,
(ii) $\frac{\partial^3}{\partial s^2 \partial t} S(t,s)y = \frac{\partial}{\partial t} S(t,s)A(s)y$,

and for all $y \in D(A)$ the function $(t,s) \mapsto A(t) \frac{\partial}{\partial s} S(t,s)y$ is continuous in D.

Definition 2. *A function $f : J \times E \times E \to E$ is said to be a Carathéodory function if it satisfies:*

(i) $t \to f(t, u, v)$ *is measurable for each $u, v \in E \times E$,*
(ii) $(u, v) \to f(t, u, v)$ *is continuous for almost each $t \in J$.*

For W, a nonempty subset of E, we denote by $\overline{W}$ and $ConvW$ the closure and the closed convex hull of W, respectively. Finally, the standard algebraic operations on sets are denoted by aW and $Y + W$, respectively. Now, we recall some basic definitions and properties about Kuratowski measure of noncompactness that will be used in the proof of our main results.

Definition 3. [35] *The Kuratowski measure of noncompactness $\alpha_E(\cdot)$ defined on bounded set W of Banach space E is*
$$\alpha_E(W) = \inf\{\varepsilon > 0 : W = \cup_{i=1}^n W_i \text{ and } diam(W_i) \leq \varepsilon \text{ for } i = 1, 2, \cdots n\}.$$

Some basic properties of $\alpha_E(\cdot)$ are given in the following lemma.

Lemma 1. *Let Y and W be bounded sets of E and a be a real number [35]. The Kuratowski measure of noncompactness satisfies some properties:*

- (p_1) *W is pre-compact if and only if $\alpha_E(W) = 0$,*
- (p_2) $\alpha_E(\overline{W}) = \alpha_E(W)$,
- (p_3) $\alpha_E(Y) \leq \alpha_E(W)$ *when* $Y \subset W$,
- (p_4) $\alpha_E(Y + W) \leq \alpha_E(Y) + \alpha_E(W)$,
- (p_5) $\alpha_E(aW) = |a|\alpha_E(W)$ *for any* $a \in \mathbf{R}$,
- (p_6) $\alpha_E(ConvW) = \alpha_E(W)$.

The map $Q : X \subset E \to E$ is said to be a α-contraction if there exists a positive constant $\lambda < 1$ such that $\alpha_E(Q(W)) \leq \lambda \alpha_E(W)$ for any bounded closed subset $W \subset E$.

Lemma 2. [40] *Let E be a Banach space, $W \subset E$ be bounded. Then there exists a countable set $W_0 \subset W$, such that*
$$\alpha_E(W) \leq 2\alpha_E(W_0).$$

Lemma 3. [41] *Let E be a Banach space, $-\infty < a_1 < a_2 < +\infty$ for constants, and let $W = \{y_n\} \subset PC([a_1, a_2], E)$, be a bounded and countable set. Then $\alpha_E(W(t))$ is Lebesgue integral on $[a_1, a_2]$, and*
$$\alpha_E\left(\left\{\int_{a_1}^{a_2} y_n(t)dt : n \in \mathbb{N}\right\}\right) \leq 2 \int_{a_1}^{a_2} \alpha_E(W(t))dt.$$

Denote by α_{PC} the Kuratowski measure of noncompactness of $PC(J, E)$. Before proving the existence results, we need the following Lemmas.

Lemma 4. [35] *If $W \subset PC(J; E)$ is bounded, then $\alpha_E(W(t)) \leq \alpha_{PC}(W)$, for all $t \in J$; here $W(t) = \{y(t); y \in W \subset E\}$. Furthermore if W is equicontinuous on J, then $\alpha_E(W(t))$ is continuous on J and*
$$\alpha_{PC}(W) = \sup_{t \in J} \alpha_E(W(t)).$$

Lemma 5. [42] *Let E, F be Banach spaces. If the map $\Psi : \mathcal{D}(\Psi) \subset E \to F$ is Lipschitz continuous with constant k, then $\alpha_F(\Psi(W)) \leq k\alpha_E(W)$ for any bounded subset $W \subset \mathcal{D}(\Psi)$.*

Theorem 1. (Darbo) [43] *Assume that W is a non-empty, closed and convex subset of a Banach space E and $0 \in W$. Let $Q : W \to W$ be a continuous mapping and α_E-contraction. If the set $\{y \in W : y = \lambda Qy\}$ is bounded for $0 < \lambda < 1$, then the map Q has at least one fixed point in W.*

Theorem 2. (Darbo-Sadovskii) [35] *Assume that W is a non-empty, closed, bounded, and convex subset of a Banach space E. Let $Q : W \to W$ be a continuous mapping and α_E-contraction. Then the map Q has at least one fixed point in W.*

3. Existence Results

In this section, we discuss the existence of mild solutions for system (1). Firstly, let us propose the definition of the mild solution of system (1).

Definition 4. *A function* $y \in PC(J, E)$ *is said to be a mild solution to the system* (1), *if it satisfies the following relations:*

$$y(0) = y_0, \quad y'(0) = y_1,$$

the non-instantaneous conditions

$$y(t) = \gamma_i(t, y(t_i^-)), \quad y'(t) = \zeta_i(t, y(t_i^-)), \quad t \in (t_i, s_i],$$

and y is the solution of the following integral equations

$$y(t) = \begin{cases} -\dfrac{\partial}{\partial s} S(t,0) y_0 + S(t,0) y_1 \\ \quad + \displaystyle\int_0^t S(t,s) f\left(s, y(s), \int_0^s g(s,\tau, y(\tau)) d\tau\right) ds, \quad t \in [0, t_1], \\ -\dfrac{\partial}{\partial s} S(t, s_i) \gamma_i(s_i, y(t_i^-)) + S(t, s_i) \zeta_i(s_i, y(t_i^-)) \\ \quad + \displaystyle\int_{s_i}^t S(t,s) f\left(s, y(s), \int_0^s g(s,\tau, y(\tau)) d\tau\right) ds, \quad t \in (s_i, t_{i+1}]. \end{cases}$$

In this manuscript, we list the following hypotheses:

(H_1) There exist a pair of constants $M \geq 1$ and $\delta > 0$, such that

$$\|S(t,s)\|_{B(E)} \leq M e^{-\delta(t-s)} \text{ for any } (t,s) \in D.$$

(H_2) There exists a constant $\tilde{M} > 0$ such that:

$$\left\|\dfrac{\partial}{\partial s} S(t,s)\right\|_{B(E)} \leq \tilde{M} e^{-\delta(t-s)}, (t,s) \in D.$$

(H_3) $f : J \times E \times E \to E$ is of Carathéodory type and satisfies:

(a) There exist $\Theta_f \in L^r(J, \mathbb{R}^+), r \in [1, \infty)$ and a continuous nondecreasing function $\psi : [0, \infty) \to (0, \infty)$ such that:

$$|f(t, y, z)| \leq \Theta_f(t) \psi(|y| + |z|) \text{ for a.a } t \in J \text{ and each } y, z \in E.$$

(b) There exist integrable functions $\sigma, \varrho : J \to \mathbb{R}^+$, such that:

$$\alpha_E(f(t, W_1, W_2)) \leq \sigma(t) \alpha_E(W_1) + \varrho(t) \alpha_E(W_2)$$

for a.a $t \in J$ and $W_1, W_2 \subset E$.

(H_4) $g : D \times E \to E$ is a continuous function that satisfies:

(a) There exist $\Theta_g \in L^1(J, \mathbb{R}^+)$, and a continuous nondecreasing function $\varphi : [0, \infty) \to (0, \infty)$ such that:

$$|g(t, s, y)| \leq \Theta_g(t) \varphi(|y|) \text{ for a.a } (t,s) \in D \text{ and each } y \in E.$$

(b) There exists constant $K^* > 0$, such that

$$\alpha_E(g(t, s, W)) \leq K^* \alpha_E(W) \text{ for a.a } (t,s) \in D \text{ and } W \subset E.$$

(H_5) The functions $\gamma_i : (t_i, s_i] \times E \to E, i = 1, \cdots, N$, are continuous, and they satisfy the following conditions:

(a) there exist positive constants $c_i, i = 1, \cdots, N$ such that

$$|\gamma_i(t, y_2) - \gamma_i(t, y_1)| \leq c_i |y_2 - y_1| \text{ for a.a } t \in (t_i, s_i] \text{ and each } y_1, y_2 \in E.$$

(b) there exist positive constants d_i, such that

$$d_i = \sup_{t \in [t_i, s_i]} \gamma_i(t, 0).$$

(H_6) The functions $\zeta_i : (t_i, s_i] \times E \to E, i = 1, \cdots, N$, are continuous, and satisfy the following conditions:

(a) There exist constants $e_i, l_i > 0, i = 1, \cdots, N$ such that

$$|\zeta_i(t, y)| \le e_i |y| + l_i \text{ for a.a } t \in (t_i; s_i] \text{ and each } y \in E.$$

(b) There exists constants $\bar{k}_i > 0, i = 1, \cdots, N$ such that

$$\alpha_E(\zeta_i(t, W)) \le \bar{k}_i \alpha_E(W) \text{ for a.a } t \in (t_i, s_i] \text{ and any } W \subset E.$$

(H_7)

$$\max_{1 \le i \le N} (k_i, 1) \left(\max_{1 \le i \le N} (\tilde{M} k_i + M \bar{k}_i) + 2M(\|\sigma\|_{L^1} + 2K^* a \|\varrho\|_{L^1}) \right) < 1.$$

Remark 1. *From Lemma 5 and (H_5), there exist constants $k_i > 0$, such that*

$$\alpha_E(\gamma_i(t, W)) \le k_i \alpha_E(W) \text{ for a.a } t \in (t_i, s_i] \text{ and each } y \in E.$$

Theorem 3. *Under the assumptions (H_1)–(H_7), the system (1) has at least one mild solution on J, provided that*

$$\int_0^a \max(\tilde{M}\Theta_f(s), \Theta_g(s)) ds \le \int_{m_i}^\infty \frac{ds}{\psi(s) + \varphi(s)}, i = 2, 3 \cdots N \tag{4}$$

with

$$\tilde{M} = \max_{2 \le i \le N} \left\{ \frac{M}{1 - L_1}, \frac{M}{1 - L_i}, \frac{Mc_i}{1 - L_{i-1}} \right\},$$

and

$$m_i = \frac{d_i}{1 - L_{i-1}} + \max_{2 \le i \le N} \left\{ \tilde{M} |y_0| + M |y_1|, \frac{\tilde{M} d_1}{1 - L_1} + \frac{M l_1}{1 - L_1}, \frac{\tilde{M} d_i}{1 - L_i} + \frac{M l_i}{1 - L_i}, \frac{\tilde{M} c_i d_{i-1}}{1 - L_{i-1}} + \frac{M c_i l_{i-1}}{1 - L_{i-1}} \right\},$$

where

$$L_i = \tilde{M} c_i + M e_i < 1.$$

Proof. Define the mapping $\Lambda : PC(J, E) \to PC(J, E)$ by

$$(\Lambda y)(t) = \begin{cases} \gamma_i \left(t, -\frac{\partial}{\partial s} S(t, s_i) \gamma_{i-1}(s_{i-1}, y(t_{i-1}^-)) + S(t, s_{i-1}) \zeta_{i-1}(s_{i-1}, y(t_{i-1}^-)) \right. \\ \left. + \int_{s_{i-1}}^{t_i} S(t, s) f \left(s, y(s), \int_0^s g(s, \tau, y(\tau)) d\tau \right) ds \right), & t \in (t_i, s_i], \\ -\frac{\partial}{\partial s} S(t, 0) y_0 + S(t, 0) y_1 \\ + \int_0^t S(t, s) f \left(s, y(s), \int_0^s g(s, \tau, y(\tau)) d\tau \right) ds, & t \in [0, t_1], \\ -\frac{\partial}{\partial s} S(t, s_i) \gamma_i(s_i, y(t_i^-)) + S(t, s_i) \zeta_i(s_i, y(t_i^-)) \\ + \int_{s_i}^t S(t, s) f \left(s, y(s), \int_0^s g(s, \tau, y(\tau)) d\tau \right) ds, & t \in (s_i, t_{i+1}]. \end{cases} \tag{5}$$

It is obvious that the fixed point of Λ is the mild solution of (1). We shall show that Λ satisfies the assumptions of Theorem 1. The proof will be given in four steps.

Step 1. A priori bounds.

Let $\lambda \in (0,1)$ and let $y \in Y$ be a possible solution of $y = \lambda \Lambda(y)$ for some $0 < \lambda < 1$. Thus,

Case 1. For each $t \in [0, t_1]$, we get

$$y(t) = -\lambda \frac{\partial}{\partial s} S(t,0) y_0 + \lambda S(t,0) y_1 + \lambda \int_0^t S(t,s) f(s, y(s), \int_0^s g(s, \tau, y(\tau)) d\tau) ds.$$

Then

$$
\begin{aligned}
|y(t)| &\leq \left\| \frac{\partial}{\partial s} S(t,0) \right\|_{B(E)} |y_0| + \|S(t,0)\|_{B(E)} |y_1| \\
&\quad + \int_0^t \|S(t,s)\|_{B(E)} \Theta_f(s) \psi\left(|y(s)| + \int_0^s \Theta_g(\tau) \varphi(|y(\tau)|) d\tau\right) ds \\
&\leq \tilde{M} |y_0| e^{-\delta t} + M |y_1| e^{-\delta t} \\
&\quad + \int_0^t M e^{-\delta(t-s)} \Theta_f(s) \psi\left(|y(s)| + \int_0^s \Theta_g(\tau) \varphi(|y(\tau)|) d\tau\right) ds. \\
&\leq (\tilde{M} |y_0| + M |y_1|) e^{-\delta t} \\
&\quad + \int_0^t M e^{-\delta(t-s)} \Theta_f(s) \psi\left(|y(s)| + \int_0^s \Theta_g(\tau) \varphi(|y(\tau)|) d\tau\right) ds.
\end{aligned}
$$

Case 2. For each $t \in (s_i, t_{i+1}]$, we have

$$
\begin{aligned}
y(t) &= -\lambda \frac{\partial}{\partial s} S(t, s_i) \gamma_i(s_i, y(s_i)) + \lambda S(t, s_i) \zeta_i(s_i, y(s_i)) \\
&\quad + \lambda \int_{s_i}^t S(t,s) f(s, y(s), \int_0^s g(s, \tau, y(\tau)) d\tau) ds,
\end{aligned}
$$

then

$$
\begin{aligned}
|y(t)| &\leq \left\| \frac{\partial}{\partial s} S(t, s_i) \right\|_{B(E)} |\gamma_i(s_i, y(s_i))| + \|S(t, s_i)\|_{B(E)} |\zeta_i(s_i, y(s_i))| \\
&\quad + \int_{s_i}^t \|S(t,s)\|_{B(E)} \Theta_f \psi\left(|y(s)| + \int_0^s \Theta_g(\tau) \varphi(|y(\tau)|) d\tau\right) ds \\
&\leq \tilde{M} c_i |y(s_i)| e^{-\delta(t-s_i)} + \tilde{M} d_i e^{-\delta(t-s_i)} \\
&\quad + M e_i |y(s_i)| e^{-\delta(t-s_i)} + M l_i e^{-\delta(t-s_i)} \\
&\quad + \int_{s_i}^t M e^{-\delta(t-s)} \Theta_f(s) \psi\left(|y(s)| + \int_0^s \Theta_g(\tau) \varphi(|y(\tau)|) d\tau\right) ds. \\
&\leq \tilde{M} c_i |y(s_i)| + \tilde{M} d_i e^{-\delta(t-s_i)} \\
&\quad + M e_i |y(s_i)| + M l_i e^{-\delta(t-s_i)} \\
&\quad + \int_{s_i}^t M e^{-\delta(t-s)} \Theta_f(s) \psi\left(|y(s)| + \int_0^s \Theta_g(\tau) \varphi(|y(\tau)|) d\tau\right) ds.
\end{aligned}
$$

It is easy to see that

$$
\begin{aligned}
\sup_{s \in [0,t]} |y(s)| &\leq \left(\frac{\tilde{M} d_i e^{s_i}}{1 - L_i} + \frac{M l_i e^{s_i}}{1 - L_i} \right) e^{-\delta t} \\
&\quad + \int_{s_i}^t \frac{M}{1 - L_i} e^{-\delta(t-s)} \Theta_f(s) \psi\left(\sup_{s \in [0,t]} |y(s)| + \int_0^s \Theta_g(\tau) \varphi(\sup_{s \in [0,t]} |y(s)|) d\tau\right) ds.
\end{aligned}
$$

Case 3. For each $t \in (s_i, t_i]$, we have,

$$\begin{aligned}
|y(t)| &= \lambda\left|\gamma_i\left(t, -\frac{\partial}{\partial s}S(t,s_{i-1})\gamma_{i-1}(s_{i-1},y(t_{i-1}^-)) + S(t,s_{i-1})\zeta_{i-1}(s_{i-1},y(t_{i-1}^-))\right.\right.\\
&\quad\left.\left. + \int_{s_{i-1}}^{t_i} S(t,s)f\left(s,y(s), \int_0^s g(s,\tau,y(\tau))d\tau\right)ds\right)\right|\\
&\leq \lambda\left|\gamma_i\left(t, -\frac{\partial}{\partial s}S(t,s_i)\gamma_{i-1}(s_{i-1},y(t_{i-1}^-)) + S(t,s_{i-1})\zeta_{i-1}(s_{i-1},y(t_{i-1}^-))\right.\right.\\
&\quad\left.\left. + \int_{s_{i-1}}^{t_i} S(t,s)f\left(s,y(s), \int_0^s g(s,\tau,y(\tau))d\tau\right)ds - \gamma_i(t,0)\right)\right|\\
&\quad + \lambda|\gamma_i(t,0)|\\
&\leq \lambda c_i\left|-\frac{\partial}{\partial s}S(t,s_i)\gamma_{i-1}(s_{i-1},y(t_{i-1}^-)) + S(t,s_{i-1})\zeta_{i-1}(s_{i-1},y(t_{i-1}^-))\right.\\
&\quad\left. + \int_{s_{i-1}}^{t_i} S(t,s)f\left(s,y(s), \int_0^s g(s,\tau,y(\tau))d\tau\right)ds\right|\\
&\quad + \lambda d_i.
\end{aligned}$$

This implies

$$\sup_{s\in[0,t]} |y(s)| \leq \left(\frac{\tilde{M}c_i d_{i-1}e^{s_{i-1}}}{1-L_{i-1}} + \frac{Mc_i l_{i-1}e^{s_{i-1}}}{1-L_{i-1}}\right)e^{-\delta t} + \frac{d_i}{1-L_{i-1}}$$

$$+ \int_{s_{i-1}}^{t_i} \frac{Mc_i}{1-L_{i-1}}e^{-\delta(t-s)}\Theta_f(s)\psi\left(\sup_{\tau\in[0,s]}|y(\tau)| + \int_0^s \Theta_g(\tau)\varphi(\sup_{z\in[0,\tau]}|y(z)|)d\tau\right)ds.$$

Then, for all $t\in J$, we have

$$|y(t)| \leq M_i^* e^{-\delta t} + \frac{d_i}{1-L_{i-1}}$$

$$+ e^{-\delta t}\int_0^t \tilde{M}e^{\delta s}\Theta_f(s)\psi\left(\sup_{\tau\in[0,s]}|y(\tau)| + \int_0^s \Theta_g(\tau)\varphi(\sup_{z\in[0,\tau]}|y(z)|)d\tau\right)ds.$$

where

$$M^* = \max_{2\leq i\leq N}\left\{\tilde{M}|y_0| + M|y_1|, \frac{\tilde{M}d_1 e^{s_1}}{1-L_1} + \frac{Ml_1 e^{s_1}}{1-L_1}, \frac{\tilde{M}d_i e^{s_i}}{1-L_i} + \frac{Ml_i e^{s_i}}{1-L_i}, \frac{\tilde{M}c_i d_{i-1}e^{s_{i-1}}}{1-L_{i-1}} + \frac{Mc_i l_{i-1}e^{s_{i-1}}}{1-L_{i-1}}\right\}.$$

Let us take the right-hand side of the above inequality as $\mu(t)$. Then

$$\mu(0) = M^* + \frac{d_i}{1-L_{i-1}},$$

$$\sup_{s\in[0,t]}|y(s)| \leq \mu(t),$$

and

$$\mu'(t) \leq -\delta\mu(t) + \tilde{M}\Theta_f(t)\psi\left(\mu(t) + \int_0^t \Theta_g(s)\varphi(\mu(s))ds\right)$$

$$\leq \tilde{M}\Theta_f(t)\psi\left(\mu(t) + \int_0^t \Theta_g(s)\varphi(\mu(s))ds\right).$$

Let

$$\beta(t) = \mu(t) + \int_0^t \Theta_g(s)\varphi(\mu(s))ds.$$

Then

$$\begin{aligned}
\beta'(t) &= \mu'(t) + \Theta_g(t)\varphi(\mu(t))\\
&\leq \tilde{M}\Theta_f(t)\psi(\beta(t)) + \Theta_g(t)\varphi(\beta(t)).
\end{aligned}$$

This implies that

$$\int_{\beta(0)}^{\beta(t)} \frac{ds}{\psi(s) + \varphi(s)} \leq \int_{m_i}^{a} \max(\tilde{M}\Theta_f(s), \Theta_g(s))ds < \int_{m_i}^{+\infty} \frac{ds}{\psi(s) + \varphi(s)}.$$

This above inequality implies that there exists a constant L such that $\beta(t) \leq L, t \in J$, and hence $\mu(t) \leq L, t \in J$. Since for every $t \in J, |y(t)| \leq \mu(t)$, we have $\|y\|_{PC} \leq L$.

Step 2. Λ is continuous.

Suppose that $(y_n)_{n \in \mathbb{N}}$ is a sequence in B_R which converges to y in B_R as $n \to \infty$. By the continuity of nonlinear term γ and ζ with respect to the second argument, for each $s \in J$, we have

$$\sup_{s \in J} |\gamma_i(s, y_n(s)) - \gamma_i(s, y(s))| \to 0 \quad \text{as} \quad n \to \infty, \tag{6}$$

$$\sup_{s \in J} |\zeta(s, y_n(s)) - \zeta(s, y(s))| \to 0 \quad \text{as} \quad n \to \infty. \tag{7}$$

By the Carathéodory character of nonlinear term f, for each $s \in J$, we have

$$\left| f\left(s, y_n(s), \int_0^s g(s, \tau, y_n(\tau))d\tau\right) - f\left(s, y(s), \int_0^s g(s, \tau, y(\tau))d\tau\right) \right| \to 0 \text{ as } n \to \infty. \tag{8}$$

Case 1. For the interval $(s_i, t_i]$, we obtain

$$|(\Lambda y_n)(t) - (\Lambda y)(t)|$$
$$\leq \gamma_i\left(t, -\frac{\partial}{\partial s}S(t, s_i)\gamma_{i-1}(s_{i-1}, y_n(t_{i-1}^-)) + S(t, s_{i-1})\zeta_{i-1}(s_{i-1}, y_n(t_{i-1}^-))\right.$$
$$\left. + \int_{s_{i-1}}^{t_i} S(t, s)f\left(s, y_n(s), \int_0^s g(s, \tau, y_n(\tau))d\tau\right) ds\right)$$
$$- \gamma_i\left(t, -\frac{\partial}{\partial s}S(t, s_i)\gamma_{i-1}(s_{i-1}, y(t_{i-1}^-)) + S(t, s_{i-1})\zeta_{i-1}(s_{i-1}, y(t_{i-1}^-))\right.$$
$$\left. + \int_{s_{i-1}}^{t_i} S(t, s)f\left(s, y(s), \int_0^s g(s, \tau, y(\tau))d\tau\right) ds\right).$$

Since the function γ_i is continuous and

$$\left| -\frac{\partial}{\partial s}S(t, s_i)\gamma_{i-1}(s_{i-1}, y_n(t_{i-1}^-)) + S(t, s_{i-1})\zeta_{i-1}(s_{i-1}, y_n(t_{i-1}^-)) \right.$$
$$+ \int_{s_{i-1}}^{t_i} S(t, s)f\left(s, y_n(s), \int_0^s g(s, \tau, y_n(\tau))d\tau\right) ds + \frac{\partial}{\partial s}S(t, s_i)\gamma_{i-1}(s_{i-1}, y(t_{i-1}^-))$$
$$\left. + S(t, s_{i-1})\zeta_{i-1}(s_{i-1}, y(t_{i-1}^-)) + \int_{s_{i-1}}^{t_i} S(t, s)f\left(s, y(s), \int_0^s g(s, \tau, y(\tau))d\tau\right) ds \right|$$
$$\leq \tilde{M}|\gamma_{i-1}(s_{i-1}, y_n(s_{i-1})) - \gamma_{i-1}(s_i, y(s_i))| + M|\zeta_{i-1}(s_{i-1}, y_n(s_{i-1})) - \zeta_{i-1}(s_{i-1}, y(s_{i-1}))|$$
$$+ M \int_{s_{i-1}}^{t} \left| f\left(s, y_n(s), \int_0^s g(s, \tau, y_n(\tau))d\tau\right) - f\left(s, y(s), \int_0^s g(s, \tau, y(\tau))d\tau\right) \right| ds.$$
$$\to 0, \quad \text{as} \quad n \to \infty.$$

We can conclude that $\Lambda y_n \to \Lambda y$, as $n \to +\infty$.

Case 2. For the interval $[0, t_1]$, we obtain

$$|(\Lambda y_n)(t) - (\Lambda y)(t)|$$
$$\leq M \int_0^t \left| f\left(s, y_n(s), \int_0^s g(s, \tau, y_n(\tau))d\tau\right) - f\left(s, y(s), \int_0^s g(s, \tau, y(\tau))d\tau\right) \right| ds$$
$$\to 0, \quad \text{as} \quad n \to \infty.$$

Case 3. For the interval $(s_i, t_{i+1}]$, we have

$$|(\Lambda y_n)(t) - (\Lambda y)(t)|$$
$$\leq \tilde{M}|\gamma_i(s_i, y_n(s_i)) - \gamma_i(s_i, y(s_i))| + M|\zeta_i(s_i, y_n(s_i)) - \zeta_i(s_i, y(s_i))|$$
$$+ M \int_{s_i}^{t} \left| f\left(s, y_n(s), \int_0^s g(s, \tau, y_n(\tau)) d\tau \right) - f\left(s, y(s), \int_0^s g(s, \tau, y(\tau)) d\tau \right) \right| ds$$
$$\to 0 \quad \text{as} \quad n \to \infty.$$

As a consequence of Case 1–3, $\Lambda y_n \to \Lambda y$, as $n \to +\infty$. Hence the Λ is continuous.

Step 3. Λ is equicontinuous.

Case 1. For the interval $[0, t_1]$, $0 \leq \tilde{t}_1 \leq \tilde{t}_2 \leq t_1$, any $y \in B_R$, we have

$$|(\Lambda y)(\tilde{t}_2) - (\Lambda y)(\tilde{t}_1)|$$
$$\leq \left\| \frac{\partial}{\partial s} S(\tilde{t}_2, 0) - \frac{\partial}{\partial s} S(\tilde{t}_1, 0) \right\|_{B(E)} |y_0|$$
$$+ \| S(\tilde{t}_2, 0) - S(\tilde{t}_1, 0) \|_{B(E)} |y_1|$$
$$+ \left| \int_0^{\tilde{t}_1} (S(\tilde{t}_2, s) - S(\tilde{t}_1, s)) f\left(s, y(s), \int_0^s g(s, \tau, y(\tau)) d\tau \right) ds \right.$$
$$\left. + \int_{\tilde{t}_1}^{\tilde{t}_2} S(\tilde{t}_2, \tau) f\left(s, y(s), \int_0^s g(s, \tau, y(\tau)) d\tau \right) ds \right|$$
$$\leq \int_0^{\tilde{t}_1} \| S(\tilde{t}_2, \tau) - S(\tilde{t}_1, \tau) \|_{B(E)} \Theta_f(\tau) \psi \left(|y(s)| + \int_0^s \Theta_g(\tau) \varphi(|y(\tau)|) d\tau \right) ds$$
$$+ M \int_{\tilde{t}_1}^{\tilde{t}_2} \Theta_f(s) \psi \left(|y(s)| + \int_0^s \Theta_g(\tau) \varphi(|y(\tau)|) d\tau \right) ds.$$

It follows from the Hölder's inequality that

$$|(\Lambda y)(\tilde{t}_2) - (\Lambda y)(\tilde{t}_1)|$$
$$\leq \left\| \frac{\partial}{\partial s} S(\tilde{t}_2, 0) - \frac{\partial}{\partial s} S(\tilde{t}_1, 0) \right\|_{B(E)} |y_0|$$
$$+ \| S(\tilde{t}_2, 0) - S(\tilde{t}_1, 0) \|_{B(E)} |y_1|$$
$$+ \psi \left(R + \varphi(R) \| \Theta_g \|_{L^1} \right) \int_0^{\tilde{t}_1} \| S(\tilde{t}_2, \tau) - S(\tilde{t}_1, \tau) \|_{B(E)} \Theta_f(\tau) d\tau$$
$$+ \frac{M \| \Theta_f \|_{L^r} \psi \left(R + \varphi(R) \| \Theta_g \|_{L^1} \right)}{\delta^{1 - \frac{1}{r}}} \left(e^{-\frac{r\delta}{r-1}(t - \tilde{t}_2)} - e^{-\frac{r\delta}{r-1}(t - \tilde{t}_1)} \right)^{1 - \frac{1}{r}}.$$

Case 2. For the interval $(s_i, t_{i+1}]$, $s_i \leq \tilde{t}_1 \leq \tilde{t}_2 \leq t_{i+1}$, any $y \in B_R$, then we get

$$|(\Lambda y)(\tilde{t}_2) - (\Lambda y)(\tilde{t}_1)| \leq \left\| \frac{\partial}{\partial s} S(\tilde{t}_2, s_i) - \frac{\partial}{\partial s} S(\tilde{t}_1, s_i) \right\|_{B(E)} |\gamma_i(s_i, y(s_i))|$$
$$+ \| S(\tilde{t}_2, s_i) - S(\tilde{t}_1, s_i) \|_{B(E)} |\zeta_i(s_i, y(s_i))|$$
$$+ \left| \int_{s_i}^{\tilde{t}_1} (S(\tilde{t}_2, s) - S(\tilde{t}_1, s)) f\left(s, y(s), \int_0^s g(s, \tau, y(\tau)) d\tau \right) ds \right.$$
$$\left. + \int_{\tilde{t}_1}^{\tilde{t}_2} S(\tilde{t}_2, \tau) f\left(s, y(s), \int_0^s g(s, \tau, y(\tau)) d\tau \right) ds \right|$$
$$\leq \int_{s_i}^{\tilde{t}_1} \| S(\tilde{t}_2, \tau) - S(\tilde{t}_1, \tau) \|_{B(E)} \Theta_f(\tau) \psi \left(|y(s)| + \int_0^s \Theta_g(\tau) \varphi(|y(\tau)|) d\tau \right) ds$$
$$+ M \int_{\tilde{t}_1}^{\tilde{t}_2} \Theta_f(s) \psi \left(|y(s)| + \int_0^s \Theta_g(\tau) \varphi(|y(\tau)|) d\tau \right) ds.$$

It follows from the Hölder's inequality that

$$
\begin{aligned}
|(\Lambda y)(\tilde{t}_2) - (\Lambda y)(\tilde{t}_1)| &\leq \left\|\frac{\partial}{\partial s}S(\tilde{t}_2,s_i) - \frac{\partial}{\partial s}S(\tilde{t}_1,s_i)\right\|_{B(E)}|\gamma_i(s_i,y(s_i))| \\
&+ \|S(\tilde{t}_2,s_i) - S(\tilde{t}_1,s_i)\|_{B(E)}|\zeta_i(s_i,y(s_i))| \\
&+ \psi\left(R + \varphi(R)\|\Theta_g\|_{L^1}\right)\int_{s_i}^{\tilde{t}_1}\|S(\tilde{t}_2,\tau) - S(\tilde{t}_1,\tau)\|_{B(E)}\,p(\tau)d\tau \\
&+ \frac{M\|\Theta_f\|_{L^r}\psi\left(R + \varphi(R)\|\Theta_g\|_{L^1}\right)}{\delta^{1-\frac{1}{r}}}\left(e^{-\frac{r\delta}{r-1}(t-\tilde{t}_2)} - e^{-\frac{r\delta}{r-1}(t-\tilde{t}_1)}\right)^{1-\frac{1}{r}}.
\end{aligned}
$$

Case 3. For the interval $(s_i, t_i]$, $s_i \leq \tilde{t}_1 \leq \tilde{t}_2 \leq t_i$, any $y \in B_R$, we have

$$
\begin{aligned}
&|(\Lambda y)(\tilde{t}_2) - (\Lambda y)(\tilde{t}_1)| \\
&= \Bigg|\gamma_i\left(\tilde{t}_2, -\frac{\partial}{\partial s}S(\tilde{t}_2,s_i)\gamma_{i-1}(s_{i-1}, y(t_{i-1}^-)) + S(\tilde{t}_2,s_{i-1})\zeta_{i-1}(s_{i-1}, y(t_{i-1}^-))\right. \\
&\quad + \int_{s_{i-1}}^{\tilde{t}_2} S(t,s)f\left(s,y(s),\int_0^s g(s,\tau,y(\tau))d\tau\right)ds\Bigg) \\
&\quad - \gamma_i\left(\tilde{t}_1, -\frac{\partial}{\partial s}S(\tilde{t}_1,s_i)\gamma_{i-1}(s_{i-1}, y(t_{i-1}^-)) + S(\tilde{t}_1,s_{i-1})\zeta_{i-1}(s_{i-1}, y(t_{i-1}^-))\right. \\
&\quad + \int_{s_{i-1}}^{\tilde{t}_1} S(\tilde{t}_1,s)f\left(s,y(s),\int_0^s g(s,\tau,y(\tau))d\tau\right)ds\Bigg)\Bigg|.
\end{aligned}
$$

then

$$
\begin{aligned}
&|(\Lambda y)(\tilde{t}_2) - (\Lambda y)(\tilde{t}_1)| \\
&\leq c_i\Bigg| -\frac{\partial}{\partial s}S(\tilde{t}_2,s_i)\gamma_{i-1}(s_{i-1}, y(t_{i-1}^-)) + S(\tilde{t}_2,s_{i-1})\zeta_{i-1}(s_{i-1}, y(t_{i-1}^-)) \\
&\quad + \int_{s_{i-1}}^{\tilde{t}_2} S(t,s)f\left(s,y(s),\int_0^s g(s,\tau,y(\tau))d\tau\right)ds \\
&\quad + \frac{\partial}{\partial s}S(\tilde{t}_1,s_i)\gamma_{i-1}(s_{i-1}, y(t_{i-1}^-)) - S(\tilde{t}_1,s_{i-1})\zeta_{i-1}(s_{i-1}, y(t_{i-1}^-)) \\
&\quad - \int_{s_{i-1}}^{\tilde{t}_1} S(\tilde{t}_1,s)f\left(s,y(s),\int_0^s g(s,\tau,y(\tau))d\tau\right)ds\Bigg|.
\end{aligned}
$$

Similarly, one can easily see that

$$
\begin{aligned}
|(\Lambda y)(\tilde{t}_2) - (\Lambda y)(\tilde{t}_1)| &\leq c_i\left\|\frac{\partial}{\partial s}S(\tilde{t}_2,s_{i-1}) - \frac{\partial}{\partial s}S(\tilde{t}_1,s_{i-1})\right\|_{B(E)}|\gamma_{i-1}(s_{i-1},y(t_{i-1}^-))| \\
&+ c_i\|S(\tilde{t}_2,s_{i-1}) - S(\tilde{t}_1,s_{i-1})\|_{B(E)}|\zeta_{i-1}(s_{i-1},y(t_{i-1}^-))| \\
&+ c_i\psi\left(R + \varphi(R)\|\Theta_g\|_{L^1}\right)\int_{s_{i-1}}^{\tilde{t}_1}\|S(\tilde{t}_2,\tau) - S(\tilde{t}_1,\tau)\|_{B(E)}\,\Theta_f(\tau)d\tau \\
&+ \frac{Mc_i\|\Theta_f\|_{L^r}\psi\left(R + \varphi(R)\|\Theta_g\|_{L^1}\right)}{\delta^{1-\frac{1}{r}}}\left(e^{-\frac{r\delta}{r-1}(t-\tilde{t}_2)} - e^{-\frac{r\delta}{r-1}(t-\tilde{t}_1)}\right)^{1-\frac{1}{r}}.
\end{aligned}
$$

In view of Case 1–3, as a result, $\|(\Lambda y)(\tilde{t}_2) - (\Lambda y)(\tilde{t}_1)\| \to 0$ as $\tilde{t}_2 \to \tilde{t}_1$, which means that Λ is equicontinuous.

Step 4. Λ is a α_{PC}-contraction operator.

For every bounded subset $B \subset PC(J, E)$, then we know that there exists a countable set $B_1 = \{y\}_{n=1}^\infty \subset B$ (see Lemma 2), such that for any $t \in J$, we have

$$\alpha_E(\Lambda(B)(t)) \leq 2\alpha_E(\Lambda(B_1)(t)). \tag{9}$$

Note that B and ΛB are equicontinuous, we can get from Lemma 2, Lemma 3, Lemma 4 and using the assumptions (H_1)–(H_6), we obtain

Case 1. For the interval $(t_i, s_i]$, we have

$$\begin{aligned}
\alpha_E(\Lambda B_1(t)) &\leq \tilde{M}k_i \left\{\alpha_E\left(\gamma_{i-1}(s_{i-1}, y_n(t_{i-1}^-))\right)\right\}_{n=0}^{\infty} \\
&+ Mk_i \left\{\alpha_E\left(\zeta_{i-1}(s_{i-1}, y_n(t_{i-1}^-))\right)\right\}_{n=0}^{\infty} \\
&+ k_i \alpha_E\left(\left\{\int_{s_{i-1}}^t S(t,s) f\left(s, y_n(s), \int_0^s g(s,\tau, y_n(s))d\tau\right) ds\right\}_{n=0}^{\infty}\right) \\
&\leq \tilde{M}k_i \bar{k}_{i-1} \left\{\alpha_E(y_n(t_i^-))\right\}_{n=0}^{\infty} + Mk_i \bar{k}_{i-1}\left\{\alpha_E(y_n(t_{i-1}^-))\right\}_{n=0}^{\infty} \\
&+ 2Mk_i \int_{s_{i-1}}^t \left\{\alpha_E\left(f\left(s, y_n(s), \int_0^s g(s,\tau, y_n(\tau))d\tau\right)\right)\right\}_{n=0}^{\infty} ds \\
&\leq \tilde{M}k_i \bar{k}_{i-1} \left\{\alpha_E(y_n(t_i^-))\right\}_{n=0}^{\infty} + Mk_i \bar{k}_{i-1}\left\{\alpha_E(y_n(t_{i-1}^-))\right\}_{n=0}^{\infty} \\
&+ 2Mk_i \int_{s_{i-1}}^t \sigma_1(s) \left\{\alpha_E(y_n(s))\right\}_{n=0}^{\infty} \\
&+ \varrho_i(s)\left\{\alpha_E\left(\int_0^s g(s,\tau, y_n(\tau))d\tau\right)\right\}_{n=0}^{\infty} ds \\
&\leq \tilde{M}k_i \bar{k}_{i-1}\left\{\alpha_E(y_n(t_i^-))\right\}_{n=0}^{\infty} + Mk_i \bar{k}_{i-1}\left\{\alpha(y_n(t_{i-1}^-))\right\}_{n=0}^{\infty} \\
&+ 2Mk_i \int_{s_{i-1}}^t \sigma_i(s) \left\{\alpha_E(y_n(s))\right\}_{n=0}^{\infty} \\
&+ 2K^*\varrho_i(s)\left\{\int_0^s \alpha_E(y_n(\tau))d\tau\right\}_{n=0}^{\infty} ds \\
&\leq (\tilde{M}k_i\bar{k}_{i-1} + Mk_i\bar{k}_{i-1})\alpha_E(B_1(t_i^-)) \\
&+ 2Mk_i \int_{s_{i-1}}^t \left(\sigma_i(s)\alpha_E(B_1(s)) + 2K^*\varrho_i(s)\int_0^s \alpha_E(B_1(\tau))d\tau\right) ds. \\
&\leq (\tilde{M}k_i\bar{k}_{i-1} + k_i\bar{k}_{i-1}) \sup\nolimits_{s\in(t_i,s_i]} \alpha_E(B(t)) \\
&+ 2Mk_i \int_{s_{i-1}}^t \left(\sigma(s)\alpha_E(B_1(s)) + 2K^*\varrho(s)s \sup_{\tau\in[0,s]} \alpha_E(B_1(\tau))\right) ds. \\
&\leq (\tilde{M}k_i\bar{k}_{i-1} + Mk_i\bar{k}_{i-1}) \sup\nolimits_{s\in(s_i,t_{i+1}]} \alpha_E(B(t)) \\
&+ 2M \int_{s_{i-1}}^t \left(\sigma(s) \sup_{s\in(s_i,t_{i+1}]} \alpha_E(B_1(s)) + 2K^*\varrho(s)s \sup_{\tau\in(t_i,s_i]} \alpha_E(B_1(\tau))\right) ds. \\
&\leq (\tilde{M}k_i\bar{k}_{i-1} + Mk_i\bar{k}_{i-1}) \sup\nolimits_{s\in(t_i,s_i]} \alpha_E(B(t)) \\
&+ 2Mk_i \int_{s_{i-1}}^t (\sigma(s) + 2K^*s\varrho(s)) \sup_{s\in(t_i,s_i]} \alpha_E(B_1(s)) ds \\
&\leq k_i(\tilde{M}k_{i-1} + M\bar{k}_{i-1} + 2M(\|\sigma\|_{L^1} + 2K^*s_i\|\varrho\|_{L^1})) \sup\nolimits_{t\in(t_i,s_i]} \alpha_E(B(t)) \\
&\leq k_i(\tilde{M}k_{i-1} + M\bar{k}_{i-1} + 2M(\|\sigma\|_{L^1} + 2K^*a\|\varrho\|_{L^1})) \sup\nolimits_{t\in(t_i,s_i]} \alpha_E(B(t)).
\end{aligned}$$

Then

$$\alpha_E(N(B(t))) \leq k_i\left(\tilde{M}k_{i-1} + M\bar{k}_{i-1} + 2M(\|\sigma\|_{L^1} + 2K^*a\|\varrho\|_{L^1})\right)\alpha_{PC}(B(t)). \tag{10}$$

Case 2. For the interval $[0, t_1]$, we have

$$\alpha_E(\Lambda B_1(t)) \leq \alpha_E\left(\left\{\int_0^t S(t,s) f\left(s, y_n(s), \int_0^s g(s,\tau, y_n(s))d\tau\right) ds\right\}_{n=0}^{\infty}\right)$$

$$\leq 2M \int_0^t \left\{ \alpha_E \left(f(s, y_n(s), \int_0^s g(s, \tau, y_n(\tau)) d\tau) \right) \right\}_{n=0}^{\infty} ds$$

$$\leq 2M \int_0^t \sigma_1(s) \{\alpha_E(y_n(s))\}_{n=0}^{\infty} ds$$
$$+ \varrho_i(s) \left\{ \alpha_E \left(\int_0^s g(s, \tau, y_n(\tau)) d\tau \right) \right\}_{n=0}^{\infty} ds$$

$$\leq 2M \int_0^t \sigma_i(s) \{\alpha_E(y_n(s))\}_{n=0}^{\infty}$$
$$+ 2K^* \varrho_i(s) \left\{ \int_0^s \alpha_E(y_n(\tau)) d\tau \right\}_{n=0}^{\infty} ds$$

$$\leq 2M \int_0^t \left(\sigma_i(s) \alpha_E(B_1(s)) + 2K^* \varrho_i(s) \int_0^s \alpha_E(B_1(\tau)) d\tau \right) ds$$

$$\leq 2M \int_0^t \left(\sigma(s) \alpha_E(B_1(s)) + 2K^* \varrho(s) s \sup_{\tau \in [0,s]} \alpha_E(B_1(\tau)) \right) ds$$

$$\leq 2M \int_0^t \left(\sigma(s) \sup_{s \in [0;t_1]} \alpha_E(B_1(s)) + 2K^* \varrho(s) s \sup_{\tau \in [0;t_1]} \alpha_E(B_1(\tau)) \right) ds$$

$$\leq 2M \int_0^t (\sigma(s) + 2K^* s \varrho(s)) \sup_{s \in [0;t_1]} \alpha_E(B_1(s)) ds.$$

$$\leq 2M(\|\sigma\|_{L^1} + 2K^* t_1 \|\varrho\|_{L^1}) \sup_{t \in [0;t_1]} \alpha_E(B(t))$$

$$\leq 2M(\|\sigma\|_{L^1} + 2K^* a \|\varrho\|_{L^1}) \sup_{t \in [0;t_1]} \alpha_E(B(t)).$$

Then

$$\alpha_E(\Lambda(B(t))) \leq 2M(\|\sigma\|_{L^1} + 2K^* a \|\varrho\|_{L^1}) \alpha_{PC}(B(t)). \tag{11}$$

Case 3. For the interval $(s_i, t_{i+1}]$, we have

$$\alpha_E(\Lambda B_1(t)) \leq \tilde{M} \{\alpha_E(\gamma_i(s, y_n(t_i^-)))\}_{n=0}^{\infty} + M \{\alpha_E(\zeta_i(s, y_n(t_i^-)))\}_{n=0}^{\infty}$$
$$+ \alpha_E \left(\left\{ \int_{s_i}^t S(t,s) f(s, y_n(s), \int_0^s g(s, \tau, y_n(s)) d\tau) ds \right\}_{n=0}^{\infty} \right)$$
$$\leq \tilde{M} k_i \{\alpha_E(y_n(t_i^-))\}_{n=0}^{\infty} + M \bar{k}_i \{\alpha_E(y_n(t_i^-))\}_{n=0}^{\infty}$$
$$+ 2M \int_{s_i}^t \left\{ \alpha_E \left(f(s, y_n(s), \int_0^s g(s, \tau, y_n(\tau)) d\tau) \right) \right\}_{n=0}^{\infty} ds$$
$$\leq \tilde{M} k_i \{\alpha_E(y_n(t_i^-))\}_{n=0}^{\infty} + M \bar{k}_i \{\alpha_E(y_n(t_i^-))\}_{n=0}^{\infty}$$
$$+ 2M \int_{s_i}^t \sigma_1(s) \{\alpha_E(y_n(s))\}_{n=0}^{\infty} ds$$
$$+ \varrho_i(s) \left\{ \alpha_E \left(\int_0^s g(s, \tau, y_n(\tau)) d\tau \right) \right\}_{n=0}^{\infty} ds$$

$$
\begin{aligned}
&\leq \tilde{M}k_i \{\alpha_E(y_n(t_i^-))\}_{n=0}^\infty + M\bar{k}_i \{\alpha_E(y_n(t_i^-))\}_{n=0}^\infty \\
&\quad + 2M \int_{s_i}^t \sigma_i(s) \{\alpha_E(y_n(s))\}_{n=0}^\infty \\
&\quad + 2K^* \varrho_i(s) \left\{ \int_0^s \alpha_E(y_n(\tau)) d\tau \right\}_{n=0}^\infty ds \\
&\leq (\tilde{M}k_i + M\bar{k}_i)\alpha_E(B(t_i^-)) \\
&\quad + 2M \int_{s_i}^t \left(\sigma(s)\alpha_E(B_1(s)) + 2K^*\varrho(s) \int_0^s \alpha_E(B_1(\tau)) d\tau \right) ds \\
&\leq (\tilde{M}k_i + M\bar{k}_i) \sup_{s\in(s_i,t_{i+1}]} \alpha_E(B_1(s)) \\
&\quad + 2M \int_{s_i}^t \left(\sigma(s)\alpha_E(B_1(s)) + 2K^*\varrho(s)s \sup_{\tau\in[0,s]} \alpha_E(B_1(\tau)) \right) ds \\
&\leq (\tilde{M}k_i + M\bar{k}_i) \sup_{s\in(s_i,t_{i+1}]} \alpha_E(B(s)) \\
&\quad + 2M \int_{s_i}^t \left(\sigma(s) \sup_{s\in(s_i,t_{i+1}]} \alpha_E(B_1(s)) + 2K^*\varrho(s)s \sup_{\tau\in(s_i,t_{i+1}]} \alpha_E(B_1(\tau)) \right) ds \\
&\leq (\tilde{M}k_i + M\bar{k}_i)\alpha_E(B_1(t_i^-)) \\
&\quad + 2M \int_{s_i}^t (\sigma(s) + 2K^*s\varrho(s)) \sup_{s\in(s_i,t_{i+1}]} \alpha_E(B_1(s)) ds \\
&\leq (\tilde{M}k_i + M\bar{k}_i + 2M(\|\sigma\|_{L^1} + 2K^* t_{i+1} \|\varrho\|_{L^1})) \sup_{t\in(s_i,t_{i+1}]} \alpha_E(B(t)) \\
&\leq (\tilde{M}k_i + M\bar{k}_i + 2M(\|\sigma\|_{L^1} + 2K^* a \|\varrho\|_{L^1})) \sup_{t\in(s_i,t_{i+1}]} \alpha_E(B(t)).
\end{aligned}
$$

Then

$$\alpha_E(\Lambda(B(t))) \leq \left(\tilde{M}k_i + M\bar{k}_i + 2M(\|\sigma\|_{L^1} + 2K^* a \|\varrho\|_{L^1}) \right) \alpha_{PC}(B(t)). \tag{12}$$

From the above cases (10)–(12), for all $t \in J$, we obtain

$$\alpha_{PC}(\Lambda(B)) \leq \max_{1\leq i\leq N}(k_i, 1) \left(\max_{1\leq i\leq N}(\tilde{M}k_i + M\bar{k}_i) + 2M(\|\sigma\|_{L^1} + 2K^* a \|\varrho\|_{L^1}) \right) \alpha_{PC}(B).$$

Thus, we find that Λ is α_{PC}-contraction operator. Applying now theorem 1, we conclude that Λ has a fixed point which is an solution of the system (1). □

Next, we present another existence result for the mild solution of the system (1).

Theorem 4. *Assume that hypotheses* (H_1)–(H_6) *are fulfilled and*

$$\lim_{R\to+\infty} \inf \frac{\psi(R + \|\Theta_g\|_{L^1}\varphi(R))\|\Theta_f\|_{L^r})}{R} = \rho < \infty,$$

and

$$\tilde{M}c_i + Me_i + \frac{M\rho\|\Theta_f\|_{L^r}}{\delta^{1-\frac{1}{r}}} \leq 1, i = 1, \cdots, N. \tag{13}$$

Then, there exists a mild solution of system (1).

Proof. Following the proof of Theorem 3 we conclude that the map $\Lambda : B_R \to B_R$ given by Equation (5) is continuous. Next, we show that there exists $R > 0$ such that $\Lambda(B_R) \subset B_R$. In fact, if it is not true,

then for each positive number R, there exists a function $\breve{y} \in B_R$ and $\breve{t} \in J$ such that $R \leq |(\Lambda y)(\breve{t})|$.
Therefore for

Case 1. For $\breve{t} \in (s_i, t_i]$, and $\breve{y} \in B_R$, we have,

$$\begin{aligned}|(\Lambda \breve{y})(\breve{t})| &\leq \left\|\frac{\partial}{\partial s}S(\breve{t}, s_{i-1})\right\|_{B(E)} |\gamma_i(s_{i-1}, \breve{y}(s_{i-1}))| \\ &+ \|S(\breve{t}, s_{i-1})\|_{B(E)} |\zeta_i(s_{i-1}, \breve{y}(s_{i-1}))| \\ &+ \int_{s_{i-1}}^{\breve{t}} \|S(t, s)\|_{B(E)} \Theta_f(s)\psi\left(|\breve{y}(s)| + \int_0^s \Theta_g(\tau)\varphi(|\breve{y}(\tau)|)d\tau\right) ds \\ &\leq \tilde{M}c_{i-1}|\breve{y}(s_{i-1})| + \tilde{M}d_{i-1} \\ &+ Me_{i-1}|\breve{y}(s_{i-1})| + Ml_{i-1} \\ &+ \int_{s_{i-1}}^{\breve{t}} Me^{-\delta(t-s)}\Theta_f(s)\psi\left(|\breve{y}(s)| + \int_0^s \Theta_g(\tau)\varphi(|\breve{y}(\tau)|)d\tau\right) ds.\end{aligned}$$

Then

$$\begin{aligned}|(\Lambda y)(\breve{t})| &\leq \tilde{M}c_{i-1}R + \tilde{M}d_{i-1} \\ &+ Me_{i-1}R + Ml_{i-1} \\ &+ \int_{s_{i-1}}^{\breve{t}} Me^{-\delta(t-s)}\Theta_f(s)\psi\left(|\breve{y}(s)| + \int_0^s \Theta_g(\tau)\varphi(|\breve{y}(\tau)|)d\tau\right) ds. \\ &\leq (\tilde{M}c_{i-1} + Me_{i-1})R + \tilde{M}l_{i-1} + Ml_{i-1} \\ &+ M\psi(R + \|\Theta_g\|_{L^1}\varphi(R)) \int_{s_{i-1}}^{\breve{t}} e^{-\delta(t-s)}\Theta_f(s)ds.\end{aligned}$$

It follows from the Hölder's inequality that

$$\begin{aligned}|(\Lambda y)(\breve{t})| &\leq (\tilde{M}c_{i-1} + Me_{i-1})R + \tilde{M}d_{i-1} + Ml_{i-1} \\ &+ \frac{M\psi(R + \|\Theta_g\|_{L^1}\varphi(R))\|\Theta_f\|_{L^r}}{\delta^{1-\frac{1}{r}}}.\end{aligned}$$

Case 2. For $\breve{t} \in [0; t_1]$, and $\breve{y} \in B_R$, we get,

$$\begin{aligned}|(\Lambda y)(\breve{t})| &\leq \left\|\frac{\partial}{\partial s}S(t,0)\right\|_{B(E)} |y_0| \\ &+ \|S(t,s)\|_{B(E)} |y_1| \\ &+ \int_0^{\breve{t}} \|S(t,s)\|_{B(E)} \Theta_f(s)\psi\left(|\breve{y}(s)| + \int_0^s \Theta_g(\tau)\varphi(|\breve{y}(\tau)|)d\tau\right) ds \\ &\leq \tilde{M}|y_1| + M|y_0| + M\psi(R + \|\Theta_g\|_{L^1}\varphi(R))\int_0^{\breve{t}} e^{-\delta(t-s)}\Theta_f(s)ds.\end{aligned}$$

It follows from the Hölder's inequality that

$$\begin{aligned}|(\Lambda y)(\breve{t})| &\leq \tilde{M}|y_0| + M|y_1| + \frac{M\psi(R + \|\Theta_g\|_{L^1}\varphi(R))\|\Theta_f\|_{L^r}}{\delta^{1-\frac{1}{r}}}(1 - e^{-\frac{r\delta}{r-1}\breve{t}})^{1-\frac{1}{r}} \\ &\leq \tilde{M}|y_0| + M|y_1| + \frac{M\psi(R + \|\Theta_g\|_{L^1}\varphi(R))\|\Theta_f\|_{L^r}}{\delta^{1-\frac{1}{r}}}.\end{aligned}$$

Case 3. For $\check{t} \in (s_i, t_{i+1}]$, and $\check{y} \in B_R$, we have,

$$\begin{aligned}
|(\Lambda y)(\check{t})| &\leq \tilde{M} c_i |y(\check{t})| + \tilde{M} d_i + M e_i |y(\check{t})| + M l_i \\
&\quad + \int_{s_i}^{\check{t}} \|S(\check{t},s)\|_{B(E)} \Theta_f(s) \psi\left(|\check{y}(s)| + \int_0^s \Theta_g(\tau)\varphi(|\check{y}(\tau)|)d\tau\right) ds. \\
&\leq \tilde{M} c_i R + \tilde{M} d_i + M e_i R + M l_i \\
&\quad + M\psi(R + \|\Theta_g\|_{L^1}\varphi(R)) \int_{s_i}^{\check{t}} e^{-\delta(\check{t}-s)} \Theta_f(s) ds.
\end{aligned}$$

It follows from the Hölder's inequality that

$$\begin{aligned}
|(\Lambda y)(t)| &\leq \tilde{M} d_i + M l_i \\
&\quad + (\tilde{M} c_i + M e_i) R \\
&\quad + \frac{M\psi(R + \|\Theta_g\|_{L^1}\varphi(R)) \|\Theta_f\|_{L^r}}{\delta^{1-\frac{1}{r}}}.
\end{aligned}$$

Therefore for all $\check{t} \in J$, we have

$$\begin{aligned}
R < |(\Lambda y)(\check{t})| &\leq (\tilde{M} c_i + M e_i) R \\
&\quad + \max(\tilde{M} d_i + M l_i, \tilde{M}|y_0| + M|y_1|) \\
&\quad + \frac{M\psi(R + \|\Theta_g\|_{L^1}\varphi(R)) \|\Theta_f\|_{L^r}}{\delta^{1-\frac{1}{r}}}.
\end{aligned}$$

Dividing both sides by R and taking the lim inf as $R \to +\infty$, we have

$$\tilde{M} c_i + M e_i + \frac{M\rho \|\Theta_f\|_{L^r}}{\delta^{1-\frac{1}{r}}} > 1, i = 0, \cdots, N.$$

which contradicts (13). Hence, the operator Λ transforms the set B_R into itself.

The proof of $\Lambda : B_R \to B_R$ is α_E-contraction is similar to those in Theorem 3. Therefore, we omit the details. By the Darbo-Sadovskii fixed point theorem 2 we deduce that Λ has a fixed point which is a mild solution of system (1). □

4. An Example

In this section, we give an example to illustrate the above theoretical result.

Set $E = L^2([0, \pi], \mathbb{R})$ be the space of all square integrable functions from $[0, \pi]$ into $\mathbb{R}$. We denote by $\mathbb{H}^2([0, \pi], \mathbb{R})$ the Sobolev space of functions $u : [0, \pi] \to \mathbb{R}$, such that $u'' \in L^2([0, \pi], \mathbb{R})$. Define the operator $\mathbb{A} : D(\mathbb{A}) \to E$ by

$$\mathbb{A} u(\tau) = u''(\tau),$$

with domain

$$D(\mathbb{A}) = \{\omega \in E : \omega, \omega' \text{ are absolutely continuous}, \omega'' \in E, \omega(0) = \omega(\pi) = 0\}.$$

It is well known that $\mathbb{A}$ is the infinitesimal generator of a C_0-semigroup and of a strongly continuous cosine function on E, which will be denoted by $(C(t))$. From [14], for all $x \in \mathbb{H}^2([0, \pi], \mathbb{R}), t \in \mathbb{R}, \|C(t)\|_{B(E)} \leq 1$. Define also the operator $\mathbb{B} : \mathbb{H}^1([0, \pi], \mathbb{R}) \to E$ by

$$\mathbb{B}(t)u(s) = a(t)u'(s),$$

where $a : [0, 1] \to \mathbb{R}$ is a Hölder continuous function.

Consider the closed linear operator $\mathcal{A}(t) = \mathbb{B}(t) + \mathbb{A}$. It has been proved by Henríquez in [44] that the family $\{\mathcal{A}(t) : t \in J\}$ generates an evolution operator $\{S(t,s)\}_{(t,s)\in D}$. Moreover, $S(\cdot,\cdot)$ is well defined and satisfies the conditions $(H1)$ and $(H2)$, with $M = \tilde{M} = 1$ and $\delta = 1$.

We consider the following system:

$$\begin{cases} \frac{\partial^2}{\partial t^2} u(t,\tau) = \frac{\partial^2}{\partial \tau^2} u(t,\tau) + a(t) \frac{\partial}{\partial t} u(t,\tau) \\ \qquad + \dfrac{u(t,\tau)}{12(\sqrt{t}+1)(1+|u(t,\tau)|)} \\ \qquad + \dfrac{e^{-t}}{(\sqrt{t}+1)(t+1)} \int_0^t \dfrac{\sqrt{t}u(s,\tau)}{8(1+s^2+t)(1+u^2(s,\tau))} ds, & t \in \left(0, \frac{1}{\sqrt{3}}\right] \cup \left(\frac{2}{\sqrt{3}}, 1\right], \\ u(t,\tau) = \frac{1}{12} \cos \pi t\, u\left(\frac{1}{\sqrt{3}}^-,\tau\right), & t \in \left[\frac{1}{\sqrt{3}}, \frac{2}{\sqrt{3}}\right], \tau \in [0,\pi], \\ \frac{\partial}{\partial t} u(t,\tau) = \frac{1}{12} \sin \pi t\, u\left(\frac{1}{\sqrt{3}}^-,\tau\right), & t \in \left[\frac{1}{\sqrt{3}}, \frac{2}{\sqrt{3}}\right], \tau \in [0,\pi], \\ u(t,0) = u(t,\pi) = 0, & t \in [0,1], \\ u(0,\tau) = y_0, & \tau \in [0,\pi], \\ \frac{\partial}{\partial t} u(0,\tau) = y_1, & \tau \in [0,\pi]. \end{cases} \quad (14)$$

Take $a = t_2 = 1$, $t_0 = s_0 = 0$, $t_1 = \frac{1}{\sqrt{3}}$, $s_1 = \frac{2}{\sqrt{3}}$. The system (14) can be written in the abstract form:

$$\begin{cases} y''(t) = A(t)y(t) + f\left(t, y(t), \int_0^t g(t,s,y(s)) ds\right), & t \in (s_i, t_{i+1}], i = 1, 2 \\ y(t) = \gamma_i(t, y(t_i^-)), & t \in (t_i, s_i], \; i = 1, \\ y'(t) = \zeta_i(t, y(t_i^-)), & t \in (t_i, s_i], \; i = 1, \\ y(0) = y_0,\; y'(0) = y_1, \end{cases} \quad (15)$$

where $y(t) = u(t,\cdot)$, that is $y(t)(\tau) = u(t,\tau)$, $\tau \in [0,\pi]$.

The function $f : J \times E \times E \to E$, is given by

$$f(t,y,z)(\tau) = \frac{|y(t)(\tau)|}{12(\sqrt{t}+1)(1+|y(t)(\tau)|)} + \frac{e^{-t}}{(\sqrt{t}+1)(t+1)} z(t)(\tau),$$

The function $g : D \times E \to E$, is given by

$$g(t,s,y)(\tau) = \frac{\sqrt{t}y(t)(\tau)}{8(1+s^2+t)(1+y^2(t)(\tau))},$$

Functions

$$\gamma_1(t, y(t_1^-))(\tau) = \frac{1}{12} \cos \pi t\, y\left(\frac{1}{\sqrt{3}}^-\right)(\tau), \quad (16)$$

and

$$\zeta_1(t, y(t_1^-))(\tau) = \frac{1}{12} \sin \pi t\, y\left(\frac{1}{\sqrt{3}}^-\right)(\tau), \quad (17)$$

represent noninstantaneous impulses during interval $\left(\frac{1}{\sqrt{3}}, \frac{2}{\sqrt{3}}\right]$. We have

$$|f(t,y,z)(\tau)| \le \frac{1}{1+\sqrt{t}} \psi(|y(t)(\tau)| + |z(t)(\tau)|), \quad (18)$$

and

$$|g(t,s,y)(\tau)| \le \frac{\sqrt{t}}{8+8t} |y(t)(\tau)|. \quad (19)$$

From the above discussion, we obtain

$$\psi(t) = t, \varphi(t) = t, \Theta_f(t) = \frac{1}{1+\sqrt{t}}, \quad \Theta_g(t) = \frac{\sqrt{t}}{8+8t}.$$

For each $t \in J$, and $W_1, W_2 \subset E$, we get

$$\alpha_E(f(t, W_1, W_2)) \leq \frac{1}{12(\sqrt{t}+1)}\alpha_E(W_1) + \frac{e^{-t}}{(\sqrt{t}+1)(t+1)}\alpha_E(W_2),$$

We shall show that condition (H_3) holds with

$$\sigma(t) = \frac{1}{12(\sqrt{t}+1)}, \quad \rho(t) = \frac{e^{-t}}{(\sqrt{t}+1)(t+1)}.$$

Moreover

$$\|\sigma\|_{L^1} \leq \frac{1}{12}, \quad \|\rho\|_{L^1} \leq 1.$$

By (19), for any $t \in J$ and $W \subset E$, we get

$$\alpha_E(g(t,s,W)) \leq \frac{1}{8} \sup_{t \in [0,1]} \frac{\sqrt{t}}{1+t} \alpha_E(W),$$

then

$$\alpha_E(g(t,s,W)) \leq \frac{\sqrt{2}}{24}\alpha_E(W).$$

Hence (H_5) is satisfied with $K^* = \frac{\sqrt{2}}{24}$.

Next, let us observe that, in view of (16) and (17), the mapping γ_1 and ζ_1 fulfil the hypothes (H_5) and (H_6) with $c_1 = e_1 = k_1 = \bar{k}_1 = \frac{1}{12}$ and $d_1 = l_1 = 0$. Furthermore, we have

$$\max(k_1, 1)\left(\tilde{M}k_1 + M\bar{k}_1 + 2M(\|\sigma\|_{L^1} + 2K^*a\|\varrho\|_{L^1})\right) = \frac{2+\sqrt{2}}{6} < 1.$$

Clearly all the conditions of theorem 3 are satisfied. Hence by the conclusion of Theorem 3, it follows that problem (14) has a solution.

Author Contributions: Investigation, M.B. and Y.Z.; Writing original draft, N.R. and B.S.

Funding: The work is supported by the Macau Science and Technology Development Fund (Grant No. 0074/2019/A2) from the Macau Special Administrative Region of the People's Republic of China and the National Natural Science Foundation of China (No. 11671339).

Acknowledgments: Y. Zhou was supported by the Macau Science and Technology Development Fund (Grant No. 0074/2019/A2) from the Macau Special Administrative Region of the People's Republic of China and the National Natural Science Foundation of China (No. 11671339). B. Samet was supported by Researchers Supporting Project RSP-2019/4, King Saud University, Saudi Arabia, Riyadh. The authors would like to thank the editor and the reviewers for their constructive comments and suggestions.

Conflicts of Interest: The authors declare no conflict of interest.

References

1. Aviles, P.; Sandefur, J. Nolinear second order equations wtih applications to partial differential equations. *J. Differ. Equ.* **1985**, *58*, 404–427. [CrossRef]
2. Azodi, H.D.; Yaghouti, M.R. Bernoulli polynomials collocation for weakly singular Volterra integro-differential equations of fractional order. *Filomat* **2018**, *32*, 3623–3635. [CrossRef]
3. Bloom, F. Asymptotic bounds for solutions to a system of damped integro-differential equations of electromagnetic theory. *J. Math. Anal. Appl.* **1980**, *73*, 524–542. [CrossRef]

4. Forbes, L.K.; Crozier, S.; Doddrell, D.M. Calculating current densities and fields produced by shielded magnetic resonance imaging probes. *SIAM J. Appl. Math.* **1997**, *57*, 401–425.
5. Keskin, B. Reconstruction of the Volterra-type integro-differential operator from nodal points. *Bound. Value Probl.* **2018**, *2018*. [CrossRef]
6. Kostić, M. Weyl-almost periodic solutions and asymptotically Weyl-almost periodic solutions of abstract Volterra integro-differential equations. *Banach J. Math. Anal.* **2019**, *13*, 64–90. [CrossRef]
7. Ren, Y.; Qin, Y.; Sakthivel, R. Existence results for fractional order semilinear integro-differential evolution equations with infinite delay. *Integral Equ. Oper. Theory* **2010**, *67*, 33–49. [CrossRef]
8. Wu, J. *Theory and Application of Partial Functional Differential Equations*; Springer: New York, NY, USA, 1996.
9. Abbas, S.; Benchohra, M. *Advanced Functional Evolution Equations and Inclusions*; Springer: Cham, Switzerland, 2015.
10. Balachandran, K.; Park, D.G.; Anthoni, S.M. Existence of solutions of abstract nonlinear second-order neutral functional integrodifferential equations. *Comput. Math. Appl.* **2003**, *46*, 1313–1324. [CrossRef]
11. Benchohra, M.; Ntouyas, S.K. Existence of mild solutions on noncompact intervals to second order initial value problems for a class of differential inclusions with nonlocal conditions. *Comput. Math. Appl.* **2000**, *39*, 11–18. [CrossRef]
12. Fattorini, H.O. *Second Order Linear Differential Equations in Banach Spaces*; North-Holland Mathematics Studies; Elsevier: Amsterdam, The Netherlands, 1985; Volume 108.
13. Mönch, H. Boundary value problems for nonlinear ordinary differential equations of second order in Banach spaces. *Nonlinear Anal.* **1980**, *4*, 985–999. [CrossRef]
14. Travis, C.C.; Webb, G.F. Second order differential equations in Banach spaces. In Proceedings of the International Symposium on Nonlinear Equations in Abstract Spaces, Arlington, TX, USA, 8–10 June 1977; Academic Press: New York, NY, USA, 1978; pp. 331–361.
15. Kozak, M. A fundamental solution of a second-order differential equation in a Banach space. *Univ. Iagel. Acta Math.* **1995**, *32*, 275–289.
16. Batty, C.J.K.; Chill, R.; Srivastava, S. Maximal regularity for second order non-autonomous Cauchy problems. *Studia Math.* **2008**, *189*, 205–223. [CrossRef]
17. Benchohra, M.; Rezoug, N.; Zhou, Y. Semilinear mixed type integro-differential evolution equations via Kuratowski measure of noncompactness. *Z. Anal. Anwend.* **2019**, *38*, 143–156
18. Cardinali, T.; Gentili, S. An existence theorem for a non-autonomous second order nonlocal multivalued problem. *Stud. Univ. Babe-Bolyai Math.* **2017**, *6291*, 101–117. [CrossRef]
19. Faraci, F.; Iannizzotto, A. A multiplicity theorem for a perturbed second-order non-autonomous system. *Proc. Edinb. Math. Soc.* **2006**, *49*, 267–275. [CrossRef]
20. Henríquez, H.; Poblete, V.; Pozo, J. Mild solutions of non-autonomous second order problems with nonlocal initial conditions. *J. Math. Anal. Appl.* **2014**, *412*, 1064–1083. [CrossRef]
21. Mahmudov, N.I.; Vijayakumar, V.; Murugesu, R. Approximate controllability of second-order evolution differential inclusions in Hilbert spaces. *Mediterr. J. Math.* **2016**, *13*, 3433–3454. [CrossRef]
22. Winiarska, T. Evolution equations of second order with operator dependent on t. *Sel. Probl. Math. Cracow Univ. Tech.* **1995**, *6*, 299–314.
23. Hernández, E.; O'Regan, D. On a new class of abstract impulsive differential equations. *Proc. Am. Math. Soc.* **2013**, *141*, 1641–1649. [CrossRef]
24. Fečkan, M.; Wang, J. A general class of impulsive evolution equations. *Topol. Methods Nonlinear Anal.* **2015**, *46*, 915–933.
25. Abbas, S.; Benchohra, M.; Darwish, M. Some existence and stability results for abstract fractional differential inclusions with not instantaneous impulses. *Math. Rep. (Bucuresti)* **2017**, *19*, 245–262.
26. Anguraj, A.; Kanjanadevi, S. Existence results for fractional non-instantaneous impulsive integro-differential equations with nonlocal conditions. *Dyn. Contin. Discret. Impuls. Syst. Ser. A Math. Anal.* **2016**, *23*, 429–445.
27. Banas, J.; Jleli, M.; Mursaleen, M.; Samet, B.; Vetro, C. *Advances in Nonlinear Analysis via the Concept of Measure of Noncompactness*; Springer: Singapore, 2017.
28. Benchohra, M.; Litimein, S. Existence results for a new class of fractional integro-differential equations with state dependent delay. *Mem. Differ. Equ. Math. Phys.* **2018**, *74*, 27–38.
29. Fečkan, M.; Wang, J.; Zhou, Y. Periodic solutions for nonlinear evolution equations with non-instantaneous impulses. *Nonauton. Dyn. Syst.* **2014**, *1*, 93–101.

30. Ganga, R.; Jaydev, D. Existence result of fractional functional integro-differential equation with not instantaneous impulse. *Int. J. Adv. Appl. Math. Mech.* **2014**, *1*, 11–21.
31. Muslim, M.; Kumar, A. Controllability of fractional differential equation of order $\alpha \in (1,2]$ with non-instantaneous impulses. *Asian J. Control* **2018**, *20*, 935–942. [CrossRef]
32. Gautam, G.; Dabas, J. Mild solution for nonlocal fractional functional differential equation with not instantaneous impulse. *Int. J. Nonlinear Sci.* **2016**, *21*, 151–160.
33. Saadati, R.; Pourhadi, E.; Samet, B. On the $\mathcal{PC}$-mild solutions of abstract fractional evolution equations with non-instantaneous impulses via the measure of noncompactness. *Bound. Value Probl.* **2019**, *2019*. [CrossRef]
34. Wang, J.; Zhou, Y.; Lin, Z. On a new class of impulsive fractional differential equations. *Appl. Math. Comput.* **2014**, *242*, 649–657. [CrossRef]
35. Banaś, J.; Goebel, K. *Measures of Noncompactness in Banach Spaces*; Lecture Notes in Pure and Applied Mathematics 60; Dekker: New York, NY, USA, 1980.
36. Li, K.; Peng, J.; Gao, J. Nonlocal fractional semilinear differential equations in separable Banach spaces. *Electron. J. Diff. Eqns.* **2013**, *2013*, 1–7.
37. Aissani, K.; Benchohra, M. Semilinear fractional order integro-differential equations with infinite delay in Banach spaces. *Arch. Math. (Brno)* **2013**, *49*, 105–117. [CrossRef]
38. Akhmerov, R.R.; Kamenskii, M.I.; Patapov, A.S.; Rodkina, A.E.; Sadovskii, B.N. *Measures of Noncompactness an Condensing Operators*; Birkhauser Verlag: Basel, Switzerland, 1992.
39. Guo, D.; Lakshmikantham, V.; Liu, X. *Nonlinear Integral Equations in Abstract Spaces*; Kluwer Academic Publishers Group: Dordrecht, The Netherlands, 1996.
40. Chen, P.; Li, Y. Monotone iterative technique for a class of semilinear evolution equations with nonlocal conditions. *Results Math.* 2013, *63*, 731–744. [CrossRef]
41. Heinz, H.P. On the behaviour of measure of noncompactness with respect to differentiation and integration of rector-valued functions. *Nonlinear Anal.* **1983**, *7*, 1351–1371. [CrossRef]
42. Deimling, K. *Nonlinear Functional Analysis*; Springer: New York, NY, USA, 1985.
43. Agarwal, R.; Meehan, M.; O'Regan, D. Fixed point theory and applications. In *Cambridge Tracts in Mathematics*; Cambridge University Press: New York, NY, USA, 2001.
44. Henríquez, H. Existence of solutions of non-autonomous second order functional differential equations with infinite delay. *Nonlinear Anal.* **2011**, *74*, 3333–3352. [CrossRef]

© 2019 by the authors. Licensee MDPI, Basel, Switzerland. This article is an open access article distributed under the terms and conditions of the Creative Commons Attribution (CC BY) license (http://creativecommons.org/licenses/by/4.0/).

Article

A New Scheme Using Cubic B-Spline to Solve Non-Linear Differential Equations Arising in Visco-Elastic Flows and Hydrodynamic Stability Problems

Asifa Tassaddiq [1], Aasma Khalid [2,3], Muhammad Nawaz Naeem [3], Abdul Ghaffar [4], Faheem Khan [5], Samsul Ariffin Abdul Karim [6] and Kottakkaran Sooppy Nisar [7,*]

1. College of Computer and Information Sciences, Majmaah University, Al-Majmaah 11952, Saudi Arabia; a.tassaddiq@mu.edu.sa
2. Department of Mathematics, Government College Women University Faisalabad, Faisalabad 38023, Pakistan; aasmakhalid@gcwuf.edu.pk
3. Department of Mathematics, Government College University Faisalabad, Faisalabad 38023, Pakistan; mnawaznaeem@yahoo.com
4. Department of Mathematical Sciences, BUITEMS, Quetta 87300, Pakistan; abdulghaffar.jaffar@gmail.com
5. Department of Mathematics, University of Sargodha, Sargodha 40100, Pakistan; fahimscholar@gmail.com
6. Fundamental and Applied Sciences Department and Centre for Smart Grid Energy Research (CSMER), Institute of Autonomous System, Universiti Teknologi PETRONAS, Bandar Seri Iskandar, Seri Iskandar 32610, Perak Darul Ridzuan, Malaysia; samsul_ariffin@utp.edu.my
7. Department of Mathematics, College of Arts and Sciences, Prince Sattam bin Abdulaziz University, Wadi Aldawaser 11991, Saudi Arabia
* Correspondence: n.sooppy@psau.edu.sa; Tel.: +966-563-456-976

Received: 4 August 2019; Accepted: 28 October 2019; Published: 8 November 2019

Abstract: This study deals with the numerical solution of the non-linear differential equations (DEs) arising in the study of hydrodynamics and hydro-magnetic stability problems using a new cubic B-spline scheme (CBS). The main idea is that we have modified the boundary value problems (BVPs) to produce a new system of linear equations. The algorithm developed here is not only for the approximation solutions of the 10th order BVPs but also estimate from 1st derivative to 10th derivative of the exact solution as well. Some examples are illustrated to show the feasibility and competence of the proposed scheme.

Keywords: non-linear differential equation; cubic B-spline; central finite difference approximations; absolute errors

MSC: 34K10; 34K28; 42A10; 65D05; 65D07

1. Introduction

Recent research in the field of hydrodynamic and hydromagnetics stability have found the presence of a family of problems in differential equations (DEs) of a high order, and which have real mathematical interest. There are various approximate (numerical) methods in the literature that have been used for the solution of boundary value problems (BVPs). The existence and uniqueness to finding the solution of higher order BVPs are systematically examined in [1]. The BVPs of higher order DEs have been examined due to their significance and the potential for applications in applied sciences. To find the analytical solutions of such BVPs analytically is very tough and are available in very few cases. Very few researchers have tried the numerical solution of 10th order BVPs. Some of the approximate techniques have been established over the years to the numerical solution for these kinds of BVPs. In [2,3], the authors has solved 10th and 12th order BVPs using the Adomian decomposition method (ADM) involving Green's function. The homotopy perturbation approach was utilized in [4] to solve BVPs of 10th order. When a uniform magnetic field is applied across the fluid in the direction of gravity, the instability sets now as ordinary convection and it is modeled by 10th order BVPs as discussed in [5]. In [6], established approximate techniques for solving the 10th order non-linear BVPs occurring in thermal instability.

Numerical methods for the solution of non-linear BVPs of order 2 m were found in [7]. An effective numerical procedure DTM for solving some linear and non-linear BVPs of 10th order is discussed in [8]. In [9,10], the BVPs of 9th and 10th order are considered by adopting homotopy perturbation technique and the modified-variational iteration technique. Also the variational iterative technique was adopted in [11] for solving the 10th order BVPs. Wazwaz [12–15] proposed modified form of ADM for solving 6th, 8th, 10th and 12th order.

The study of non-polynomial spline [16] of 11th degree is a key element to solve 10th order BVPs. In [17], it is depicted that the DEs that describe the 10th order model to incorporate a 3rd order model of enlistment machine, two equations for dynamic power control, two equations for receptive power control, and three equations for edge pitch control. A 10th order nonlinear dynamic model was developed in [18] to turn mobile robots that incorporate slip between the driven wheels and the ground. Based on binary six-point and eight-point approximating subdivision scheme, two collocation algorithms are constructed by [19,20] to find the solution of BVPs. The 4th order linear BVPs using a new cubic B-spline were solved in [21]. Authors explained the 10th and 12th order BVPs by using the Galerkin weighted residual technique in [22]. The 5th, 6th and 8th order linear and non-linear BVPs by using the cubic B-spline scheme (CBS) method were solved in [23–25]. The higher (10th and 11th) degree splines were tested in [26,27] for solving 10th order BVPs. In [28] they practiced 2nd order finite difference schemes for the mathematical solutions of the 8th, 10th and 12th order Eigen-value problems. Galerkin method with septic B-spline and quintic B-spline was adopted in [29,30] for solving 10th order BVPs. Quintic B-spline and septic-B spline collocation methods was discussed in [31,32] to find solution of a 10th order BVPs.

For discrete methods, e.g., Adomian decomposition, shooting, homotopy perturbation, finite differences and variational-iterative technique, only give discrete approximate values of the unknown $y(x)$. For fitting curve to data we require further data processing methods. To overcome these disadvantages, we introduced a new CBS scheme for the solution of 10th order BVPs. The algorithm developed here is not only for the approximation solutions of the 10th order boundary value problems(BVPs) employing CBS but also estimate derivatives of 1st order to 10th order (where boundary conditions (BCs) are defined) of the exact solution as well.

The rest of the paper is organized as follows. The construction of CBS is presented in Section 2. In Section 3, the CBS scheme is utilized as an interpolating function in the solution of 10th order nonlinear BVPs. The results and discussion are presented in Section 4. Also some problems are considered in this section to show the efficiency of the CBS scheme. Finally, the concluding remarks are given in the final section.

2. The Construction of CBS

In this section, we construct the CBS basis functions for solving numerically the non-linear equations arising in the study of hydrodynamics and hydro-magnetic stability problems. To find the approximate solution at nodal points defined in the region $[a,b]$. For an interval $\Omega = [a,b]$, we divide it into n sub-intervals $\Omega_i = [\kappa_i, \kappa_{i+1}]$; $i = 0,1,2,...,n-1$, by the equidistant knots. For this range, we select equidistant points such that

$$\Omega_i = \kappa_i = a + \imath h, \tag{1}$$

such that

$$\Omega = \{a = \kappa_0, ..., \kappa_n = b\}, \tag{2}$$

i.e., $\kappa_\iota = a + \iota h, (\iota = 0, ..., n)$ and $h = \frac{b-a}{n}$.

Assume $S_3(\Omega) = \{p(t) \in C^2[a,b]\}$ such that $p(t)$ converted to to cubic-polynomial on separately sub interval$(\kappa_\iota, \kappa_{\iota+1})$. The basis function is defined as

$$M_\iota(\kappa) = \frac{1}{6h^3} \begin{cases} (\kappa - \kappa_{\iota-2})^3, & \text{if } \kappa \in [\kappa_{\iota-2}, \kappa_{\iota-1}], \\ h^3 + 3h^2(\kappa - \kappa_{\iota-1}) + 3h(\kappa - \kappa_{\iota-1})^2 - 3(\kappa - \kappa_{\iota-1})^3, & \text{if } \kappa \in [\kappa_{\iota-1}, \kappa_\iota], \\ h^3 + 3h^2(\kappa_{\iota+1} - \kappa) + 3h(\kappa_{\iota+1} - \kappa)^2 - 3(\kappa_{\iota+1} - \kappa)^3, & \text{if } \kappa \in [\kappa_\iota, \kappa_{\iota+1}], \\ (\kappa_{\iota+2} - \kappa)^3, & \text{if } \kappa \in [\kappa_{\iota+1}, \kappa_{\iota+2}], \\ 0, & \text{otherwise,} \end{cases}$$

for $(\iota = 2, 3, 4, ..., n-2)$. Considering one and all $M_\iota(\kappa)$ is also a piece-wise cubic with knots at Ω, simultaneously $M_\iota(\kappa) \in S_3(\Omega)$.

Assume $\Psi = \{M_\iota\}$; $(\iota = -1, 0, 1, 2 ... n, n+1)$ be linearly independent and let $M_3(\Omega) = \text{span}\Psi$. Thus $M_3(\Omega)$ is $(n+3)$ dimensional and $M_3(\Omega) = S_3(\Omega)$. Let $s(\kappa)$ be the cubic-B spline function interpolating at the nodal points and $s(\kappa) \in S_3(\Omega)$. Then $s(\kappa)$ can be written as

$$s(\kappa) = \sum_{\iota=-1}^{n+1} j_\iota M_\iota(\kappa).$$

Consequently now for a function $w(\kappa)$, there happened to be a distinctive cubic-B spline $s(\kappa) = \sum_{\iota=-1}^{n+1} j_\iota M_\iota(\kappa)$, satisfying the interpolating conditions:

$$w(\kappa_\iota) = s(\kappa_\iota) = \frac{j_{\iota-1} + 4j_\iota + j_{\iota+1}}{6}, \tag{3}$$

for $\iota = 0, ..., n$.

The values of $M_\iota(\kappa)$, and its derivatives $M_\iota^{(1)}(\kappa)$, $M_\iota^{(2)}(\kappa)$ at nodal points are required and these derivatives are tabulated in Table 1.

Table 1. Values of $M_\iota(\kappa)$ and its derivatives.

	$M_\iota(\kappa)$	$M_\iota^{(1)}(\kappa)$	$M_\iota^{(2)}(\kappa)$
$\kappa_{\iota-2}, \kappa_{\iota+2}$	0	0	0
$\kappa_{\iota-1}$	1/6	1/2h	$1/h^2$
κ_ι	4/6	0	$-2/h^2$
$\kappa_{\iota+1}$	1/6	$-1/2h$	$1/h^2$
otherwise	0	0	0

Assume $m_\iota = s^{(1)}(\kappa_\iota)$ and $\aleph_\iota = s^{(2)}(\kappa_\iota)$ then from

$$m_\iota = s^{(1)}(\kappa_\iota) = w^{(1)}(\kappa_\iota) - \frac{1}{180}h^4 w^{(5)}(\kappa_\iota) + O(h^6) \tag{4}$$

$$w^{(1)}(\kappa) = s^{(1)}(\kappa_\iota) = \frac{j_{\iota+1} - j_{\iota-1}}{2h} \tag{5}$$

$$\aleph_\iota = s^{(2)}(\kappa_\iota) = w^{(2)}(\kappa_\iota) - \frac{1}{12}h^2 w^{(4)}(\kappa_\iota) + \frac{1}{360}h^4 w^{(6)}(\kappa_\iota) + O(h^6) \tag{6}$$

$$w^{(2)}(\kappa) = s^{(2)}(\kappa_i) = \frac{J_{i+1} - 2J_i + J_{i-1}}{h^2}, \quad (7)$$

$\aleph_i$ may be used to determine numerical-difference formulas for $w^{(3)}(\kappa_i), w^{(4)}(\kappa_i)$ such that ($i = 1$ to $n - 1$), for $w^{(5)}(\kappa_i), w^{(6)}(\kappa_i)$ such that ($i = 2$ to $n - 2$), for $w^{(7)}(\kappa_i), w^{(8)}(\kappa_i)$ such that ($i = 3$ to $n - 3$) and $w^{(9)}(\kappa_i), w^{(10)}(\kappa_i)$ such that ($i = 4$ to $n - 4$) like so the errors can be obtained by using Taylor-series

$$\begin{cases} \frac{\aleph_{i+1} - \aleph_{i-1}}{2h} = \frac{s^{(3)}(\kappa_{i-}) + s^{(3)}(\kappa_{i+})}{2} = w^{(3)}(\kappa_i) + \frac{1}{12}h^2 w^{(5)}(\kappa_i) + O(h^4); \\ w^{(3)}(\kappa) = s^{(3)}(\kappa_i) = \frac{J_{i+2} - 2J_{i+1} + 2J_{i-1} - J_{i-2}}{2h^3}, \\ \frac{\aleph_{i+1} - 2\aleph_i + \aleph_{i-1}}{h^2} = \frac{s^{(3)}(\kappa_{i-}) - s^{(3)}(\kappa_{i+})}{h} = w^{(4)}(\kappa_i) - \frac{1}{720}h^4 w^{(8)}(\kappa_i) + O(h^6); \\ w^{(4)}(\kappa) = s^{(4)}(\kappa_i) = \frac{J_{i+2} - 4J_{i+1} + 6J_i - 4J_{i-1} + J_{i-2}}{h^4}, \\ \frac{\aleph_{i+2} - 2\aleph_{i+1} + 2\aleph_{i-1} - \aleph_{i-2}}{2h^3} = w^{(5)}(\kappa_i) + O(h^2); \\ w^{(5)}(\kappa) = s^{(5)}(\kappa_i) = \frac{J_{i+3} - 4J_{i+2} + 5J_{i+1} + 5J_{i-1} + 4J_{i-2} - J_{i-3}}{2h^5}. \end{cases} \quad (8)$$

Similarly (see [31]),

$$\begin{cases} w^{(6)}(\kappa_i) = s^{(6)}(\kappa_i) = \frac{J_{i+3} - 6J_{i+2} + 15J_{i+1} - 20J_i + 15J_{i-1} - 6J_{i-2} + J_{i-3}}{h^6}, \\ w^{(7)}(\kappa_i) = s^{(7)}(\kappa_i) = \frac{J_{i+4} - 6J_{i+3} + 14J_{i+2} - 14J_{i+1} + 14J_{i-1} - 14J_{i-2} + 6J_{i-3} - J_{i-4}}{2h^7}, \\ w^{(8)}(\kappa_i) = s^{(8)}(\kappa_i) = \frac{1}{h^8}(J_{i+4} - 8J_{i+3} + 28J_{i+2} - 56J_{i+1} + 70J_i - 56J_{i-1} + 28J_{i-2} - 8J_{i-3} + J_{i-4}), \\ w^{(9)}(\kappa_i) = s^{(9)}(\kappa_i) = \frac{1}{2h^9}(J_{i+5} - 8J_{i+4} + 27J_{i+3} - 48J_{i+2} + 42J_{i+1} - 42J_{i-1} + 48J_{i-2} - 27J_{i-3} + 8J_{i-4} - J_{i-5}). \end{cases} \quad (9)$$

3. The 10th Order Nonlinear BVPs

In this section, we consider the 10th order nonlinear BVPs arising in the study of hydrodynamics stability and visco-elastic flows.

$$w^{(10)}(\kappa) = f(\kappa, w(\kappa), w^{(1)}(\kappa), w^{(2)}(\kappa), w^{(3)}(\kappa), w^{(4)}(\kappa), w^{(5)}(\kappa), w^{(6)}(\kappa), w^{(7)}(\kappa), w^{(8)}(\kappa), w^{(9)}(\kappa)), \kappa \in [a,b], \quad (10)$$

with BCs

$$\begin{aligned} w(a) &= \lambda_0, & w^{(1)}(a) &= \lambda_1, & w^{(2)}(a) &= \lambda_2, \\ w^{(3)}(a) &= \lambda_3, & w^{(4)}(a) &= \lambda_4, & w(b) &= \chi_0, \\ w^{(1)}(b) &= \chi_1, & w^{(2)}(b) &= \chi_2, & w^{(3)}(b) &= \chi_3, \\ w^{(4)}(b) &= \chi_4, \end{aligned} \quad (11)$$

where $\lambda_0, \lambda_1, \lambda_2, \lambda_3, \lambda_4$ and $\chi_0, \chi_1, \chi_2, \chi_3, \chi_4$ are given real constants, $(a_i(\kappa); i = 1, 2, ..., 10)$ and f is continuous in interval $[a, b]$.

The Taylor, series for $w^{(10)}(\kappa_i)$ at the preferred collocation points alongside central difference (see [31]), we have

$$w^{(10)}(\kappa_i) = \frac{1}{h^6} \left(w_{i+3}^{(4)}(\kappa_i) - 6w_{i+2}^{(4)}(\kappa_i) + 15w_{i+1}^{(4)}(\kappa_i) - 20w_i^{(4)}(\kappa_i) + 15 \right. \\ \left. w_{i-1}^{(4)}(\kappa_i) - 6w_{i-2}^{(4)}(\kappa_i) + w_{i-3}^{(4)}(\kappa_i) \right). \quad (12)$$

Equation (9) can be written as

$$\frac{\aleph_{I-2} - 2\aleph_{I-3} + \aleph_{I-4}}{h^2} = w^{(4)}(\kappa_{I-3}) - \frac{1}{720}h^4 w^{(8)}(\kappa_{I-3}) + O(h^6),$$

$$\frac{\aleph_{I-1} - 2\aleph_{I-2} + \aleph_{I-3}}{h^2} = w^{(4)}(\kappa_{I-2}) - \frac{1}{720}h^4 w^{(8)}(\kappa_{I-2}) + O(h^6),$$

$$\frac{\aleph_{I} - 2\aleph_{I-1} + \aleph_{I-2}}{h^2} = w^{(4)}(\kappa_{I-1}) - \frac{1}{720}h^4 w^{(8)}(\kappa_{I-1}) + O(h^6), \quad (13)$$

$$\frac{\aleph_{I+2} - 2\aleph_{I+1} + \aleph_{I}}{h^2} = w^{(4)}(\kappa_{I+1}) - \frac{1}{720}h^4 w^{(8)}(\kappa_{I+1}) + O(h^6),$$

$$\frac{\aleph_{I+3} - 2\aleph_{I+2} + \aleph_{I+1}}{h^2} = w^{(4)}(\kappa_{I+2}) - \frac{1}{720}h^4 w^{(8)}(\kappa_{I+2}) + O(h^6),$$

$$\frac{\aleph_{I+4} - 2\aleph_{I+3} + \aleph_{I+2}}{h^2} = w^{(4)}(\kappa_{I+3}) - \frac{1}{720}h^4 w^{(8)}(\kappa_{I+3}) + O(h^6).$$

Substituting Equation (13) into Equation (12), we obtain

$$\frac{1}{h^8}(\aleph_{I+4} - 8\aleph_{I+3} + 28\aleph_{I+2} - 56\aleph_{I+1} + 70\aleph_{I} - 56\aleph_{I-1} + 28\aleph_{I-2} - 8\aleph_{I-3} + \aleph_{I-4}) \quad (14)$$
$$= w^{(10)}(\kappa_I) + O(h^2).$$

Since $\aleph_I = \frac{J_{I+1} - 2J_I + J_{I-1}}{h^2}$ so, Equation (14) becomes

$$w^{(10)}(\kappa_I) = \frac{1}{h^8} \left(\frac{J_{I+5} - 2J_{I+4} + J_{I+3}}{h^2} - 8\left(\frac{J_{I+4} - 2J_{I+3} + J_{I+2}}{h^2}\right) + 28\left(\frac{J_{I+3} - 2J_{I+2} + J_{I+1}}{h^2}\right) - 56\left(\frac{J_{I+2} - 2J_{I+1} + J_I}{h^2}\right) \right.$$
$$+ 70\left(\frac{J_{I+1} - 2J_I + J_{I-1}}{h^2}\right) - 56\left(\frac{J_I - 2J_{I-1} + J_{I-2}}{h^2}\right) + 28\left(\frac{J_{I-1} - 2J_{I-2} + J_{I-3}}{h^2}\right) - 8\left(\frac{J_{I-2} - 2J_{I-3} + J_{I-4}}{h^2}\right) + \quad (15)$$
$$\left. \frac{J_{I-3} - 2J_{I-4} + J_{I-5}}{h^2} \right).$$

After some simplifications the above equation becomes

$$w^{(10)}(\kappa_I) = s^{(10)}(\kappa_I) = \frac{1}{h^{10}} \left(J_{I+5} - 10J_{I+4} + 45J_{I+3} - 120J_{I+2} + 210J_{I+1} - 252J_I \right.$$
$$\left. + 210J_{I-1} - 120J_{I-2} + 45J_{I-3} - 10J_{I-4} + J_{I-5} \right). \quad (16)$$

Let $w(\kappa_I) = s(\kappa_I) = \sum_{l=-1}^{n+1} J_l M_l(\kappa_I)$ be the accurate solution of non-linear 10th order BVPs

$$w^{(10)}(\kappa_I) = f(\kappa_I, w(\kappa_I), w^{(1)}(\kappa_I), w^{(2)}(\kappa_I), w^{(3)}(\kappa_I), w^{(4)}(\kappa_I),$$
$$w^{(5)}(\kappa_I), w^{(6)}(\kappa_I), w^{(7)}(\kappa_I), w^{(8)}(\kappa_I), w^{(9)}(\kappa_I)), \kappa_I \in [a,b]. \quad (17)$$

Imposing Equations (3), (5), (7), (8) and (9) into Equation (17), we have

$$\frac{1}{h^{10}} \left(J_{I+5} - 10J_{I+4} + 45J_{I+3} - 120J_{I+2} + 210J_{I+1} - 252J_I + 210J_{I-1} - 120J_{I-2} + 45J_{I-3} \right.$$
$$- 10J_{I-4} + J_{I-5} \right) = f_I \left(\kappa_I, \frac{1}{6}(J_{I-1} + 4J_I + J_{I+1}), \frac{1}{2h}(J_{I+1} - J_{I-1}), \frac{1}{h^2}(J_{I+1} - 2J_I + J_{I-1}), \right.$$
$$\frac{1}{2h^3}(J_{I+2} - 2J_{I+1} + 2J_{I-1} - J_{I-2}), \frac{1}{h^4}(J_{I+2} - 4J_{I+1} + 6J_I - 4J_{I-1} + J_{I-2}), \frac{1}{2h^5}(J_{I+3}$$
$$- 4J_{I+2} + 5J_{I+1} + 5J_{I-1} + 4J_{I-2} - J_{I-3}), \frac{1}{h^6}(J_{I+3} - 6J_{I+2} + 15J_{I+1} - 20J_I + 15J_{I-1} \quad (18)$$
$$- 6J_{I-2} + J_{I-3}), \frac{1}{2h^7}(J_{I+4} - 6J_{I+3} + 14J_{I+2} - 14J_{I+1} + 14J_{I-1} - 14J_{I-2} + 6J_{I-3} - J_{I-4}),$$
$$\frac{1}{h^8}(J_{I+4} - 8J_{I+3} + 28J_{I+2} - 56J_{I+1} + 70J_I - 56J_{I-1} + 28J_{I-2} - 8J_{I-3} + J_{I-4}), \frac{1}{2h^9}(J_{I+5} - 8J_{I+4}$$
$$\left. + 27J_{I+3} - 48J_{I+2} + 42J_{I+1} - 42J_{I-1} + 48J_{I-2} - 27J_{I-3} + 8J_{I-4} - J_{I-5}) \right), \kappa \in [a,b].$$

Equation (18) we will produce a new system consisting of $(n-7)$ linear equations $(\imath = 4, 5, \ldots, n-4)$ with $(n+3)$ unknowns $\jmath_\imath$ where $(\imath = -1, 0, \ldots, n+1)$, therefore ten further equations are required. From given BCs at $\kappa = a$, we have five equations:

$$w(a) = \lambda_0 \Rightarrow \jmath_{-1} + 4\jmath_0 + \jmath_1 = 6\lambda_0$$
$$w^{(1)}(a) = \lambda_1 \Rightarrow -\jmath_{-1} + \jmath_1 = 2\lambda_1 h$$
$$w^{(2)}(a) = \lambda_2 \Rightarrow \jmath_{-1} - 2\jmath_0 + \jmath_1 = \lambda_2 h^2 \tag{19}$$
$$w^{(3)}(a) = \lambda_3 \Rightarrow \jmath_2 - 2\jmath_1 + 2\jmath_{-1} - \jmath_{-2} = 2\lambda_3 h^3$$
$$w^{(4)}(a) = \lambda_4 \Rightarrow \jmath_2 - 4\jmath_1 + 6\jmath_0 - 4\jmath_{-1} + \jmath_{-2} = \lambda_4 h^4,$$

similarly from $\kappa = b$ there will be other five equations

$$w(b) = \chi_0 \Rightarrow \jmath_{n-1} + 4\jmath_n + \jmath_{n+1} = 6\chi_0$$
$$w^{(1)}(b) = \chi_1 \Rightarrow -\jmath_{n-1} + \jmath_{n+1} = 2\chi_1 h$$
$$w^{(2)}(b) = \chi_2 \Rightarrow \jmath_{n-1} - 2\jmath_n + \jmath_{n+1} = \chi_2 h^2 \tag{20}$$
$$w^{(3)}(b) = \chi_3 \Rightarrow \jmath_{n+2} - 2\jmath_{n+1} + 2\jmath_{n-1} - \jmath_{n-2} = 2\chi_3 h^3$$
$$w^{(4)}(b) = \chi_4 \Rightarrow \jmath_{n+2} - 4\jmath_{n+1} + 6\jmath_n - 4\jmath_{n-1} + \jmath_{n-2} = \chi_4 h^4.$$

Omitting the order of the error of terms, the exact solution $w(\kappa_\imath) = s(\kappa_\imath) = \sum_{\imath=-1}^{n+1} \jmath_\imath M_\imath(\kappa_\imath)$ is accomplished by finding solution of the discussed above linear system of $(n+3)$ equations in $(n+3)$ unknowns considering the Equations (18)–(20).

4. Convergence Analysis

Let $\hat{w}(\kappa)$ be the exact solution of the Equations (10)–(12) and also $\hat{s}(\kappa)$ be the CBS approximation to $\hat{w}(\kappa)$. Therefore, we have

$$\hat{w}(\kappa_\imath) = \hat{s}(\kappa_\imath) = \sum_{\imath=-1}^{n+1} \hat{\jmath}_\imath M_\imath(\kappa_\imath), \tag{21}$$

where

$$\hat{\jmath} = \hat{\jmath}_{imath} = \left[\hat{\jmath}_{-1}, \hat{\jmath}_0, \hat{\jmath}_1, \ldots, \hat{\jmath}_{n+1}\right]^T.$$

Also, we have assume that $s'(\kappa)$ be the computed cubic B spline approximation to $\hat{s}(\kappa)$, namely

$$w'(\kappa_\imath) = s'(\kappa_\imath) = \sum_{i=-1}^{n+1} \jmath'_i M_\imath(\kappa_\imath),$$

$$\jmath' = \jmath'_i = \left[\jmath'_{-1}, \jmath'_0, \jmath'_1, \ldots, \jmath'_{n+1}\right]^T.$$

To approximate the error $\|\hat{w}(\kappa_\imath)) - \hat{s}(\kappa_\imath))\|_\infty$ we have to estimate error $\|\hat{w}(\kappa_\imath)) - s'(\kappa_\imath))\|_\infty$ and $\|w'(\kappa_\imath)) - \hat{s}(\kappa_\imath))\|_\infty$ seperately
The system of $(n+3) \times (n+3)$ matrix can be written as:

$$B\jmath = G.$$

Then, we have

$$B\hat{\jmath} = \hat{G} \tag{22}$$

and

$$B\jmath' = G'. \tag{23}$$

Now, by subtracting Equations (22) and (23), we obtain

$$B(J' - \hat{J}) = G' - \hat{G},$$

where B is an $(n+3) \times (n+3)$-dimensional band matrix, and

$$G = \begin{bmatrix} G_{-1}, & G_0, & G_1, & ..., & G_{n+1} \end{bmatrix}^T,$$

where T denoting transpose.

We can write

$$(J' - \hat{J}) = B^{-1}(G' - \hat{G}). \tag{24}$$

Taking the infinity norm from Equation (24), we obtain

$$\|(J' - \hat{J})\|_\infty = \|B^{-1}\|_\infty \|G' - \hat{G}\|_\infty.$$

The B-spline $M = M_t = \{M_{-1}, M_0, M_1, ..., M_{n+1}\}$ satisfy the following property

$$\left| \sum_{i=-1}^{n+1} J'_i M_t(\kappa_t) \right| \leq 1.$$

Using [24]

$$\|B^{-1}\|_\infty \|G' - \hat{G}\|_\infty \leq bh^2.$$

$$\|(J' - \hat{J})\|_\infty \leq bh^2. \tag{25}$$

$$s'(\kappa_t) - \hat{s}(\kappa_t) = (J' - \hat{J}) \sum_{i=-1}^{n+1} M_t(\kappa_t).$$

$$\|s'(\kappa_t) - \hat{s}(\kappa_t)\|_\infty = \|(J' - \hat{J}) \sum_{i=-1}^{n+1} M_t(\kappa_t)\|_\infty.$$

$$\|s'(\kappa_t) - \hat{s}(\kappa_t)\|_\infty \leq \|(J' - \hat{J})\|_\infty \left| \sum_{i=-1}^{n+1} M_t(\kappa_t) \right| \leq bh^2. \tag{26}$$

$$\|\hat{w}(\kappa_t) - s'(\kappa_t)\|_\infty \leq \rho h^4. \tag{27}$$

$$\|\hat{w}(\kappa_t) - \hat{s}(\kappa_t)\|_\infty \leq \|\hat{w}(\kappa_t) - s'(\kappa_t)\|_\infty + \|s'(\kappa_t) - \hat{s}(\kappa_t)\|_\infty. \tag{28}$$

Using Equations (26) and (27) in Equation (28)

$$\|\hat{w}(\kappa_t) - \hat{s}(\kappa_t)\|_\infty \leq bh^2 + \rho h^4 = \ell h^2.$$

which proves that this method is second order convergent and $\|\hat{w}(\kappa) - \hat{s}(\kappa)\|_\infty \leq \ell h^2$.

5. Results and Discussions

To test the accuracy of CBS method, three problems are discussed and compared with the existing methods in this section.

5.1. Problem 1

We consider the following DEs arising in viscoelastic flows and hydrodynamic stability problems as given in [29,31]

$$w^{(10)}(\kappa) = \frac{14175}{4}(J + w(\kappa) + 1)^{11}; \; 0 \leq \kappa \leq 1;$$

subject to BCs;

$$w(0) = w(1) = 0, \quad w^{(1)}(0) = -\frac{1}{2} = -w^{(2)}(0), \quad w^{(1)}(1) = 1,$$

$$w^{(2)}(1) = 4, \quad w^{(3)}(0) = \frac{3}{4}, \quad w^{(3)}(1) = 12, \quad w^{(4)}(0) = \frac{3}{2}, \quad w^{(4)}(1) = 48.$$

the exact solution of given equation is $w(\kappa) = \frac{2}{2-\kappa} - \kappa - 1$. The values of fifteen unknowns J_i from the Equations (18)–(20) are

$J_{-2} = 0.10849167,$ $J_3 = -0.12456626,$ $J_8 = -0.13626667,$
$J_{-1} = 0.05166667,$ $J_4 = -0.15061957,$ $J_9 = -0.08666667,$
$J_0 = -0.00083333,$ $J_5 = -0.16684713,$ $J_{10} = -0.0066667,$
$J_1 = -0.04833333,$ $J_6 = -0.17169449,$ $J_{11} = 0.11333333,$
$J_2 = -0.09000833,$ $J_7 = -0.16277005,$ $J_{12} = 0.28773333.$

Tables 2 and 3 analyzed the exact solution and cubic B-spline scheme (CBS) solution of problem 1 at $h = \frac{1}{10}$ and $h = \frac{1}{5}$ respectively. Figures 1–3 analyze the exact solution with cubic B-spline scheme (CBS) solution of problem 1 at $h = \frac{1}{10}$ and $h = \frac{1}{5}$ graphically. Table 4 analyze the errors at those derivatives where boundary conditions (BCs) are defined in problem 1 at $h = \frac{1}{10}$.

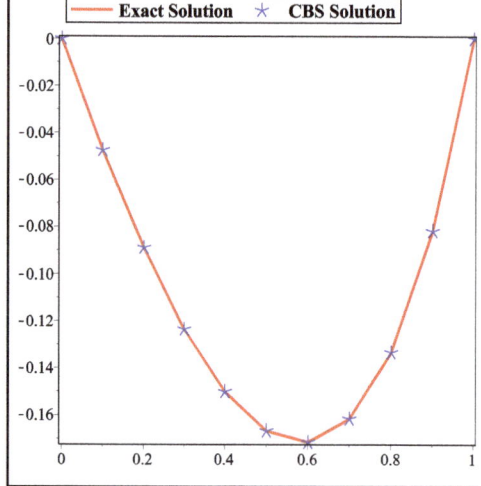

Figure 1. Problem 1 at $h = \frac{1}{10}$.

Table 2. Analyzing exact solution and cubic B-spline scheme (CBS) solution of problem 1 at $h = \frac{1}{10}$.

κ	Exact Solution	CBS Solution	Absolute Error
0	0	0	0×10^0
0.1	−0.0473684	−0.0473665	1.900×10^{-06}
0.2	−0.0888889	−0.0888822	6.670×10^{-05}
0.3	−0.1235294	−0.1235488	3.810×10^{-05}
0.4	−0.1500000	−0.1509819	1.020×10^{-04}
0.5	−0.1666667	−0.1669504	1.720×10^{-04}
0.6	−0.1714286	−0.1714992	2.030×10^{-05}
0.7	−0.1615385	−0.1615302	1.700×10^{-06}
0.8	−0.1333333	−0.1333172	9.160×10^{-05}
0.9	−0.0818182	−0.0818000	2.180×10^{-05}
1	0	0	0×10^0

Table 3. Analyzing exact solution and CBS solution of problem 1 at $h = \frac{1}{5}$.

κ	Exact Solution	CBS Solution	Absolute Error of CBS
0	0	0	0×10^0
0.2	−0.0888889	−0.0888000	8.890×10^{-05}
0.4	−0.1500000	−0.1500222	2.980×10^{-05}
0.6	−0.1714286	−0.1714778	1.150×10^{-05}
0.8	−0.1333333	−0.1333000	3.730×10^{-05}
1	0	0	0×10^0

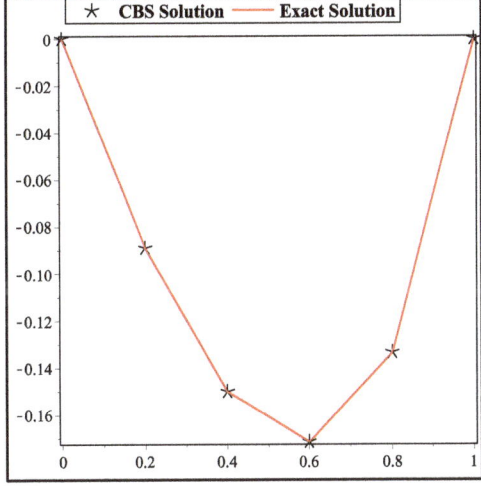

Figure 2. Problem 1 at $h = \frac{1}{5}$.

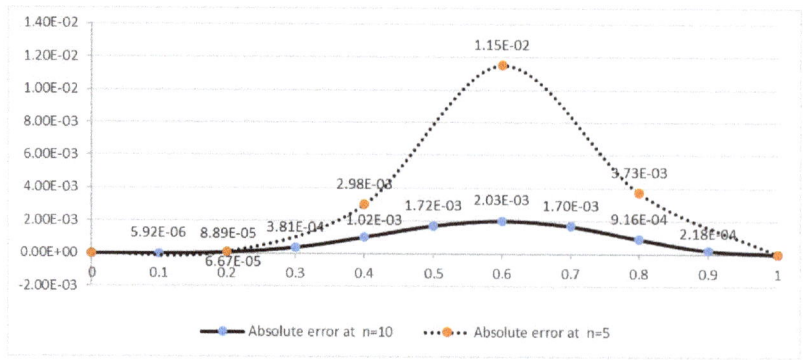

Figure 3. Problem 1 at $h = \frac{1}{10}$ and $h = \frac{1}{5}$.

Table 4. Errors at derivatives where boundary conditions (BCs) are defined in problem 1 at $h = \frac{1}{10}$.

κ	CBS-Solution of $w^{(1)}(\kappa)$	CBS-Solution of $w^{(2)}(\kappa)$	CBS-Solution of $w^{(3)}(\kappa)$	CBS-Solution of $w^{(4)}(\kappa)$
0	−0.5	0.5	0.75	1.5
0.1	−0.4459	0.5825	1.0585	1.93853
0.2	−0.3812	0.7117	1.3398	2.54026
0.3	−0.3031	0.8505	1.3543	3.38062
0.4	−0.2114	0.9826	1.4378	4.57764
0.5	−0.1054	1.1380	1.9730	6.32099
0.6	0.0204	1.3772	3.0994	8.92485
0.7	0.1771	1.7579	4.6624	12.92780
0.8	0.3805	2.3097	6.4105	19.29012
0.9	0.6480	3.0400	8.4517	29.80422
1	1	4	12	48

5.2. Problem 2

We consider the following problem as given in [16]

$$w^{(10)}(\kappa) = 9!(e^{-10w(\kappa)} - \frac{2}{(1+\kappa)^{10}}); \ 0 \leq \kappa \leq e^{1/2-1}$$

subject to BCs;

$$w(0) = 0, \quad w\left(e^{1/2-1}\right) = \frac{1}{2}, \quad w^{(1)}(0) = -w^{(2)}(0) = 1, \quad w^{(1)}\left(e^{1/2-1}\right) = e^{\left(\frac{-1}{2}\right)},$$

$$w^{(2)}\left(e^{1/2-1}\right) = -e^{(-1)}, \quad w^{(3)}(0) = 2, \quad w^{(3)}\left(e^{1/2-1}\right) = 2e^{\left(-\frac{3}{2}\right)},$$

$$w^{(4)}(0) = -6, \quad w^{(4)}\left(e^{1/2-1}\right) = -6e^{(-2)},$$

the exact solution of a given equation is $w(\kappa) = \ln(1+\kappa)$ where the domain $[0, e^{1/2-1}]$ for $h = 2^{-i}e^{1/2-1}$.

The values of fifteen unknowns J_i from Equations (18)–(20) are

$J_{-2} = -0.13805879$,
$J_{-1} = -0.0662749$,
$J_0 = 0.0007014$,
$J_1 = 0.0634693$,
$J_2 = 0.1225218$,

$J_3 = 0.1782805$,
$J_4 = 0.2311044$,
$J_5 = 0.2812925$,
$J_6 = 0.3290924$,
$J_7 = 0.3747130$,

$J_8 = 0.4183388$,
$J_9 = 0.4601370$,
$J_{10} = 0.5002580$,
$J_{11} = 0.5388309$,
$J_{12} = 0.57597018$.

Tables 5 and 6 analyzed the exact solution and cubic B-spline scheme (CBS) solution of problem 2 at $h = 0.064872$ and $h = 0.12974426$ respectively. Figures 4–6 analyze the exact solution with cubic B-spline scheme (CBS) solution of problem 2 at $h = 0.064872$ and $h = 0.12974426$ graphically. Table 7 analyze the errors at those derivatives where boundary conditions (BCs) are defined in problem 2 at $h = 0.064872$.

Table 5. Analyzing exact solution and CBS-solution of problem 2 at $h = 0.064872$.

κ	Exact Solution	CBS Solution	Absolute Error of CBS
0	0	0	0×10^0
0.065	0.06285473	0.0628501	4.650×10^{-06}
0.13	0.12199129	0.1219728	1.850×10^{-05}
0.195	0.17782512	0.1777914	3.370×10^{-05}
0.259	0.23070570	0.2306651	4.060×10^{-05}
0.324	0.28092982	0.2808945	3.530×10^{-05}
0.389	0.32875164	0.3287292	2.250×10^{-05}
0.454	0.37439053	0.3743805	1.000×10^{-05}
0.519	0.41803711	0.4180342	2.920×10^{-06}
0.584	0.45985807	0.4598575	5.980×10^{-07}
0.648	0.5	0.5	0×10^0

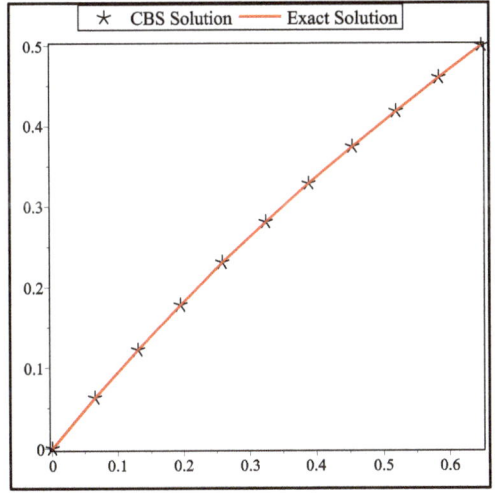

Figure 4. Problem 1 at $h = 0.064872$.

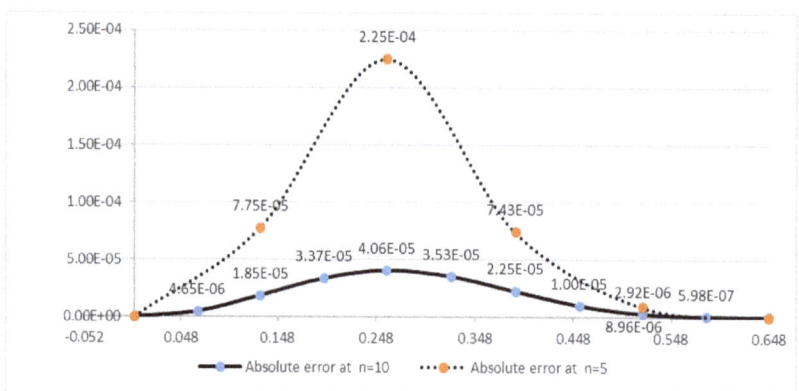

Figure 5. Problem 2 at $h = 0.064872$ and $h = 0.12974426$.

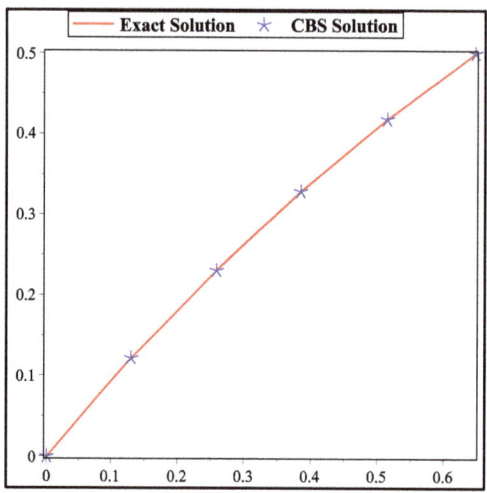

Figure 6. Problem 1 at $h = 0.12974426$.

Table 6. Analyzing the exact solution and CBS solution of problem 2 at $h = 0.12974426$.

κ	Exact Solution	CBS Solution	Absolute Error of CBS
0	0	0	0×10^0
0.130	0.1219912	0.1219138	7.750×10^{-05}
0.259	0.2307057	0.2304804	2.250×10^{-04}
0.389	0.3287516	0.3286773	7.430×10^{-05}
0.519	0.41803711	0.4180281	8.960×10^{-06}
0.649	0.5	0.5	0×10^0

Table 7. Errors at derivatives where BCs are defined in problem 2 at $h = 0.064872$.

κ	CBS Solution of $w^{(1)}(\kappa)$	CBS Solution of $w^{(2)}(\kappa)$	CBS Solution of $w^{(3)}(\kappa)$	CBS Solution of $w^{(4)}(\kappa)$
0	1	−1	2	−6
0.065	0.93893	−0.88288	1.67530	−4.66533031
0.13	0.88490	−0.78264	1.42971	−3.68168892
0.195	0.83690	−0.69738	1.20485	−2.94393959
0.259	0.79396	−0.62632	1.00104	−2.38195984
0.324	0.75524	−0.56751	0.83630	−1.94791017
0.389	0.72004	−0.51781	0.72045	−1.60848492
0.454	0.68786	−0.47403	0.64400	−1.34006674
0.519	0.65840	−0.43426	0.58187	−1.12562522
0.584	0.63139	−0.39854	0.51160	−0.95268994
0.648	0.60653	−0.36788	0.44626	−0.81200035

5.3. Problem 3

We consider the following equation as given in [29,33]

$$w^{(10)}(\kappa) + e^{-\kappa}(w(\kappa))^2 = e^{-3\kappa} + e^{-\kappa}; \ 0 \leq z' \leq 1$$

subject to BCs;

$$w(0) = w^{(2)}(0) = w^{(4)}(0) = -w^{(1)}(0) = -w^{(3)}(0) = 1,$$

$$w(0) = w^{(2)}(0) = w^{(4)}(0) = -w^{(1)}(0) = -w^{(3)}(0) = e^{-1}$$

the exact solution of given equation is $w(\kappa) = e^{-\kappa}$. The values of fifteen unknowns J_i the Equations (18)–(20) are

$J_{-2} = 1.21938333,$ $J_3 = -0.73961579,$ $J_8 = -0.44858605,$
$J_{-1} = 1.10333333,$ $J_4 = -0.66924328,$ $J_9 = 0.405893650,$
$J_0 = 0.99833333,$ $J_5 = 0.605557470,$ $J_{10} = 0.36726630,$
$J_1 = -0.9033333,$ $J_6 = 0.547923909,$ $J_{11} = 0.33231776,$
$J_2 = -0.5333053,$ $J_7 = 0.495772367,$ $J_{12} = 0.30069852$.

Tables 8 and 9 analyzed the exact solution and cubic B-spline scheme (CBS) solution of problem 3 at $h = \frac{1}{10}$ and $h = \frac{1}{5}$ respectively. Figures 7–9 analyze the exact solution with cubic B-spline scheme (CBS) solution of problem 3 at $h = \frac{1}{10}$ and $h = \frac{1}{5}$ graphically. Table 10 analyze the errors at those derivatives where boundary conditions (BCs) are defined in problem 3 at $h = \frac{1}{10}$.

Table 8. Analyzing exact solution and CBS solution of problem 3 at $h = \frac{1}{10}$.

κ	Exact Solution	CBS Solution	Absolute Error of CBS
0	1	1	0
0.1	0.9048374	0.9048417	4.250×10^{-06}
0.2	0.8187308	0.8187471	1.630×10^{-05}
0.3	0.7408182	0.7408483	3.010×10^{-05}
0.4	0.6703200	0.6703577	3.770×10^{-05}
0.5	0.6065307	0.6065662	3.550×10^{-05}
0.6	0.5488116	0.5488376	2.590×10^{-05}
0.7	0.4965853	0.4965999	1.460×10^{-05}
0.8	0.4493290	0.4493350	6.080×10^{-06}
0.9	0.4065697	0.4065712	1.500×10^{-06}
1	0.3678794	0.3678794	0

213

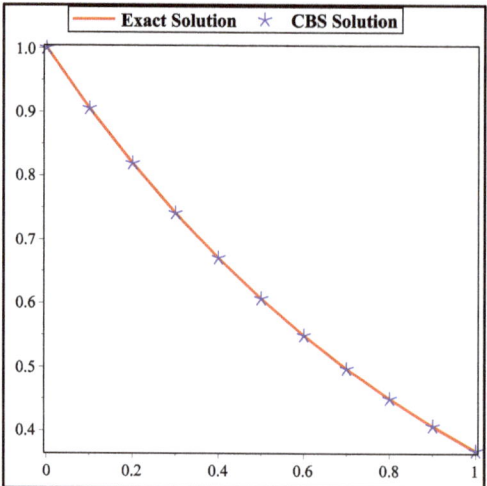

Figure 7. Problem 1 at $h = \frac{1}{10}$.

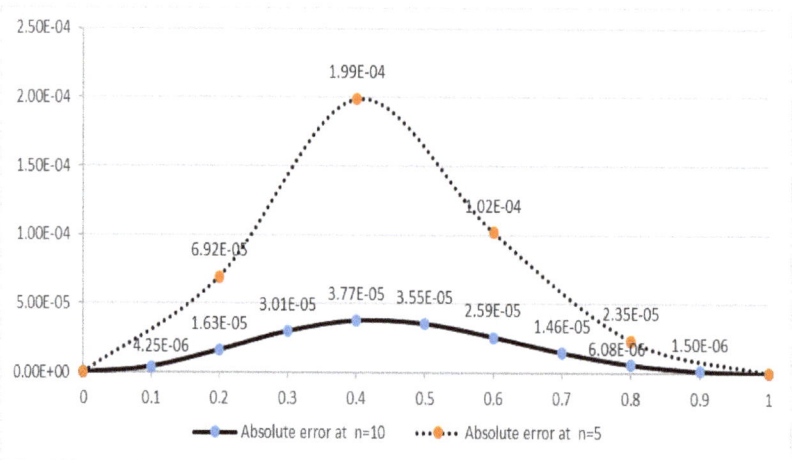

Figure 8. Problem 3 at $h = \frac{1}{10}$ and $h = \frac{1}{5}$.

Table 9. Analyzing exact solution and CBS solution of problem 3 at $h = \frac{1}{5}$.

κ	Exact Solution	CBS	Absolute Error of CBS
0	1	1	0×10^0
0.2	0.8187308	0.8188000	6.920×10^{-05}
0.4	0.6703200	0.6705188	1.990×10^{-04}
0.6	0.5488116	0.5489132	1.020×10^{-04}
0.8	0.4493290	0.4493525	2.350×10^{-05}
1	0.3678794	0.3678794	0×10^0

Table 10. Errors at derivatives where BCs are defined in problem 3 at $h = \frac{1}{10}$.

κ	CBS Solution of $w^{(2)}(\kappa), w^{(4)}(\kappa)$	CBS Solution of $w^{(1)}(\kappa), w^{(3)}(\kappa)$
0	1	−1
0.1	0.90482409	−0.90484731
0.2	0.81870008	−0.81872330
0.3	0.74077662	−0.74079984
0.4	0.67027318	−0.67029640
0.5	0.60648354	−0.60650676
0.6	0.54876876	−0.54879198
0.7	0.49655070	−0.49657392
0.8	0.44930633	−0.44932955
0.9	0.40656241	−0.40658563
1	0.36787944	−0.36787944

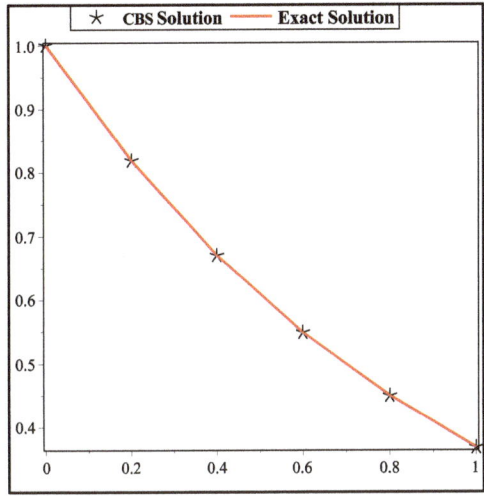

Figure 9. Problem 1 at $h = \frac{1}{10}$ and $h = \frac{1}{5}$.

6. Conclusions

In this study, we present new scheme using CBS of some non-linear differential equations arising in visco-elastic flows and hydrodynamic stability problems. The proper selection for the choice of the scheme and an appropriate of adjustment BCs may cause elasticity for the betterment of the results. The new CBS scheme proposed in this study is very simple to apply in solving the non-linear DEs compared with some existing schemes. An advantage of using the CBS scheme is that it gives a spline function on each new time line which can be applied to achieve the numerical solutions at any stage in the space direction.

Author Contributions: Conceptualization, A.T., A.G. and A.K.; methodology, A.T., A.K. and M.N.N.; software, A.T., A.K., F.K. and K.S.N; formal analysis, S.A.A.K., K.S.N. and F.K.; writing-original draft preparation, A.K., A.G. and S.A.A.K.; writing-review and editing, A.T., K.S.N. and A.G.; visualization, F.K., M.N.N. and A.G.; supervision, M.N.N. and S.A.A.K.; funding acquisition, A.T.

Funding: This research is funded by Deanship of Scientific Research at Majmaah University, Project Number (R-1441-25).

Acknowledgments: The first author Asifa Tassaddiq (A.T) would like to thank Deanship of Scientific Research at Majmaah University, for supporting this work under Project Number (R-1441-25).

Conflicts of Interest: The authors declare no conflict of interest.

References

1. Agarwal, R.P. *Boundary Value Problems from Higher Order Differential Equations*; World Scientific Singapore: Singapore, 1986.
2. Al-Hayani, W. Adomian decomposition method with green's function for solving tenth-order boundary value problems. *Appl. Math.* **2014**, *5*, 1437. [CrossRef]
3. Al-Hayani, W. Adomian decomposition method with green's function for solving twelfth-order boundary value problems. *Appl. Math. Sci.* **2015**, *9*, 353–368. [CrossRef]
4. Barari, A.; Omidvar, M.; Najafi, T.; Ghotbi, A.R. Homotopy perturbation method for solving tenth order boundary value problems. *Int. J. Math. Comput.* **2009**, *3*, 15–27.
5. Chandrasekhar, S. *Hydrodynamic and Hydromagnetic Stability*; Oxford University Press: London, UK, 1961.
6. Burns, W.W.; Owen, H.A.; Wilson, T.G.; Rodriguez, G.E.; Paulkovich, J. A digital computer simulation and study of a direct-energy-transfer power-conditioning system. In Proceedings of the IEEE Power Electronics Specialists Conference, Culver City, CA, USA, 9–11 June 1975; pp. 138–149.
7. Djidjeli, K.; Twizell, E.; Boutayeb, A. Numerical methods for special nonlinear boundary-value problems of order 2 m. *J. Comput. Appl. Math.* **1993**, *47*, 35–45. [CrossRef]
8. Erturk, V.S.; Momani, S. A reliable algorithm for solving tenth-order boundary value problems. *Numer. Algorithms* **2007**, *44*, 147–158. [CrossRef]
9. Mohyud-Din, S.T.; Yildirim, A. Solution of tenth and ninth-order boundary value problems by homotopy perturbation method. *J. Korean Soc. Ind. Appl. Math.* **2010**, *14*, 17–27.
10. Mohyud-Din, S.T.; Yildirim, A. Solutions of tenth and ninth-order boundary value problems by modified variational iteration method. *Appl. Appl. Math.* **2010**, *5*, 11–25.
11. Noor, M.A.; Al-Said, E.; Mohyud-Din, S.T. A reliable algorithm for solving tenth-order boundary value problems. *Appl. Math.* **2012**, *6*, 103–107.
12. Wazwaz, A.-M. Approximate solutions to boundary value problems of higher order by the modified decomposition method. *Comput. Math. Appl.* **2000**, *40*, 679–691. [CrossRef]
13. Wazwaz, A.-M. The modified adomian decomposition method for solving linear and nonlinear boundary value problems of tenth-order and twelfth-order. *Int. J. Nonlinear Sci. Numer. Simul.* **2000**, *1*, 17–24. [CrossRef]
14. Wazwaz, A.-M. The numerical solution of special eighth-order boundary value problems by the modified decomposition method. *Neural Parallel Sci. Comput.* **2000**, *8*, 133–146.
15. Wazwaz, A.-M. The numerical solution of sixth-order boundary value problems by the modified decomposition method. *Appl. Math. Comput.* **2001**, *118*, 311–325. [CrossRef]
16. Taiwo, O.A.; Ogunlaran, O.M. A non-polynomial spline method for solving linear fourth-order boundary-value problems. *Int. J. Phys. Sci.* **2011**, *13*, 3246–3254.
17. Hamdy, A.; Badreddin, E. Dynamic modeling of a wheeled mobile robot for identification, navigation and control. In *IMACS Conference on Modeling and Control of Technological Systems*; IMAS: Lille, France, 1992; pp. 119–128.
18. Hughes, J.T. Type-C Wind Turbine Model Order Reduction and Parameter Identification. Master's Thesis, University of Illinois at Urbana-Champaign, Champaign, IL, USA, 2013.
19. Kanwal, G.; Ghaffar, A.; Hafeezullah, M.M.; Manan, S.A.; Rizwan, M.; Rahman, G. Numerical solution of 2-point boundary value problem by subdivision scheme. *Commun. Math. Appl.* **2019**, *10*, 1–11.
20. Manan, S.A.; Ghaffar, A.; Rizwan, M.; Rahman, G.; Kanwal, G. A subdivision approach to the approximate solution of 3rd order boundary value problem. *Commun. Math. Appl.* **2018**, *9*, 499–512.
21. Gupta, Y.; Kumar, M. B-spline method for solution of linear fourth order boundary value problem. *System* **2011**, *1*, 4.
22. Islam, M.S.; Hossain, M.B.; Rahman, M.A. Numerical approaches for tenth and twelfth order linear and nonlinear differential equations. *Br. J. Math. Comput. Sci.* **2015**, *5*, 637. [CrossRef]
23. Khalid, A.; Naeem, M.N. Cubic B-spline solution of nonlinear sixth order boundary value problems. *J. Math.* **2018**, *50*, 91–103.
24. Khalid, A.; Naeem, M.N.; Ullah, Z.; Ghaffar, A.; Baleanu, D.; Nisar, K.S.; Al-Qurashi, M.M. Numerical solution of the boundary value problems arising in magnetic fields and cylindrical shells. *Mathematics* **2019**, *7*, 508. [CrossRef]

25. Lang, F.-G.; Xu, X.-P. A new CBS method for linear fifth order boundary value problems. *J. Appl. Math. Comput.* **2011**, *36*, 101–116. [CrossRef]
26. Siddiqi, S.S.; Akram, G. Solution of 10th-order boundary value problems using non-polynomial spline technique. *Appl. Math. Comput.* **2007**, *190*, 641–651. [CrossRef]
27. Siddiqi, S.S.; Akram, G. Solutions of tenth-order boundary value problems using eleventh degree spline. *Appl. Math. Comput.* **2007**, *185*, 115–127. [CrossRef]
28. Twizell, E.H.; Boutayeb, A.; Djidjeli, K. Numerical methods for eighth-, tenth-and twelfth-order eigenvalue problems arising in thermal instability. *Adv. Comput. Math.* **1994**, *2*, 407–436. [CrossRef]
29. Kasi Viswanadham, K.N.S.; Ballem, S. Numerical solution of tenth order boundary value problems by galerkin method with quintic B-splines. *Int. J. Innov. Sci. Math.* **2014**, *2*, 288–294.
30. Kasi Viswanadham, K.N.S.; Ballem, S. Numerical solution of tenth order boundary value problems by galerkin method with septic B-spline. *Int. J. Appl. Sci. Eng.* **2015**, *13*, 247–260.
31. Kasi Viswanadham, K.N.S.; Raju, Y.S. Quintic B-spline collocation method for tenth order boundary value problems. *Bound. Value Probl.* **2012**, *51*, 7–13.
32. Kasi Viswanadham, K.N.S.; Raju, Y.S. Sextic B-spline collocation method for eighth order boundary value problems. *Int. J. Appl. Sci. Eng.* **2014**, *12*, 43–57.
33. Kasi Viswanadham, K.N.S.; Ballem, S. Numerical solution of eighth order boundary value problems by galerkin method with septic B-splines. *Procedia Eng.* **2015**, *127*, 1370–1377.

© 2019 by the authors. Licensee MDPI, Basel, Switzerland. This article is an open access article distributed under the terms and conditions of the Creative Commons Attribution (CC BY) license (http://creativecommons.org/licenses/by/4.0/).

Article

Identification of Source Term for the Time-Fractional Diffusion-Wave Equation by Fractional Tikhonov Method

Le Dinh Long [1,†], Nguyen Hoang Luc [2,†], Yong Zhou [3,4,†], and Can Nguyen [5,*,†]

[1] Faculty of Natural Sciences, Thu Dau Mot University, Thu Dau Mot City 820000, Binh Duong Province, Vietnam; ledinhlong@tdmu.edu.vn
[2] Institute of Research and Development, Duy Tan University, Da Nang 550000, Vietnam; hoangluctt@gmail.com
[3] Faculty of Mathematics and Computational Science, Xiangtan University, Xiangtan 411105, China; yzhou@xtu.edu.cn
[4] Nonlinear Analysis and Applied Mathematics (NAAM) Research Group, Faculty of Science, King Abdulaziz University, Jeddah 21589, Saudi Arabia
[5] Applied Analysis Research Group, Faculty of Mathematics and Statistics, Ton Duc Thang University, Ho Chi Minh City 700000, Vietnam
* Correspondence: nguyenhuucan@tdtu.edu.vn
† These authors contributed equally to this work.

Received: 17 August 2019; Accepted: 4 October 2019; Published: 10 October 2019

Abstract: In this article, we consider an inverse problem to determine an unknown source term in a space-time-fractional diffusion equation. The inverse problems are often ill-posed. By an example, we show that this problem is NOT well-posed in the Hadamard sense, i.e., this problem does not satisfy the last condition-the solution's behavior changes continuously with the input data. It leads to having a regularization model for this problem. We use the Tikhonov method to solve the problem. In the theoretical results, we also propose *a priori* and *a posteriori* parameter choice rules and analyze them.

Keywords: fractional diffusion-wave equation; fractional derivative; ill-posed problem; Tikhonov regularization method

MSC: 26A33; 35K05; 35R11; 35K99; 47H10

---

1. Introduction

Let Ω be a bounded domain in $\mathbb{R}^d$ with sufficiently smooth boundary $\partial\Omega$, $\beta \in (1,2)$. In this paper, we consider the inverse source problem of the time-fractional diffusion-wave equation:

$$\begin{cases} \partial^\beta_{0+} u(x,t) = \Delta u(x,t) + \Xi(x), & (x,t) \in \Omega \times (0,T), \\ u(x,t) = 0, & (x,t) \in \partial\Omega \times (0,T], \\ u(x,0) = f(x), & x \in \Omega, \\ u_t(x,0) = g(x), & x \in \Omega, \\ u(x,T) = h(x), & x \in \Omega, \end{cases} \quad (1)$$

where $\partial^\beta_{0+} u(x,t)$ is the Caputo fractional derivative of order β defined as [1]

$$\partial^\beta_{0+} u(x,t) = \frac{1}{\Gamma(2-\alpha)} \int_0^t \frac{\partial^2 u(x,s)}{\partial s^2} \frac{ds}{(t-s)^{\beta-1}}, \quad 1 < \beta < 2, \quad (2)$$

where $\Gamma(.)$ is the Gamma function.

It is known that the inverse source problem mentioned above is ill-posed in general, i.e., a solution does not always exist and, in the case of existence of a solution, it does not depend continuously on the given data. In fact, from a small noise of physical measurement, for example, (h, f, g) is noised by observation data $(h^{\varepsilon_1}, g^{\varepsilon_2}, h^{\varepsilon_3})$ with order of $\varepsilon_1 > 0, \varepsilon_2 > 0$, and $\varepsilon_3 > 0$.

$$\|h - h^{\varepsilon_1}\|_{\mathcal{L}^2(\Omega)} \leq \varepsilon_1, \quad \|f - f^{\varepsilon_2}\|_{\mathcal{L}^2(\Omega)} \leq \varepsilon_2, \text{ and } \|g - g^{\varepsilon_3}\|_{\mathcal{L}^2(\Omega)} \leq \varepsilon_3. \tag{3}$$

In all functions $f(x), g(x)$, and $h(x)$ are given data. It is well-known that if $\varepsilon_1, \varepsilon_2$, and ε_3 are small enough, the sought solution $\Xi(x)$ may have a large error. The backward problem is to find $\Xi(x)$ from Ξ^{ε} and g^{ε} which satisfies (3), where $\|\cdot\|_{\mathcal{L}^2(\Omega)}$ denotes the $\mathcal{L}^2$ norm.

It is known that the inverse source problem mentioned above is ill-posed in general, i.e., a solution does not always exist, and in the case of existence of a solution, it does not depend continuously on the given data. In fact, from a small noise of physical measurement, the corresponding solutions may have a large error. Hence, a regularization is required. Inverse source problems for a time-fractional diffusion equation for $0 < \beta < 1$ have been studied. Tuan et al. [2] used the Tikhonov regularization method to solve the inverse source problem with the later time and show the estimation for the exact solution and regularized solution by a priori and a posteriori parameter choices rules. Wei et al. [3–5] studied an inverse source problem in a spatial fractional diffusion equation by quasi-boundary value and truncation methods. Fan Yang et al., see [6], used the Landweber iteration regularization method for determining the unknown source for the modified Helmholtz equation. Nevertheless, to our best knowledge, Salir Tarta et al. [7] used these properties and analytic Fredholm theorem to prove that the inverse source problem is well-posed, i.e., $f(t, x)$ can be determined uniquely and depends continuously on additional data $u(T, x)$, $x \in \Omega$, see [8,9]—the authors studied the inverse source problem in the case of nonlocal inverse problem in a one-dimensional time-space and numerical algorithm. Furthermore, the research of backward problems for the diffusion-wave equation is an open problem and still receives attention. In 2017, Tuan et al. [10] considered

$$\begin{cases} \dfrac{\partial^{\beta}}{\partial t^{\beta}} u(x,t) = -r^{\beta}(-\Delta)^{\frac{\alpha}{2}} u(x,t) + h(t)f(x), & (x,t) \in \Omega_T, \\ u(-1,t) = u(1,t) = 0, & 0 < t < T, \\ u(x,0) = 0, & x \in \Omega, \\ u(x,T) = g(x), & x \in \Omega, \end{cases} \tag{4}$$

where $\Omega_T = (-1, 1) \times (0, T)$; $r > 0$ is a parameter; $h \in C[0, T]$ is a given function; $\beta \in (0, 1)$; $\alpha \in (1, 2)$ are fractional order of the time and the space derivatives, respectively; and $T > 0$ is a final time. The function $u = u(x, t)$ denotes a concentration of contaminant at a position x and time t with $(-\Delta)^{\frac{\alpha}{2}}$ as the fractional Laplacian. If α tends to 2, the fractional Laplacian tends to the Laplacian normal operator, see [1,2,7–16]. In this paper, we use the fractional Tikhonov regularization method to solve the identification of source term of the fractional diffusion-wave equation inverse source problem with variable coefficients in a general bounded domain. However, a fractional Tikhonov is not a new method for mathematicians in the world. In [16], Zhi Quan and Xiao Li Feng used this method for considering the Helmholtz equation. Here, we estimate a convergence rate under an a priori bound assumption of the exact solution and a priori parameter choice rule and estimate a convergence rate under the a posteriori parameter choice rule.

In several papers, many authors have shown that the fractional diffusion-wave equation plays a very important role in describing physical phenomena, such as the diffusion process in media with fractional geometry, see [17]. Nowadays, fractional calculus receives increasing attention in the scientific community, with a growing number of applications in physics, electrochemistry, biophysics, viscoelasticity, biomedicine, control theory, signal processing, etc., see [18]. In a lot of papers, the Mittag–Leffler function and its properties are researched and the results are used to model the different physical phenomena, see [19,20].

The rest of this article is organized as follows. In Section 2, we introduce some preliminary results. The ill-posedness of the fractional inverse source problem (1) and conditional stability are provided in Section 3. We propose a Tikhonov regularization method and give two convergence estimates under an a priori assumption for the exact solution and two regularization parameter choice rules: Section 4 (a priori parameter choice) and Section 5 (a posteriori parameter choice).

2. Preliminary Results

In this section, we introduce a few properties of the eigenvalues of the operator $(-\Delta)$, see [21].

Definition 1 (Eigenvalues of the Laplacian operator).

1. Each eigenvalues of $(-\Delta)$ is real. The family of eigenvalues $\{\tilde{b}_i\}_{i=1}^{\infty}$ satisfy $0 \leq \tilde{b}_1 \leq \tilde{b}_2 \leq \tilde{b}_3 \leq \ldots$, and $\tilde{b}_i \to \infty$ as $i \to \infty$.
2. We take $\{\tilde{b}_i, e_i\}$ the eigenvalues and corresponding eigenvectors of the fractional Laplacian operator in Ω with Dirichlet boundary conditions on $\partial\Omega$:

$$-\Delta e_i(x) = \tilde{b}_i e_i(x), \quad x \in \Omega, \tag{5}$$
$$e_i(x) = 0, \quad \text{on } \partial\Omega,$$

for $i = 1, 2, \ldots$. Then, we define the operator $(-\Delta)$ by

$$-\Delta u := \sum_{i=0}^{\infty} c_i(-\Delta e_i(x)) = \sum_{i=0}^{\infty} c_i \tilde{b}_i e_i(x), \tag{6}$$

which maps $H_0^{\kappa}(\Omega)$ into $\mathcal{L}_2(\Omega)$. Let $0 \neq \kappa < \infty$. By $H^{\kappa}(\Omega)$, we denote the space of all functions $g \in \mathcal{L}_2(\Omega)$ with the property

$$\sum_{i=1}^{\infty}(1+\tilde{b}_i)^{2\kappa}|g_i|^2 < \infty, \tag{7}$$

where $g_i = \int_{\Omega} g(x) e_i(x) dx$. Then, we also define $\|g\|_{H^{\kappa}(\Omega)} = \sqrt{\sum_{i=1}^{\infty}(1+\tilde{b}_i)^{2\kappa}|g_i|^2}$. If $\kappa = 0$, then $H^{\kappa}(\Omega)$ is $\mathcal{L}_2(\Omega)$.

Definition 2 (See [1]). The Mittag–Leffler function is:

$$E_{\beta,\gamma}(z) = \sum_{i=0}^{\infty} \frac{z^i}{\Gamma(\beta i + \gamma)}, \quad z \in \mathbb{C},$$

where $\beta > 0$ and $\gamma \in \mathbb{R}$ are arbitrary constants.

Lemma 1 (See [21]). For $1 < \beta < 2$, $\gamma \in \mathbb{R}$, and $\omega > 0$, we get

$$E_{\beta,\gamma}(-\omega) = \frac{1}{\Gamma(\gamma - \beta)\omega} + o\left(\frac{1}{\omega^2}\right), \quad \omega \to \infty. \tag{8}$$

Lemma 2 (See [1]). If $\beta < 2$ and $\gamma \in \mathbb{R}$, suppose ζ satisfies $\frac{\pi\beta}{2} < \zeta < \min\{\pi, \pi\beta\}$, $\zeta \leq |\arg(y)| \leq \pi$. Then, there exists a constant $\tilde{A}$ as follows:

$$|E_{\beta,\gamma}(y)| \leq \frac{\tilde{A}}{1+|y|}. \tag{9}$$

Lemma 3 (See [22]). The following equality holds for $\tilde{b} > 0$, $\alpha > 0$ and $m \in \mathbb{N}$

$$\frac{d^m}{dt^m} E_{\alpha,1}(-\tilde{b}t^{\alpha}) = -\tilde{b}t^{\alpha-m} E_{\alpha,\alpha-m+1}(-\tilde{b}t^{\alpha}), \quad t > 0. \tag{10}$$

Lemma 4. For $\tilde{b}_i > 0$, $\beta > 0$, and positive integer $i \in \mathbb{N}$, we have

$$(1) \frac{d}{dt}(tE_{\beta,2}(\tilde{b}_i t^{\beta})) = E_{\beta,1}(-\tilde{b}_i t^{\beta}),$$
$$(2) \frac{d}{dt}(E_{\beta,1}(-\tilde{b}_i t^{\beta})) = -\tilde{b}_i t^{\beta-1} E_{\beta,\beta}(-\tilde{b}_i t^{\beta}). \tag{11}$$

Lemma 5. *For any $\tilde{b}_i$ satisfying $\tilde{b}_i \geq \tilde{b}_1 > 0$, there exists positive constants A, B such that*

$$\frac{A}{\tilde{b}_i T^\beta} \leq \left| E_{\beta,\beta+1}(-\tilde{b}_i T^\beta) \right| \leq \frac{B}{\tilde{b}_i T^\beta}. \tag{12}$$

Lemma 6 (See [21]). *Let $\tilde{b} > 0$, we have*

$$\int_0^\infty e^{-pt} t^{\beta i + \gamma - 1} E_{\beta,\gamma}^{(i)}(\pm at^\beta) dt = \frac{i! p^{\beta - \gamma}}{(p^\beta \mp a)^{i+1}}, \quad \Re(a) > \|a\|^{\frac{1}{\beta}}, \tag{13}$$

where $E_{\beta,\gamma}^{(i)}(y) := \frac{d^i}{dy^i} E_{\beta,\gamma}(y)$.

Lemma 7. *For constant $\xi \geq \tilde{b}_1$ and $\frac{1}{2} \leq \tau \leq 1$, one has*

$$\mathcal{C}(\xi) = \frac{\xi}{A^{2\tau} + \alpha^2 \xi^{2\tau}} \leq \overline{C}(\tau, A) \alpha^{-\frac{1}{\tau}}, \tag{14}$$

where $\overline{C} = \overline{C}(\tau, A)$ are independent on α, ξ.

Proof. Let $\frac{1}{2} \leq \tau \leq 1$, we solve the equation $\mathcal{C}'(\xi_0) = 0$, then there exists a unique $\xi_0 = A(2\tau - 1)^{-\frac{1}{2\tau}} \alpha^{-\frac{1}{\tau}}$, it gives

$$\mathcal{C}(\xi) \leq \mathcal{C}(\xi_0) \leq \frac{A^{1-2\tau}}{2\tau}(2\tau - 1)^{\frac{2\tau-1}{2\tau}} \alpha^{-\frac{1}{\tau}} := \overline{C}(\tau, A) \alpha^{-\frac{1}{\tau}}.$$

□

Lemma 8. *Let the constant $\xi \geq \tilde{b}_1$ and $\frac{1}{2} \leq \tau \leq 1$, we get*

$$\mathcal{D}(\xi) = \frac{\alpha^2 \xi^{2\tau - j}}{A^{2\tau} + \alpha^2 \xi^{2\tau}} \leq \begin{cases} \mathcal{B}_1(j, \tau, A) \alpha^{\frac{j}{\tau}}, & 0 < j < 2\tau, \\ \mathcal{B}_2(j, \tau, A, \tilde{b}_1) \alpha^2, & j \geq 2\tau. \end{cases}$$

Proof.

- If $j \geq 2\tau$, then from $\xi \geq \tilde{b}_1$, we get

$$\mathcal{D}(\xi) = \frac{\alpha^2 \xi^{2\tau - j}}{A^{2\tau} + \alpha^2 \xi^{2\tau}} \leq \frac{\alpha^2 \xi^{2\tau - j}}{A^{2\tau}} \leq \frac{\alpha^2}{A^{2\tau} \tilde{b}_1^{j-2\tau}} \leq \frac{1}{A^{2\tau} \tilde{b}_1^{j-2\tau}} \alpha^2. \tag{15}$$

- If $0 < j < 2\tau$, then it can be seen that $\lim_{\xi \to 0} \mathcal{D}(\xi) = \lim_{\xi \to +\infty} \mathcal{D}(\xi) = 0$. Taking the derivative of $\mathcal{D}$ with respect to ξ, we know that

$$\mathcal{D}'(\xi) = \frac{\alpha^2 (2\tau - j) \xi^{2\tau - j - 1}(A^{2\tau} + \alpha^2 \xi^{2\tau}) - \alpha^4 2\tau \xi^{4\tau - j - 1}}{(A^{2\tau} + \alpha^2 \xi^{2\tau})^2}. \tag{16}$$

From (16), a simple transformation gives

$$\mathcal{D}'(\xi) = \frac{\alpha^2 (2\tau - j) A^{2\tau} \xi^{2\tau - j - 1} - \alpha^4 j \xi^{4\tau - j - 1}}{(A^{2\tau} + \alpha^2 \xi^{2\tau})^2}. \tag{17}$$

$\mathcal{D}(\xi)$ attains maximum value at $\xi = \xi_0$ such that it satisfies $\mathcal{D}'(\xi) = 0$. Solving $\mathcal{D}'(\xi_0) = 0$, we know that $\xi_0 = A(2\tau - j)^{\frac{1}{2\tau}} \alpha^{-\frac{1}{\tau}} j^{-\frac{1}{2\tau}}$.

Hence, we conclude

$$\mathcal{D}(\xi) \leq \mathcal{D}(\xi_0) = \mathcal{D}\left(A(2\tau-j)^{\frac{1}{2\tau}}\alpha^{-\frac{1}{\tau}}j^{-\frac{1}{2\tau}}\right) = \frac{(2\tau-j)^{\frac{2\tau-j}{2\tau}}A^{-j}j^{\frac{j}{2\tau}}}{2\tau}\alpha^{\frac{j}{\tau}}. \tag{18}$$

□

Lemma 9. *Let $\xi > \widetilde{b}_1 > 0$ and $\frac{1}{2} \leq \tau < 1$, and $\mathcal{F}(\xi)$ be a function defined by*

$$\mathcal{F}(\xi) = \frac{\alpha^2 \xi^{2\tau-(j+1)}}{A^{2\tau} + \alpha^2 \xi^{2\tau}} \leq \begin{cases} \mathcal{B}_3(j,\tau,A)\alpha^{\frac{j+1}{\tau}}, & 0 < j < 2\tau-1, \\ \mathcal{B}_4(j,\tau,A,\widetilde{b}_1)\alpha^2, & j \geq 2\tau-1, \end{cases}$$

where $\mathcal{B}_3(j,\tau,A,\widetilde{b}_1) = \frac{2\tau-j-1}{2\tau A^{2\tau}}\left(\frac{j+1}{2\tau-j-1}\right)^{\frac{j+1}{2\tau}}$ and $\mathcal{B}_4(j,\tau,A,\widetilde{b}_1) = \frac{1}{A^{2\tau}\widetilde{b}_1^{(j+1)-2\tau}}$.

Proof.

- If $j \geq 2\tau - 1$, then for $\xi \geq \widetilde{b}_1$ we know that

$$\mathcal{F}(\xi) \leq \frac{\alpha^2 \xi^{2\tau-(j+1)}}{A^{2\tau}} \leq \frac{1}{A^{2\tau}\widetilde{b}_1^{(j+1)-2\tau}}\alpha^2 = \mathcal{B}_4(j,\tau,A,\widetilde{b}_1)\alpha^2. \tag{19}$$

- If $0 < j < 2\tau - 1$, then we have $\lim_{\xi \to 0} \mathcal{F}(\xi) = \lim_{\xi \to \infty} \mathcal{F}(\xi) = 0$, then we know

$$\mathcal{F}(\xi) \leq \sup_{\xi \in (0,+\infty)} \mathcal{F}(\xi) \leq \mathcal{F}(\xi_0).$$

By taking the derivative of $\mathcal{F}$ with respect to ξ, we know that

$$(\mathcal{F})'(\xi) = \frac{A^{2\tau}\alpha^2(2\tau-j-1)\xi^{2\tau-j-2} + \alpha^4(-j-1)\xi^{4\tau-j-2}}{(A^{2\tau} + \alpha^2\xi^{2\tau})^2}. \tag{20}$$

The function $\mathcal{F}(\xi)$ attains maximum at value $\xi = \xi_0$, whereby $\xi_0 \in (0,+\infty)$, which satisfies $(\mathcal{F})'(\xi_0) = 0$. Solving $(\mathcal{F})'(\xi_0) = 0$, we obtain that $\xi_0 = \frac{A(2\tau-j-1)^{\frac{1}{2\tau}}}{\alpha^{\frac{1}{\tau}}(j+1)^{\frac{1}{2\tau}}} > 0$, then we have

$$\mathcal{F}(\xi) \leq \mathcal{F}(\xi_0) = \mathcal{F}\left(\frac{A(2\tau-j-1)^{\frac{1}{2\tau}}}{\alpha^{\frac{1}{\tau}}(j+1)^{\frac{1}{2\tau}}}\right) = \frac{2\tau-j-1}{2\tau A^{2\tau}}\left(\frac{j+1}{2\tau-j-1}\right)^{\frac{j+1}{2\tau}}\alpha^{\frac{j+1}{\tau}}.$$

The proof of Lemma 9 is completed. Our main results are described in the following Theorem. □

Now, we use the separation of variables to yield the solution of (1). Suppose that the solution of (1) is defined by Fourier series

$$u(x,t) = \sum_{i=1}^{\infty} u_i(t)e_i(x), \quad \text{with } u_i(t) = \langle u(\cdot,t), e_i(\cdot) \rangle. \tag{21}$$

Next, we apply the separating variables method and suppose that problem (1) has a solution of the form $u(x,t) = \sum_{i=1}^{\infty} u_i(t)e_i(x)$. Then, $u_i(t)$ is the solution of the following fractional ordinary differential equation with initial conditions as follows:

$$\begin{cases} \frac{\partial^\beta}{\partial t^\beta}u_i(t) = \Delta u_i(t) + \Xi_i(x), & (x,t) \in \Omega \times (0,T), \\ u_i(0) = \langle f(x), e_i(x) \rangle, & x \in \Omega, \\ u_{it}(0) = \langle g(x), e_i(x) \rangle, & x \in \Omega. \end{cases} \tag{22}$$

As Sakamoto and Yamamoto [22], the formula of solution corresponding to the initial value problem for (22) is obtained as follows:

$$u_i(t) = t^\beta E_{\beta,\beta+1}(-\widetilde{b}_i t^\beta) \langle \Xi, e_i \rangle + E_{\beta,1}(-\widetilde{b}_i t^\beta) \langle f, e_i \rangle + t E_{\beta,2}(-\widetilde{b}_i t^\beta) \langle g, e_i \rangle. \quad (23)$$

Hence, we get

$$u(x,t) = \sum_{i=1}^{\infty} \left[t^\beta E_{\beta,\beta+1}(-\widetilde{b}_i t^\beta) \langle \Xi(x), e_i(x) \rangle + E_{\beta,1}(-\widetilde{b}_i t^\beta) \langle f(x), e_i(x) \rangle \right.$$
$$\left. + t E_{\beta,2}(-\widetilde{b}_i t^\beta) \langle g(x), e_i(x) \rangle \right] e_i(x). \quad (24)$$

Letting $t = T$, we obtain

$$u(x,T) = \sum_{i=1}^{\infty} \left[T^\beta E_{\beta,\beta+1}(-\widetilde{b}_i T^\beta) \langle \Xi(x), e_i(x) \rangle + E_{\beta,1}(-\widetilde{b}_i T^\beta) \langle f(x), e_i(x) \rangle \right.$$
$$\left. + T E_{\beta,2}(-\widetilde{b}_i T^\beta) \langle g(x), e_i(x) \rangle \right] e_i(x). \quad (25)$$

From (25) and using final condition $u(x,T) = h(x)$, we get

$$h(x) = \sum_{i=1}^{\infty} \left[T^\beta E_{\beta,\beta+1}(-\widetilde{b}_i T^\beta) \langle \Xi(x), e_i(x) \rangle + E_{\beta,1}(-\widetilde{b}_i T^\beta) \langle f(x), e_i(x) \rangle \right.$$
$$\left. + T E_{\beta,2}(-\widetilde{b}_i T^\beta) \langle g(x), e_i(x) \rangle \right] e_i(x). \quad (26)$$

By denoting $h_i = \langle h(x), e_i(x) \rangle$, $f_i = \langle f(x), e_i(x) \rangle$, $g_i = \langle g(x), e_i(x) \rangle$, and $\Xi_i = \langle \Xi(x), e_i(x) \rangle$, using a simple transformation, we have

$$\Xi_i = \frac{h_i - E_{\beta,1}(-\widetilde{b}_i T^\beta) f_i - T E_{\beta,2}(-\widetilde{b}_i T^\beta) g_i}{T^\beta E_{\beta,\beta+1}(-\widetilde{b}_i T^\beta)}. \quad (27)$$

Then, we receive the formula of the source function $\Xi(x)$

$$\Xi(x) = \sum_{i=1}^{\infty} \frac{\mathcal{R}_i}{T^\beta E_{\beta,\beta+1}(-\widetilde{b}_i T^\beta)} e_i(x), \quad (28)$$

where $\mathcal{R}_i = h_i - E_{\beta,1}(-\widetilde{b}_i T^\beta) f_i - T E_{\beta,2}(-\widetilde{b}_i T^\beta) g_i$.

In the following Theorem, we provide the uniqueness property of the inverse source problem.

Theorem 1. *The couple solution $(u(x,t), \Xi(x))$ of problem (1) is unique.*

Proof. We assume Ξ_1 and Ξ_2 to be the source functions corresponding to the final values $\mathcal{R}_1$ and $\mathcal{R}_2$ in form (27) and (28), respectively, whereby

$$\mathcal{R}_1 = h_1 - E_{\beta,1}(-\widetilde{b}_i T^\beta) f_1 - T E_{\beta,2}(-\widetilde{b}_i T^\beta) g_1,$$
$$\mathcal{R}_2 = h_2 - E_{\beta,1}(-\widetilde{b}_i T^\beta) f_2 - T E_{\beta,2}(-\widetilde{b}_i T^\beta) g_2. \quad (29)$$

Suppose that $h_1 = h_2$, $f_1 = f_2$, and $g_1 = g_2$, then we prove that $\Xi_1 = \Xi_2$. In fact, using the inequality $(a+b+c)^2 \le 3(a^2+b^2+c^2)$, we get

$$\|\mathcal{R}_1 - \mathcal{R}_2\|_{\mathcal{L}^2(\Omega)}^2 \le 3\|h_1 - h_2\|_{\mathcal{L}^2(\Omega)}^2 + 3\big[E_{\beta,1}(-\tilde{b}_iT^\beta)\big]^2\|f_1 - f_2\|_{\mathcal{L}^2(\Omega)}^2$$
$$+ 3T^2\big[E_{\beta,2}(-\tilde{b}_iT^\beta)\big]^2\|g_1 - g_2\|_{\mathcal{L}^2(\Omega)}^2$$
$$\le 3\|h_1 - h_2\|_{\mathcal{L}^2(\Omega)}^2 + \frac{3B^2}{\tilde{b}_1 T^{2\beta}}\|f_1 - f_2\|_{\mathcal{L}^2(\Omega)}^2 + \frac{3T^2 B^2}{\tilde{b}_1 T^{2\beta}}\|g_1 - g_2\|_{\mathcal{L}^2(\Omega)}^2. \quad (30)$$

From (30), we can see that if the right hand side tends to 0, then $\|\mathcal{R}_1 - \mathcal{R}_2\|_{\mathcal{L}^2(\Omega)}^2 \to 0$. Therefore, we have $\mathcal{R}_1 = \mathcal{R}_2$. The proof is completed. □

2.1. The Ill-Posedness of Inverse Source Problem

Theorem 2. *The inverse source problem is ill-posed.*

Define a linear operator $\mathcal{K} : \mathcal{L}^2(\Omega) \to \mathcal{L}^2(\Omega)$ as follows:

$$\mathcal{K}\Xi(x) = \int_\Omega k(x,\omega)\Xi(\omega)d\omega = \mathcal{R}(x), \quad x \in \Omega, \quad (31)$$

where $k(x,\omega)$ is the kernel

$$k(x,\omega) = \sum_{i=1}^\infty T^\beta E_{\beta,\beta+1}(-\tilde{b}_i T^\beta)e_i(x)e_i(\omega). \quad (32)$$

Due to $k(x,\omega) = k(\omega,x)$, we know $\mathcal{K}$ is a self-adjoint operator. Next, we are going to prove its compactness. We use the fractional Tikhonov regularization method to rehabilitate it, where $e_i(x)$ is an orthogonal basis in $\mathcal{L}^2(\Omega)$ and

$$\xi_i = T^\beta E_{\beta,\beta+1}(-\tilde{b}_i T^\beta). \quad (33)$$

Proof. Due to $k(x,\omega) = k(\omega,x)$, we know $\mathcal{K}$ is a self-adjoint operator. Next, we are going to prove its compactness. Defining the finite rank operators $\mathcal{K}_N$ as follows:

$$\mathcal{K}_N\Xi(x) = \sum_{i=1}^N \big[T^\beta E_{\beta,\beta+1}(-\tilde{b}_i T^\beta)\big]\langle\Xi(x), e_i(x)\rangle e_i(x). \quad (34)$$

Then, from (31) and (34) and combining Lemma 5, we have

$$\|\mathcal{K}_N\Xi - \mathcal{K}\Xi\|_{\mathcal{L}^2(\Omega)}^2 = \sum_{i=N+1}^\infty \big[T^\beta E_{\beta,\beta+1}(-\tilde{b}_i T^\beta)\big]^2 \big|\langle\Xi(x), e_i(x)\rangle\big|^2$$
$$\le \sum_{i=N+1}^\infty \frac{B^2}{\tilde{b}_i}\big|\langle\Xi(x), e_i(x)\rangle\big|^2$$
$$\le \frac{B^2}{\tilde{b}_N} \sum_{i=N+1}^\infty \big|\langle\Xi(x), e_i(x)\rangle\big|^2. \quad (35)$$

Therefore, $\|\mathcal{K}_N - \mathcal{K}\|_{\mathcal{L}^2(\Omega)} \to 0$ in the sense of operator norm in $\mathcal{L}(\mathcal{L}^2(\Omega); \mathcal{L}^2(\Omega))$ as $N \to \infty$. Additionally, $\mathcal{K}$ is a compact and self-adjoint operator. Therefore, $\mathcal{K}$ admits an orthonormal eigenbasis e_i in $\mathcal{L}^2(\Omega)$. From (31), the inverse source problem we introduced above can be formulated as an operator equation

$$\mathcal{K}\Xi(x) = \mathcal{R}(x), \quad (36)$$

and by Kirsch [23], we conclude that it is ill-posed. To illustrate an ill-posed problem, we present an example. To perform this example ill-posed, we fix β and let us choose the input data

$$h^m(x) = \frac{e_m(x)}{\sqrt{\widetilde{b}_m}}, \quad g^m = \left(\frac{T^{2-2\beta}}{B^2}\right)\frac{e_m(x)}{\sqrt{\widetilde{b}_m}}, \quad \text{and } f^m = \left(\frac{T^{2\beta}}{B^2}\right)\frac{e_m(x)}{\sqrt{\widetilde{b}_m}}. \tag{37}$$

Due to (28) and combining (37), by (23), the source term corresponding to Ξ^m is

$$\Xi^m(x) = \sum_{i=1}^{\infty} \frac{\langle \mathcal{R}^m(x), e_i(x) \rangle}{T^\beta E_{\beta,\beta+1}(-\widetilde{b}_i T^\beta)} e_i(x)$$

$$= \sum_{i=1}^{\infty} \frac{\langle \frac{e_m(x)}{\sqrt{\widetilde{b}_m}}, e_i(x) \rangle - E_{\beta,1}(-\widetilde{b}_i T^\beta) \langle \frac{e_m}{\sqrt{\widetilde{b}_m}}, e_i(x) \rangle - T E_{\beta,2}(-\widetilde{b}_i T^\beta) \langle \frac{e_m}{\sqrt{\widetilde{b}_m}}, e_i(x) \rangle}{T^\beta E_{\beta,\beta+1}(-\widetilde{b}_i T^\beta)} e_i(x)$$

$$= \frac{e_m(x)}{\sqrt{\widetilde{b}_m}} \frac{\left(1 - E_{\beta,1}(-\widetilde{b}_i T^\beta)\right) - T E_{\beta,2}(-\widetilde{b}_i T^\beta)}{T^\beta E_{\beta,\beta+1}(-\widetilde{b}_i T^\beta)}, \tag{38}$$

where $\mathcal{R}^m = h^m - E_{\beta,1}(-\widetilde{b}_i T^\beta) f^m - T E_{\beta,2}(-\widetilde{b}_i T^\beta) g^m$.

Let us choose other input data $h, f, g = 0$. By (28), the source term corresponding to h, f, g is $\Xi = 0$. An error in $\mathcal{L}^2(\Omega)$ norm between two input final data is

$$\|h^m - h\|_{\mathcal{L}^2(\Omega)} = \left\|\frac{e_m(x)}{\sqrt{\widetilde{b}_m}}\right\|_{\mathcal{L}^2(\Omega)} = \frac{1}{\sqrt{\widetilde{b}_m}},$$

$$\|g^m - g\|_{\mathcal{L}^2(\Omega)} = \left(\frac{T^{2-2\beta}}{B^2}\right)\left\|\frac{e_m(x)}{\sqrt{\widetilde{b}_m}}\right\|_{\mathcal{L}^2(\Omega)} = \left(\frac{T^{2-2\beta}}{B^2}\right)\frac{1}{\sqrt{\widetilde{b}_m}},$$

$$\|f^m - f\|_{\mathcal{L}^2(\Omega)} = \left(\frac{T^{2\beta}}{B^2}\right)\left\|\frac{e_m(x)}{\sqrt{\widetilde{b}_m}}\right\|_{\mathcal{L}^2(\Omega)} = \left(\frac{T^2}{B^2}\right)\frac{1}{\sqrt{\widetilde{b}_m}}, \tag{39}$$

with B as defined in Lemma 5. Therefore,

$$\lim_{m \to +\infty} \|h^m - h\|_{\mathcal{L}^2(\Omega)} = \lim_{m \to +\infty} \frac{1}{\sqrt{\widetilde{b}_m}} = 0,$$

$$\lim_{m \to +\infty} \|g^m - g\|_{\mathcal{L}^2(\Omega)} = \left(\frac{T^{2-2\beta}}{B^2}\right)\lim_{m \to +\infty} \frac{1}{\sqrt{\widetilde{b}_m}} = 0,$$

$$\lim_{m \to +\infty} \|f^m - f\|_{\mathcal{L}^2(\Omega)} = \left(\frac{T^{2\beta}}{B^2}\right)\lim_{m \to +\infty} \frac{1}{\sqrt{\widetilde{b}_m}} = 0. \tag{40}$$

An error in $\mathcal{L}^2$ norm between two corresponding source terms is

$$\|\Xi^m - \Xi\|_{\mathcal{L}^2(\Omega)} = \left\|\frac{e_m(x)\left(1 - E_{\beta,1}(-\widetilde{b}_m T^\beta) - T^\beta E_{\beta,2}(-\widetilde{b}_m T^\beta)\right)}{\sqrt{\widetilde{b}_m} T^\beta E_{\beta,\beta+1}(-\widetilde{b}_i T^\beta)}\right\|_{\mathcal{L}^2(\Omega)}$$

$$= \frac{\left(1 - E_{\beta,1}(-\widetilde{b}_m T^\beta) - T^\beta E_{\beta,2}(-\widetilde{b}_m T^\beta)\right)}{\sqrt{\widetilde{b}_m}\, T^\beta E_{\beta,\beta+1}(-\widetilde{b}_i T^\beta)}. \tag{41}$$

From (41) and using the inequality in Lemma 5, we obtain

$$\|\Xi^m - \Xi\|_{\mathcal{L}^2(\Omega)} \geq \frac{\sqrt{\tilde{b}_m}}{B}\left(1 - E_{\beta,1}(-\tilde{b}_m T^\beta) - T^\beta E_{\beta,2}(-\tilde{b}_m T^\beta)\right). \tag{42}$$

From (42), we have

$$\lim_{m\to+\infty} \|\Xi^m - \Xi\|_{\mathcal{L}^2(\Omega)} > \lim_{m\to+\infty} \frac{\sqrt{\tilde{b}_m}}{B}\left(1 - \frac{\tilde{A}}{\tilde{b}_m T^\beta} - \frac{\tilde{A}}{\tilde{b}_m}\right)$$

$$> \lim_{m\to+\infty} \frac{\sqrt{\tilde{b}_m}}{B}\left(1 - \frac{1}{\sqrt{\tilde{b}_m}}\left(\frac{\tilde{A}}{T^\beta} + \tilde{A}\right)\right) = +\infty. \tag{43}$$

Combining (40) and (43), we conclude that the inverse source problem is ill-posed. □

2.2. Conditional Stability of Source Term $\Xi(x)$

In this section, we show a conditional stability of source function $\Xi(x)$.

Theorem 3. *If* $\|\Xi\|_{H^{\gamma j}(\Omega)} \leq M_1$ *for* $M_1 > 0$, *then*

$$\|\Xi\|_{\mathcal{L}^2(\Omega)} \leq M_1 \left(\frac{T^\beta}{A}\right)^{\frac{j}{j+1}} \|\mathcal{R}\|_{\mathcal{L}^2(\Omega)}^{\frac{j}{j+1}}. \tag{44}$$

Proof. By using the (28) and Hölder inequality, we have

$$\|\Xi\|_{\mathcal{L}^2(\Omega)}^2 = \sum_{i=1}^\infty \Xi_i^2 = \sum_{i=1}^\infty \frac{\mathcal{R}_i^2}{\left|T^\beta E_{\beta,\beta+1}(-\tilde{b}_i T^\beta)\right|^2}$$

$$\leq \left[\sum_{i=1}^\infty \frac{|\mathcal{R}_i|^2}{\left|T^\beta E_{\beta,\beta+1}(-\tilde{b}_i T^\beta)\right|^{2j+2}}\right]^{\frac{1}{j+1}} \left[\sum_{i=1}^\infty |\mathcal{R}_i|^2\right]^{\frac{j}{j+1}}$$

$$\leq \left[\sum_{i=1}^\infty \frac{|\mathcal{R}_i|^2}{\left|T^\beta E_{\beta,\beta+1}(-\tilde{b}_i T^\beta)\right|^{2j}\left|T^\beta E_{\beta,\beta+1}(-\tilde{b}_i T^\beta)\right|^2}\right]^{\frac{1}{j+1}} \|\mathcal{R}_i\|_{\mathcal{L}^2(\Omega)}^{\frac{2j}{j+1}}. \tag{45}$$

Using Lemma (5) leads to

$$\frac{1}{\left|T^\beta E_{\beta,\beta+1}(-\tilde{b}_i T^\beta)\right|^{2j}} \leq \frac{\left|\tilde{b}_i T^\beta\right|^{2j}}{\left|A^{2j}\right|}. \tag{46}$$

Combining (45) and (46), we get

$$\|\Xi\|_{\mathcal{L}^2(\Omega)}^2 \leq \sum_{i=1}^\infty \frac{\left|\tilde{b}_i^j T^{\beta j}\right|^{\frac{2}{j+1}} \|\Xi_i\|^{\frac{2j}{j+1}}}{A^{\frac{2j}{j+1}}} \|\mathcal{R}\|_{\mathcal{L}^2(\Omega)}^{\frac{2j}{j+1}} \leq \|\Xi\|_{H^{\gamma j}(\Omega)}^2 \left(\frac{T^\beta}{A}\right)^{\frac{2j}{j+1}} \|\mathcal{R}\|_{\mathcal{L}^2(\Omega)}^{\frac{2j}{j+1}}. \tag{47}$$

Taking square root in both sides, we have (44). □

3. Regularization of the Inverse Source Problem for the Time-Fractional Diffusion-Wave Equation by the Fractional Tikhonov Method

As mentioned above, applying the fractional Tikhonov regularization method we solve the inverse source problem. Due to singular value decomposition for compact self-adjoint operator $\mathcal{K}$, as in (33). If the measured data $(h^{\varepsilon_1}(x), f^{\varepsilon_2}(x), g^{\varepsilon_3}(x))$ and $(h(x), f(x)), g(x))$ with a noise level of $\varepsilon_1, \varepsilon_2,$ and ε_3 satisfy

$$\|h - h^{\varepsilon_1}\|_{\mathcal{L}^2(\Omega)} \leq \varepsilon_1, \quad \|f - f^{\varepsilon_2}\|_{\mathcal{L}^2(\Omega)} \leq \varepsilon_2, \text{and } \|g - g^{\varepsilon_3}\|_{\mathcal{L}^2(\Omega)} \leq \varepsilon_3, \tag{48}$$

then we can present a regularized solution as follows:

$$\Xi_{\alpha,\tau}^{\varepsilon_1,\varepsilon_2,\varepsilon_3}(x) = \sum_{i=1}^{\infty} \frac{\left(T^{\beta} E_{\beta,\beta+1}(-\widetilde{b}_i T^{\beta})\right)^{2\tau-1}}{\alpha^2 + \left(T^{\beta} E_{\beta,\beta+1}(-\widetilde{b}_i T^{\beta})\right)^{2\tau}} \langle \mathcal{R}^{\varepsilon_1,\varepsilon_2,\varepsilon_3}(x), e_i(x)\rangle e_i(x), \quad \frac{1}{2} \leq \tau \leq 1, \tag{49}$$

where α is a parameter regularization.

$$\Xi_{\alpha,\tau}(x) = \sum_{i=1}^{\infty} \frac{\left(T^{\beta} E_{\beta,\beta+1}(-\widetilde{b}_i T^{\beta})\right)^{2\tau-1}}{\alpha^2 + \left(T^{\beta} E_{\beta,\beta+1}(-\widetilde{b}_i T^{\beta})\right)^{2\tau}} \langle \mathcal{R}(x), e_i(x)\rangle e_i(x), \quad \frac{1}{2} \leq \tau \leq 1, \tag{50}$$

where

$$\mathcal{R}_i^{\varepsilon_1,\varepsilon_2,\varepsilon_3} = h_i^{\varepsilon_1} - E_{\beta,1}(-\widetilde{b}_i T^{\beta}) f_i^{\varepsilon_2} - T E_{\beta,2}(-\widetilde{b}_i T^{\beta}) g_i^{\varepsilon_3},$$

$$\mathcal{R}_i = h_i - E_{\beta,1}(-\widetilde{b}_i T^{\beta}) f_i - T E_{\beta,2}(-\widetilde{b}_i T^{\beta}) g_i. \tag{51}$$

4. A Priori Parameter Choice

Afterwards, we will give an error estimation for $\left\|\Xi(x) - \Xi_{\alpha,\tau}^{\varepsilon_1,\varepsilon_2,\varepsilon_3}(x)\right\|_{\mathcal{L}^2(\Omega)}$ and show convergence rate under a suitable choice for the regularization parameter.

Theorem 4. *Let Ξ be as (28) and the noise assumption (48) hold. Then, we have the following estimate:*

- If $0 \leq j \leq 2\tau$, since $\alpha = \left(\frac{(\max\{\varepsilon_1^2,\varepsilon_2^2,\varepsilon_3^2\})^{\frac{1}{2}}}{M_1}\right)^{\frac{\tau}{j+2}}$ we have

$$\|\Xi - \Xi_{\alpha,\tau}^{\varepsilon_1,\varepsilon_2,\varepsilon_3}\|_{\mathcal{L}^2(\Omega)} \leq \left((\max\{\varepsilon_1^2,\varepsilon_2^2,\varepsilon_3^2\})^{\frac{1}{2}}\right)^{\frac{j}{j+2}} M_1^{\frac{1}{j+2}} \left(\overline{C}(\tau,A)(\mathcal{P}(B,\widetilde{b}_1,T,\beta))^{\frac{1}{2}} + \mathcal{B}_1(j,\tau,A)\right). \tag{52}$$

- If $j \geq 2\tau$, by choosing $\alpha = \left(\frac{(\max\{\varepsilon_1^2,\varepsilon_2^2,\varepsilon_3^2\})^{\frac{1}{2}}}{M_1}\right)^{\frac{\tau}{\tau+2}}$ we have

$$\|\Xi - \Xi_{\alpha,\tau}^{\varepsilon_1,\varepsilon_2,\varepsilon_3}\|_{\mathcal{L}^2(\Omega)}$$

$$\leq \left((\max\{\varepsilon_1^2,\varepsilon_2^2,\varepsilon_3^2\})^{\frac{1}{2}}\right)^{\frac{\tau}{\tau+1}} M_1^{\frac{1}{\tau+1}} \left(\overline{C}(\tau,A)(\mathcal{P}(B,\widetilde{b}_1,T,\beta))^{\frac{1}{2}} + \mathcal{B}_2(j,\tau,A,\widetilde{b}_1)\right), \tag{53}$$

where

$$\mathcal{P}(B,\widetilde{b}_1,T,\beta) = \left(1 + \frac{B^2}{|\widetilde{b}_1 T^{\beta}|^2} + \frac{B^2 T^{2-2\beta}}{|\widetilde{b}_1|^2}\right), \tag{54}$$

M_1 is a positive number satisfies $\|\Xi\|_{H^{\gamma_j}(\Omega)} \leq M_1,$ \hfill (55)

$$\mathcal{B}_1(j,\tau,A) = \frac{(2\tau-j)^{\frac{2\tau-j}{2\tau}} A^{-j} j^{\frac{j}{2\tau}}}{2\tau}, \quad \mathcal{B}_2(j,\tau,A) = \frac{1}{A^{2\tau} \widetilde{b}_1^{j-2\tau}}. \tag{56}$$

Proof. By the triangle inequality, we know

$$\left\| \Xi - \Xi_{\alpha,\tau}^{\varepsilon_1,\varepsilon_2,\varepsilon_3} \right\|_{\mathcal{L}^2(\Omega)} \leq \underbrace{\left\| \Xi_{\alpha,\tau} - \Xi_{\alpha,\tau}^{\varepsilon_1,\varepsilon_2,\varepsilon_3} \right\|_{\mathcal{L}^2(\Omega)}}_{\|\mathcal{K}_1\|_{\mathcal{L}^2(\Omega)}} + \underbrace{\left\| \Xi - \Xi_{\alpha,\tau} \right\|_{\mathcal{L}^2(\Omega)}}_{\|\mathcal{K}_2\|_{\mathcal{L}^2(\Omega)}}. \tag{57}$$

The proof falls naturally into two steps.
Step 1: Estimation for $\|\mathcal{K}_1\|_{\mathcal{L}^2(\Omega)}$, we receive

$$\Xi_{\alpha,\tau}^{\varepsilon_1,\varepsilon_2,\varepsilon_3}(x) - \Xi_{\alpha,\tau}(x) = \sum_{i=1}^{\infty} \frac{\left(T^\beta E_{\beta,\beta+1}(-\tilde{b}_i T^\beta)\right)^{2\tau-1}}{\alpha^2 + \left(T^\beta E_{\beta,\beta+1}(-\tilde{b}_i T^\beta)\right)^{2\tau}} \Big(\langle \mathcal{R}^{\varepsilon_1,\varepsilon_2,\varepsilon_3}(x) - \mathcal{R}(x), e_i(x)\rangle\Big) e_i(x)$$

$$= \sum_{i=1}^{\infty} \frac{\left(T^\beta E_{\beta,\beta+1}(-\tilde{b}_i T^\beta)\right)^{2\tau-1}}{\alpha^2 + \left(T^\beta E_{\beta,\beta+1}(-\tilde{b}_i T^\beta)\right)^{2\tau}} \Big(\langle h_i^{\varepsilon_1} - h_i, e_i(x)\rangle$$

$$+ E_{\beta,1}(-\tilde{b}_i T^\beta)\langle f_i^{\varepsilon_2} - f_i, e_i(x)\rangle$$

$$+ TE_{\beta,2}(-\tilde{b}_i T^\beta)\langle g_i^{\varepsilon_3} - g_i, e_i(x)\rangle\Big) e_i(x). \tag{58}$$

Combining (50) to (51), and Lemma 5, it is easily seen that $\left|T^\beta E_{\beta,\beta}(-\tilde{b}_i T^\beta)\right| \geq \frac{A}{\tilde{b}_i}$. From (58), applying the inequality $(a+b+c)^2 \leq 3a^2 + 3b^2 + 3c^2$ and combining Lemma 7, we know that

$$\|\mathcal{K}_1\|_{\mathcal{L}^2(\Omega)}^2 \leq \sup_{i \in \mathbb{N}} \left(\frac{\tilde{b}_i}{A^{2\tau} + \alpha^2|\tilde{b}_i|^{2\tau}}\right)^2 \left(\sum_{i=1}^{\infty} 3\left|\langle h_i^{\varepsilon_1} - h_i, e_i(x)\rangle\right|^2\right.$$

$$+ 3\sum_{i=1}^{\infty} \left|E_{\beta,1}(-\tilde{b}_i T^\beta)\right|^2 \left|\langle f_i^{\varepsilon_2} - f_i, e_i(x)\rangle\right|^2$$

$$\left. + 3T^2 \sum_{i=1}^{\infty} \left|E_{\beta,2}(-\tilde{b}_i T^\beta)\right|^2 \left|\langle g_i^{\varepsilon_3} - g_i, e_i(x)\rangle\right|^2\right)$$

$$\leq \sup_{i \in \mathbb{N}} \left(\frac{\tilde{b}_i}{A^{2\tau} + \alpha^2|\tilde{b}_i|^{2\tau}}\right)^2 \left(3\varepsilon_1^2 + 3\sum_{i=1}^{\infty} \left|E_{\beta,1}(-\tilde{b}_i T^\beta)\right|^2 \varepsilon_2^2\right.$$

$$\left. + 3T^2 \sum_{i=1}^{\infty} \left|E_{\beta,2}(-\tilde{b}_i T^\beta)\right|^2 \varepsilon_3^2\right). \tag{59}$$

Using the result of Lemma 1 in above, we receive

$$\|\mathcal{K}_1\|_{\mathcal{L}^2(\Omega)}^2 \leq \left(\overline{C}(\tau,A)\alpha^{-\frac{1}{\tau}}\right)^2 \left(3\varepsilon_1^2 + \sum_{i=1}^{\infty} \frac{3B^2\varepsilon_2^2}{|\tilde{b}_i T^\beta|^2} + \sum_{i=1}^{\infty} \frac{3B^2 T^2 \varepsilon_3^2}{|\tilde{b}_i T^\beta|^2}\right)$$

$$\leq \left(\overline{C}(\tau,A)\alpha^{-\frac{1}{\tau}}\right)^2 \left(3\varepsilon_1^2 + \frac{3B^2\varepsilon_2^2}{|\tilde{b}_1 T^\beta|^2} + \frac{3B^2 T^2 \varepsilon_3^2}{|\tilde{b}_1 T^\beta|^2}\right)$$

$$\leq \left(\overline{C}(\tau,A)\alpha^{-\frac{1}{\tau}}\right)^2 \max\{\varepsilon_1^2,\varepsilon_2^2,\varepsilon_3^2\} \left(3 + \frac{B^2}{|\tilde{b}_1 T^\beta|^2} + \frac{B^2 T^{2-2\beta}}{|\tilde{b}_1|^2}\right). \tag{60}$$

Therefore, we have concluded

$$\|\mathcal{K}_1\|_{\mathcal{L}^2(\Omega)} \leq \overline{C}(\tau,A)\alpha^{-\frac{1}{\tau}} \left(\max\{\varepsilon_1^2,\varepsilon_2^2,\varepsilon_3^2\} \mathcal{P}(B,\tilde{b}_1,T,\beta)\right)^{\frac{1}{2}}, \tag{61}$$

where
$$P(B, \widetilde{b}_1, T, \beta) = \left(3 + \frac{B^2}{|\widetilde{b}_1 T^\beta|^2} + \frac{B^2 T^{2-2\beta}}{|\widetilde{b}_1|^2}\right). \tag{62}$$

□

Step 2: Next, we have to estimate $\|\mathcal{K}_2\|_{\mathcal{L}^2(\Omega)}^2$. From (28) and (50), and using Parseval equality, we get

$$\|\mathcal{K}_2\|_{\mathcal{L}^2(\Omega)}^2 \le \sum_{i=1}^{+\infty} \left(\frac{(T^\beta E_{\beta,\beta+1}(-\widetilde{b}_i T^\beta))^{2\tau-1}}{\alpha^2 + (T^\beta E_{\beta,\beta+1}(-\widetilde{b}_i T^\beta))^{2\tau}} - \frac{1}{|T^\beta E_{\beta,\beta+1}(-\widetilde{b}_i T^\beta)|}\right)^2 |\langle \mathcal{R}(x), e_i(x)\rangle|^2$$

$$\le \sum_{i=1}^{+\infty} \left(\frac{\alpha^2}{\left|T^\beta E_{\beta,\beta+1}(-\widetilde{b}_i T^\beta)\right|\left(\alpha^2 + \left|T^\beta E_{\beta,\beta+1}(-\widetilde{b}_i T^\beta)\right|^{2\tau}\right)}\right)^2 |\langle \mathcal{R}(x), e_i(x)\rangle|^2. \tag{63}$$

From (63), we have estimation for $\|\mathcal{K}_2\|_{\mathcal{L}^2(\Omega)}^2$

$$\|\mathcal{K}_2\|_{\mathcal{L}^2(\Omega)}^2 = \sum_{i=1}^{+\infty} \frac{\alpha^4 |\langle \mathcal{R}(x), e_i(x)\rangle|^2}{\left|T^\beta E_{\beta,\beta+1}(-\widetilde{b}_i T^\beta)\right|^2 \left(\alpha^2 + \left|T^\beta E_{\beta,\beta+1}(-\widetilde{b}_i T^\beta)\right|^{2\tau}\right)^2}$$

$$\le \sum_{i=1}^{+\infty} \frac{\alpha^4 \widetilde{b}_i^{2j}\widetilde{b}_i^{-2j} |\langle \mathcal{R}(x), e_i(x)\rangle|^2}{\left|T^\beta E_{\beta,\beta+1}(-\widetilde{b}_i T^\beta)\right|^2 \left(\alpha^2 + \left|T^\beta E_{\beta,\beta+1}(-\widetilde{b}_i T^\beta)\right|^{2\tau}\right)^2}$$

$$\le \sup_{i \in \mathbb{N}} |\mathcal{D}(i)|^2 \sum_{i=1}^{+\infty} \frac{\widetilde{b}_i^{2j} |\langle \mathcal{R}(x), e_i(x)\rangle|^2}{\left|T^\beta E_{\beta,\beta+1}(-\widetilde{b}_i T^\beta)\right|^2} = \sup_{i \in \mathbb{N}} |\mathcal{D}(i)|^2 \|\Xi\|_{H^j(\Omega)}^2. \tag{64}$$

Hence, $\mathcal{D}(i)$ has been estimated

$$\mathcal{D}(i) = \frac{\alpha^2 \widetilde{b}_i^{-j}}{\alpha^2 + \left|T^\beta E_{\beta,\beta+1}(-\widetilde{b}_i T^\beta)\right|^{2\tau}}. \tag{65}$$

Next, using the Lemmas 5 and 8, we continue to estimate $\mathcal{D}(i)$. In fact, we get

$$\mathcal{D}(i) \le \frac{\alpha^2 \widetilde{b}_i^{2\tau-j}}{A^{2\tau} + \alpha^2 \widetilde{b}_i^{2\tau}} \le \begin{cases} \mathcal{B}_1(j,\tau,A)\alpha^{\frac{j}{\tau}}, & 0 < j < 2\tau, \\ \mathcal{B}_2(j,\tau,A,\widetilde{b}_1)\alpha^2, & j \ge 2\tau. \end{cases} \tag{66}$$

Combining (64) to (66), we receive

$$\|\mathcal{K}_2\|_{\mathcal{L}^2(\Omega)}^2 \le \begin{cases} \mathcal{B}_1(j,\tau,A) M_1 \alpha^{\frac{j}{\tau}}, & 0 < j < 2\tau, \\ \mathcal{B}_2(j,\tau,A,\widetilde{b}_1) M_1 \alpha^2, & j \ge 2\tau. \end{cases} \tag{67}$$

Next, combining the above two inequalities, we obtain

$$\|\Xi(x) - \Xi_{\alpha,\tau}^{\varepsilon_1,\varepsilon_2,\varepsilon_3}(x)\|_{\mathcal{L}^2(\Omega)} \le \overline{C}(\tau,A)\alpha^{-\frac{1}{\tau}} \left(\max\{\varepsilon_1^2,\varepsilon_2^2,\varepsilon_3^2\} P(B,\widetilde{b}_1,T,\beta,\gamma)\right)^{\frac{1}{2}}$$

$$+ \begin{cases} \mathcal{B}_1(j,\tau,A) M_1 \alpha^{\frac{j}{\tau}}, & 0 < j < 2\tau, \\ \mathcal{B}_2(j,\tau,A,\widetilde{b}_1) M_1 \alpha^2, & j \ge 2\tau. \end{cases} \tag{68}$$

Choose the regularization parameter α as follows:

$$\alpha = \begin{cases} \left(\dfrac{(\max\{\varepsilon_1^2,\varepsilon_2^2,\varepsilon_3^2\})^{\frac{1}{2}}}{M_1}\right)^{\frac{T}{j+2}}, & 0 < j < 2\tau, \\ \left(\dfrac{(\max\{\varepsilon_1^2,\varepsilon_2^2,\varepsilon_3^2\})^{\frac{1}{2}}}{M_1}\right)^{\frac{T}{\tau+1}}, & j \geq 2\tau. \end{cases} \qquad (69)$$

Hence, we conclude that

Case 1: If $0 \leq j \leq 2\tau$, since $\alpha = \left(\dfrac{(\max\{\varepsilon_1^2,\varepsilon_2^2,\varepsilon_3^2\})^{\frac{1}{2}}}{M_1}\right)^{\frac{T}{j+2}}$ we have

$$\|\Xi(x) - \Xi_{\alpha,\tau}^{\varepsilon_1,\varepsilon_2,\varepsilon_3}(x)\|_{L^2(\Omega)}$$

$$\leq \left((\max\{\varepsilon_1^2,\varepsilon_2^2,\varepsilon_3^2\})^{\frac{1}{2}}\right)^{\frac{j}{j+2}} M_1^{\frac{1}{j+2}} \left(\overline{C}(\tau,A)(P(B,\tilde{b}_1,T,\beta,\gamma))^{\frac{1}{2}} + \mathscr{B}_1(j,\tau,A)\right). \qquad (70)$$

Case 2: If $j \geq 2\tau$, since $\alpha = \left(\dfrac{(\max\{\varepsilon_1^2,\varepsilon_2^2,\varepsilon_3^2\})^{\frac{1}{2}}}{M_1}\right)^{\frac{T}{\tau+2}}$ we have

$$\|\Xi(x) - \Xi_{\alpha,\tau}^{\varepsilon_1,\varepsilon_2,\varepsilon_3}(x)\|_{L^2(\Omega)}$$

$$\leq \left((\max\{\varepsilon_1^2,\varepsilon_2^2,\varepsilon_3^2\})^{\frac{1}{2}}\right)^{\frac{\tau}{\tau+1}} M_1^{\frac{1}{\tau+1}} \left(\overline{C}(\tau,A)(P(B,\tilde{b}_1,T,\beta,\gamma))^{\frac{1}{2}} + \mathscr{B}_2(j,\tau,A,\tilde{b}_1)\right). \qquad (71)$$

5. A Posteriori Parameter Choice

In this section, we consider an a posteriori regularization parameter choice in Morozov's discrepancy principle (see in [21]). We use the discrepancy principle in the following form:

$$\left\|\dfrac{(T^\beta E_{\beta,\beta+1}(-\tilde{b}_i T^\beta))^{2\tau}}{\alpha^2 + \left(T^\beta E_{\beta,\beta+1}(-\tilde{b}_i T^\beta)\right)^{2\tau}} \mathcal{R}^{\varepsilon_1,\varepsilon_2,\varepsilon_3}(x) - \mathcal{R}^{\varepsilon_1,\varepsilon_2,\varepsilon_3}(x)\right\|_{L^2(\Omega)} = k(\max\{\varepsilon_1^2,\varepsilon_2^2,\varepsilon_3^2\})^{\frac{1}{2}}, \qquad (72)$$

whereby $\frac{1}{2} \leq \tau \leq 1$, $k > 1$, and α is the regularization parameter.

Lemma 10. *Let*

$$\rho(\alpha) = \sqrt{\sum_{i=1}^{\infty}\left(\dfrac{\alpha^2}{\alpha^2 + \left(T^\beta E_{\beta,\beta+1}(-\tilde{b}_i T^\beta)\right)^{2\tau}}\right)^2 \left|\langle \mathcal{R}^{\varepsilon_1,\varepsilon_2,\varepsilon_3}(x), e_i(x)\rangle\right|^2}. \qquad (73)$$

If $0 < k(\max\{\varepsilon_1^2,\varepsilon_2^2,\varepsilon_3^2\})^{\frac{1}{2}} < \|\mathcal{R}^{\varepsilon_1,\varepsilon_2,\varepsilon_3}\|_{L^2(\Omega)}$, then the following results hold:

(a) $\rho(\alpha)$ is a continuous function;
(b) $\rho(\alpha) \to 0$ as $\alpha \to 0$;
(c) $\rho(\alpha) \to \|\mathcal{R}^{\varepsilon_1,\varepsilon_2,\varepsilon_3}\|_{L^2(\Omega)}$ as $\alpha \to \infty$;
(d) $\rho(\alpha)$ is a strictly increasing function.

Lemma 11. *Let α be the solution of (72), it gives*

$$\dfrac{1}{\alpha^{\frac{1}{\tau}}} \leq \begin{cases} \dfrac{(\sqrt{2}\mathscr{B}_3(j,\tau,A))^{\frac{1}{j+1}}}{(k^2 - 6P(B,\tilde{b}_1,T,\beta))^{\frac{1}{2(j+1)}}} \dfrac{M_1^{\frac{1}{j+1}}}{(\max\{\varepsilon_1^2,\varepsilon_2^2,\varepsilon_3^2\})^{\frac{1}{2(j+1)}}}, & 0 < j < 2\tau - 1, \\ \dfrac{(\sqrt{2}\mathscr{B}_4(j,\tau,A,\tilde{b}_1))^{\frac{1}{2\tau}}}{(k^2 - 6P(B,\tilde{b}_1,T,\beta))^{\frac{1}{4\tau}}} \dfrac{M_1^{\frac{1}{2\tau}}}{(\max\{\varepsilon_1^2,\varepsilon_2^2,\varepsilon_3^2\})^{\frac{1}{4\tau}}}, & j \geq 2\tau - 1, \end{cases} \qquad (74)$$

which gives the required results.

Proof. Step 1: First of all, we have the error estimation between $\mathcal{R}^{\varepsilon_1,\varepsilon_2,\varepsilon_3}$ and $\mathcal{R}$. Indeed, using the inequality $(a+b+c)^2 \leq 3(a^2+b^2+c^2), \forall a,b,c \geq 0$, we get

$$\|\mathcal{R}^{\varepsilon_1,\varepsilon_2,\varepsilon_3} - \mathcal{R}\|_{\mathcal{L}^2(\Omega)}^2 \leq \left(3\varepsilon_1^2 + 3\sum_{i=1}^{\infty}\left|E_{\beta,1}(-\widetilde{b}_iT^\beta)\right|^2\varepsilon_2^2 + 3T^2\sum_{i=1}^{\infty}\left|E_{\beta,2}(-\widetilde{b}_iT^\beta)\right|^2\varepsilon_3^2\right),$$

$$\leq \left(3\varepsilon_1^2 + \frac{3B^2\varepsilon_2^2}{|\widetilde{b}_1T^\beta|^2} + \frac{3B^2T^2\varepsilon_3^2}{|\widetilde{b}_1T^\beta|^2}\right) \leq 3\left(\max\left\{\varepsilon_1^2,\varepsilon_2^2,\varepsilon_3^2\right\}\right)\mathcal{P}(B,\widetilde{b}_1,T,\beta). \quad (75)$$

Step 2: Using the inequality $(a+b)^2 \leq 2(a^2+b^2), \forall a,b \geq 0$, we can receive the following estimation

$$k^2\left(\max\left\{\varepsilon_1^2,\varepsilon_2^2,\varepsilon_3^2\right\}\right) = \sum_{i=1}^{\infty}\left(\frac{\alpha^2}{\alpha^2 + \left|T^\beta E_{\beta,\beta+1}(-\widetilde{b}_iT^\beta)\right|^{2\tau}}\right)^2 \left|\left\langle\mathcal{R}_i^{\varepsilon_1,\varepsilon_2,\varepsilon_3}(x),e_i(x)\right\rangle\right|^2$$

$$\leq 2\sum_{i=1}^{\infty}\left(\frac{\alpha^2}{\alpha^2 + \left|T^\beta E_{\beta,\beta+1}(-\widetilde{b}_iT^\beta)\right|^{2\tau}}\right)^2 \left|\left\langle\mathcal{R}_i^{\varepsilon_1,\varepsilon_2,\varepsilon_3}(x) - \mathcal{R}(x),e_i(x)\right\rangle\right|^2$$

$$+ 2\sum_{i=1}^{\infty}\left(\frac{\alpha^2\left|T^\beta E_{\beta,\beta+1}(-\widetilde{b}_iT^\beta)\right|}{\left[\alpha^2 + \left|T^\beta E_{\beta,\beta+1}(-\widetilde{b}_iT^\beta)\right|^{2\tau}\right]\widetilde{b}_i^j}\right)^2 \frac{\widetilde{b}_i^{2j}\left|\langle\mathcal{R}(x),e_i(x)\rangle\right|^2}{\left|T^\beta E_{\beta,\beta+1}(-\widetilde{b}_iT^\beta)\right|^2}. \quad (76)$$

From (76), we get

$$k^2\left(\max\left\{\varepsilon_1^2,\varepsilon_2^2,\varepsilon_3^2\right\}\right) \leq 2\sum_{i=1}^{\infty}\left|\left\langle\mathcal{R}_i^{\varepsilon_1,\varepsilon_2,\varepsilon_3}(x) - \mathcal{R}(x),e_i(x)\right\rangle\right|^2$$

$$+ 2\sum_{i=1}^{\infty}\left(\frac{\alpha^2\left|T^\beta E_{\beta,\beta+1}(-\widetilde{b}_iT^\beta)\right|}{\left[\alpha^2 + \left|T^\beta E_{\beta,\beta+1}(-\widetilde{b}_iT^\beta)\right|^{2\tau}\right]\widetilde{b}_i^j}\right)^2 \frac{\widetilde{b}_i^{2j}\left|\langle\mathcal{R}(x),e_i(x)\rangle\right|^2}{\left|T^\beta E_{\beta,\beta+1}(-\widetilde{b}_iT^\beta)\right|^2}$$

$$\leq 6\left(\max\left\{\varepsilon_1^2,\varepsilon_2^2,\varepsilon_3^2\right\}\right)\mathcal{P}(B,\widetilde{b}_1,T,\beta,\gamma)$$

$$+ 2\sum_{i=1}^{\infty}\left(\frac{\alpha^2\left|T^\beta E_{\beta,\beta+1}(-\widetilde{b}_iT^\beta)\right|}{\left[\alpha^2 + \left|T^\beta E_{\beta,\beta+1}(-\widetilde{b}_iT^\beta)\right|^{2\tau}\right]\widetilde{b}_i^j}\right)^2 \frac{\widetilde{b}_i^{2j}\left|\langle\mathcal{R}(x),e_i(x)\rangle\right|^2}{\left|T^\beta E_{\beta,\beta+1}(-\widetilde{b}_iT^\beta)\right|^2}$$

$$\leq 6\left(\max\left\{\varepsilon_1^2,\varepsilon_2^2,\varepsilon_3^2\right\}\right)\mathcal{P}(B,\widetilde{b}_1,T,\beta)$$

$$+ 2\sum_{i=1}^{\infty}|\mathcal{H}_i|^2 \frac{\widetilde{b}_i^{2j}\left|\langle\mathcal{R}(x),e_i(x)\rangle\right|^2}{\left|T^\beta E_{\beta,\beta+1}(-\widetilde{b}_iT^\beta)\right|^2}, \quad (77)$$

whereby

$$\mathcal{H}_i = \frac{\alpha^2\left|T^\beta E_{\beta,\beta+1}(-\widetilde{b}_iT^\beta)\right|}{\left[\alpha^2 + \left|T^\beta E_{\beta,\beta+1}(-\widetilde{b}_iT^\beta)\right|^{2\tau}\right]\widetilde{b}_i^j}. \quad (78)$$

From (78), we get $\mathcal{H}_i$ as follows:

$$\mathcal{H}_i = \frac{\alpha^2 T^\beta E_{\beta,\beta+1}(-\widetilde{b}_iT^\beta)}{\left[\alpha^2 + \left|T^\beta E_{\beta,\beta+1}(-\widetilde{b}_iT^\beta)\right|^{2\tau}\right]\widetilde{b}_i^j} \leq \frac{\alpha^2 T^\beta \frac{B}{T^\beta \widetilde{b}_i^j}}{\left[\alpha^2 + \left|\frac{T^\beta A}{T^\beta \widetilde{b}_i^j}\right|^{2\tau}\right]\widetilde{b}_i^j} \leq \frac{\alpha^2 B}{A^{2\tau} + \alpha^2 \widetilde{b}_i^{2\tau}}\widetilde{b}_i^{2\tau-(j+1)}. \quad (79)$$

From (79), using Lemma 9, we have

$$\mathcal{H}_i \leq \begin{cases} \mathcal{B}_3(j,\tau,A)\alpha^{\frac{j+1}{\tau}}, & 0 < j < 2\tau - 1, \\ \mathcal{B}_4(j,\tau,A,\widetilde{b}_1)\alpha^2, & j \geq 2\tau - 1. \end{cases}$$

Therefore, combining (77) to (79), we know that

$$k^2(\max\{\varepsilon_1^2, \varepsilon_2^2, \varepsilon_3^2\}) \leq 6(\max\{\varepsilon_1^2, \varepsilon_2^2, \varepsilon_3^2\})\mathcal{P}(B,\widetilde{b}_1,T,\beta) \\ + \begin{cases} 2\mathcal{B}_3^2(j,\tau,A)M_1^2\alpha^{2\frac{j+1}{\tau}}, & 0 < j < 2\tau - 1, \\ 2\mathcal{B}_4^2(j,\tau,A,\widetilde{b}_1)M_1^2\alpha^4, & j \geq 2\tau - 1. \end{cases} \qquad (80)$$

From (80), it is very easy to see that

$$(k^2 - 6\mathcal{P}(B,\widetilde{b}_1,T,\beta))(\max\{\varepsilon_1^2, \varepsilon_2^2, \varepsilon_3^2\}) \leq \begin{cases} 2\mathcal{B}_3^2(j,\tau,A)M_1^2\alpha^{2\frac{j+1}{\tau}}, & 0 < j < 2\tau - 1, \\ 2\mathcal{B}_4^2(j,\tau,A,\widetilde{b}_1)M_1^2\alpha^4, & j \geq 2\tau - 1. \end{cases} \qquad (81)$$

So,

$$\frac{1}{\alpha^{\frac{1}{\tau}}} \leq \begin{cases} \dfrac{(\sqrt{2}\mathcal{B}_3(j,\tau,A))^{\frac{1}{j+1}}}{(k^2 - 6\mathcal{P}(B,\widetilde{b}_1,T,\beta))^{\frac{1}{2(j+1)}}} \dfrac{M_1^{\frac{1}{j+1}}}{(\max\{\varepsilon_1^2,\varepsilon_2^2,\varepsilon_3^2\})^{\frac{1}{2(j+1)}}}, & 0 < j < 2\tau - 1, \\ \dfrac{(\sqrt{2}\mathcal{B}_4(j,\tau,A,\widetilde{b}_1))^{\frac{1}{2\tau}}}{(k^2 - 6\mathcal{P}(B,\widetilde{b}_1,T,\beta))^{\frac{1}{4\tau}}} \dfrac{M_1^{\frac{1}{2\tau}}}{(\max\{\varepsilon_1^2,\varepsilon_2^2,\varepsilon_3^2\})^{\frac{1}{4\tau}}}, & j \geq 2\tau - 1, \end{cases} \qquad (82)$$

which gives the required results. The estimation of $\|\Xi(x) - \Xi_{\alpha,\tau}^{\varepsilon_1,\varepsilon_2,\varepsilon_3}(x)\|_{\mathcal{L}^2(\Omega)}$ is established by our next Theorem. □

Theorem 5. *Assume the a priori condition and the noise assumption hold, and there exists $\tau > 1$ such that $0 < k\left(\max\{\varepsilon_1^2, \varepsilon_2^2, \varepsilon_3^2\}\right)^{\frac{1}{2}} < \|\mathcal{R}^{\varepsilon_1,\varepsilon_2,\varepsilon_3}\|_{\mathcal{L}^2(\Omega)}$. This Theorem now shows the convergent estimate between the exact solution and the regularized solution such that*

- *If $0 \leq j \leq 2\tau - 1$, we have the convergence estimate*

$$\|\Xi(x) - \Xi_{\alpha,\tau}^{\varepsilon_1,\varepsilon_2,\varepsilon_3}(x)\|_{\mathcal{L}^2(\Omega)} \leq \mathcal{Q}(j,T,A,B,\beta,k,\widetilde{b}_1)\left(\max\{\varepsilon_1^2,\varepsilon_2^2,\varepsilon_3^2\}\right)^{\frac{j}{2(j+1)}} M_1^{\frac{1}{j+1}}. \qquad (83)$$

- *If $j \geq 2\tau - 1$, we have the convergence estimate*

$$\|\Xi(x) - \Xi_{\alpha,\tau}^{\varepsilon_1,\varepsilon_2,\varepsilon_3}(x)\|_{\mathcal{L}^2(\Omega)} \leq \mathscr{L}(\overline{C},T,A,B,\beta,k,j,\widetilde{b}_1) \times \left(\max\{\varepsilon_1^2,\varepsilon_2^2,\varepsilon_3^2\}\right)^{\frac{1}{2}\left(1-\frac{1}{2\tau}\right)} M_1^{\frac{1}{2\tau}}, \qquad (84)$$

whereby

$$Q(j,T,A,B,\beta,k,\widetilde{b}_1) = \frac{\overline{C}(\tau,A)\left(\sqrt{2}\mathcal{B}_3(j,\tau,A)\right)^{\frac{1}{j+1}}\mathcal{P}(B,\widetilde{b}_1,T,\beta))^{\frac{1}{2}}}{(k^2 - 6\mathcal{P}(B,\widetilde{b}_1,T,\beta))^{\frac{1}{2(j+1)}}}$$
$$+ \left(\frac{\left(\sqrt{3}(\mathcal{P}(B,\widetilde{b}_1,T,\beta))^{\frac{1}{2}} + k\right)}{A}\right)^{\frac{j}{j+1}},$$

$$\mathcal{P}(B,\widetilde{b}_1,T,\beta,\gamma) = \left(1 + \frac{B^2}{|\widetilde{b}_1 T^\beta|^2} + \frac{B^2 T^{2-2\beta}}{|\widetilde{b}_1|^2}\right),$$

$$\mathcal{L}(\overline{C},T,A,B,\beta,k,j,\widetilde{b}_1) = \left[\frac{\overline{C}(\tau,A)\left(\sqrt{2}\mathcal{B}_4(j,\tau,A,\widetilde{b}_1)\right)^{\frac{1}{2\tau}}(\mathcal{P}(B,\widetilde{b}_1,T,\beta))^{\frac{1}{2}}}{(k^2 - 6\mathcal{P}(B,\widetilde{b}_1,T,\beta))^{\frac{1}{4\tau}}}\right.$$
$$\left. + \left(\frac{\sqrt{3}(\mathcal{P}(B,\widetilde{b}_1,T,\beta))^{\frac{1}{2}} + k}{A}\right)^{1-\frac{1}{2\tau}}\widetilde{b}_1^{2\tau-j-1}\right]. \quad (85)$$

Proof. Applying the triangle inequality, we get

$$\left\|\Xi(x) - \Xi_{\alpha,\tau}^{\varepsilon_1,\varepsilon_2,\varepsilon_3}(x)\right\|_{\mathcal{L}^2(\Omega)} \le \left\|\Xi(x) - \Xi_{\alpha,\tau}(x)\right\|_{\mathcal{L}^2(\Omega)} + \left\|\Xi_{\alpha,\tau}(x) - \Xi_{\alpha,\tau}^{\varepsilon_1,\varepsilon_2,\varepsilon_3}(x)\right\|_{\mathcal{L}^2(\Omega)}. \quad (86)$$

Case 1: If $0 < j \le 2\tau - 1$. First of all, we recalled estimation from (61) and, by Lemma 9 Part (a), we have

$$\left\|\Xi_{\alpha,\tau}(x) - \Xi_{\alpha,\tau}^{\varepsilon_1,\varepsilon_2,\varepsilon_3}(x)\right\|_{\mathcal{L}^2(\Omega)} \le Q(j,T,A,B,\beta,k,\widetilde{b}_1)\left(\max\{\varepsilon_1^2,\varepsilon_2^2,\varepsilon_3^2\}\right)^{\frac{j}{2(j+1)}} M_1^{\frac{1}{j+1}}. \quad (87)$$

Next, we have estimate $\left\|\Xi(x) - \Xi_{\alpha,\tau}(x)\right\|_{\mathcal{L}^2(\Omega)}$. From (28) and (50), and using Parseval equality, we get

$$\left\|\Xi(x) - \Xi_{\alpha,\tau}(x)\right\|_{\mathcal{L}^2(\Omega)}$$
$$= \left\|\sum_{i=1}^{+\infty}\left(\frac{(T^\beta E_{\beta,\beta+1}(-\widetilde{b}_i T^\beta))^{2\tau-1}}{\alpha^2 + (T^\beta E_{\beta,\beta+1}(-\widetilde{b}_i T^\beta))^{2\tau}} - \frac{1}{(T^\beta E_{\beta,\beta+1}(-\widetilde{b}_i T^\beta))}\right)\langle \mathcal{R}(x),e_i(x)\rangle e_i(x)\right\|_{\mathcal{L}^2(\Omega)}$$
$$= \left\|\sum_{i=1}^{+\infty}\frac{\alpha^2}{\alpha^2 + (T^\beta E_{\beta,\beta+1}(-\widetilde{b}_i T^\beta))^{2\tau}}\langle \Xi(x),e_i(x)\rangle e_i(x)\right\|_{\mathcal{L}^2(\Omega)}$$
$$= \left\|\sum_{i=1}^{+\infty}\frac{\alpha^2 T^\beta E_{\beta,\beta+1}(-\widetilde{b}_i T^\beta)}{\alpha^2 + (T^\beta E_{\beta,\beta+1}(-\widetilde{b}_i T^\beta))^{2\tau}}\frac{\langle \Xi(x),e_i(x)\rangle}{T^\beta E_{\beta,\beta+1}(-\widetilde{b}_i T^\beta)}e_i(x)\right\|_{\mathcal{L}^2(\Omega)}. \quad (88)$$

Using the Hölder's inequality, we obtain

$$\|\Xi(x) - \Xi_{\alpha,\tau}(x)\|_{\mathcal{L}^2(\Omega)}$$

$$\leq \left\| \sum_{i=1}^{+\infty} \frac{\alpha^2 T^\beta E_{\beta,\beta+1}(-\widetilde{b}_i T^\beta)}{\alpha^2 + (T^\beta E_{\beta,\beta+1}(-\widetilde{b}_i T^\beta))^{2\tau}} \langle \Xi(x), e_i(x) \rangle e_i(x) \right\|_{\mathcal{L}^2(\Omega)}^{\frac{j}{j+1}}$$

$$\times \left\| \sum_{i=1}^{+\infty} \frac{\alpha^2 T^\beta E_{\beta,\beta+1}(-\widetilde{b}_i T^\beta)}{\alpha^2 + (T^\beta E_{\beta,\beta+1}(-\widetilde{b}_i T^\beta))^{2\tau}} \frac{\langle \Xi(x), e_i(x) \rangle}{(T^\beta E_{\beta,\beta+1}(-\widetilde{b}_i T^\beta))^{j+1}} e_i(x) \right\|_{\mathcal{L}^2(\Omega)}^{\frac{1}{j+1}}$$

$$\leq \underbrace{\left\| \sum_{i=1}^{+\infty} \frac{\alpha^2}{\alpha^2 + (T^\beta E_{\beta,\beta+1}(-\widetilde{b}_i T^\beta))^{2\tau}} \langle \mathcal{R}(x), e_i(x) \rangle e_i(x) \right\|_{\mathcal{L}^2(\Omega)}^{\frac{j}{j+1}}}_{\mathcal{A}_1}$$

$$\times \underbrace{\left\| \sum_{i=1}^{+\infty} \frac{\alpha^2}{\alpha^2 + (T^\beta E_{\beta,\beta+1}(-\widetilde{b}_i T^\beta))^{2\tau}} \frac{\langle \Xi(x), e_i(x) \rangle}{(T^\beta E_{\beta,\beta+1}(-\widetilde{b}_i T^\beta))^j} e_i(x) \right\|_{\mathcal{L}^2(\Omega)}^{\frac{1}{j+1}}}_{\mathcal{A}_2}. \tag{89}$$

From (89) and (75), using Lemma 11, one has

$$\mathcal{A}_1 \leq \left(\left\| \sum_{i=1}^{+\infty} \frac{\alpha^2}{(T^\beta E_{\beta,\beta+1}(-\widetilde{b}_i T^\beta))^{2\tau} + \alpha^2} \langle \mathcal{R}(x) - \mathcal{R}^{\varepsilon_1,\varepsilon_2,\varepsilon_3}(x), e_i(x) \rangle e_i(x) \right\|_{\mathcal{L}^2(\Omega)} \right.$$

$$\left. + \left\| \sum_{i=1}^{+\infty} \frac{\alpha^2}{(T^\beta E_{\beta,\beta+1}(-\widetilde{b}_i T^\beta))^{2\tau} + \alpha^2} \langle \mathcal{R}^{\varepsilon_1,\varepsilon_2,\varepsilon_3}(x), e_i(x) \rangle e_i(x) \right\|_{\mathcal{L}^2(\Omega)} \right)^{\frac{j}{j+1}}$$

$$\leq \left(\sqrt{3} \left(\max\{\varepsilon_1^2, \varepsilon_2^2, \varepsilon_3^2\} \right)^{\frac{1}{2}} (\mathcal{P}(B, \widetilde{b}_1, T, \beta))^{\frac{1}{2}} + k \left(\max\{\varepsilon_1^2, \varepsilon_2^2, \varepsilon_3^2\} \right)^{\frac{1}{2}} \right)^{\frac{j}{j+1}}$$

$$\leq \left(\sqrt{3} (\mathcal{P}(B, \widetilde{b}_1, T, \beta))^{\frac{1}{2}} + k \right)^{\frac{j}{j+1}} \left(\max\{\varepsilon_1^2, \varepsilon_2^2, \varepsilon_3^2\} \right)^{\frac{j}{2(j+1)}}. \tag{90}$$

Next, using the priori condition a, we have

$$\mathcal{A}_2 = \left\| \sum_{i=1}^{+\infty} \frac{\alpha^2}{\alpha^2 + (T^\beta E_{\beta,\beta+1}(-\widetilde{b}_i T^\beta))^{2\tau}} \frac{\langle \Xi(x), e_i(x) \rangle}{(T^\beta E_{\beta,\beta+1}(-\widetilde{b}_i T^\beta))^j} e_i(x) \right\|_{\mathcal{L}^2(\Omega)}^{\frac{1}{j+1}}$$

$$\leq \left\| \sum_{i=1}^{+\infty} \frac{\langle \Xi(x), e_i(x) \rangle}{(T^\beta E_{\beta,\beta+1}(-\widetilde{b}_i T^\beta))^j} e_i(x) \right\|_{\mathcal{L}^2(\Omega)}^{\frac{1}{j+1}}$$

$$\leq \left\| \sum_{i=1}^{+\infty} \left(\frac{\widetilde{b}_i}{A} \right)^j \Xi_i e_i(x) \right\|_{\mathcal{L}^2(\Omega)}^{\frac{1}{j+1}} \leq \frac{1}{A^{\frac{j}{j+1}}} M_1^{\frac{1}{j+1}}. \tag{91}$$

Combining (88) to (91), we conclude that

$$\|\Xi(x) - \Xi_{\alpha,\tau}(x)\|_{\mathcal{L}^2(\Omega)} \leq \left(\frac{(\sqrt{3}(\mathcal{P}(B, \widetilde{b}_1, T, \beta))^{\frac{1}{2}} + k)}{A} \right)^{\frac{j}{j+1}} \left(\max\{\varepsilon_1^2, \varepsilon_2^2, \varepsilon_3^2\} \right)^{\frac{j}{2(j+1)}} M_1^{\frac{1}{j+1}}. \tag{92}$$

Combining (87) to (92), we know that

$$\|\Xi(x) - \Xi_{\alpha,\tau}^{\varepsilon_1,\varepsilon_2,\varepsilon_3}(x)\|_{\mathcal{L}^2(\Omega)} \leq \mathcal{Q}(j, T, A, B, \beta, k, \widetilde{b}_1) \left(\max\{\varepsilon_1^2, \varepsilon_2^2, \varepsilon_3^2\} \right)^{\frac{j}{2(j+1)}} M_1^{\frac{1}{j+1}}, \tag{93}$$

whereby

$$Q(j,T,A,B,\beta,k,\widetilde{b}_1) = \frac{\overline{C}(\tau,A)\left(\sqrt{2}\mathscr{B}_3(j,\tau,A)\right)^{\frac{1}{j+1}}\mathcal{P}(B,\widetilde{b}_1,T,\beta))^{\frac{1}{2}}}{\left(k^2 - 6\mathcal{P}(B,\widetilde{b}_1,T,\beta)\right)^{\frac{1}{2(j+1)}}}$$

$$+ \left(\frac{(\sqrt{3}(\mathcal{P}(B,\widetilde{b}_1,T,\beta))^{\frac{1}{2}} + k)}{A}\right)^{\frac{j}{j+1}},$$

$$\mathcal{P}(B,\widetilde{b}_1,T,\beta,\gamma) = \left(1 + \frac{B^2}{|\widetilde{b}_1 T^\beta|^2} + \frac{B^2 T^{2-2\beta}}{|\widetilde{b}_1|^2}\right). \tag{94}$$

Case 2: Our next goal is to determine the estimation of $\left\|\Xi_{\alpha,\tau}(x) - \Xi_{\alpha,\tau}^{\varepsilon_1,\varepsilon_2,\varepsilon_3}(x)\right\|_{\mathcal{L}^2(\Omega)}$ when $j \geq 2\tau - 1$, we get

$$\left\|\Xi_{\alpha,\tau}(x) - \Xi_{\alpha,\tau}^{\varepsilon_1,\varepsilon_2,\varepsilon_3}(x)\right\|_{\mathcal{L}^2(\Omega)} \leq \frac{\overline{C}(\tau,A)\left(\sqrt{2}\mathscr{B}_4(j,\tau,A,\widetilde{b}_1)\right)^{\frac{1}{2\tau}}(\mathcal{P}(B,\widetilde{b}_1,T,\beta))^{\frac{1}{2}}}{\left(k^2 - 6\mathcal{P}(B,\widetilde{b}_1,T,\beta)\right)^{\frac{1}{4\tau}}}$$

$$\times M_1^{\frac{1}{2\tau}}\left(\max\{\varepsilon_1^2,\varepsilon_2^2,\varepsilon_3^2\}\right)^{\frac{1}{2}(1-\frac{1}{2\tau})}. \tag{95}$$

Next, for $\|\Xi(x) - \Xi_{\alpha,\tau}(x)\|_{\mathcal{L}^2(\Omega)}$, we get

$$\|\Xi(x) - \Xi_{\alpha,\tau}(x)\|_{\mathcal{L}^2(\Omega)} = \left\|\sum_{i=1}^{+\infty} \frac{\alpha^2}{\alpha^2 + (T^\beta E_{\beta,\beta+1}(-\widetilde{b}_i T^\beta))^{2\tau}} \langle\Xi(x),e_i(x)\rangle e_i(x)\right\|_{\mathcal{L}^2(\Omega)}$$

$$= \left\|\sum_{i=1}^{+\infty} \frac{\alpha^2 T^\beta E_{\beta,\beta+1}(-\widetilde{b}_i T^\beta)}{\alpha^2 + (T^\beta E_{\beta,\beta+1}(-\widetilde{b}_i T^\beta))^{2\tau}} \frac{\langle\Xi(x),e_i(x)\rangle}{T^\beta E_{\beta,\beta+1}(-\widetilde{b}_i T^\beta)} e_i(x)\right\|_{\mathcal{L}^2(\Omega)}$$

$$\leq \underbrace{\left\|\sum_{i=1}^{+\infty} \frac{\alpha^2 T^\beta E_{\beta,\beta+1}(-\widetilde{b}_i T^\beta)}{\alpha^2 + (T^\beta E_{\beta,\beta+1}(-\widetilde{b}_i T^\beta))^{2\tau}} \langle\Xi(x),e_i(x)\rangle e_i(x)\right\|_{\mathcal{L}^2(\Omega)}^{1-\frac{1}{2\tau}}}_{\mathcal{B}_1}$$

$$\times \underbrace{\left\|\sum_{i=1}^{+\infty} \frac{\alpha^2 T^\beta E_{\beta,\beta+1}(-\widetilde{b}_i T^\beta)}{\alpha^2 + (T^\beta E_{\beta,\beta+1}(-\widetilde{b}_i T^\beta))^{2\tau}} \frac{\langle\Xi(x),e_i(x)\rangle}{(T^\beta E_{\beta,\beta+1}(-\widetilde{b}_i T^\beta))^{2\tau}} e_i(x)\right\|_{\mathcal{L}^2(\Omega)}^{\frac{1}{2\tau}}}_{\mathcal{B}_2}. \tag{96}$$

From (96), repeated application of Lemma 11 Part (b) enables us to write $\mathcal{B}_1$, it is easy to check that

$$\mathcal{B}_1 \leq \left(\sqrt{3}\left(\max\{\varepsilon_1^2,\varepsilon_2^2,\varepsilon_3^2\}\right)^{\frac{1}{2}}(\mathcal{P}(B,\widetilde{b}_1,T,\beta))^{\frac{1}{2}} + k\left(\max\{\varepsilon_1^2,\varepsilon_2^2,\varepsilon_3^2\}\right)^{\frac{1}{2}}\right)^{1-\frac{1}{2\tau}}$$

$$\leq \left(\sqrt{3}(\mathcal{P}(B,\widetilde{b}_1,T,\beta))^{\frac{1}{2}} + k\right)^{1-\frac{1}{2\tau}}\left(\max\{\varepsilon_1^2,\varepsilon_2^2,\varepsilon_3^2\}\right)^{\frac{1}{2}(1-\frac{1}{2\tau})}. \tag{97}$$

In the same way as in $\mathcal{A}_2$, it follows easily that $\dfrac{\alpha^2}{\alpha^2 + (T^\beta E_{\beta,\beta+1}(-\tilde{b}_i T^\beta))^{2\tau}} < 1$, we now proceed by induction

$$B_2 \leq \left\|\sum_{i=1}^{+\infty} \frac{\langle \Xi(x), e_i(x)\rangle}{(T^\beta E_{\beta,\beta+1}(-\tilde{b}_i T^\beta))^{2\tau-1}} e_i(x)\right\|_{L^2(\Omega)}^{\frac{1}{2\tau}}$$

$$\leq \left\|\sum_{i=1}^{+\infty} \left(\frac{\tilde{b}_i}{A}\right)^{2\tau-1} \tilde{b}_i^{-j} \tilde{b}_i^j \Xi_i e_i(x)\right\|_{L^2(\Omega)}^{\frac{1}{2\tau}} \leq A^{\frac{1}{2\tau}-1} \tilde{b}_1^{2\tau-j-1} M_1^{\frac{1}{2\tau}}. \tag{98}$$

Combining (86) and (95)–(98), it may be concluded that

$$\|\Xi(x) - \Xi_{\alpha,\tau}^{\varepsilon_1,\varepsilon_2,\varepsilon_3}(x)\|_{L^2(\Omega)} \leq \mathscr{L}(\overline{C}, T, A, B, \beta, k, j, \tilde{b}_1) \times \left(\max\left\{\varepsilon_1^2, \varepsilon_2^2, \varepsilon_3^2\right\}\right)^{\frac{1}{2}\left(1-\frac{1}{2\tau}\right)} M_1^{\frac{1}{2\tau}}, \tag{99}$$

whereby

$$\mathscr{L}(\overline{C}, T, A, B, \beta, k, j, \tilde{b}_1) = \left[\frac{\overline{C}(\tau, A)(\sqrt{2}\mathcal{B}_4(j,\tau,A,\tilde{b}_1))^{\frac{1}{2\tau}}(\mathcal{P}(B,\tilde{b}_1,T,\beta))^{\frac{1}{2}}}{(k^2 - 6\mathcal{P}(B,\tilde{b}_1,T,\beta))^{\frac{1}{4\tau}}}\right.$$

$$\left. + \left(\frac{\sqrt{3}(\mathcal{P}(B,\tilde{b}_1,T,\beta))^{\frac{1}{2}} + k}{A}\right)^{1-\frac{1}{2\tau}} \tilde{b}_1^{2\tau-j-1}\right]. \tag{100}$$

The proof is completed. □

6. Simulation Example

In this section, we are going to show an example to simulate the theory. In order to do this, we consider the problem as follows:

$$\partial_{0+}^\beta u(x,t) = \frac{\partial^2}{\partial x^2} u(x,t) + \Xi(x), \ (x,t) \in (0,\pi) \times (0,1), \tag{101}$$

where the Caputo fractional derivative of order β is defined as

$$\partial_{0+}^\beta u(x,t) = \frac{1}{\Gamma(2-\alpha)} \int_0^t \frac{\partial^2 u(x,s)}{\partial s^2} \frac{ds}{(t-s)^{\beta-1}}, \ 1 < \beta < 2, \tag{102}$$

where $\Gamma(.)$ is the Gamma function.

We chose the operator $\Delta u = \frac{\partial^2}{\partial x^2} u$ on the domain $\Omega = (0,\pi)$ with the Dirichlet boundary condition $u(0,t) = u(\pi,t) = 0$ for $t \in (0,1)$, we have the eigenvalues and corresponding eigenvectors given by $\tilde{b}_i = i^2$, $i = 1,2,...$ and $e_i(x) = \sqrt{\dfrac{2}{\pi}} \sin(ix)$, respectively.

In addition, problem (101) satisfies the conditions

$$u(x,0) = f(x), \ \frac{\partial}{\partial t}u(x,0) = g(x), \ x \in (0,\pi)$$

and the final condition

$$u(x,1) = h(x), x \in (0,\pi). \tag{103}$$

We consider the following assumptions:

$$f(x) = \sqrt{\frac{2}{\pi}} \sin(2x),$$

$$g(x) = \sqrt{\frac{2}{\pi}} \sin(x),$$

$$h(x) = \sqrt{\frac{2}{\pi}} \Big[E_{\beta,2}(-1) \sin(x) + E_{\beta,1}(-2) \sin(2x) + E_{\beta,\beta+1}(-3) \sin(3x) \Big].$$

In this example, we choose the following solution

$$u(x,t) = \sqrt{\frac{2}{\pi}} \Big[t E_{\beta,2}(-t^\beta) \sin(x) + E_{\beta,1}(-2t^\beta) \sin(2x) + t^\beta E_{\beta,\beta+1}(-3t^\beta) \sin(3x) \Big]. \quad (104)$$

Before giving the main results of this section, we present some of the following numerical approximation methods.

- Composite Simpson's rule: Suppose that the interval $[a,b]$ is split up into n sub-intervals, with n being an even number. Then, the composite Simpson's rule is given by

$$\int_a^b \varphi(z)\, dz \approx \frac{h}{3} \sum_{j=1}^{n/2} \Big[\varphi(z_{2j-2}) + 4\varphi(z_{2j-1}) + \varphi(z_{2j}) \Big]$$

$$= \frac{h}{3} \Big[\varphi(z_0) + 2 \sum_{j=1}^{n/2-1} \varphi(z_{2j}) + 4 \sum_{j=1}^{n/2} \varphi(z_{2j-1}) + \varphi(z_n) \Big], \quad (105)$$

where $z_j = a + jh$ for $j = 0,1,...,n-1,n$ with $h = \dfrac{b-a}{n}$, in particular, $z_0 = a$ and $z_n = b$.

- For a,b are two positive integers given. We use the finite difference method to discretize the time and spatial variable for $(x,t) \in (0,\pi) \times (0,1)$ as follows:

$$x_p = p\Delta x, \quad t_q = q\Delta t, \quad 0 \le p \le X, \quad 0 \le q \le T,$$

$$\Delta x = \frac{\pi}{X}, \quad \Delta t = \frac{1}{T}.$$

- Explicit forward Euler method: Let $u_p^q = u(x_p, t_q)$, then the finite difference approximations are given by

$$\frac{\partial^2 u(x_p,t_q)}{\partial x^2} = \frac{u_{p+1}^q - 2u_p^q + u_{p-1}^q}{\Delta x^2}, \quad (106)$$

$$\frac{\partial^2 u(x_p,t_q)}{\partial t^2} = \frac{u_p^{q+1} - 2u_p^q + u_p^{q-1}}{\Delta t^2}. \quad (107)$$

Instead of getting accurate data (h,f,g), we get approximated data of (h,f,g), i.e., the input data (h,f,g) is noised by observation data $(h^{\varepsilon_1}, g^{\varepsilon_2}, h^{\varepsilon_3})$ with order of $\varepsilon_1, \varepsilon_2, \varepsilon_3 > 0$ which satisfies

$$h^{\varepsilon_1} = h + \varepsilon_1(\text{rand}(\cdot) - 1),$$
$$f^{\varepsilon_2} = f + \varepsilon_2(2\,\text{rand}(\cdot) + 1),$$
$$g^{\varepsilon_3} = g + \varepsilon_3(\text{rand}(\cdot) - 2),$$

where, in Matlab software, the $\text{rand}(\cdot)$ function generates arrays of random numbers whose elements are uniformly distributed in the interval $(0,1)$.

The absolute error estimation is defined by

$$\text{Error}^{\beta,\varepsilon,\alpha,\tau} = \left[\frac{1}{X-1}\sum_{p=1}^{X-1}\left(\Xi(x_p) - \Xi_{\alpha,\tau}^{\varepsilon_1,\varepsilon_2,\varepsilon_3}(x_p)\right)^2\right]^{1/2},$$

where $\frac{1}{2} \leq \tau \leq 1$ and $\alpha = \left(\frac{(\max\{\varepsilon_1^2,\varepsilon_2^2,\varepsilon_3^2\})^{\frac{1}{2}}}{M_1}\right)^{\frac{T}{\tau+2}}$.

From the above analysis, we present some results as follows.

In Table 1, we show the convergent estimate between Ξ and $\Xi_{\alpha,\tau}^{\varepsilon_1,\varepsilon_2,\varepsilon_3}$ with a priori and a posteriori parameter choice rules. From the observations on this table, we can conclude that the approximation result is acceptable. Moreover, we also present the graph of the source functions with cases of the input data noise and the corresponding errors, respectively (see Figures 1–3). In addition, the solution $u(x,t)$ is also shown in Figure 4 for $0 \leq x \leq \pi$ and $0 \leq t \leq 1$.

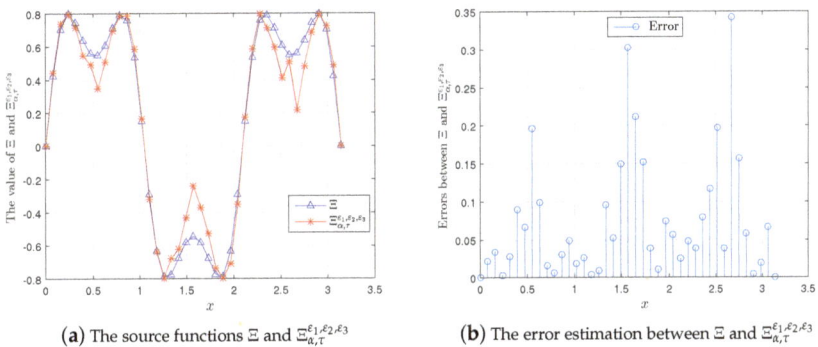

Figure 1. A comparison between Ξ and $\Xi_{\alpha,\tau}^{\varepsilon_1,\varepsilon_2,\varepsilon_3}$ for $\beta = 1.5$, $X = T = 40$, $\{\varepsilon_1, \varepsilon_2, \varepsilon_3\} := \{9 \times 10^{-2}, 2 \times 10^{-2}, 1 \times 10^{-3}\}$, $\tau = \frac{4}{5}$.

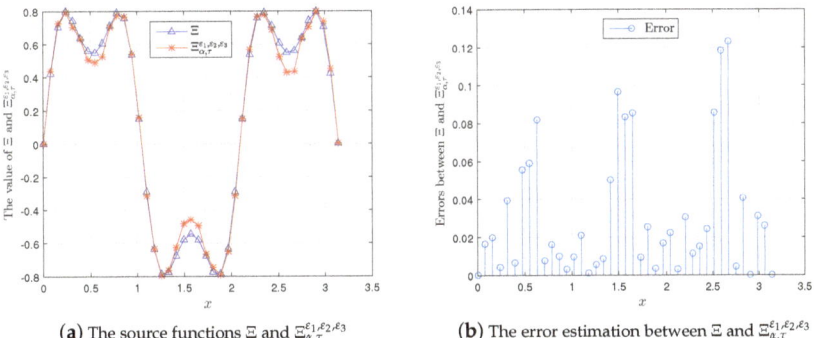

Figure 2. A comparison between Ξ and $\Xi_{\alpha,\tau}^{\varepsilon_1,\varepsilon_2,\varepsilon_3}$ for $\beta = 1.5$, $X = T = 40$, $\{\varepsilon_1, \varepsilon_2, \varepsilon_3\} := \{1 \times 10^{-3}, 2 \times 10^{-3}, 3 \times 10^{-3}\}$, $\tau = \frac{4}{5}$.

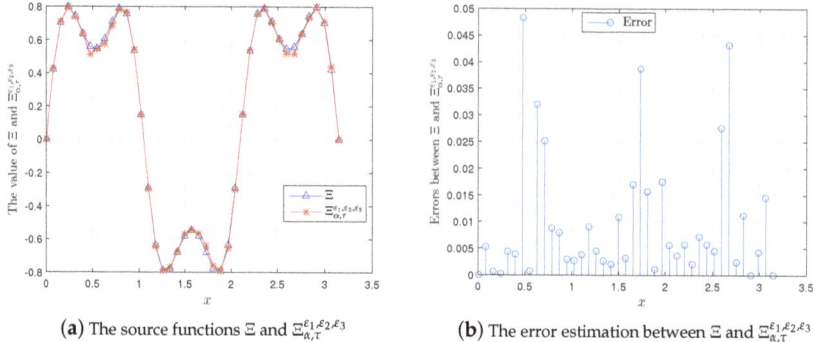

(a) The source functions Ξ and $\Xi_{\alpha,\tau}^{\varepsilon_1,\varepsilon_2,\varepsilon_3}$

(b) The error estimation between Ξ and $\Xi_{\alpha,\tau}^{\varepsilon_1,\varepsilon_2,\varepsilon_3}$

Figure 3. A comparison between Ξ and $\Xi_{\alpha,\tau}^{\varepsilon_1,\varepsilon_2,\varepsilon_3}$ for $\beta = 1.5$, $X = T = 40$, $\{\varepsilon_1,\varepsilon_2,\varepsilon_3\} := \{3 \times 10^{-1}, 2 \times 10^{-2}, 5 \times 10^{-1}\}$, $\tau = \dfrac{4}{5}$.

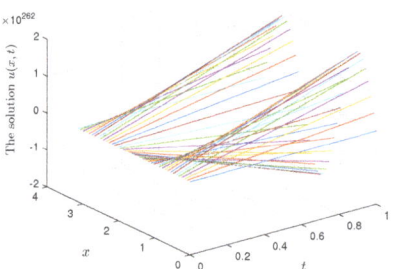

Figure 4. The solution $u(x,t)$ for $(x,t) \in (0,\pi) \times (0,1)$.

Table 1. The errors estimation between Ξ and $\Xi_{\alpha,\tau}^{\varepsilon_1,\varepsilon_2,\varepsilon_3}$ at $\beta = 1.5$ with $X = T = 40$, $\tau = \dfrac{4}{5}$.

$\{\varepsilon_1, \varepsilon_2, \varepsilon_3\}$	$X = 40, T = 40$	
	$\text{Error}_{priori}^{\beta,\varepsilon,\alpha,\tau}$	$\text{Error}_{posteriori}^{\beta,\varepsilon,\alpha,\tau}$
$\{3 \times 10^{-1}, 2 \times 10^{-2}, 5 \times 10^{-1}\}$	0.164478172012052	0.182258736154960
$\{9 \times 10^{-2}, 2 \times 10^{-2}, 1 \times 10^{-3}\}$	0.031066747441897	0.030595088570760
$\{1 \times 10^{-3}, 2 \times 10^{-3}, 3 \times 10^{-3}\}$	0.014676586512256	0.015071362259137

7. Conclusions

In this study, we use the Tikhonov method to regularize the inverse problem to determine an unknown source term in a space-time-fractional diffusion equation. By an example, we prove that this problem is ill-posed in the sense of Hadamard. Under *a priori* and *a posteriori* parameter choice rules, we show the results about the convergent estimate between the exact solution and the regularized solution. In addition, we show an example to illustrate our proposed regularization.

Author Contributions: Project administration, Y.Z.; Resources, L.D.L.; Methodology, N.H.L.; Writing—review, editing and software, C.N.

Funding: This research received no external funding.

Conflicts of Interest: The authors declare no conflict of interest.

References

1. Podlubny, I. Fractional Differential Equations. In *Mathematics in Science and Engineering*; Academic Press Inc.: San Diego, CA, USA, 1990; Volume 198.
2. Nguyen, H.T.; Le, D.L.; Nguyen, T.V. Regularized solution of an inverse source problem for a time fractional diffusion equation. *Appl. Math. Model.* **2016**, *40*, 8244–8264. [CrossRef]
3. Wei, T.; Wang, J. A modified quasi-boundary value method for an inverse source problem of the time-fractional diffusion equation. *Appl. Numer. Math.* **2014**, *78*, 95–111. [CrossRef]
4. Wang, J.G.; Zhou, Y.B.; Wei, T. Two regularization methods to identify a space-dependent source for the time-fractional diffusion equation. *Appl. Numer. Math.* **2013**, *68*, 39–57. [CrossRef]
5. Zhang, Z.Q.; Wei, T. Identifying an unknown source in time-fractional diffusion equation by a truncation method. *Appl. Math. Comput.* **2013**, *219*, 5972–5983. [CrossRef]
6. Yang, F.; Liu, X.; Li, X.X. Landweber iterative regularization method for identifying the unknown source of the modified Helmholtz equation. *Bound. Value Probl.* **2017**, *91*. [CrossRef]
7. Tatar, S.; Ulusoy, S. An inverse source problem for a one-dimensional space-time fractional diffusion equation. *Appl. Anal.* **2015**, *94*, 2233–2244. [CrossRef]
8. Tatar, S.; Tinaztepe, R.; Ulusoy, S. Determination of an unknown source term in a space-time fractional diffusion equation. *J. Frac. Calc. Appl.* **2015**, *6*, 83–90.
9. Tatar, S.; Tinaztepe, R.; Ulusoy, S. Simultaneous inversion for the exponents of the fractional time and space derivatives in the space-time fractional diffusion equation. *Appl. Anal.* **2016**, *95*, 1–23. [CrossRef]
10. Tuan, N.H.; Long, L.D. Fourier truncation method for an inverse source problem for space-time fractional diffusion equation. *Electron. J. Differ. Equ.* **2017**, *2017*, 1–16.
11. Mehrdad, L.; Dehghan, M. The use of Chebyshev cardinal functions for the solution of a partial differential equation with an unknown time-dependent coefficient subject to an extra measurement. *J. Comput. Appl. Math.* **2010**, *235*, 669–678.
12. Pollard, H. The completely monotonic character of the Mittag-Leffler function $E_\alpha(-x)$. *Bull. Am. Math. Soc.* **1948**, *54*, 1115–1116. [CrossRef]
13. Yang, M.; Liu, J.J. Solving a final value fractional diffusion problem by boundary condition regularization. *Appl. Numer. Math.* **2013**, *66*, 45–58. [CrossRef]
14. Kilbas, A.A.; Srivastava, H.M.; Trujillo, J.J. *Theory and Applications of Fractional Differential Equations*; Elsevier Science Limited: Amsterdam, The Netherlands, 2006.
15. Luchko, Y. Initial-boundary-value problems for the one-dimensional time-fractional diffusion equation. *Fract. Calc. Appl. Anal.* **2012**, *15*, 141–160. [CrossRef]
16. Quan, Z.; Feng, X.L. A fractional Tikhonov method for solving a Cauchy problem of Helmholtz equation. *Appl. Anal.* **2017**, *96*, 1656–1668. [CrossRef]
17. Trifce, S.; Tomovski, Z. The general time fractional wave equation for a vibrating string. *J. Phys. A Math. Theor.* **2010**, *43*, 055204.
18. Trifce, S.; Ralf, M.; Zivorad, T. Fractional diffusion equation with a generalized Riemann–Liouville time fractional derivative. *J. Phys. A Math. Theor.* **2011**, *44*, 255203.
19. Hilfer, R.; Seybold, H.J. Computation of the generalized Mittag-Leffler function and its inverse in the complex plane. *Integral Transform Spec. Funct.* **2006**, *17*, 37–652.
20. Seybold, H.; Hilfer, R. Numerical Algorithm for Calculating the Generalized Mittag-Leffler Function. *SIAM J. Numer. Anal.* **2008**, *47*, 69–88. [CrossRef]
21. Kilbas, A.A.; Srivastava, H.M.; Trujillo, J.J. Theory and Application of Fractional differential equations. In *North—Holland Mathematics Studies*; Elsevier Science B.V.: Amsterdam, The Netherlands, 2006; Volume 204. [CrossRef]

22. Sakamoto, K.; Yamamoto, M. Initial value/boundary value problems for fractional diffusion-wave equations and applications to some inverse problems. *J. Math. Anal. Appl.* **2011**, *382*, 426–447.
23. Kirsch, A. *An Introduction to the Mathematical Theory of Inverse Problem*; Springer: Berlin, Germany, 1996. [CrossRef]

© 2019 by the authors. Licensee MDPI, Basel, Switzerland. This article is an open access article distributed under the terms and conditions of the Creative Commons Attribution (CC BY) license (http://creativecommons.org/licenses/by/4.0/).

Article

Approximate Solutions of Time Fractional Diffusion Wave Models

Abdul Ghafoor [1], Sirajul Haq [2], Manzoor Hussain [2], Poom Kumam [3,4,5],* and Muhammad Asif Jan [1]

[1] Institute of Numerical Sciences, Kohat University of Science and Technology, Kohat 26000, KP, Pakistan; abdulghafoor@kust.edu.pk (A.G.); majan@kust.edu.pk (M.A.J.)
[2] Faculty of Engineering Sciences, GIK Institute, Topi 23640, KP, Pakistan; siraj@giki.edu.pk (S.H.); ges1612@giki.edu.pk (M.H.)
[3] Theoretical and Computational Science (TaCS) Center Department of Mathematics, Faculty of Science, King Mongkuts University of Technology Thonburi (KMUTT), 126 Pracha Uthit Rd., Bang Mod, Thung Khru, Bangkok 10140, Thailand
[4] KMUTT-Fixed Point Research Laboratory, Room SCL 802 Fixed Point Laboratory, Science Laboratory Building, Department of Mathematics, Faculty of Science, King Mongkut's University of Technology Thonburi (KMUTT), 126 Pracha-Uthit Road, Bang Mod, Thrung Khru, Bangkok 10140, Thailand
[5] Department of Medical Research, China Medical University Hospital, China Medical University, Taichung 40402, Taiwan
* Correspondence: poom.kum@kmutt.ac.th

Received: 11 July 2019; Accepted: 27 August 2019; Published: 3 October 2019

Abstract: In this paper, a wavelet based collocation method is formulated for an approximate solution of (1 + 1)- and (1 + 2)-dimensional time fractional diffusion wave equations. The main objective of this study is to combine the finite difference method with Haar wavelets. One and two dimensional Haar wavelets are used for the discretization of a spatial operator while time fractional derivative is approximated using second order finite difference and quadrature rule. The scheme has an excellent feature that converts a time fractional partial differential equation to a system of algebraic equations which can be solved easily. The suggested technique is applied to solve some test problems. The obtained results have been compared with existing results in the literature. Also, the accuracy of the scheme has been checked by computing L_2 and L_∞ error norms. Computations validate that the proposed method produces good results, which are comparable with exact solutions and those presented before.

Keywords: fractional differential equations; two-dimensional wavelets; finite differences

---

1. Introduction

The theory of fractional calculus is an ancient topic that has many applications. However, practical work in this direction has been recently started (see References [1–3]). Most of the physical phenomena in chemistry, physics, engineering and other fields of science can be modeled using parameters of fractional calculus [4,5], means fractional derivative and integral operators. Amongst these are electrolyte polarization [6], viscoelastic systems [7], dielectric polarization [8] and so forth. Fractional models in different circumstances lead towards more accurate behaviour than those of integer order models.

The time fractional diffusion wave equation (TFDWE) is such an important model which has extensive uses. The TFDWE is actually a wave equation [9] with a fractional time derivative which describes universal acoustic, electromagnetic and mechanical responses [10,11] with an enhanced method. Over the past few decades, extensive attention has been paid to the closed form solution of time fractional diffusion wave equations (TFDWEs) and is still an open area of research. The closed form solution of such problems is not an easy job and needs herculean efforts. Owing to the fact several authors proposed numerical methods for the solution of fractional models, Tadjeran et al. [12] used second order accurate approximation for fractional diffusion equations. Zhuang et al. [13] applied an implicit numerical method for the anomalous sub-diffusion equation. Yuste and Acedo [14] studied fractional diffusion equations via an explicit finite difference method. Chen et al. [15] proposed the Fourier method for fractional diffusion equations. Hosseini et al. [16] solved the fractional telegraph equation with the help of radial basis functions. Zhou and Xu [17] applied the Chebyshev wavelets collocation method for the solution of time fractional diffusion wave equations. Bhrawya [18] used the spectral Tau algorithm based on the Jacobi operational matrix for the numerical solution of time fractional diffusion-wave equations. Yaseen et al. [19] solved fractional diffusion wave equations with reaction terms using finite differences and a trigonometric B-splines technique. Khader [20] and his co-author applied the finite difference method coupled with the Hermite formula for solutions of fractional diffusion wave equations. Kanwal et al. [21] implemented two-dimensional Genocchi Polynomials combined with the Ritz-Galerkin Method for solutions of fractional diffusion wave and Klein-Gordon equations. Datsko et al. [22] studied time-fractional diffusion-wave equation with mass absorption in a sphere under harmonic impact.

Recently, numerical methods using wavelets have been given more emphasis because of their simple applicability. These methods also have some other interesting properties such as the ability to detect singularities and express the function in different resolution levels, which improves the accuracy. Amongst different classes of wavelets, Haar wavelets deserve special consideration. Haar wavelets consist of piece wise constant functions. The integration of these wavelets in different times is one of the best features. Also, Haar wavelets have orthogonality and normalization properties with compact support. For more discussion on Haar wavelets one can see References [23,24].

In the present study, we propose a hybrid numerical scheme, based on Haar wavelets and finite differences, to solve (1 + 1)- and (1 + 2)-dimensional TFDWEs. The stability of the proposed method is discussed with the matrix method which is an essential part of the manuscript. The models which will be under consideration are characterized in the following types:

(1 + 1)-Dimensional Equation:

$$^cD_t^\delta w(x,t) = -w_t(x,t) + w_{xx}(x,t) + \mathcal{A}(x,t), \quad x \in \Omega, \quad t \in [0,T], \quad 1 < \delta \leq 2, \tag{1}$$

$$\begin{cases} w(x,0) = f(x), \quad w_t(x,0) = g(x) \quad x \in \tilde{\Omega} = \Omega \cup \partial\Omega, \\ w(x,t) = \alpha(t), \quad x \in \partial\Omega \quad t \in [0,T]. \end{cases} \tag{2}$$

(1 + 2)-Dimensional Equation:

$$^cD_t^\delta w(x,y,t) = \Delta w(x,y,t) + \mathcal{B}(x,y,t), \quad (x,y) \in \Phi, \quad t \in [0,T], \quad 1 < \delta \leq 2, \tag{3}$$

$$\begin{cases} w(x,y,0) = \chi(x,y), \quad w_t(x,y,0) = \kappa(x,y), \quad (x,y) \in \tilde{\Phi} = \Phi \cup \partial\Phi, \\ w(x,y,t) = \chi_1(x,y,t), \quad (x,y) \in \partial\Phi, \quad t \in [0,T]. \end{cases} \tag{4}$$

In Equations (1)–(4), Δ is two-dimensional Laplacian; $\mathcal{A}$, $\mathcal{B}$, f, g, α, χ, κ, χ_1 are known functions and w is unknown function. Equations (2) and (4) are the corresponding initial and boundary conditions. The symbols, Ω and $\partial\Omega$, Φ and $\partial\Phi$ represent the domain and boundary of the domain respectively for

(1 + 1)- and (1 + 2)-dimensional problems. Also ${}^cD_t^\delta w$ denotes the time fractional derivative of w with respect to t in the Caputo sense which is given by

$$
{}^cD_t^\delta w = \begin{cases} \frac{1}{\Gamma(2-\delta)} \int_0^t \frac{w_{\zeta\zeta}(x,\zeta)}{(t-\zeta)^{\delta-1}} d\zeta, & 1 < \delta < 2, \\ \frac{\partial^2 w(x,t)}{\partial t^2}, & \delta = 2. \end{cases} \tag{5}
$$

2. Ground Work

In this section, some basic definitions of fractional calculus and Haar wavelets are presented, which will be required for the demonstration of our results. For a basic definition of Haar wavelets and its integrals we refer to Reference [23]. Let us consider $x \in [a,b]$ where a and b are the limits of the interval. Next, the interval is subdivided into $2M$ intervals where $M = 2^J$ and J denote the maximal level of resolution. Further, the two parameters $j = 0, \cdots, J$ and $k = 0, \cdots, 2^j - 1$ are introduced. These parameters show the integer decomposition of wavelet number $i = m + k + 1$, where $m = 2^j$. The first and ith wavelets are defined as follows:

$$
\mathcal{H}_1(x) = \begin{cases} 1, & x \in [a,b] \\ 0, & \text{otherwise}. \end{cases} \tag{6}
$$

$$
\mathcal{H}_i(x) = \begin{cases} 1, & x \in [\zeta_1(i), \zeta_2(i)) \\ -1, & x \in [\zeta_2(i), \zeta_3(i)) \\ 0, & \text{otherwise}, \end{cases} \tag{7}
$$

where

$$
\zeta_1(i) = a + 2kv\delta x, \quad \zeta_2(i) = a + (2k+1)v\delta x, \quad \zeta_3(i) = a + 2(k+1)v\delta x, \quad v = \frac{M}{m}, \quad \delta x = \frac{b-a}{2M}.
$$

To solve nth order time fractional PDEs the following repeated integrals are needed:

$$
\mathcal{P}_{i,\beta}(x) = \int_a^x \int_a^x \cdots \int_a^x \mathcal{H}_i(z) dz^\beta = \frac{1}{(\beta-1)!} \int_a^x (x-z)^{\beta-1} \mathcal{H}_i(z) dz, \tag{8}
$$

where

$$
\beta = 1, 2, \ldots n, \quad i = 1, 2, \ldots 2M.
$$

Keeping in view Equations (6) and (7) the close form expressions of these integrals are given by

$$
\mathcal{P}_{1,\beta}(x) = \frac{(x-a)^\beta}{\beta!}. \tag{9}
$$

$$
\mathcal{P}_{i,\beta}(x) = \begin{cases} 0, & x < \zeta_1(x) \\ \frac{1}{\beta!}(x - \zeta_1(i))^\beta & x \in [\zeta_1(i), \zeta_2(i)) \\ \frac{1}{\beta!}[(x - \zeta_1(i))^\beta - 2((x - \zeta_2(i))^\beta)] & x \in [\zeta_2(i), \zeta_3(i)) \\ \frac{1}{\beta!}[(x - \zeta_1(i))^\beta - 2((x - \zeta_2(i))^\beta + (x - \zeta_3(i))^\beta)] & x \geq \zeta_3(i). \end{cases} \tag{10}
$$

3. Description of the Method

This section is devoted to discussing the scheme for Equations (1) and (3) separately. In both cases, the fractional order time derivative has been approximated by the quadrature formula [16]

$$
\begin{aligned}
{}^c D_t^\delta w(x, t^{j+1}) &= \frac{1}{\Gamma(2-\delta)} \int_0^{t^{j+1}} w^{(2)}(x, \zeta) \left(t^{j+1} - \zeta\right)^{1-\delta} d\zeta \\
&= \frac{1}{\Gamma(2-\delta)} \sum_{k=0}^{j} \int_{t^j}^{t^{j+1}} \left[\frac{w^{k+1} - 2w^k + w^{k-1}}{\tau^2} \right] \left(t^{j+1} - \zeta\right)^{1-\delta} d\zeta \\
&= \frac{1}{\Gamma(2-\delta)} \sum_{k=0}^{j} \left[\frac{w^{k+1} - 2w^k + w^{k-1}}{\tau^2} \right] \int_{t^j}^{t^{j+1}} \left[(j+1)\tau - \zeta\right]^{1-\delta} d\zeta \\
&= \frac{1}{\Gamma(2-\delta)} \sum_{k=0}^{j} \left[\frac{w^{k+1} - 2w^k + w^{k-1}}{\tau^2} \right] \frac{(j-k+1)^{2-\delta} - (j-k)^{2-\delta}}{(2-\delta)(\tau^{\delta-2})} \\
&= \frac{\tau^{-\delta}}{\Gamma(3-\delta)} \sum_{k=0}^{j} \left[w^{j-k+1} - 2w^{j-k} + w^{j-k-1} \right] \left[(k+1)^{2-\delta} - (k)^{2-\delta} \right] \\
&= A_\delta \left[w^{j+1} - 2w^j + w^{j-1} \right] + A_\delta \sum_{k=1}^{j} \left[w^{j-k+1} - 2w^{j-k} + w^{j-k-1} \right] B(k),
\end{aligned}
\quad (11)
$$

where $A_\delta = \frac{\tau^{-\delta}}{\Gamma(3-\delta)}$, τ is time step size and $B(k) = (k+1)^{2-\delta} - (k)^{2-\delta}$.

Case i:

Using Equation (11) and θ−weighted scheme ($0 \le \theta \le 1$) in Equation (1), we obtain

$$
A_\delta \left[w^{j+1} - 2w^j + w^{j-1} \right] + A_\delta \sum_{k=1}^{j} \left[w^{j-k+1} - 2w^{j-k} + w^{j-k-1} \right] B(k) + \frac{1}{\tau} \left\{ w^{j+1} - w^j \right\} \\
= \theta w_{xx}^{j+1} + (1-\theta) w_{xx}^j + \mathcal{A}(x, t^{j+1}).
\quad (12)
$$

After simplification, the above equation transforms to

$$
(\tau A_\delta + 1) w^{j+1} - \tau \theta w_{xx}^{j+1} = 2\tau A_\delta w^j - \tau A_\delta w^{j-1} - \tau A_\delta \sum_{k=1}^{j} \left[w^{j-k+1} - 2w^{j-k} + w^{j-k-1} \right] B(k) \\
+ w^j + \tau(1-\theta) w_{xx}^j + \tau \mathcal{A}(x, t^{j+1}).
\quad (13)
$$

In our analysis we take $\theta = 1/2$. Now approximating the highest order derivative by a truncated Haar wavelets series as:

$$
w_{xx}^{j+1}(x) = \sum_{i=1}^{2M} a_i^{j+1} \mathcal{H}_i(x).
\quad (14)
$$

Integrating Equation (14) from 0 to x

$$
w_x^{j+1}(x) = \sum_{i=1}^{2M} a_i^{j+1} \mathcal{P}_{i,1}(x) + w_x^{j+1}(0).
\quad (15)
$$

Integrating Equation (15) from 0 to 1, we get

$$
w_x^{j+1}(0) = w^{j+1}(1) - w^{j+1}(0) - \sum_{i=1}^{2M} a_i^{j+1} \mathcal{P}_{i,2}(1).
\quad (16)
$$

Substituting Equation (16) in Equation (15), the resultant equation reduces to

$$w_x^{j+1}(x) = \sum_{i=1}^{2M} a_i^{j+1} [\mathcal{P}_{i,1}(x) - \mathcal{P}_{i,2}(1)] + w^{j+1}(1) - w^{j+1}(0). \tag{17}$$

Integration of Equation (17) from 0 to x yields

$$w^{j+1}(x) = \sum_{i=1}^{2M} a_i^{j+1} [\mathcal{P}_{i,2}(x) - x\mathcal{P}_{i,2}(1)] + x\left[w^{j+1}(1) - w^{j+1}(0)\right] + w^{j+1}(0). \tag{18}$$

Substituting values from Equations (14), (17) and (18) in Equation (13) and using collocation points $x_m = \frac{m-0.5}{2M}$, $m = 1, 2, \ldots 2M$, leads to the following system of algebraic equation

$$\sum_{i=1}^{2M} a_i^{j+1} \left[(\tau A_\delta + 1) \{\mathcal{P}_{i,2}(x) - x\mathcal{P}_{i,2}(1)\} - \tau \theta \mathcal{H}_i(x) \right]_{x=x_m} = \mathcal{R}(m), \tag{19}$$

where

$$\mathcal{R}(m) = 2\tau A_\delta w^j - \tau A_\delta w^{j-1} - \tau A_\delta \sum_{k=1}^{j} \left[w^{j-k+1} - 2w^{j-k} + w^{j-k-1} \right] B(k) + w^j$$

$$+ \tau(1-\theta) w_{xx}^j + \tau \mathcal{A}(x_m, t^{j+1}) - (\tau A_\delta + 1) \left\{ x_m \left(w^{j+1}(1) - w^{j+1}(0) \right) + w^{j+1}(0) \right\}.$$

Equation (19) contains $2M$ equations. The unknown wavelet coefficients can be computed from this system. After determination of these unknown constants, the required solution at each time can be calculated from Equation (18).

Case ii:

Following a similar approach, as discussed earlier, Equation (3) gives

$$A_\delta w^{j+1} - \theta \left[w_{xx}^{j+1} + w_{yy}^{j+1} \right] = (1-\theta) \left[w_{xx}^j + w_{yy}^j \right] + \mathcal{B}(x, y, t^{j+1}) + 2A_\delta w^j - A_\delta w^{j-1}$$

$$- A_\delta \sum_{k=1}^{j} \left[w^{j-k+1} - 2w^{j-k} + w^{j-k-1} \right] B(k). \tag{20}$$

Now we approximate $w_{xxyy}^{j+1}(x, y)$ with a two dimensional truncated Haar wavelets series as:

$$w_{xxyy}^{j+1}(x, y) = \sum_{i=1}^{2M} \sum_{l=1}^{2M} a_{i,l}^{j+1} \mathcal{H}_i(x) \mathcal{H}_i(y), \tag{21}$$

where $a_{i,l}^{j+1}$ are unknowns to be determined. Integration of Equation (21) w.r.t. to y, between 0 and y, gives

$$w_{xxy}^{j+1}(x, y) = \sum_{i=1}^{2M} \sum_{l=1}^{2M} a_{i,l}^{j+1} \mathcal{H}_i(x) \mathcal{P}_{l,1}(y) + w_{xxy}^{j+1}(x, 0). \tag{22}$$

Integrating Equation (22) w.r.t y from 0 to 1, the unknown term $w_{xxy}^{j+1}(x, 0)$ is given by

$$w_{xxy}^{j+1}(x, 0) = w_{xx}^{j+1}(x, 1) - w_{xx}^{j+1}(x, 0) - \sum_{i=1}^{2M} \sum_{l=1}^{2M} a_{i,l}^{j+1} \mathcal{H}_i(x) \mathcal{P}_{l,2}(1). \tag{23}$$

Substituting Equation (23) in Equation (22), the obtained result is

$$w_{xxy}^{j+1}(x,y) = \sum_{i=1}^{2M}\sum_{l=1}^{2M} a_{i,l}^{j+1} \mathcal{H}_i(x)\left[\mathcal{P}_{l,1}(y) - \mathcal{P}_{l,2}(1)\right] + w_{xx}^{j+1}(x,1) - w_{xx}^{j+1}(x,0). \tag{24}$$

Integrating Equation (24) from 0 to y, we get

$$w_{xx}^{j+1}(x,y) = \sum_{i=1}^{2M}\sum_{l=1}^{2M} a_{i,l}^{j+1} \mathcal{H}_i(x)\left[\mathcal{P}_{l,2}(y) - y\mathcal{P}_{l,2}(1)\right] + yw_{xx}^{j+1}(x,1) + (1-y)w_{xx}^{j+1}(x,0). \tag{25}$$

Repeating the same procedure one can easily derive the subsequent expressions

$$w_{yy}^{j+1}(x,y) = \sum_{i=1}^{2M}\sum_{l=1}^{2M} a_{i,l}^{j+1} \left[\mathcal{P}_{i,2}(x) - x\mathcal{P}_{i,2}(1)\right]\mathcal{H}_l(y) + xw_{yy}^{j+1}(1,y) + (1-x)w_{yy}^{j+1}(0,y). \tag{26}$$

$$w_x^{j+1}(x,y) = \sum_{i=1}^{2M}\sum_{l=1}^{2M} a_{i,l}^{j+1}\left[\mathcal{P}_{i,1}(x) - \mathcal{P}_{i,2}(1)\right]\left[\mathcal{P}_{l,2}(y) - y\mathcal{P}_{l,2}(1)\right] + yw_x^{j+1}(x,1)$$
$$+ (1-y)w_x^{j+1}(x,0) + w^{j+1}(1,y) - w^{j+1}(0,y) - yw^{j+1}(1,1) + yw^{j+1}(0,1)$$
$$+ (y-1)w^{j+1}(1,0) + (1-y)w^{j+1}(0,0). \tag{27}$$

$$w_y^{j+1}(x,y) = \sum_{i=1}^{2M}\sum_{l=1}^{2M} a_{i,l}^{j+1}\left[\mathcal{P}_{i,2}(x) - x\mathcal{P}_{i,2}(1)\right]\left[\mathcal{P}_{l,1}(y) - \mathcal{P}_{l,2}(1)\right] + xw_y^{j+1}(1,y)$$
$$+ (1-x)w_y^{j+1}(0,y) + w^{j+1}(x,1) - w^{j+1}(x,0) - xw^{j+1}(1,1) + xw^{j+1}(1,0)$$
$$+ (x-1)w^{j+1}(0,1) + (1-x)w^{j+1}(0,0). \tag{28}$$

$$w^{j+1}(x,y) = \sum_{i=1}^{2M}\sum_{l=1}^{2M} a_{i,l}^{j+1}\left[\mathcal{P}_{i,2}(x) - x\mathcal{P}_{i,2}(1)\right]\left[\mathcal{P}_{l,2}(y) - y\mathcal{P}_{l,2}(1)\right] + yw^{j+1}(x,1)$$
$$- yw^{j+1}(0,1) + (1-y)\left[w^{j+1}(x,0) - w^{j+1}(0,0)\right] + xw^{j+1}(1,y)$$
$$- xw^{j+1}(0,y) - xy\left[w^{j+1}(1,1) - w^{j+1}(0,1)\right] + x(y-1)w^{j+1}(1,0)$$
$$+ x(1-y)w^{j+1}(0,0) + w^{j+1}(0,y). \tag{29}$$

Substitution of Equations (25), (26) and (29) in Equation (20) and using the collocation points, $x_m = \frac{m-0.5}{2M}$, $y_n = \frac{n-0.5}{2M}$, $m,n = 1,2,\ldots 2M$, produces the following system of equations

$$\sum_{i=1}^{2M}\sum_{l=1}^{2M} a_{i,l}^{j+1}\left[A_\delta \mathcal{D}(i,l,m,n) - \theta\mathcal{E}(i,l,m,n) - \theta\mathcal{F}(i,l,m,n)\right] = \mathcal{L}(m,n) + \mathcal{M}(m,n), \tag{30}$$

where

$$D(i,l,m,n) = [\mathcal{P}_{i,2}(x_m) - x_m\mathcal{P}_{i,2}(1)][\mathcal{P}_{l,2}(y_n) - y_n\mathcal{P}_{l,2}(1)],$$
$$\mathcal{E}(i,l,m,n) = \mathcal{H}_i(x_m)[\mathcal{P}_{l,2}(y_n) - y_n\mathcal{P}_{l,2}(1)],$$
$$\mathcal{F}(i,l,m,n) = [\mathcal{P}_{i,2}(x_m) - x_m\mathcal{P}_{i,2}(1)]\mathcal{H}_l(y_n),$$
$$\mathcal{L}(m,n) = (1-\theta)\left[w^j_{xx} + w^j_{yy}\right] + \mathcal{B}(x_m, y_n, t^{j+1}) + 2A_\delta w^j - A_\delta w^{j-1}$$
$$- A_\delta \sum_{k=1}^{j}\left[w^{j-k+1} - 2w^{j-k} + w^{j-k-1}\right] B(k),$$
$$\mathcal{M}(m,n) = -A_\delta\left[y_n w^{j+1}_x(x_m,1) - y_n w^{j+1}(0,1) + (1-y_n)\left\{w^{j+1}(x_m,0) - w^{j+1}(0,0)\right\}\right.$$
$$+ x_m w^{j+1}(1,y_n) - x_m w^{j+1}(0,y_n) - x_m y_n\left\{w^{j+1}(1,1) - w^{j+1}(0,1)\right\}$$
$$+ x_m(y_n - 1)w^{j+1}(1,0) + x_m(1-y_n)w^{j+1}(0,0) + w^{j+1}(0,y_n)\right] + \theta\left[y_n w^{j+1}_{xx}(x_m,1)\right.$$
$$+ (1-y_n)w^{j+1}_{xx}(x_m,0) + x_m w^{j+1}_{yy}(1,y_n) + (1-x_m)w^{j+1}_{yy}(0,y_n)\right].$$

Equation (30) represents $2M \times 2M$ equations in so many unknowns which can be solved easily. After calculation of these unknowns, an approximate solution can be obtained from Equation (29).

4. Stability Analysis

Here we present the stability analysis of the proposed scheme for (1 + 2)-dimensional problems; a similar result can be proved for (1 + 1)-dimensional problems. In matrix form Equations (25), (26) and (29) can be written as

$$w^{j+1}_{xx} = \mathcal{U}\alpha^{j+1} + \tilde{\mathcal{U}}^{j+1}, \tag{31}$$
$$w^{j+1}_{yy} = \mathcal{V}\alpha^{j+1} + \tilde{\mathcal{V}}^{j+1}, \tag{32}$$
$$w^{j+1} = \mathcal{Z}\alpha^{j+1} + \tilde{\mathcal{Z}}^{j+1}, \tag{33}$$

where $\alpha^{j+1} = \alpha^{j+1}(i,l)$, $\mathcal{U}, \mathcal{V}, \mathcal{Z}$ and $\tilde{\mathcal{U}}^{j+1}, \tilde{\mathcal{V}}^{j+1}, \tilde{\mathcal{Z}}^{j+1}$ are interpolation matrices of $w^{j+1}_{xx}, w^{j+1}_{yy}, w^{j+1}$ at collocation points and boundary terms, respectively. Now using Equations (31), (32) and (33) in Equation (20), we get

$$[A_\delta \mathcal{Z} - \theta(\mathcal{U} + \mathcal{V})]\alpha^{j+1} = [2A_\delta \mathcal{Z} + (1-\theta)(\mathcal{U} + \mathcal{V})]\alpha^j + \mathcal{G}^{j+1}, \tag{34}$$

where $\mathcal{G}^{j+1} = -A_\delta \tilde{\mathcal{Z}}^{j+1} + \theta(\tilde{\mathcal{U}}^{j+1} + \tilde{\mathcal{V}}^{j+1}) + 2A_\delta \tilde{\mathcal{Z}}^j + (1-\theta)(\tilde{\mathcal{U}}^j + \tilde{\mathcal{V}}^j) + \mathcal{B}^{j+1} - A_\delta w^{j-1} - A_\delta \sum_{k=1}^{j}\left[w^{j-k+1} - 2w^{j-k} + w^{j-k-1}\right] B(k)$.

Now From Equation (34) one can write

$$\alpha^{j+1} = \mathcal{C}^{-1}\mathcal{T}\alpha^j + \mathcal{C}^{-1}\mathcal{G}^{j+1}, \tag{35}$$

where $\mathcal{C} = [A_\delta \mathcal{Z} - \theta(\mathcal{U} + \mathcal{V})]$, $\mathcal{T} = 2A_\delta \mathcal{Z} + (1-\theta)[\mathcal{U} + \mathcal{V}]$. Putting Equation (35) in Equation (33) we get

$$w^{j+1} = \mathcal{Z}\mathcal{C}^{-1}\mathcal{T}\alpha^j + \mathcal{Z}\mathcal{C}^{-1}\mathcal{G}^{j+1} + \tilde{\mathcal{Z}}^{j+1}. \tag{36}$$

Using Equation (33) in Equation (36) we have

$$w^{j+1} = \mathcal{Z}\mathcal{C}^{-1}\mathcal{T}\mathcal{Z}^{-1}w^j - \mathcal{Z}\mathcal{C}^{-1}\mathcal{T}\mathcal{Z}^{-1}\tilde{\mathcal{Z}}^j + \mathcal{Z}\mathcal{C}^{-1}\mathcal{G}^{j+1} + \tilde{\mathcal{Z}}^{j+1}. \tag{37}$$

The above equation shows a recurrence relation of a full discretization scheme which allow us refinement in time. If $\tilde{w}^{j+1}$ is numerical solution then

$$\tilde{w}^{j+1} = \mathcal{Z}\mathcal{C}^{-1}\mathcal{T}\mathcal{Z}^{-1}\tilde{w}^j - \mathcal{Z}\mathcal{C}^{-1}\mathcal{T}\mathcal{C}^{-1}\tilde{\mathcal{Z}}^j + \mathcal{Z}\mathcal{C}^{-1}\mathcal{G}^{j+1} + \tilde{\mathcal{Z}}^{j+1}. \tag{38}$$

Let $e^{j+1} = w^{j+1} - \tilde{w}^{j+1}$ be the error at $(j+1)^{\text{th}}$ time level. Subtracting Equation (37) from Equation (38) then

$$e^{j+1} = \Lambda e^j,$$

where $\Lambda = \mathcal{Z}\mathcal{C}^{-1}\mathcal{T}\mathcal{Z}^{-1}$ is the amplification matrix. According to Lax-Richtmyer criterion, the scheme will be stable if $\|\Lambda\| \leq 1$. It has been verified computationally that $\|\Lambda\| \leq 1$. For $J = 1$ the spectral radius is 0.01025 which lies in the stability domain.

5. Convergence Analysis

The convergence analysis of scheme (18) and (29) is similar to the following theorems, therefore the proofs are omitted.

Lemma 1 (see [24]). *If $w(x) \in L^2(R)$ with $|w'(x)| \leq \rho$, for all $x \in (0,1)$, $\rho > 0$ and $w(x) = \sum_{i=0}^{\infty} a_i \mathcal{H}_i(x)$ then $|a_i| \leq \frac{\rho}{2^{j+1}}$.*

Lemma 2 (see [25]). *If $f(x,y)$ satisfies a Lipschitz condition on $[0,1] \times [0,1]$, that is, there exists a positive L such that for all $(x_1, y), (x_2, y) \in [0,1] \times [0,1]$ we have $|f(x_1, y) - f(x_2, y)| \leq L |x_1 - x_2|$ then*

$$a_{i,l}^2 \leq \frac{L^2}{2^{4j+4}m^2}$$

Theorem 1. *If $w(x)$ and $w_{2M}(x)$ are the exact and approximate solution of Equation (1), then the error norm $\|E_J\|$ at J^{th} resolution level is*

$$\|E_J\| \leq \frac{4\rho}{3}\left(\frac{1}{2^{J+1}}\right)^2. \tag{39}$$

Proof. See [26]. □

Theorem 2. *Assume $w(x,y)$ and $w_{2M}(x,y)$ be the exact and approximate solution of Equation (3), then*

$$\|E_J\| \leq \frac{L}{4\sqrt{255}} \frac{1}{2^{4J}}. \tag{40}$$

Proof. See [27]. □

6. Illustrative Test Problems

In this part, we chose some test problem to confirm the reliability and efficiency of the present scheme. For validation of our results L_∞ and L_2 error norm are figured out which are defined as follows:

$$L_2 = \sqrt{\sum_{i=1}^{2M}\left(w^{ext} - w^{app}\right)^2}, \quad L_\infty = \max_{1 \leq i \leq 2M}\left|w^{ext} - w^{app}\right|, \tag{41}$$

where w^{app} and w^{ext} are respectively approximate and exact solutions.

Problem 5.1

Let us take the following (1 + 1)-dimensional TFDWE with damping

$$^cD_t^\delta w(x,t) = -w_t(x,t) + w_{xx}(x,t) + \mathcal{A}(x,t), \quad x \in [0,1], \quad t \in [0,1], \quad 1 < \delta \le 2, \quad (42)$$

with $\mathcal{A}(x,t) = \frac{2x(1-x)t^{2-\delta}}{\Gamma(3-\delta)} + 2tx(1-x) + 2t^2$. Initial and boundary conditions are derived from the exact solution $w(x,t) = t^2 x(1-x)$. This problem has been solved for parameters $J = 4$, $t = 0.01$, 0.1, 1, $\delta = 1.1, 1.3, 1.5, 1.7, 1.9$. The obtained error norms are shown in Table 1. From table it is obvious that results of the present scheme match well with exact solution. Also in Table 2 it has been observed that accuracy increases with increasing resolution level which shows the convergence in the spatial direction. In the same table, the results have been matched with existing results in the literature which clarify that computed solutions are in good agreement with the work of Chen et al. [28]. Table 3 shows convergence in time for fixed $dx = 1/32$. The convergence rate of the proposed scheme has been addressed in Table 4. the graphical solution and error plot are given in Figure 1. From this Figure it is clear that approximate solutions are matchable with exact.

Table 1. Error norms of problem 5.1 for at $J = 4$.

δ	$t = 0.01, \tau = 0.0001$		$t = 0.1, \tau = 0.001$		$t = 1, \tau = 0.01$	
	L_∞	L_2	L_∞	L_2	L_∞	L_2
1.1	7.0694×10^{-8}	2.9496×10^{-7}	1.0799×10^{-5}	4.5921×10^{-5}	2.1556×10^{-3}	8.9537×10^{-3}
1.3	3.1776×10^{-8}	1.3294×10^{-7}	7.6592×10^{-6}	3.2979×10^{-5}	2.1082×10^{-3}	8.7615×10^{-3}
1.5	1.1646×10^{-8}	4.8890×10^{-8}	4.8457×10^{-6}	2.1318×10^{-5}	2.0653×10^{-3}	8.5871×10^{-3}
1.7	5.2296×10^{-9}	2.1899×10^{-8}	3.3635×10^{-6}	1.4944×10^{-5}	2.1431×10^{-3}	8.8989×10^{-3}
1.9	2.2087×10^{-9}	9.2078×10^{-9}	2.1386×10^{-6}	9.3949×10^{-5}	2.4382×10^{-3}	1.0094×10^{-2}

Table 2. Comparison of maximum error of problem 5.1 with previous work at $t = 1$ and $\delta = 1.7$.

[28]			Present Method		
h	τ	Error	dx	τ	Error
0.05	0.05	4.4333×10^{-3}	1/4	0.05	8.2306×10^{-4}
0.025	0.0125	7.7368×10^{-4}	1/8	...	5.4184×10^{-4}
0.0125	0.00625	3.1827×10^{-4}	1/16	...	4.9195×10^{-4}

Table 3. Error norms of problem 5.1 for different values of τ and δ.

	$\delta = 1.5$		$\delta = 1.7$	
τ	L_∞	L_2	L_∞	L_2
1/4	5.4216×10^{-2}	2.2515×10^{-1}	5.4689×10^{-2}	2.2708×10^{-1}
1/8	2.7891×10^{-2}	1.1571×10^{-1}	2.8861×10^{-2}	1.1961×10^{-1}
1/16	1.3645×10^{-2}	5.6647×10^{-2}	1.4343×10^{-2}	5.9443×10^{-2}
1/32	6.6683×10^{-3}	2.7699×10^{-2}	7.0022×10^{-3}	2.9034×10^{-2}
1/64	3.2674×10^{-3}	1.3580×10^{-2}	3.4061×10^{-3}	1.4135×10^{-2}

Table 4. Convergence rate of maximum error of problem 5.1 at $t = 1$ and $\delta = 1.7$.

J	τ	Error	Rate
1	1/10	2.6194×10^{-3}	-
2	1/20	1.3086×10^{-3}	1.0012
3	1/40	5.5920×10^{-4}	1.2265
4	1/80	2.2921×10^{-4}	1.2866
5	1/160	9.3261×10^{-5}	1.2973
6	1/320	3.7377×10^{-5}	1.3191

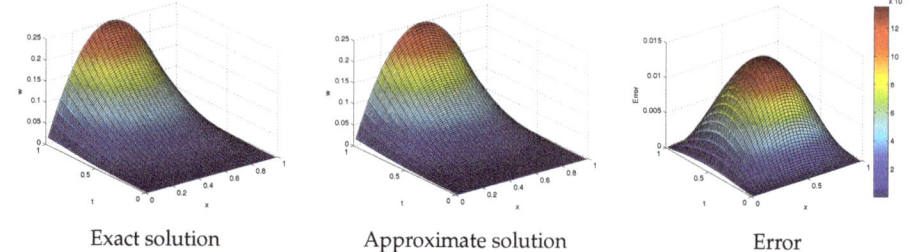

Exact solution   Approximate solution   Error

Figure 1. Graphical behaviour of problem 5.1 when $t = 1$, $\delta = 1.5$.

Problem 5.2:

Consider the following TFDWE with damping

$$^cD_t^\delta w(x,t) = -w_t(x,t) + w_{xx}(x,t) + \mathcal{A}(x,t), \quad x \in [0,1], \quad t \in [0,1], \quad 1 < \delta \le 2, \quad (43)$$

coupled with initial and boundary conditions

$$\begin{cases} w(x,0) = 0, & w_t(x,0) = 0 \quad x \in (0,1) \\ w(0,t) = t^3, & w(1,t) = et^3, \quad t \in [0,1]. \end{cases} \quad (44)$$

The exact solution and source term are given by $w(x,t) = e^x t^3$ and $\mathcal{A}(x,t) = \frac{6t^{3-\delta}e^x}{\Gamma(4-\delta)} + 3t^2 e^x - t^3 e^x$. In Table 5 the obtained error norms are shown for parameters $t = 0.01, 0.1$, $\delta = 1.1, 1.3, 1.5, 1.7, 1.9$, $J = 4$. Table 5 shows that exact and approximate solutions agree with each other. The solution profile and absolute error are displayed Figure 2. From the Figure, the coincidence of both solutions are visible.

Table 5. Error norms of problem 5.2 at $J = 4$.

δ	$t = 0.01$, $\tau = 0.0001$		$t = 0.1$, $\tau = 0.001$	
	L_∞	L_2	L_∞	L_2
1.1	1.7079×10^{-7}	6.8446×10^{-7}	1.2504×10^{-4}	5.3397×10^{-4}
1.3	6.5331×10^{-7}	2.5683×10^{-6}	4.4278×10^{-4}	1.8777×10^{-3}
1.5	1.2494×10^{-6}	4.7989×10^{-6}	8.7071×10^{-4}	3.6354×10^{-3}
1.7	1.3386×10^{-6}	5.0827×10^{-6}	1.0489×10^{-3}	4.2541×10^{-3}
1.9	5.6739×10^{-7}	2.1936×10^{-6}	5.1085×10^{-4}	2.0046×10^{-3}

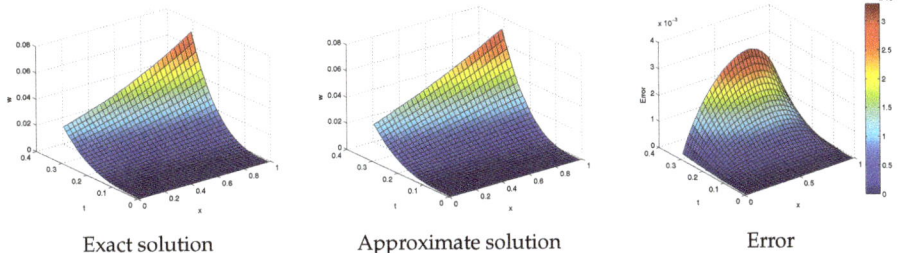

Exact solution   Approximate solution   Error

Figure 2. Graphical behaviour of problem 5.2 at $t = 0.3$, $\delta = 1.1$.

Problem 5.3:

Now we consider (1+2)-dimensional TFDWE [29]

$$^cD_t^\delta w(x,y,t) = \Delta w(x,y,t) + \mathcal{B}(x,y,t), \quad (x,y) \in [0,1] \times [0,1], \quad t \in [0,1], \quad 1 < \delta \leq 2, \quad (45)$$

with exact solution $w(x,y,t) = \sin(\pi x)\sin(\pi y)t^{\delta+3}$, and source term

$$\mathcal{B}(x,y,t) = \sin(\pi x)\sin(\pi y)\left[\frac{\Gamma(\delta+3)t^2}{2} - 2t^{\delta+2}\right].$$

We solved this problem for resolution level $J = 4$ and the obtained results are recorded in Table 6 for different values of time and τ. From Table 6 it is clear that the proposed scheme works well for the solution of two dimensional problems. Table 7 shows the comparison of the computed results with the previous work of Zhang [29]. One can see that our results are matchable with existing results. The same table presents convergence in time for $(1+2)$-dimensional problems. The graphical solution and absolute error of the problem are shown in Figure 3. It is obvious from Figure 3 that the exact and approximate solutions have strong agreement.

Table 6. Comparison of problem 5.4 at $t = 1$ and δ with previous results.

δ	$t = 0.1$, $\tau = 0.001$		$t = 0.2$, $\tau = 0.01$		$t = 0.5$, $\tau = 0.05$	
	L_∞	L_2	L_∞	L_2	L_∞	L_2
1.5	1.6049×10^{-4}	8.0439×10^{-5}	4.3534×10^{-4}	2.1819×10^{-4}	2.5390×10^{-2}	1.2725×10^{-2}
1.7	1.1635×10^{-4}	5.8320×10^{-5}	5.9673×10^{-4}	2.9908×10^{-4}	7.5824×10^{-3}	3.8003×10^{-3}
1.9	3.0965×10^{-5}	1.5519×10^{-5}	3.9390×10^{-4}	1.9742×10^{-4}	4.8842×10^{-3}	2.4480×10^{-3}

Table 7. Error norms of problem 5.3 for different values of τ and δ.

		L_∞	
δ	τ	Present	[29]
1.25	1/10	8.1748×10^{-3}	8.1577×10^{-2}
	1/20	6.5092×10^{-3}	3.4379×10^{-2}
	1/40	5.7150×10^{-3}	1.4484×10^{-2}
1.5	1/10	6.7087×10^{-3}	2.9942×10^{-2}
	1/20	4.8922×10^{-3}	1.0749×10^{-2}
	1/40	4.1390×10^{-3}	3.8291×10^{-3}
1.75	1/10	6.7087×10^{-3}	8.4482×10^{-3}
	1/20	4.8922×10^{-3}	2.5877×10^{-3}
	1/40	4.1390×10^{-3}	7.8500×10^{-4}

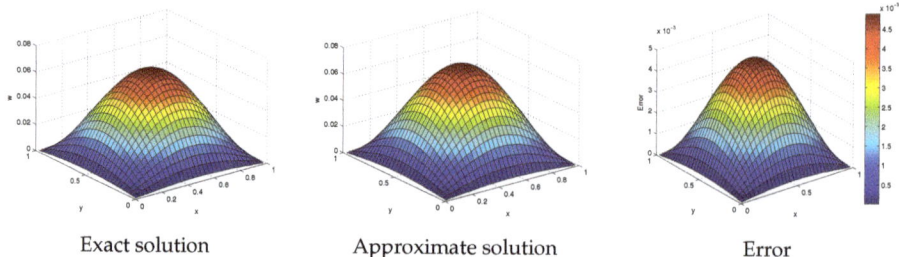

Figure 3. Graphical behaviour of problem 5.3 when $t = 0.5$, $\delta = 1.9$.

Problem 5.4:

Consider the following TFDWE with reaction term [19]

$$^cD_t^\delta w(x,t) + w(x,t) = w_{xx}(x,t) + \mathcal{A}(x,t), \quad x \in [0,1], \quad t \in [0,1], \quad 1 < \delta \leq 2, \tag{46}$$

coupled with initial and boundary conditions

$$\begin{cases} w(x,0) = 0, & w_t(x,0) = 0 \quad x \in (0,1) \\ w(0,t) = o, & w(1,t) = 0, \quad t \in [0,1], \end{cases} \tag{47}$$

where the forcing terms are $\mathcal{A}(x,t) = \frac{2t^{2-\delta}x(1-x)}{\Gamma(3-\delta)} + t^2 x(1-x) - 2t^2$. This problem has been solved with the help of the proposed scheme. In Table 8 we presented the solutions at different points. Also the obtained results have been compared with the work presented in Reference [19]. It is clear from table that our results are more accurate. From the table it is also obvious that the exact and numerical solutions are in good agreement. Exact verses numerical solutions are plotted in Figure 4. Graphical solutions also indicate that the proposed scheme works in the case where the reaction term exists.

Table 8. Absolute error at different points of example 5.4 at $\tau = 0.001$.

(x,t)	$\delta = 1.1$		$\delta = 1.3$		$\delta = 1.5$		$\delta = 1.9$	
	L_∞	L_∞ [19]	L_∞	L_∞ [19]	L_∞	L_∞ [19]	L_∞	L_∞ [19]
(0.1,0.1)	2.1684×10^{-19}	9.5133×10^{-9}	1.0842×10^{-19}	6.6004×10^{-9}	3.2526×10^{-19}	4.4920×10^{-9}	5.8546×10^{-18}	1.9326×10^{-9}
(0.2,0.2)	9.5409×10^{-18}	1.0530×10^{-7}	2.1684×10^{-17}	7.9127×10^{-8}	3.4694×10^{-18}	5.7844×10^{-8}	1.8735×10^{-16}	2.8903×10^{-8}
(0.3,0.3)	4.1633×10^{-17}	9.6665×10^{-7}	4.1633×10^{-17}	3.3461×10^{-7}	1.0755×10^{-16}	2.5678×10^{-7}	1.3634×10^{-15}	1.4105×10^{-7}
(0.4,0.4)	1.3877×10^{-17}	1.0813×10^{-6}	1.8735×10^{-16}	9.1574×10^{-7}	7.0776×10^{-16}	7.3594×10^{-7}	4.5033×10^{-15}	4.3402×10^{-7}
(0.5,0.5)	1.3877×10^{-16}	2.2190×10^{-6}	3.9551×10^{-16}	1.6516×10^{-6}	2.0261×10^{-15}	1.6516×10^{-6}	1.4231×10^{-14}	1.0367×10^{-6}

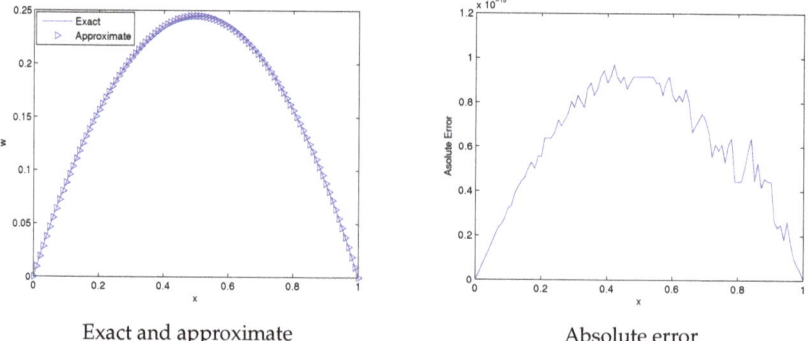

Figure 4. Graphical behaviour of problem 5.4 at $\delta = 1.1$, $t = 1$.

Problem 5.5:

Now we consider the following equation

$$^cD_t^\delta w(x,y,t) = a_1 \Delta w(x,y,t) - b_1 \sin(w(x,y,t)), \quad (x,y) \in \Phi, \quad t \in [0,T], \quad 1 < \delta \leq 2, \quad (48)$$

where a_1 and b_1 are constants and the initial and boundary conditions are

$$\begin{cases} w(x,y,0) = \arctan\left(\exp(\tfrac{1}{2} - \sqrt{15x^2 + 15y^2})\right), \quad w_t(x,y,0) = 0, \quad (x,y) \in \tilde{\Phi} = \Phi \cup \partial\Phi, \\ w(x,y,t) = 0, \quad (x,y) \in \partial\Phi, \quad t \in [0,T]. \end{cases} \quad (49)$$

Here, we examine the behaviour of circular ring soliton numerically. Due to pulsating behaviour, such waves are also known as pulsons. We choose different values of parameters a_1, b_1 to present surface plots to study the time evolution of the circular ring soliton. We observe the effect of a_1 and b_1 on solutions. In Figure 5, numerical solutions for different values of a_1 and b_1 have been plotted. Figure 6 shows the numerical solution for $a_1 = 0.05$ while varying b_1. In Figure 7 the results are plotted for $b_1 = 10$, in which the wave peak value at the centre becomes lower as a_1 increases. This reveals that the solitary wave moves in a stable way up to a large time under finite initial condition.

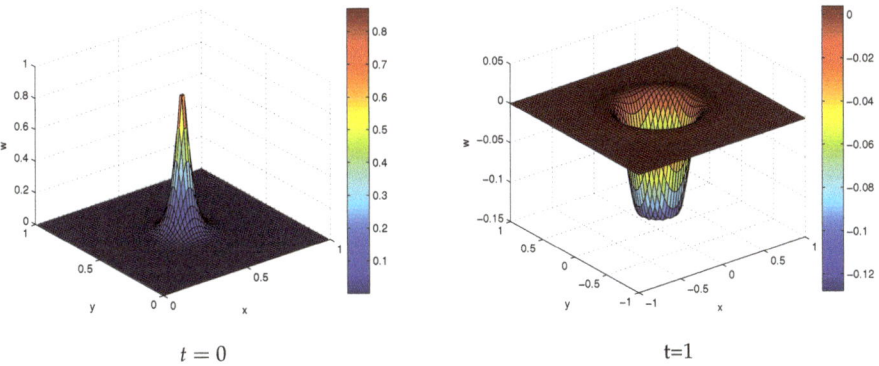

Figure 5. Graphical behaviour of problem 5.5 at $\delta = 1.9$, $a_1 = 0.1$, $b_1 = 10$.

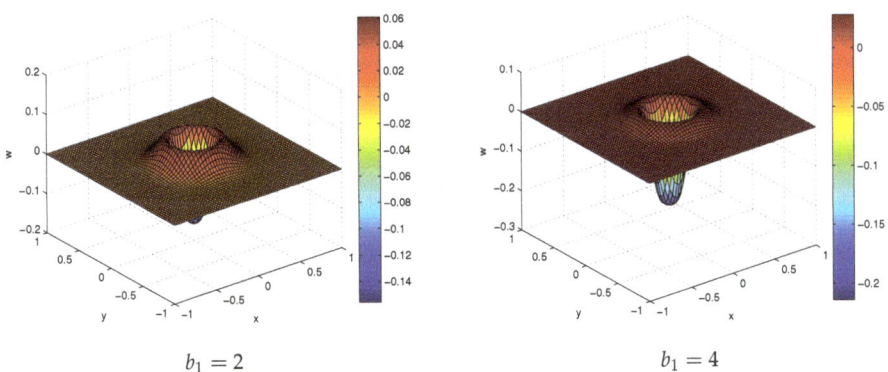

Figure 6. Graphical behaviour of problem 5.5 at $\delta = 1.9$, $a_1 = 0.05$.

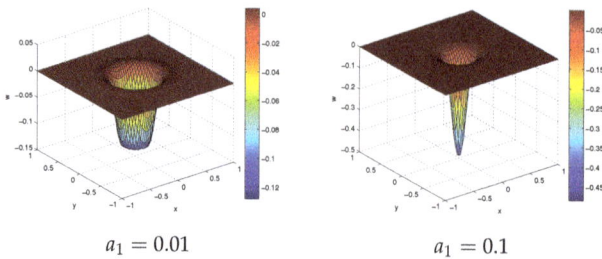

$a_1 = 0.01 \qquad a_1 = 0.1$

Figure 7. Graphical behaviour of problem 5.5 at $\delta = 1.9$, $b_1 = 10$.

7. Conclusions

In this paper, we proposed a hybrid method based on finite difference and Haar wavelets approximations. The scheme is applied for the numerical solution of (1 + 1)- and (1 + 2)-dimensional time fraction partial differential equations. The accuracy and applicability of the scheme is validated through some test problems. The tabulated data and graphical solution show that the scheme works very well for time fractional problems.

Author Contributions: Conceptualization, A.G. and S.H.; Methodology, A.G.; Software, A.G.; Validation, S.H., M.H. and M.A.J.; Formal Analysis, A.G.; Investigation, M.A.J.; Resources, P.K.; Writing–Original Draft Preparation, A.G.; Writing–Review and Editing, M.H.; Visualization, P.K.; Supervision, S.H. and P.K.; Project Administration, P.K.; Funding Acquisition, P.K.

Funding: The project was supported by the Center of Excellence in Theoretical and Computational Science (TaCS-CoE), Faculty of Science, King Mongkut's University of Technology Thonburi (KMUTT).

Acknowledgments: The authors are thankful to anonymous reviewers for their fruitful suggestion which improved the quality of the manuscript. Also we are thankful for the financial support of the Center of Excellence in Theoretical and Computational Science(TaCS-CoE), Faculty of Science, King Mongkut's University of Technology Thonburi (KMUTT).

Conflicts of Interest: The authors declare no conflict of interest.

References

1. Kilbas, A.A.; Srivastava, H.M.; Trujillo, J.J. *Theory and Applications of the Fractional Differential Equations*; Elsevier: Amsterdam, The Netherlands, 2006; Volume 204.
2. Oldham. K.B. Fractional differential equations in electrochemistry. *Adv. Eng. Softw.* **2010**, *41*, 9–12. [CrossRef]
3. Hilfer, R. *Applications of Fractional Calculus in Physics*; World Scientific Pub. Company: Singapore, 2000.
4. Machado, J.A.T. Analysis and design of fractional-order digital control systems. *Syst. Aanl. Model. Simul.* **1997**, *27*, 107–122.
5. Baleanu, D.; Defterli, O.; Agrawal, O.P. A central difference numerical scheme for fractional optimal control problems. *J. Vib. Control.* **2009**, *15*, 583–597. [CrossRef]
6. Ichise, M.; Nagayanagi, Y.; Kojima, T. An analog simulation of noninteger order transfer functions for analysis of electrode process. *J. Electroanal. Chem.* **1971**, *33*, 253–265. [CrossRef]
7. Koeller, R.C. Applications of fractional calculus to the theory of viscoelasticity. *J. Appl. Mech.* **1984**, *51*, 299–307. [CrossRef]
8. Sun, H.H. Abdelwahad, A.A.; Onaral, B. Linear approximation of transfer function with a pole of fractional order. *EEE Trans. Autom. Control* **1984**, *29*, 441–444. [CrossRef]
9. Mainardi, F. Some basic problems in continuum and statistical mechanics. In *Fractals and Fractional Calculus in Continuum Mechanics*; Carpinteri, A., Mainardi, F., Eds.; Springer: Vienna, Austria, 1997; pp. 291–348.
10. Nigmatullin, R.R. To the theoretical explanation of the universal response. *Phys. Status Solidi (B) Basic Res.* **1984**, *123*, 739–745. [CrossRef]
11. Nigmatullin, R.R. Realization of the generalized transfer equation in a medium with fractal geometry. *Phys. Status Solidi (B) Basic Res.* **1986**, *133*, 425–430. [CrossRef]

12. Tadjeran, C.; Meerschaert, M.M.; Scheffler, H.-P. A second-order accurate numerical approximation for the fractional diffusion equation. *J. Comput. Phys.* **2006**, *213*, 205–213. [CrossRef]
13. Zhuang, P.; Liu, F.; Anh, V.; Turner, I. New solution and analytical techniques of the implicit numerical methods for the anomalous sub-diffusion equation. *SIAM J. Numer. Anal.* **2008**, *46*, 1079–1095. [CrossRef]
14. Yuste, S.B.; Acedo, L. An explicit finite difference method and a new von Neumann-type stability analysis for fractional diffusion equations. *SIAM J. Numer. Anal.* **2005**, *43*, 1862–1874. [CrossRef]
15. Chen, C.M.; Liu, F.; Turner, I.; Anh, V. A Fourier method for the fractional diffusion equation describing sub-diffusion. *J. Comput. Phys.* **2007**, *227*, 886–897. [CrossRef]
16. Hosseini, V.R.; Chen,W.; Avazzadeh, Z. Numerical solution of fractional telegraph equation by using radial basis functions. *Anal. Bound. Elem.* **2014**, *38*, 31–39. [CrossRef]
17. Zhou, F.; Xu, X. Numerical solution of time-fractional diffusion-wave equations via Chebyshev wavelets collocation method. *Adv. Math. Phys.* **2017**, *2017*, 2610804. [CrossRef]
18. Bhrawya, A.H.; Doha, E.H.; Baleanu, D.; Ezz-Eldien, S.S. A spectral tau algorithm based on Jacobi operational matrix for numerical solution of time fractional diffusion-wave equations. *J. Comput. Phys.* **2015**, *293*, 142–156. [CrossRef]
19. Yaseen, M.; Abbas, M.; Nazir, T.; Baleanu, D. A finite difference scheme based on cubic trigonometric B-splines for a time fractional diffusion-wave equation. *Adv. Differ. Equ.* **2017**, *274*, 2–18. [CrossRef]
20. Khader, M.M.; Adel, M.H. Numerical solutions of fractional wave equations using an efficient class of FDM based on the Hermite formula. *Adv. Differ. Equ.* **2016**, *34*, 2–10. [CrossRef]
21. Kanwal, A.; Phang, C.; Iqbal, U. Numerical Solution of Fractional Diffusion Wave Equation and Fractional Klein–Gordon Equation via Two-Dimensional Genocchi Polynomials with a Ritz–Galerkin Method. *Computation* **2018**, *40*, 2–12. [CrossRef]
22. Datsko, B.; Podlubny, I.; Povstenko, Y. Time-Fractional Diffusion-Wave Equation with Mass Absorption in a Sphere under Harmonic Impact. *Mathematics* **2019**, *433*, 2–11. [CrossRef]
23. Lepik, U. Solving PDEs with the aid of two-dimensional Haar wavelets. *Comput. Math. Appl.* **2011**, *61*, 1873–1879. [CrossRef]
24. Majak, M.; Shvartsman, B.S.; Kirs, M.; Herranen, H. Convergence theorem for the Haar wavelet based discretization method. *Compos. Struct.* **2015**, *126*, 227–232. [CrossRef]
25. Arbabi, S.; Nazari, A.; Darvishi, M.T. A two-dimensional Haar wavelets method for solving systems of PDEs. *Appl. Math. Comput.* **2017**, *292*, 33–46. [CrossRef]
26. Haq, S.; Ghafoor, A.; Hussain, M. Numerical solutions of variable order time fractional (1 + 1)- and (1 + 2)-dimensional advection dispersion and diffusion models. *Appl. Math. Comp.* **2019**, *360*, 107–121. [CrossRef]
27. Haq, S.; Ghafoor, A. An Efficient Numerical Algorithm for Multi-Dimensional Time Dependent Partial Differential Equations. *Comput. Math. Appl.* **2018**, *75*, 2723–2734. [CrossRef]
28. Chen, J.; Liu, F.; Anh, V.; Shen, S.; Liu, Q.; Liao, C. The analytical solution and numerical solution of the fractional diffusion-wave equation with damping. *Appl. Math. Comput.* **2012**, *219*, 1737–1748. [CrossRef]
29. Zhang, Y.; Sun, Z.; Zhao, X.; Compact alternating direction implicit scheme for the two-dimensional fractional diffusion-wave equation. *SIAM J. Numer. Anal.* **2012**, *50*, 535–1555. [CrossRef]

© 2019 by the authors. Licensee MDPI, Basel, Switzerland. This article is an open access article distributed under the terms and conditions of the Creative Commons Attribution (CC BY) license (http://creativecommons.org/licenses/by/4.0/).

Article

Numerical Study for Darcy–Forchheimer Flow of Nanofluid due to a Rotating Disk with Binary Chemical Reaction and Arrhenius Activation Energy

Mir Asma [1], W.A.M. Othman [1,*] and Taseer Muhammad [2]

[1] Institute of Mathematical Sciences, Faculty of Science, University of Malaya, Kuala Lumpur 50603, Malaysia
[2] Department of Mathematics, Government College Women University, Sialkot 51310, Pakistan
* Correspondence: wanainun@um.edu.my

Received: 16 July 2019; Accepted: 11 September 2019; Published: 2 October 2019

Abstract: The present article investigates Darcy–Forchheimer 3D nanoliquid flow because of a rotating disk with Arrhenius activation energy. Flow is created by rotating disk. Impacts of thermophoresis and Brownian dispersion are accounted for. Convective states of thermal and mass transport at surface of a rotating disk are imposed. The nonlinear systems have been deduced by transformation technique. Shooting method is employed to construct the numerical arrangement of subsequent problem. Plots are organized just to investigate how velocities, concentration, and temperature are influenced by distinct emerging flow variables. Surface drag coefficients and local Sherwood and Nusselt numbers are also plotted and discussed. Our results indicate that the temperature and concentration are enhanced for larger values of porosity parameter and Forchheimer number.

Keywords: Arrhenius activation energy; rotating disk; Darcy–Forchheimer flow; binary chemical reaction; nanoparticles; numerical solution

---

1. Introduction

Nanofluid is the blend of nanometer-measured particles and the conventional base liquid. Nanofluids are generally used to conquer the low warm exhibition of normal base liquids such as oil, water, ethylene glycol, and propylene glycol. Because of intriguing physical characteristics, the nanofluids have potential use in earthenware production, metal working procedures, covering related applications, atomic reactor cooling, cooling, transportation, attractive medication, and a few others. Choi and Eastman [1] are credited with the word nanofluid. They established that nanomaterials are remarkable candidates for development in warmth transport of ordinary fluids. Regarding the convective vehicle of nanofluid, a numerical relation is accounted by Buongiorno [2]. Here, thermophoresis and Brownian movement are viewed as the most significant slip instruments. A few ongoing progressions in nanofluid streams can be found in references [3–25].

The present examiners are associated with breaking down the liquid stream due to a turning disk because of its tremendous applications in rotational air cleaners, diffusive siphons, nourishment handling advances, turbomachinery, PC stockpiling gadgets, therapeutic hardware, gas turbine rotors, greases, pivoting plate cathodes, and numerous other examples. Initially, the pivoting plate issue was tended to by von Karman [26]. Cochran [27] created asymptotic answer for the von Karman issue. Stewartson [28] broke down liquid stream between pivoting co-axial plates. Chappel and Stirs [29] talked about the liquid stream among turning and stationary plate. Ackroyd [30] thought about suction/infusion impacts in the Karman issue and created arrangements containing exponentially rotting coefficients. Shaky progression of thick fluid instigated by noncoaxial turns of a disk was explained by Erdogan [31]. Attia [32] talked about liquid stream by turning circles submerged in a permeable space using Wrench Nicolson strategy. Warmth and mass exchange attributed to pivoting streams of thick fluid because of a permeable circle was analyzed by Turkyilmazoglu and Senel [33]. They registered the numerical arrangement of the overseeing stream issue. Rashidi et al. [34] inspected the impact of entropy in a hydromagnetic stream of viscous liquid by pivoting plate. Mustafa et al. [35] investigated the progression of nanoliquid actuated by an extending circle. They inferred that constant extending of disk is a significant part of lessening limit-layer thickness. Hydromagnetic stream of a turning plate by taking slip and nanoparticles impacts was examined by Hayat et al. [36]. Mustafa [37] analyzed MHD nanoliquid flow by turning disk subjects to slip impacts. Hayat et al. [38] discussed the Darcy–Forchheimer stream of CNTs instigated by turning disk.

Concentration difference of species exists in a blend, subject to mass exchange. By fluctuating the grouping of species in a blend, they move from a high-fixation area to low-focus locale. The least compulsory vitality that is needed by reactants before synthetic response occurs is characterized as enactment vitality. A mass exchange mechanism alongside substance response with enactment vitality for the most part discovers applications in concoction building, mechanics of oil, and water emulsions, nourishment preparation etc. The regular convection stream of double-blend in a permeable medium with initiation vitality was proposed by Bestman [39]. Makinde et al. [40] explored temperamental characteristic convection stream subject to nth-request response and initiation vitality. Maleque [41] studied exothermic/endothermic response in blended convection streams subject to initiation vitality. Adjusted Arrhenius capacity was used by Awad et al. [42] to examine shaky pivoting streams of two-fold liquid past an indiscreet twisted surface. Abbas et al. [43] explored casson liquid streams subject to actuation vitality. Shafique et al. [44] inspected turning visco-elastic streams joining artificially receptive species with initiation vitality. Further recent attempts on binary chemical reaction and Arrhenius activation energy can be seen in the studies [45–47].

Darcy–Forchheimer nanoliquid flow because of rotating disk subject to binary chemical reaction and Arrhenius activation energy is investigated. Thermophoretic dispersion and arbitrary movement viewpoints are held. Heat and mass exchange highlights are broken down via convective factors. The administrative frameworks are comprehended numerically through shooting procedure. Additionally, velocities, concentration, temperature, surface drag coefficients, and local Sherwood and Nusselt numbers are discussed graphically.

2. Statement

Here, steady, laminar Darcy–Forchheimer 3D flow of viscous nanoliquid because of a rotating disk with binary chemical reaction and Arrhenius activation energy is examined. The disk at $z = 0$ rotates with constant angular velocity Ω (see Figure 1). Effects of thermophoresis and Brownian dissemination are additionally accounted for. Convection factors for warmth and mass exchange are employed. It is additionally accepted that the surface is warmed by hot liquid with concentration C_f and temperature T_f that give mass and warmth exchange coefficients k_{m^*} and h_f respectively. Velocities are (u, v, w) in directions of (r, φ, z) separately. Ensuing boundary layer articulations are [22,38,44]:

$$\frac{\partial u}{\partial r} + \frac{u}{r} + \frac{\partial w}{\partial z} = 0, \tag{1}$$

$$u\frac{\partial u}{\partial r} - \frac{v^2}{r} + w\frac{\partial u}{\partial z} = \nu \left(\frac{\partial^2 u}{\partial z^2} + \frac{\partial^2 u}{\partial r^2} + \frac{1}{r}\frac{\partial u}{\partial r} - \frac{u}{r^2} \right) - \frac{\nu}{k^*}u - Fu^2, \tag{2}$$

$$u\frac{\partial v}{\partial r} + \frac{uv}{r} + w\frac{\partial v}{\partial z} = \nu \left(\frac{\partial^2 v}{\partial z^2} + \frac{\partial^2 v}{\partial r^2} + \frac{1}{r}\frac{\partial v}{\partial r} - \frac{v}{r^2} \right) - \frac{\nu}{k^*}v - Fv^2, \tag{3}$$

$$u\frac{\partial w}{\partial r} + w\frac{\partial w}{\partial z} = \nu \left(\frac{\partial^2 w}{\partial z^2} + \frac{\partial^2 w}{\partial r^2} + \frac{1}{r}\frac{\partial w}{\partial r} \right) - \frac{\nu}{k^*}w - Fw^2, \tag{4}$$

$$u\frac{\partial T}{\partial r} + w\frac{\partial T}{\partial z} = \alpha^* \left(\frac{\partial^2 T}{\partial z^2} + \frac{\partial^2 T}{\partial r^2} + \frac{1}{r}\frac{\partial T}{\partial r} \right)$$
$$+ \frac{(\rho c)_p}{(\rho c)_f} \left(D_B \left(\frac{\partial T}{\partial r}\frac{\partial C}{\partial r} + \frac{\partial T}{\partial z}\frac{\partial C}{\partial z} \right) + \frac{D_T}{T_\infty} \left(\left(\frac{\partial T}{\partial z}\right)^2 + \left(\frac{\partial T}{\partial r}\right)^2 \right) \right), \quad (5)$$

$$u\frac{\partial C}{\partial r} + w\frac{\partial C}{\partial z} = D_B \left(\frac{\partial^2 C}{\partial z^2} + \frac{\partial^2 C}{\partial r^2} + \frac{1}{r}\frac{\partial C}{\partial r} \right)$$
$$+ \frac{D_T}{T_\infty} \left(\frac{\partial^2 T}{\partial z^2} + \frac{\partial^2 T}{\partial r^2} + \frac{1}{r}\frac{\partial T}{\partial r} \right) - k_r^2 (C - C_\infty) \left(\frac{T}{T_\infty} \right)^n \exp\left(-\frac{E_a}{\kappa T} \right). \quad (6)$$

Subjected boundary conditions are

$$u = 0, \; v = r\Omega, \; w = 0, \; -k\frac{\partial T}{\partial z} = h_f(T_f - T), \; -D_B\frac{\partial C}{\partial z} = k_{m^*}(C_f - C) \text{ at } z = 0, \quad (7)$$

$$u \to 0, \; v \to 0, \; T \to T_\infty, \; C \to C_\infty \text{ as } z \to \infty. \quad (8)$$

Here u, v and w represent velocities in directions of r, ϕ and z while ρ_f, ν ($= \mu/\rho_f$) and μ show density, kinematic and dynamic viscosities respectively, $(\rho c)_p$ effective heat capacity of nanoparticles, E_a the activation energy, $(\rho c)_f$ heat capacity of liquid, k^* the permeability of porous space, C the concentration, n the fitted rate constant, C_∞ the ambient concentration, $F = C_b/rk^{*1/2}$ the non-uniform inertia factor, D_T the thermophoretic factor, C_b the drag factor, h_f the uniform heat transfer factor, $\alpha^* = k/(\rho c)_f$ and k the thermal diffusivity and thermal conductivity respectively, T the fluid temperature, k_r the reaction rate, D_B the Brownian factor, κ the Boltzmann constant, k_{m^*} the uniform mass transfer factor and T_∞ the ambient temperature. Selecting

$$\left. \begin{array}{l} u = r\Omega f'(\zeta), \; w = -(2\Omega \nu)^{1/2} f(\zeta), \; v = r\Omega g(\zeta), \\ \phi(\zeta) = \frac{C - C_\infty}{C_f - C_\infty}, \; \zeta = \left(\frac{2\Omega}{\nu}\right)^{1/2} z, \; \theta(\zeta) = \frac{T - T_\infty}{T_f - T_\infty}. \end{array} \right\} \quad (9)$$

Continuity expression (1) is verified while Equations (2)–(8) yield

$$2f''' + 2ff'' - f'^2 + g^2 - \lambda f' - Frf'^2 = 0, \quad (10)$$

$$2g'' + 2fg' - 2f'g - \lambda g - Frg^2 = 0, \quad (11)$$

$$\frac{1}{\Pr}\theta'' + f\theta' + N_b\theta'\phi' + N_t\theta'^2 = 0, \quad (12)$$

$$\frac{1}{Sc}\phi'' + f\phi' + \frac{1}{Sc}\frac{N_t}{N_b}\theta'' - \sigma(1 + \delta\theta)^n \phi \exp\left(-\frac{E}{1 + \delta\theta}\right) = 0, \quad (13)$$

$$f(0) = 0, \; f'(0) = 0, \; g(0) = 1, \; \theta'(0) = -\gamma_1(1 - \theta(0)), \; \phi'(0) = -\gamma_2(1 - \phi(0)), \quad (14)$$

$$f'(\infty) \to 0, \; g(\infty) \to 0, \; \theta(\infty) \to 0, \; \phi(\infty) \to 0. \quad (15)$$

Here Fr stands for Forchheimer number, γ_2 for concentration Biot number, λ for porosity parameter, γ_1 for thermal Biot number, N_t thermophoresis parameter, $\Pr$ Prandtl number, σ for chemical reaction parameter, N_b for Brownian motion, δ for temperature difference parameter, Sc Schmidt number, and E for nondimensional activation energy. Nondimensional variables are defined by

$$\left. \begin{array}{l} \lambda = \frac{\nu}{k^*\Omega}, \; Fr = \frac{C_b}{k^{*1/2}}, \; \gamma_1 = \frac{h_f}{k}\sqrt{\frac{\nu}{2\Omega}}, \; \gamma_2 = \frac{k_{m^*}}{D_B}\sqrt{\frac{\nu}{2\Omega}}, \; N_b = \frac{(\rho c)_p D_B (C_f - C_\infty)}{(\rho c)_f \nu}, \\ \Pr = \frac{\nu}{\alpha^*}, \; N_t = \frac{(\rho c)_p D_T (T_f - T_\infty)}{(\rho c)_f \nu T_\infty}, \; Sc = \frac{\nu}{D_B}, \; \sigma = \frac{k_r^2}{\Omega}, \; \delta = \frac{T_f - T_\infty}{T_\infty}, \; E = \frac{E_a}{\kappa T_\infty}. \end{array} \right\} \quad (16)$$

The coefficients of skin-friction and Nusselt and Sherwood expressions are

$$\left.\begin{array}{l} \text{Re}_r^{1/2} C_f = f''(0), \ \text{Re}_r^{1/2} C_g = g'(0), \\ \text{Re}_r^{-1/2} Nu = -\theta'(0), \ \text{Re}_r^{-1/2} Sh = -\phi'(0), \end{array}\right\} \quad (17)$$

where $\text{Re}_r = 2(\Omega r)r/\nu$ represents local rotational Reynolds number.

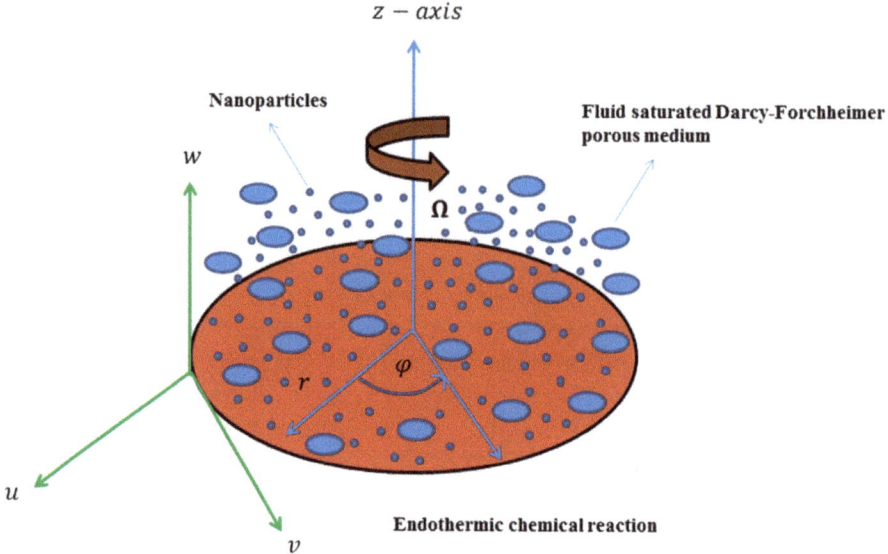

Figure 1. Flow configuration.

3. Numerical Results and Discussion

The present section outlines the commitment of various relevant parameters including Schmidt number Sc, porosity parameter λ, thermophoresis parameter N_t, Prandtl number Pr, Forchheimer number Fr, nondimensional activation energy E, thermal Biot γ_1, chemical reaction parameter σ, concentration Biot γ_2 and Brownian number N_b on velocities $f'(\zeta)$ and $g(\zeta)$, concentration $\phi(\zeta)$ and temperature $\theta(\zeta)$ distributions. Figure 2 portrays how porosity parameter λ influences the speed appropriation $f'(\zeta)$. It has been discovered that the speed profile $f'(\zeta)$ and its related energy layer are devalued by upgrading porosity λ. The presence of permeable space improves the protection from liquid stream which relates to bringing down liquid speed and its related energy layer. Figure 3 delineates the impact of Forchheimer variable Fr on $f'(\zeta)$. Higher estimations of Forchheimer variable Fr establish lower speed profile $f'(\zeta)$. Figure 4 shows how the speed conveyance $g(\zeta)$ is influenced by porosity parameter λ. Here the speed dissemination is rotted by expanding λ. Figure 5 delineates a variety of speed circulation $g(\zeta)$ for unmistakable Fr. By expanding Fr, a decrease showed up in speed dissemination and related layer. Figure 6 shows warm Biot γ_1 impact on temperature $\theta(\zeta)$. More grounded convection is delivered by upgrading warm Biot number γ_1. Thus, temperature and warm layer are raised by expanding warm Biot number γ_1. Figure 7 presents a variety in temperature field $\theta(\zeta)$ for Pr. Here, temperature is rotted for bigger Pr. The proportion of force diffusivity to warm diffusivity is termed as the Prandtl number. Higher estimations of Pr depict more fragile warm diffusivity, which compares to diminishing in the warm layer. Figure 8 is shown to investigate N_t impact on temperature field $\theta(\zeta)$. Bigger thermophoresis parameter N_t establishes a higher temperature field and progressively warm layer thickness. The purpose of such contention is that augmentation in N_t yields high grounded thermophoresis power which further permits motion of the nanoparticles in liquid zone. Far from surface in this way shapes a more grounded temperature dispersion $\theta(\zeta)$ and progressively warm layer. The effect of N_b on temperature profile $\theta(\zeta)$ is depicted in Figure 9. From a physical perspective, an unpredictable movement of nanoparticles increments by improving Brownian movement parameter N_b causes a crash of particle. As a result, the active vitality is changed into warmth vitality which causes upgrade in $\theta(\zeta)$ and associated warm layer. Figure 10 shows how concentration $\phi(\zeta)$ is influenced by concentration Biot number γ_2. Concentration is upgraded for higher estimations of γ_2. From Figure 11 we can see that bigger Sc rots concentration $\phi(\zeta)$. Schmidt number Sc is conversely relative to Brownian diffusivity. Higher Sc yields a more fragile Brownian diffusivity. Such Brownian diffusivity prompts low concentration $\phi(\zeta)$. Figure 12 demonstrates how the thermophoresis parameter N_t influences the concentration $\phi(\zeta)$. By improving thermophoresis parameter N_t, concentration $\phi(\zeta)$ and related concentration layers are upgraded. Figure 13 depicts the Brownian movement N_b and minor departure from concentration $\phi(\zeta)$. It can be seen that a more fragile concentration $\phi(\zeta)$ is produced by using higher N_b. Figure 14 explains the impact of nondimensional initiation vitality E on concentration $\phi(\zeta)$. An improvement in E rots altered Arrhenius work $\left(\frac{T}{T_\infty}\right)^n \exp\left(-\frac{E_a}{\kappa T}\right)$. Such inevitably builds up the generative synthetic response because of which concentration $\phi(\zeta)$ upgrades. Figure 15 shows that an improvement in σ shows a rot in concentration $\phi(\zeta)$ and its related layer. Highlights of N_t and N_b on $Nu(Re_r)^{-1/2}$ are revealed through Figures 16 and 17 respectively. True to form, $Nu(Re_r)^{-1/2}$ reduces for N_t and N_b. Effects of N_t and N_b on $Sh(Re_r)^{-1/2}$ have been portrayed in Figures 18 and 19 respectively. Here $Sh(Re_r)^{-1/2}$ is an expanding capacity of N_t, while the inverse pattern is seen for N_b. Table 1 is developed to validate the present results with the previously published results in a limiting case. Here, we demonstrate that the present numerical solution has good agreement with the previous solution by Naqvi et al. [48] in a limiting case.

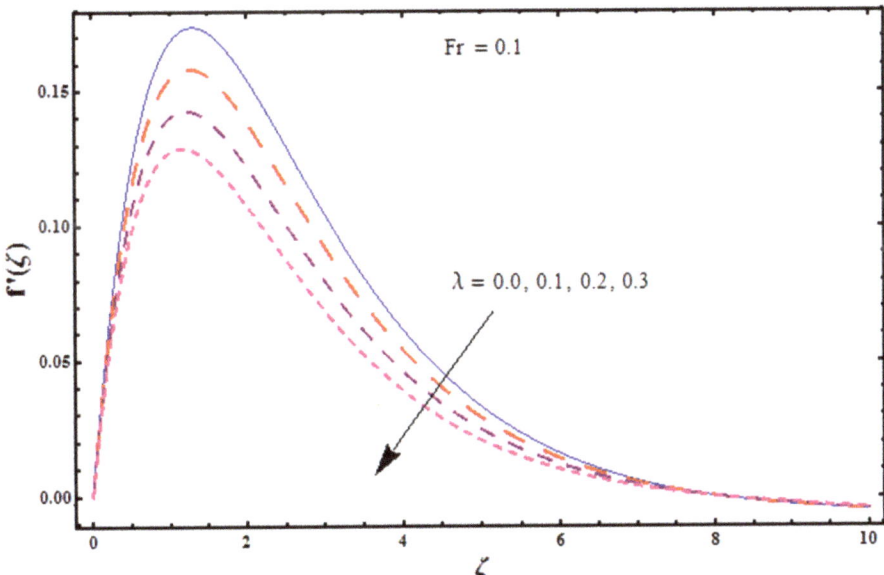

Figure 2. Curves of $f'(\zeta)$ for λ.

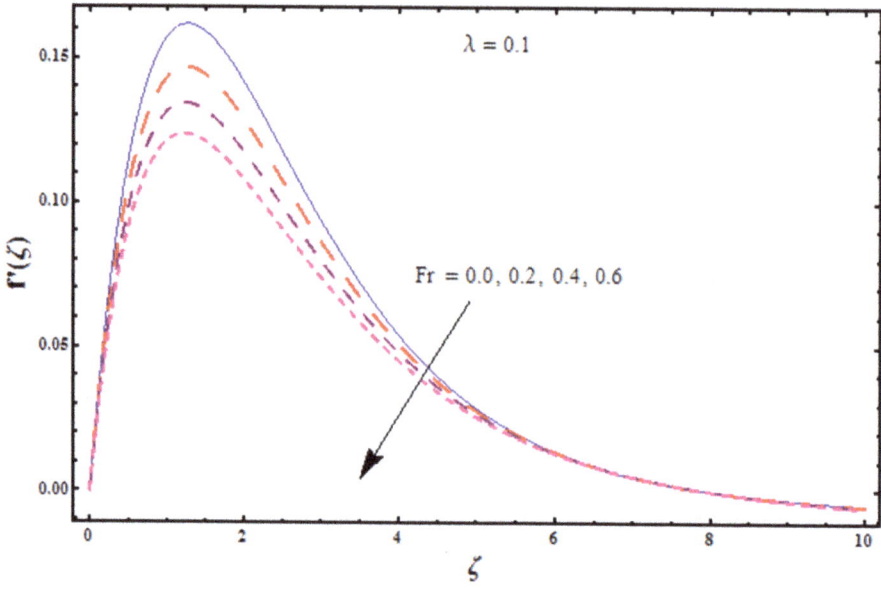

Figure 3. Curves of $f'(\zeta)$ for Fr.

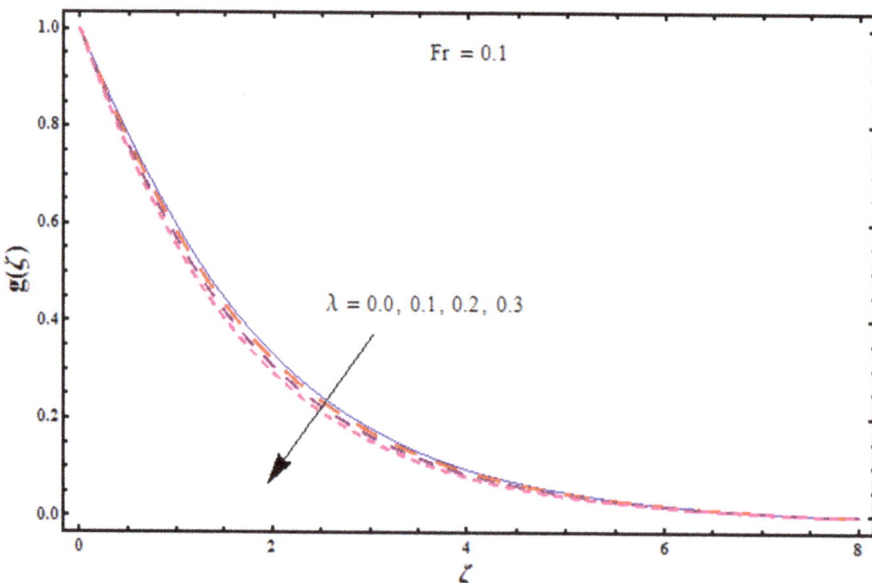

Figure 4. Curves of $g(\zeta)$ for λ.

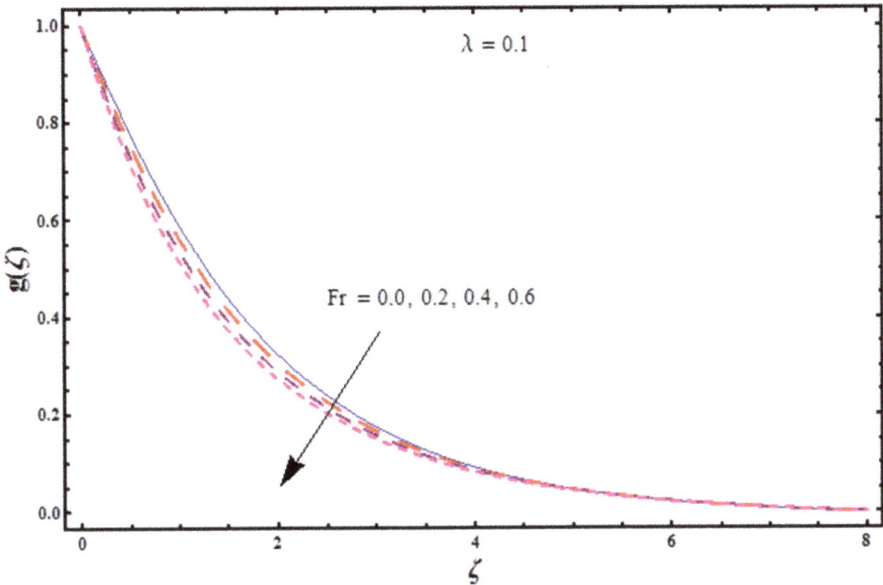

Figure 5. Curves of $g(\zeta)$ for Fr.

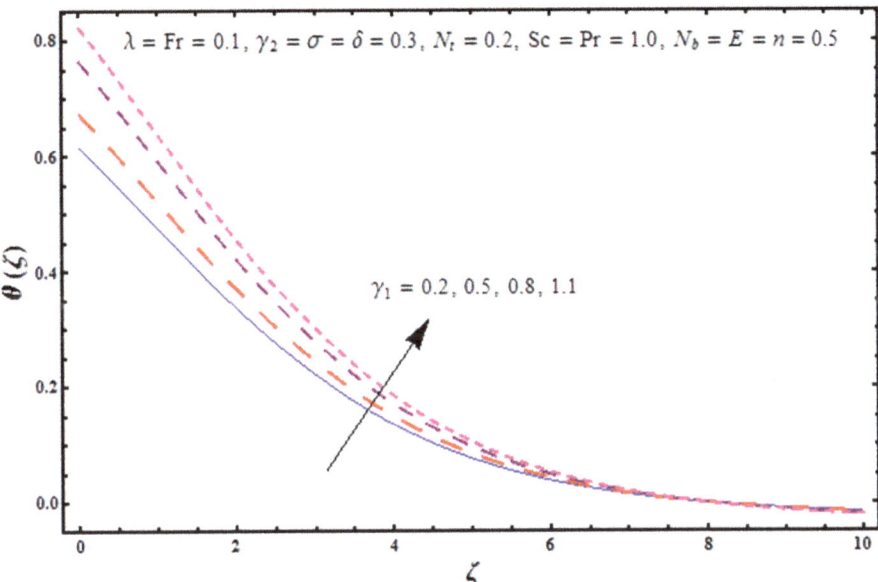

Figure 6. Curves of $\theta(\zeta)$ for γ_1.

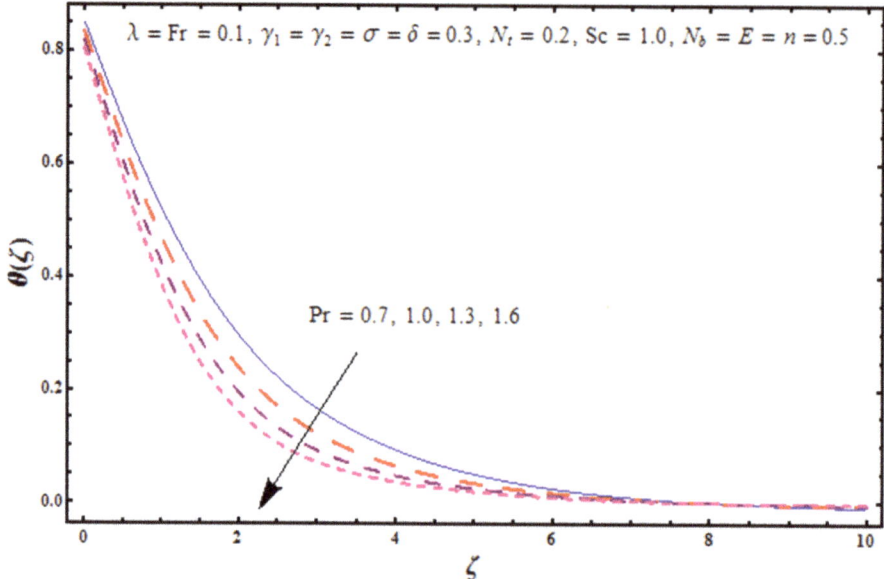

Figure 7. Curves of $\theta(\zeta)$ for Pr.

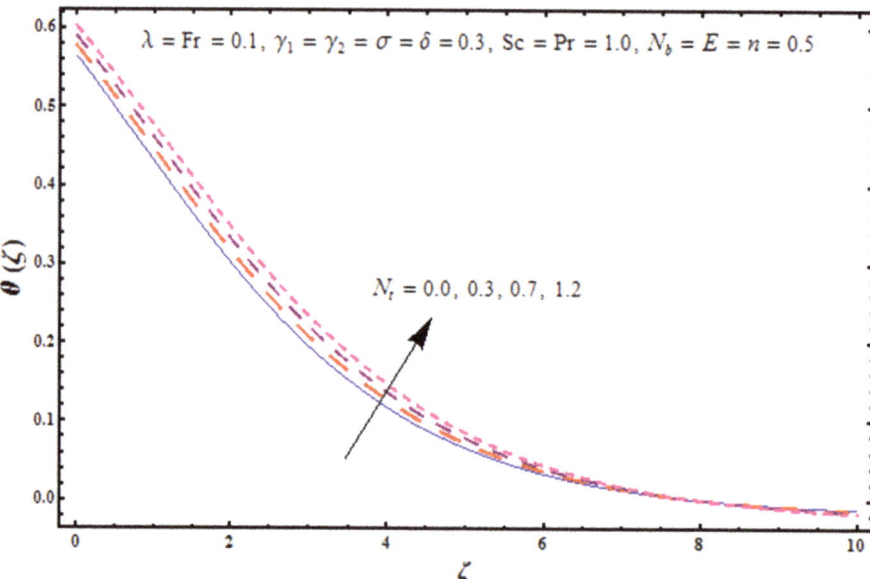

Figure 8. Curves of $\theta(\zeta)$ for N_t.

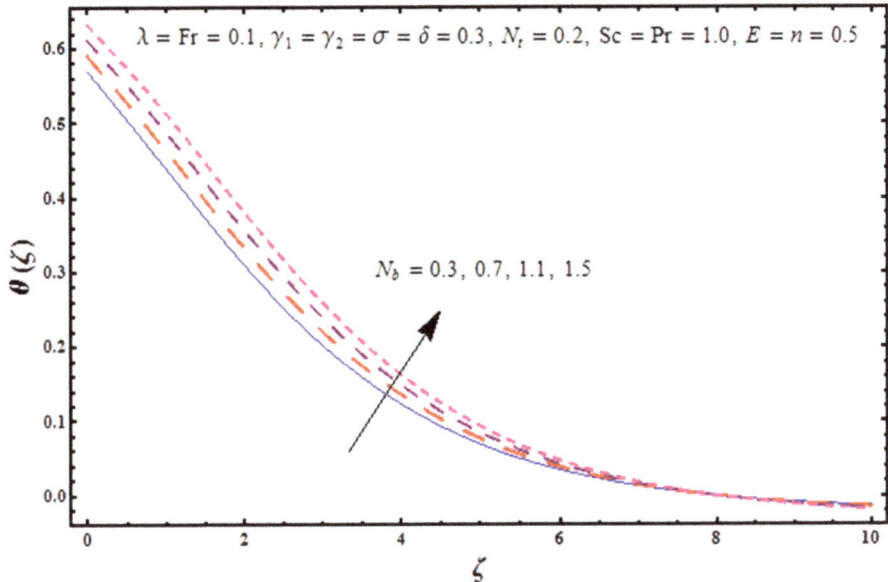

Figure 9. Curves of $\theta(\zeta)$ for N_b.

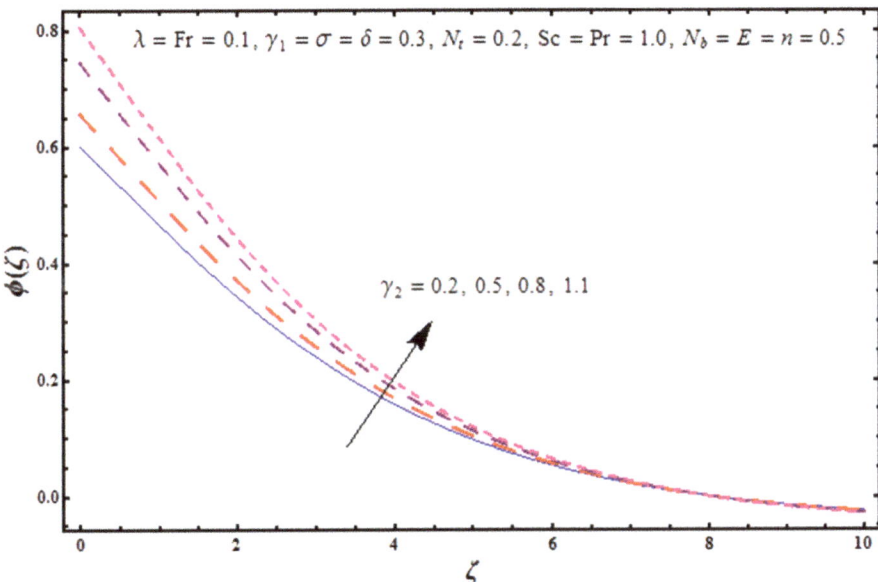

Figure 10. Curves of $\phi(\zeta)$ for γ_2.

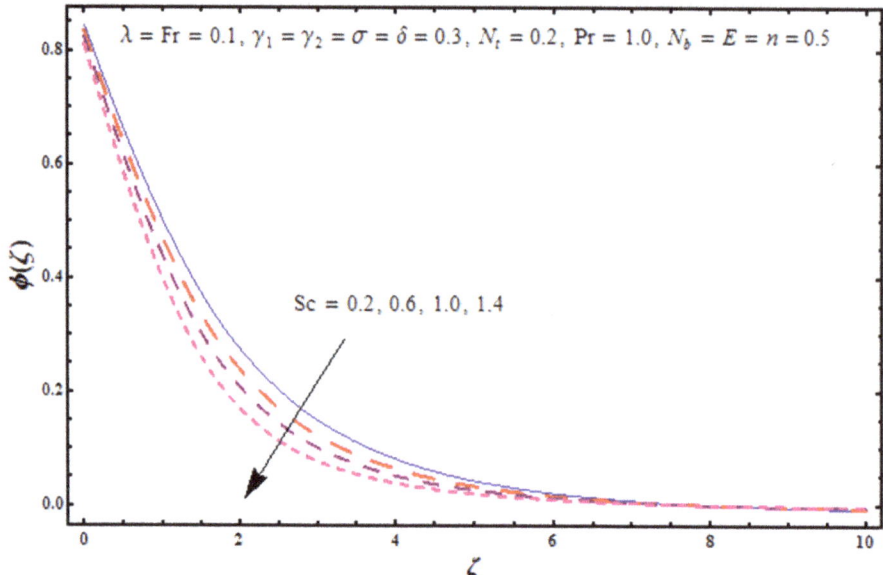

Figure 11. Curves of $\phi(\zeta)$ for Sc.

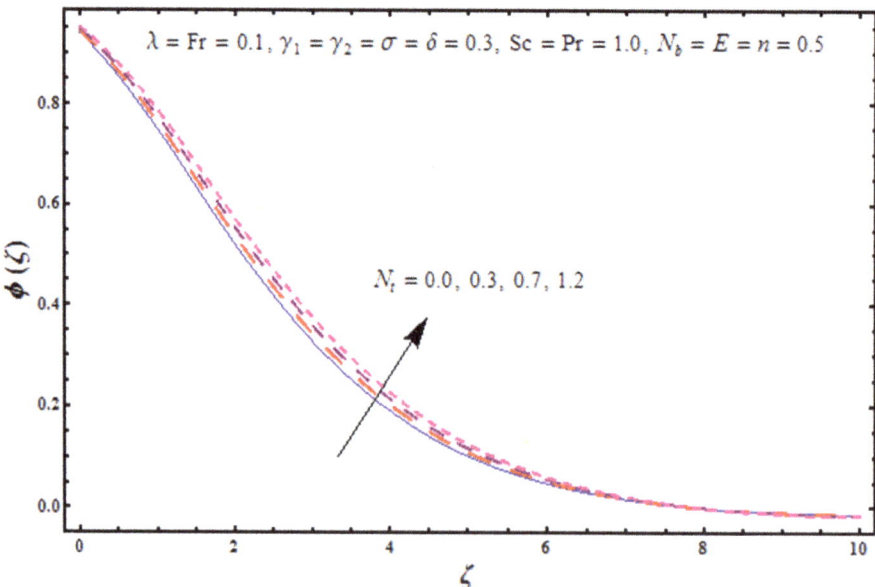

Figure 12. Curves of $\phi(\zeta)$ for N_t.

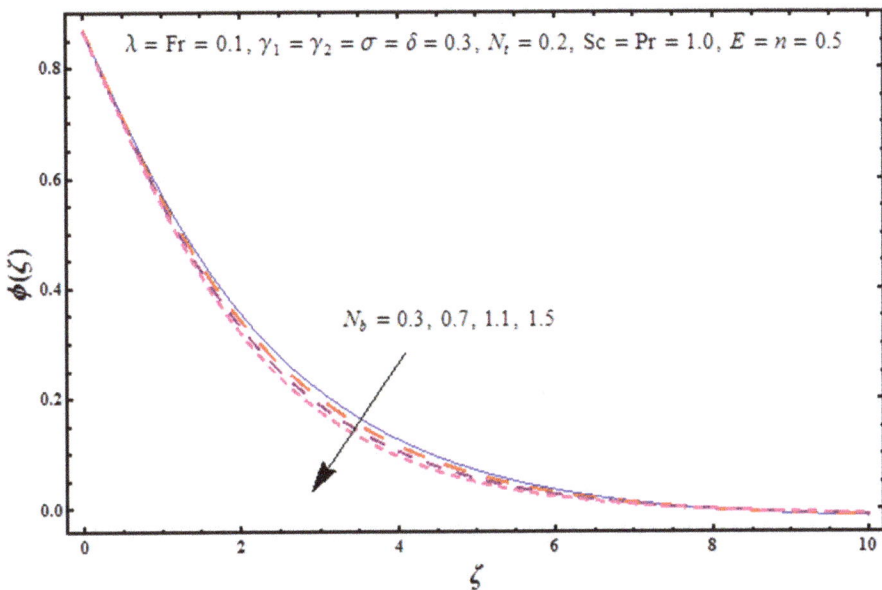

Figure 13. Curves of $\phi(\zeta)$ for N_b.

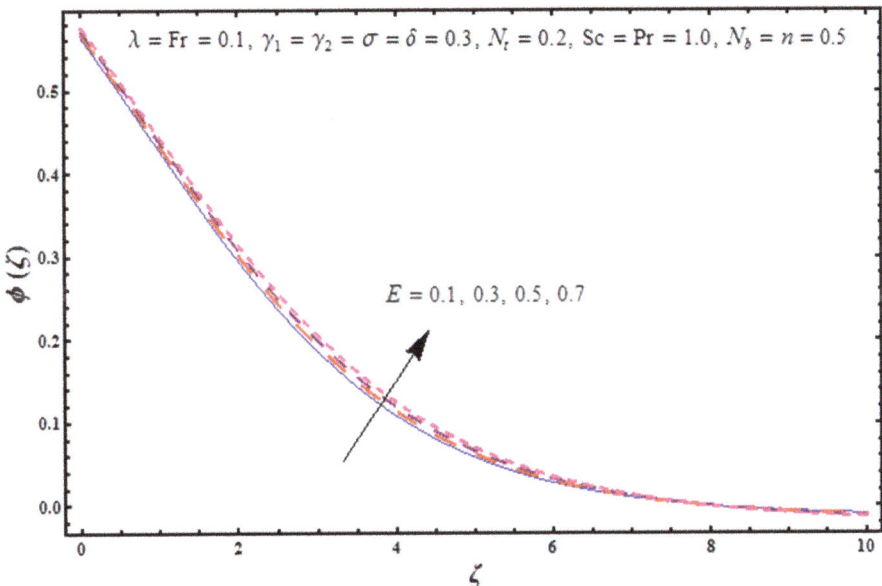

Figure 14. Curves of $\phi(\zeta)$ for E.

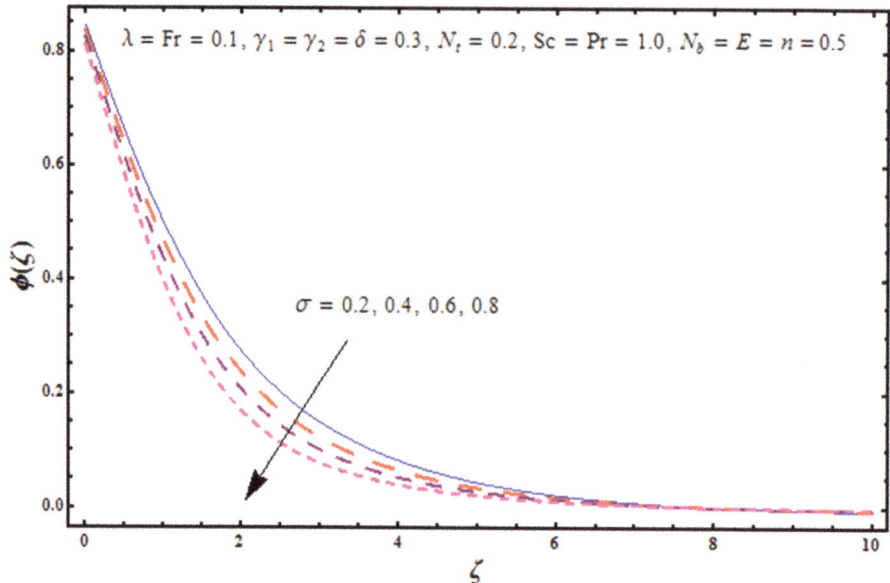

Figure 15. Curves of $\phi(\zeta)$ for σ.

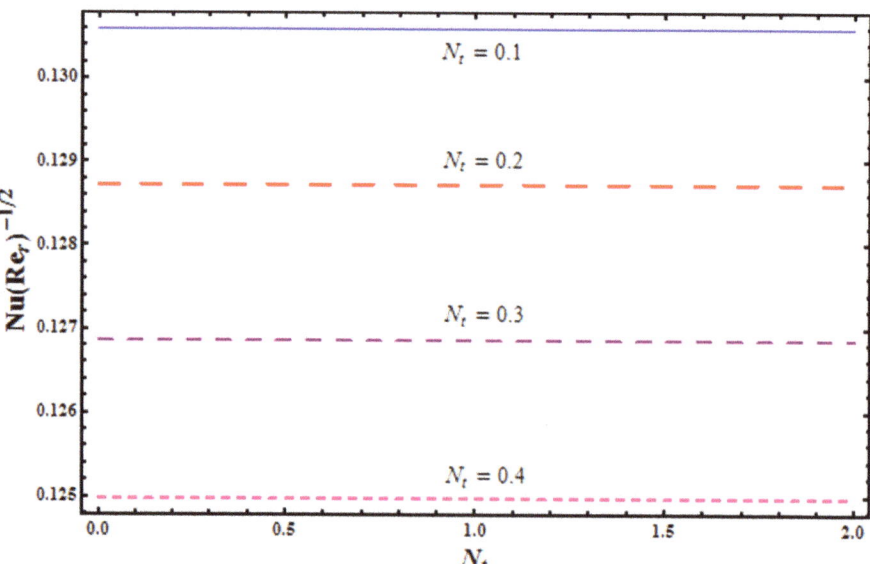

Figure 16. Curves of $Nu(Re_r)^{-1/2}$ for N_t.

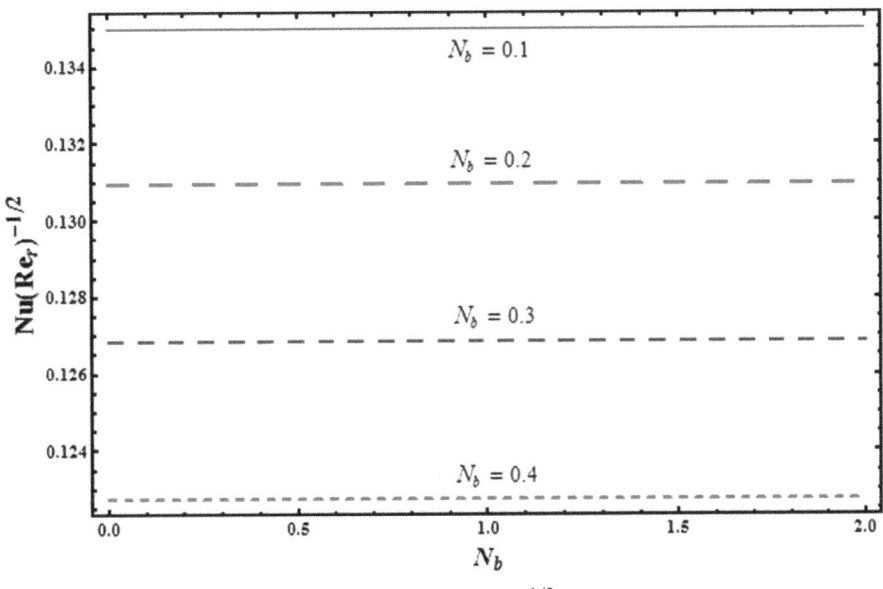

Figure 17. Curves of $Nu(\text{Re}_r)^{-1/2}$ for N_b.

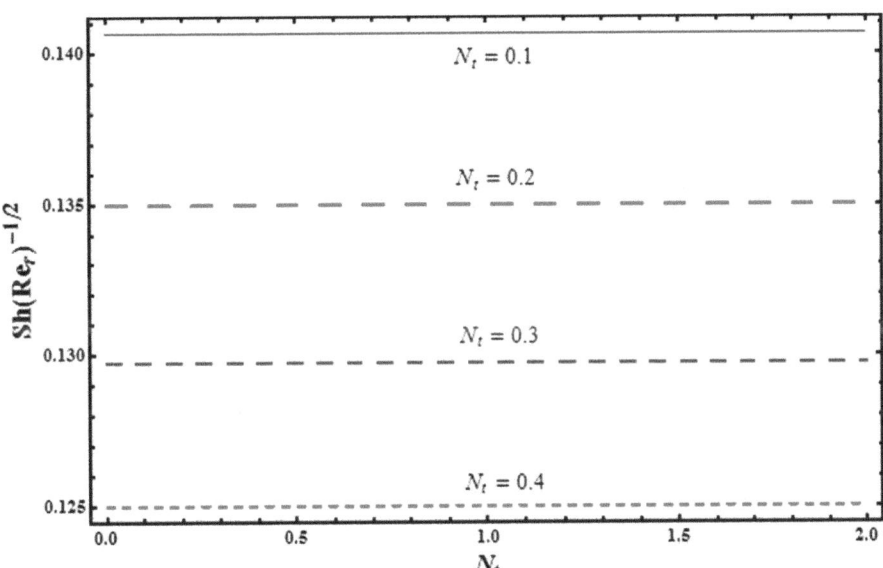

Figure 18. Curves of $Sh(\text{Re}_r)^{-1/2}$ for N_t.

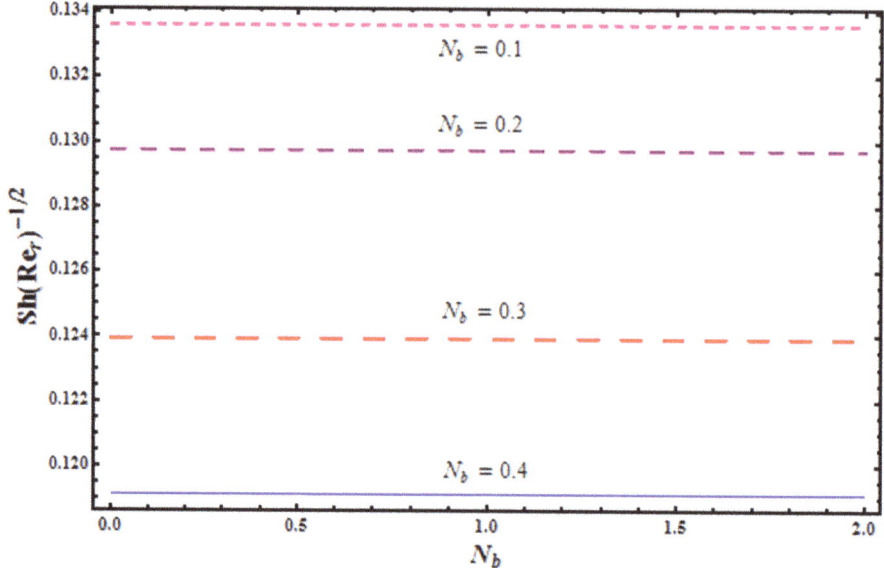

Figure 19. Curves of $Sh(\text{Re}_r)^{-1/2}$ for N_b.

Table 1. Comparative values of $f''(0)$ and $g'(0)$ for value of Fr when $\lambda = 0.2$.

	Present Results		Naqvi et al. [48]	
Fr	$f''(0)$	$g'(0)$	$f''(0)$	$g'(0)$
0.2	0.43478	−0.78139	0.4347813	−0.7813904

4. Conclusions

Darcy–Forchheimer flow of viscous nanofluid due to a rotating disk with binary chemical reaction and Arrhenius activation energy was studied. The shooting algorithm leads to the solutions of dimensionless quantities. We noticed that temperature rises for larger thermal Biot number. Temperature is less in the absence of thermal Biot number. Enhancing concentration Biot number leads to higher concentration and thickness of concentration boundary layer. An increase in activation energy leads to higher temperature. We further demonstrated that enhancement in chemical reaction parameter gives a reduction in the curves of concentration.

Author Contributions: All the authors have contributed equally in all parts.

Funding: This research was funded by University of Malaya grant number IIRG 001C-2019.

Acknowledgments: The project is partially sponsored by University of Malaya grant No: IIRG 001C-2019.

Conflicts of Interest: The authors declare no conflict of interest.

References

1. Choi, S.U.S. *Enhancing Thermal Conductivity of Fluids with Nanoparticles*; FED 231/MD; ASME: New York, NY, USA, 1995; pp. 99–105.
2. Buongiorno, J. Convective transport in nanofluids. *ASME J. Heat Transf.* **2006**, *128*, 240–250. [CrossRef]
3. Tiwari, R.K.; Das, M.K. Heat transfer augmentation in a two-sided lid-driven differentially heated square cavity utilizing nanofluid. *Int. J. Heat Mass Transf.* **2007**, *50*, 2002–2018. [CrossRef]
4. Pantzali, M.N.; Mouza, A.A.; Paras, S.V. Investigating the efficacy of nanofluids as coolants in plate heat exchangers (PHE). *Chem. Eng. Sci.* **2009**, *64*, 3290–3300. [CrossRef]

5. Sheikholeslami, M.; Bandpy, M.G.; Ganji, D.D.; Soleimani, S. Effect of a magnetic field on natural convection in an inclined half-anulus enclosure filled with Cu-water nanofluid using CVFEM. *Adv. Powder Technol.* **2013**, *24*, 980–991. [CrossRef]
6. Togun, H.; Safaei, M.R.; Sadri, R.; Kazi, S.N.; Badarudin, A.; Hooman, K.; Sadeghinezhad, E. Numerical simulation of laminar to turbulent nanofluid flow and heat transfer over a backward-facing step. *Appl. Math. Comput.* **2014**, *239*, 153–170. [CrossRef]
7. Hsiao, K.L. Nanofluid flow with multimedia physical features for conjugate mixed convection and radiation. *Comp. Fluids* **2014**, *104*, 1–8. [CrossRef]
8. Hayat, T.; Muhammad, T.; Alsaedi, A.; Alhuthali, M.S. Magnetohydrodynamic three-dimensional flow of viscoelastic nanofluid in the presence of nonlinear thermal radiation. *J. Magn. Magn. Mater.* **2015**, *385*, 222–229. [CrossRef]
9. Lin, Y.; Zheng, L.; Zhang, X.; Ma, L.; Chen, G. MHD pseudo-plastic nanofluid unsteady flow and heat transfer in a finite thin film over stretching surface with internal heat generation. *Int. J. Heat Mass Transf.* **2015**, *84*, 903–911. [CrossRef]
10. Hsiao, K.L. Stagnation electrical MHD nanofluid mixed convection with slip boundary on a stretching sheet. *Appl. Thermal Eng.* **2016**, *98*, 850–861. [CrossRef]
11. Hayat, T.; Aziz, A.; Muhammad, T.; Alsaedi, A. On magnetohydrodynamic three-dimensional flow of nanofluid over a convectively heated nonlinear stretching surface. *Int. J. Heat Mass Transf.* **2016**, *100*, 566–572. [CrossRef]
12. Hayat, T.; Muhammad, T.; Shehzad, S.A.; Alsaedi, A. Three-dimensional flow of Jeffrey nanofluid with a new mass flux condition. *J. Aerospace Eng.* **2016**, *29*, 04015054. [CrossRef]
13. Hsiao, K.L. Micropolar nanofluid flow with MHD and viscous dissipation effects towards a stretching sheet with multimedia feature. *Int. J. Heat Mass Transf.* **2017**, *112*, 983–990. [CrossRef]
14. Muhammad, T.; Alsaedi, A.; Shehzad, S.A.; Hayat, T. A revised model for Darcy-Forchheimer flow of Maxwell nanofluid subject to convective boundary condition. *Chin. J. Phys.* **2017**, *55*, 963–976. [CrossRef]
15. Hayat, T.; Muhammad, T.; Shehzad, S.A.; Alsaedi, A. An analytical solution for magnetohydrodynamic Oldroyd-B nanofluid flow induced by a stretching sheet with heat generation/absorption. *Int. J. Thermal Sci.* **2017**, *111*, 274–288. [CrossRef]
16. Hsiao, K.L. To promote radiation electrical MHD activation energy thermal extrusion manufacturing system efficiency by using Carreau-Nanofluid with parameters control method. *Energy* **2017**, *130*, 486–499. [CrossRef]
17. Hayat, T.; Sajjad, R.; Alsaedi, A.; Muhammad, T.; Ellahi, R. On squeezed flow of couple stress nanofluid between two parallel plates. *Results Phys.* **2017**, *7*, 553–561. [CrossRef]
18. Muhammad, T.; Alsaedi, A.; Hayat, T.; Shehzad, S.A. A revised model for Darcy-Forchheimer three-dimensional flow of nanofluid subject to convective boundary condition. *Results Phys.* **2017**, *7*, 2791–2797. [CrossRef]
19. Moshizi, S.A.; Zamani, M.; Hosseini, S.J.; Malvandi, A. Mixed convection of magnetohydrodynamic nanofluids inside microtubes at constant wall temperature. *J. Magn. Magn. Mater.* **2017**, *430*, 36–46. [CrossRef]
20. Hayat, T.; Hussain, Z.; Alsaedi, A.; Muhammad, T. An optimal solution for magnetohydrodynamic nanofluid flow over a stretching surface with constant heat flux and zero nanoparticles flux. *Neural Comp. Appl.* **2018**, *29*, 1555–1562. [CrossRef]
21. Eid, M.R.; Mahny, K.L.; Muhammad, T.; Sheikholeslami, M. Numerical treatment for Carreau nanofluid flow over a porous nonlinear stretching surface. *Results Phys.* **2018**, *8*, 1185–1193. [CrossRef]
22. Aziz, A.; Alsaedi, A.; Muhammad, T.; Hayat, T. Numerical study for heat generation/absorption in flow of nanofluid by a rotating disk. *Results Phys.* **2018**, *8*, 785–792. [CrossRef]
23. Muhammad, T.; Lu, D.C.; Mahanthesh, B.; Eid, M.R.; Ramzan, M.; Dar, A. Significance of Darcy-Forchheimer porous medium in nanofluid through carbon nanotubes. *Commun. Theoret. Phys.* **2018**, *70*, 361. [CrossRef]
24. Saif, R.S.; Hayat, T.; Ellahi, R.; Muhammad, T.; Alsaedi, A. Darcy-Forchheimer flow of nanofluid due to a curved stretching surface. *Int. J. Numer. Methods Heat Fluid Flow* **2019**, *29*, 2–20. [CrossRef]
25. Mahanthesh, B.; Gireesha, B.J.; Animasaun, I.L.; Muhammad, T.; Shashikumar, N.S. MHD flow of SWCNT and MWCNT nanoliquids past a rotating stretchable disk with thermal and exponential space dependent heat source. *Phys. Scr.* **2019**, *94*, 085214. [CrossRef]
26. von Karman, T. Uberlaminare und turbulente Reibung. *Z. Angew. Math. Mech.* **1921**, *1*, 233–252. [CrossRef]
27. Cochran, W.G. The flow due to a rotating disk. *Proc. Camb. Philos. Soc.* **1934**, *30*, 365–375. [CrossRef]

28. Stewartson, K. On the flow between two rotating coaxial disks. *Proc. Comb. Phil. Soc.* **1953**, *49*, 333–341. [CrossRef]
29. Chapple, P.J.; Stokes, V.K. *On the Flow Between a Rotating and a Stationary Disk*; Report No. FLD 8; Dept. Mech. Eng. Princeton University: Princeton, NJ, USA, 1962.
30. Ackroyd, J.A.D. On the steady flow produced by a rotating disk with either surface suction or injection. *J. Eng. Math.* **1978**, *12*, 207–220. [CrossRef]
31. Erdogan, M.E. Unsteady flow of a viscous fluid due to non-coaxial rotations of a disk and a fluid at infinity. *Int. J. Non-Linear Mech.* **1997**, *32*, 285–290. [CrossRef]
32. Attia, H.A. Steady flow over a rotating disk in porous medium with heat transfer. *Nonlinear Anal. Model. Control* **2009**, *14*, 21–26.
33. Turkyilmazoglu, M.; Senel, P. Heat and mass transfer of the flow due to a rotating rough and porous disk. *Int. J. Thermal Sci.* **2013**, *63*, 146–158. [CrossRef]
34. Rashidi, M.M.; Kavyani, N.; Abelman, S. Investigation of entropy generation in MHD and slip flow over a rotating porous disk with variable properties. *Int. J. Heat Mass Transf.* **2014**, *70*, 892–917. [CrossRef]
35. Mustafa, M.; Khan, J.A.; Hayat, T.; Alsaedi, A. On Bodewadt flow and heat transfer of nanofluids over a stretching stationary disk. *J. Mol. Liq.* **2015**, *211*, 119–125. [CrossRef]
36. Hayat, T.; Muhammad, T.; Shehzad, S.A.; Alsaedi, A. On magnetohydrodynamic flow of nanofluid due to a rotating disk with slip effect: A numerical study. *Comput. Methods Appl. Mech. Eng.* **2017**, *315*, 467–477. [CrossRef]
37. Mustafa, M. MHD nanofluid flow over a rotating disk with partial slip effects: Buongiorno model. *Int. J. Heat Mass Transf.* **2017**, *108*, 1910–1916. [CrossRef]
38. Hayat, T.; Haider, F.; Muhammad, T.; Alsaedi, A. On Darcy-Forchheimer flow of carbon nanotubes due to a rotating disk. *Int. J. Heat Mass Transf.* **2017**, *112*, 248–254. [CrossRef]
39. Bestman, A.R. Natural convection boundary layer with suction and mass transfer in a porous medium. *Int. J. Energy Res.* **1990**, *14*, 389–396. [CrossRef]
40. Makinde, O.D.; Olanrewaju, P.O.; Charles, W.M. Unsteady convection with chemical reaction and radiative heat transfer past a flat porous plate moving through a binary mixture. *Africka Matematika* **2011**, *22*, 65–78. [CrossRef]
41. Maleque, K.A. Effects of exothermic/endothermic chemical reactions with Arrhenius activation energy on MHD free convection and mass transfer flow in presence of thermal radiation. *J. Thermodyn.* **2013**, *2013*, 692516. [CrossRef]
42. Awad, F.G.; Motsa, S.; Khumalo, M. Heat and mass transfer in unsteady rotating fluid flow with binary chemical reaction and activation energy. *PLoS ONE* **2014**, *9*, e107622. [CrossRef] [PubMed]
43. Abbas, Z.; Sheikh, M.; Motsa, S.S. Numerical solution of binary chemical reaction on stagnation point flow of Casson fluid over a stretching/shrinking sheet with thermal radiation. *Energy* **2016**, *95*, 12–20. [CrossRef]
44. Shafique, Z.; Mustafa, M.; Mushtaq, A. Boundary layer flow of Maxwell fluid in rotating frame with binary chemical reaction and activation energy. *Results Phys.* **2016**, *6*, 627–633. [CrossRef]
45. Hayat, T.; Aziz, A.; Muhammad, T.; Alsaedi, A. Effects of binary chemical reaction and Arrhenius activation energy in Darcy-Forchheimer three-dimensional flow of nanofluid subject to rotating frame. *J. Thermal Anal. Calorimet.* **2019**, *136*, 1769–1779. [CrossRef]
46. Irfan, M.; Khan, W.A.; Khan, M.; Gulzar, M.M. Influence of Arrhenius activation energy in chemically reactive radiative flow of 3D Carreau nanofluid with nonlinear mixed convection. *J. Phys. Chem. Solids* **2019**, *125*, 141–152. [CrossRef]
47. Hayat, T.; Aziz, A.; Muhammad, T.; Alsaedi, A. Numerical simulation for Darcy-Forchheimer 3D rotating flow subject to binary chemical reaction and Arrhenius activation energy. *J. Cent. South Univ.* **2019**, *26*, 1250–1259. [CrossRef]
48. Naqvi, S.M.R.S.; Muhammad, T.; Kim, H.M.; Mahmood, T.; Saeed, A.; Khan, B.S. Numerical treatment for Darcy–Forchheimer flow of nanofluid due to a rotating disk with slip effects. *Can. J. Phys.* **2019**, *97*, 856–863. [CrossRef]

© 2019 by the authors. Licensee MDPI, Basel, Switzerland. This article is an open access article distributed under the terms and conditions of the Creative Commons Attribution (CC BY) license (http://creativecommons.org/licenses/by/4.0/).

Article
On Fractional Symmetric Hahn Calculus

Nichaphat Patanarapeelert [1] and Thanin Sitthiwirattham [2,*]

[1] Department of Mathematics, Faculty of Applied Science, King Mongkut's University of Technology North Bangkok, Bangkok 10800, Thailand; nichaphat.p@sci.kmutnb.ac.th
[2] Mathematics Department, Faculty of Science and Technology, Suan Dusit University, Bangkok 10300, Thailand
* Correspondence: thanin_sit@dusit.ac.th

Received: 14 August 2019; Accepted: 17 September 2019; Published: 20 September 2019

Abstract: In this paper, we study fractional symmetric Hahn difference calculus. The new idea of the symmetric Hahn difference operator, the fractional symmetric Hahn integral, and the fractional symmetric Hahn operators of Riemann–Liouville and Caputo types are presented. In addition, we formulate some fundamental properties based on these fractional symmetric Hahn operators.

Keywords: fractional symmetric Hahn integral; fractional symmetric Hahn difference operator

JEL Classification: 39A10; 39A13; 39A70

1. Introduction

The Hahn difference operator, one type of quantum difference operator, has been studied by many researchers. It is used to construct families of orthogonal polynomials and to study certain approximation problems (see [1–3]).

Hahn [4] is the first researcher who introduced the Hahn difference operator $D_{q,\omega}$ based on the forward difference operator and the Jackson q-difference operator where

$$D_{q,\omega}f(t) := \frac{f(qt+\omega) - f(t)}{t(q-1) + \omega}, \quad t \neq \omega_0 := \frac{\omega}{1-q}.$$

Later, the right inverse of Hahn's operator and its properties were presented (see [5,6]). There are other works related to the Hahn difference operator such as the study of Hahn quantum variational calculus [7–9], and the existence and uniqueness results for the initial value problems [10–12] and boundary value problems [13,14].

Recently, Brikshavana and Sitthiwirattham [15] introduced fractional Hahn difference operators. The boundary value problems for fractional Hahn difference equations were subsequently studied by many researchers (see [16–19]).

In 2013, Artur et al. [20] introduced the symmetric Hahn difference operator $\tilde{D}_{q,\omega}$ as

$$\tilde{D}_{q,\omega}f(t) := \frac{f(qt+\omega) - f(q^{-1}(t-\omega))}{(q-q^{-1})t + (1+q^{-1})\omega} \quad \text{for } t \neq \omega_0.$$

However, we observe from the literature that fractional symmetric Hahn difference calculus has not been studied. In order to give a rigorous analysis of symmetric Hahn calculus, this paper is devoted to presenting the new concepts of the symmetric Hahn difference operator, the fractional symmetric Hahn integral, and the fractional symmetric Hahn difference operators of the Riemann–Liouville and Caputo types. Particularly, the results from this study can be used as a tool in some applications such as approximation problems, and initial and boundary value problems associated with symmetric

Hahn operators. We first introduce some basic definitions and properties of Hahn's difference operators in Section 2. In Section 3, we present the fractional symmetric Hahn integral and its properties. Finally, we propose the fractional symmetric Hahn difference operators of the Riemann–Liouville and Caputo types and their properties in Sections 4 and 5, respectively.

2. Preliminary Definitions and Properties

In order to study the fractional symmetric Hahn difference calculus, we first introduce some notations, definitions, lemmas as follows. (see [4–8,20,21]).

For $0 < q < 1, \omega > 0, \omega_0 = \frac{\omega}{1-q}$, we define

$$\widetilde{[k]}_q := \begin{cases} \frac{1-q^{2k}}{1-q^2} = [k]_{q^2}, & k \in \mathbb{N} \\ 1, & k = 0, \end{cases}$$

$$\widetilde{[k]}_q! := \begin{cases} \widetilde{[k]}_q \widetilde{[k-1]}_q \cdots \widetilde{[1]}_q = \prod_{i=1}^{k} \frac{1-q^{2i}}{1-q^2}, & k \in \mathbb{N} \\ 1, & k = 0. \end{cases}$$

The q, ω-forward jump operator is defined by

$$\sigma_{q,\omega}^k(t) := q^k t + \omega[k]_q,$$

and the q, ω-backward jump operator is defined by

$$\rho_{q,\omega}^k(t) := \frac{t - \omega[k]_q}{q^k},$$

where $k \in \mathbb{N}$.

Letting $n \in \mathbb{N}_0 := \{0, 1, 2, \ldots\}, a, b \in \mathbb{R}$, we define the power functions as follows:

- The q-analogue of the power function

$$(a-b)_q^0 := 1, \qquad (a-b)_q^n := \prod_{i=0}^{n-1}(a - bq^i),$$

- The q-symmetric analogue of the power function

$$(a \widetilde{-} b)_q^0 := 1, \qquad (a \widetilde{-} b)_q^n := \prod_{i=0}^{n-1}(a - bq^{2i+1}),$$

- The q, ω-symmetric analogue of the power function

$$(a \widetilde{-} b)_{q,\omega}^0 := 1, \qquad (a \widetilde{-} b)_{q,\omega}^n := \prod_{i=0}^{n-1}\left[a - \sigma_{q,\omega}^{2i+1}(b)\right].$$

In general, for $\alpha \in \mathbb{R}$, we have

$$(a-b)_q^\alpha = a^\alpha \prod_{i=0}^{\infty} \frac{1 - \left(\frac{b}{a}\right) q^i}{1 - \left(\frac{b}{a}\right) q^{\alpha+i}}, \quad a \neq 0,$$

$$(a \widetilde{-} b)_q^\alpha = a^\alpha \prod_{i=0}^{\infty} \frac{1 - \left(\frac{b}{a}\right) q^{2i+1}}{1 - \left(\frac{b}{a}\right) q^{2(\alpha+i)+1}}, \quad a \neq 0.$$

Since
$$(a \widetilde{-} b)_{q,\omega}^n = \prod_{i=0}^{n-1}\left[a - \sigma_{q,\omega}^{2i+1}(b)\right] = \prod_{i=0}^{n-1}\left[(a-\omega_0) - (b-\omega_0)q^{2i+1}\right]$$
$$= \left((a-\omega_0)\widetilde{-}(b-\omega_0)\right)_q^n$$
$$= (a-\omega_0)^n \prod_{i=0}^{n-1}\left[1 - \left(\frac{b-\omega_0}{a-\omega_0}\right)q^{2i+1}\right] \cdot \frac{\prod_{i=n}^{\infty}\left[1 - \left(\frac{b-\omega_0}{a-\omega_0}\right)q^{2i+1}\right]}{\prod_{i=n}^{\infty}\left[1 - \left(\frac{b-\omega_0}{a-\omega_0}\right)q^{2i+1}\right]}$$
$$= (a-\omega_0)^n \prod_{i=0}^{\infty} \frac{1 - \left(\frac{b-\omega_0}{a-\omega_0}\right)q^{2i+1}}{1 - \left(\frac{b-\omega_0}{a-\omega_0}\right)q^{2(n+i)+1}},$$

so, we obtain
$$(a\widetilde{-}b)_{q,\omega}^{\alpha} = \left((a-\omega_0)\widetilde{-}(b-\omega_0)\right)_q^{\alpha} = (a-\omega_0)^{\alpha}\prod_{i=0}^{\infty}\frac{1 - \left(\frac{b-\omega_0}{a-\omega_0}\right)q^{2i+1}}{1 - \left(\frac{b-\omega_0}{a-\omega_0}\right)q^{2(\alpha+i)+1}}, \quad a \neq \omega_0.$$

In particular, if $a \neq b = 0$, we have $a_q^{\alpha} = \tilde{a}_q^{\alpha} = a^{\alpha}$. If $a \neq b = \omega_0$, we have $(a\widetilde{-}\omega_0)_{q,\omega}^{\alpha} = (a-\omega_0)^{\alpha}$.
Furthermore, if $a = b = 0$, we define $(0)_q^{\alpha} = \widetilde{(0)}_q^{\alpha} = \widetilde{(0)}_{q,\omega}^{\alpha} := 0$ for $\alpha > 0$.

Next, we define q-symmetric gamma and q-symmetric beta functions as
$$\tilde{\Gamma}_q(x) := \begin{cases} \frac{(1-q^2)_q^{x-1}}{(1-q^2)^{x-1}} = \frac{(\widetilde{1-q})_q^{x-1}}{(1-q^2)^{x-1}}, & x \in \mathbb{R}\setminus\{0,-1,-2,\ldots\} \\ [x-1]_q!, & x \in \mathbb{N} \end{cases}$$
$$\tilde{B}_q(x,y) := \frac{\tilde{\Gamma}_q(x)\tilde{\Gamma}_q(y)}{\tilde{\Gamma}_q(x+y)},$$

respectively.

Lemma 1. *For* $m, n \in \mathbb{N}_0$ *and* $\alpha \in \mathbb{R}$,
(a) $(x \widetilde{-} \sigma_{q,\omega}^n(x))_{q,\omega}^{\alpha} = (x-\omega_0)^k(1\widetilde{-}q^n)_q^{\alpha},$
(b) $(\sigma_{q,\omega}^m(x) \widetilde{-} \sigma_{q,\omega}^n(x))_{q,\omega}^{\alpha} = q^{m\alpha}(x-\omega_0)^{\alpha}(1\widetilde{-}q^{n-m})_q^{\alpha}.$

Proof. For $m, n \in \mathbb{N}_0$ and $\alpha \in \mathbb{R}$, we have
$$(x \widetilde{-} \sigma_{q,\omega}^n(x))_{q,\omega}^{\alpha} = \left((x-\omega_0) \widetilde{-} (\sigma_{q,\omega}^n(x) - \omega_0)\right)_q^k$$
$$= (x-\omega_0)^{\alpha}\prod_{i=0}^{\infty}\frac{1 - \left(\frac{\sigma_{q,\omega}^n(x)-\omega_0}{x-\omega_0}\right)q^{2i+1}}{1 - \left(\frac{\sigma_{q,\omega}^n(x)-\omega_0}{x-\omega_0}\right)q^{2(i+\alpha)+1}}$$
$$= (x-\omega_0)^{\alpha}\prod_{i=0}^{\infty}\frac{1 - q^n q^{2i+1}}{1 - q^n q^{2(i+\alpha)+1}}$$
$$= (x-\omega_0)^{\alpha}(1\widetilde{-}q^n)_q^{\alpha}.$$

and

$$((\widetilde{\sigma_{q,\omega}^m(x)}) - \sigma_{q,\omega}^n(x))_{q,\omega}^{\alpha} = \left(\widetilde{(\sigma_{q,\omega}^m(x) - \omega_0) - (\sigma_{q,\omega}^n(x) - \omega_0)}\right)_q^{\alpha}$$

$$= ((\sigma_{q,\omega}^m(x) - \omega_0)^{\alpha} \prod_{i=0}^{\infty} \frac{1 - \left(\frac{\sigma_{q,\omega}^n(x) - \omega_0}{\sigma_{q,\omega}^m(x) - \omega_0}\right) q^{2i+1}}{1 - \left(\frac{\sigma_{q,\omega}^n(x) - \omega_0}{\sigma_{q,\omega}^m(x) - \omega_0}\right) q^{2(i+\alpha)+1}}$$

$$= (q^m(x - \omega_0))^{\alpha} \prod_{i=0}^{\infty} \frac{1 - q^{n-m} q^{2i+1}}{1 - q^{n-m} q^{2(i+\alpha)+1}}$$

$$= q^{m\alpha}(x - \omega_0)^{\alpha} (1 - \widetilde{q^{n-m}})_q^{\alpha}.$$

So, Lemma 1 (a) and Lemma 1 (b) hold. The proof is complete. □

Lemma 2. *Let* $t, s \in I_{q,\omega}^T := \left\{ q^k T + \omega[k]_q : k \in \mathbb{N}_0 \right\} \cup \{\omega_0\}$, $T > \omega_0$. *Then,*

$$(\widetilde{t - s})_{q,\omega}^{\alpha} = 0$$

where $t \geq s$ *and* $\alpha \notin \mathbb{N}_0$.

Proof. Since $t, s \in I_{q,\omega}^T$, we have $t = \sigma_{q,\omega}^m(T)$, $s = \sigma_{q,\omega}^n(T)$ where $m, n \in \mathbb{N}$. For $t \geq s$, we find that

$$(\widetilde{t - s})_{q,\omega}^{\alpha} = \left(\widetilde{\sigma_{q,\omega}^m(T) - \sigma_{q,\omega}^n(T)}\right)_{q,\omega}^{\alpha}$$

$$= q^{m\alpha}(T - \omega_0)^{\alpha} (1 - \widetilde{q^{n-m}})_q^{\alpha}$$

$$= q^{\alpha m}(T - \omega_0)^{\alpha} \prod_{i=0}^{\infty} \left[\frac{1 - q^{2i+n-m+1}}{1 - q^{2i+n-m+1+2\alpha}}\right] = 0.$$

The proof is complete. □

Definition 1 ([20]). *For* $q \in (0, 1)$, $\omega > 0$, *we let* f *be the function defined on* $I_{q,\omega}^T \subseteq \mathbb{R}$. *The symmetric Hahn difference of* f *is defined by*

$$\tilde{D}_{q,\omega} f(t) := \frac{f(\sigma_{q,\omega}(t)) - f(\rho_{q,\omega}(t))}{\sigma_{q,\omega}(t) - \rho_{q,\omega}(t)} \quad t \in I_{q,\omega}^T - \{\omega_0\},$$

$$\tilde{D}_{q,\omega} f(\omega_0) = f'(\omega_0) \text{ where } f \text{ is differentiable at } \omega_0.$$

$\tilde{D}_{q,\omega} f$ *is called* q, ω-*symmetric derivative of* f, *and* f *is* q, ω-*symmetric differentiable on* $I_{q,\omega}^T$.

From the above definition, we note that

$$\tilde{D}_{q,\omega}^0 f(x) = f(x) \text{ and } \tilde{D}_{q,\omega}^N f(x) = \tilde{D}_{q,\omega} \tilde{D}_{q,\omega}^{N-1} f(x) \text{ where } N \in \mathbb{N}.$$

Lemma 3 ([20]). *Properties of symmetric Hahn difference operators*
If f *and* g *are* q, ω-*symmetric differentiable on* $I_{q,\omega}^T$. *Then*

(a) $\tilde{D}_{q,\omega}[f(t) + g(t)] = \tilde{D}_{q,\omega} f(t) + \tilde{D}_{q,\omega} g(t)$,
(b) $\tilde{D}_{q,\omega}[f(t) g(t)] = f(\rho_{q,\omega}(t)) \tilde{D}_{q,\omega} g(t) + g(\sigma_{q,\omega}(t)) \tilde{D}_{q,\omega} f(t)$,
(c) $\tilde{D}_{q,\omega} \left[\frac{f(t)}{g(t)}\right] = \frac{g(\rho_{q,\omega}(t)) \tilde{D}_{q,\omega} f(t) - f(\rho_{q,\omega}(t)) \tilde{D}_{q,\omega} g(t)}{g(\rho_{q,\omega}(t)) g(\sigma_{q,\omega}(t))}$ *for* $g(\rho_{q,\omega}(t)) g(\sigma_{q,\omega}(t)) \neq 0$,
(d) $\tilde{D}_{q,\omega}[C] = 0$ *where* C *is constant.*

Lemma 4. Let $0 < q < 1$, $\omega > 0$, $t \in I_{q,\omega}^T$, and $\alpha, \beta \in \mathbb{R}$. Then,

(a) $\tilde{D}_{q,\omega}\widetilde{(t-\beta)}_{q,\omega}^\alpha = [\widetilde{\alpha}]_q \widetilde{(\rho_{q,\omega}(t)-\beta)}_{q,\omega}^{\alpha-1}$,

(b) $\tilde{D}_{q,\omega}\widetilde{(\beta-t)}_{q,\omega}^\alpha = -[\widetilde{\alpha}]_q \widetilde{(\beta-\sigma_{q,\omega}(t))}_{q,\omega}^{\alpha-1}$.

Proof. By Lemma 1 and Definition 1, we find that

$$\tilde{D}_{q,\omega}(t-\beta)_{q,\omega}^\alpha = \tilde{D}_{q,\omega}\left[(t-\omega_0)^\alpha \prod_{i=0}^\infty \left(\frac{1-\left(\frac{\beta-\omega_0}{t-\omega_0}\right)q^{2i+1}}{1-\left(\frac{\beta-\omega_0}{t-\omega_0}\right)q^{2(i+\alpha)+1}}\right)\right]$$

$$= \frac{1}{\sigma_{q,\omega}(t)-\rho_{q,\omega}(t)}\left\{(\sigma_{q,\omega}(t)-\omega_0)^\alpha \prod_{i=0}^\infty \left(\frac{1-\left(\frac{\beta-\omega_0}{\sigma_{q,\omega}(t)-\omega_0}\right)q^{2i+1}}{1-\left(\frac{\beta-\omega_0}{\sigma_{q,\omega}(t)-\omega_0}\right)q^{2(i+\alpha)+1}}\right)\right.$$

$$\left. -(\rho_{q,\omega}(t)-\omega_0)^\alpha \prod_{i=0}^\infty \left(\frac{1-\left(\frac{\beta-\omega_0}{\rho_{q,\omega}(t)-\omega_0}\right)q^{2i+1}}{1-\left(\frac{\beta-\omega_0}{\rho_{q,\omega}(t)-\omega_0}\right)q^{2(i+\alpha)+1}}\right)\right\}$$

$$= -\frac{q}{(1-q^2)(t-\omega_0)}\left\{q^\alpha(t-\omega_0)^\alpha \prod_{i=0}^\infty \left(\frac{1-\left(\frac{\beta-\omega_0}{t-\omega_0}\right)q^{2i}}{1-\left(\frac{\beta-\omega_0}{t-\omega_0}\right)q^{2(i+\alpha)}}\right)\right.$$

$$\left. -\frac{(t-\omega_0)^\alpha}{q^\alpha}\prod_{i=0}^\infty \left(\frac{1-\left(\frac{\beta-\omega_0}{t-\omega_0}\right)q^{2(i+1)}}{1-\left(\frac{\beta-\omega_0}{t-\omega_0}\right)q^{2(i+\alpha+1)}}\right)\right\}$$

$$= \frac{q^{1-\alpha}}{(1-q^2)}(t-\omega_0)^{\alpha-1}\left\{\frac{\prod_{i=0}^\infty\left(1-\left(\frac{\beta-\omega_0}{t-\omega_0}\right)q^{2(i+1)}\right)}{\prod_{i=0}^\infty\left(1-\left(\frac{\beta-\omega_0}{t-\omega_0}\right)q^{2(i+\alpha+1)}\right)}\right.$$

$$\left. -q^{2\alpha}\frac{\prod_{i=0}^\infty\left(1-\left(\frac{\beta-\omega_0}{t-\omega_0}\right)q^{2i}\right)}{\prod_{i=0}^\infty\left(1-\left(\frac{\beta-\omega_0}{t-\omega_0}\right)q^{2(i+\alpha)}\right)}\right\}$$

$$= \left(\frac{1-q^{2\alpha}}{1-q^2}\right)q^{1-\alpha}(t-\omega_0)^{\alpha-1}\cdot\frac{1}{1-q^{2\alpha}}\left\{\prod_{i=0}^\infty\left(\frac{1-\left(\frac{\beta-\omega_0}{t-\omega_0}\right)q^{2(i+1)}}{1-\left(\frac{\beta-\omega_0}{t-\omega_0}\right)q^{2(i+\alpha+1)}}\right)\right.$$

$$\left. -q^{2\alpha}\prod_{i=0}^\infty\left(\frac{1-\left(\frac{\beta-\omega_0}{t-\omega_0}\right)q^{2i}}{1-\left(\frac{\beta-\omega_0}{t-\omega_0}\right)q^{2(i+\alpha)}}\right)\right\}$$

$$= [\widetilde{\alpha}]_q (t-\omega_0)^{\alpha-1}q^{1-\alpha}\frac{\prod_{i=0}^\infty\left(1-\left(\frac{\beta-\omega_0}{t-\omega_0}\right)q^{2(i+1)}\right)}{\prod_{i=0}^\infty\left(1-\left(\frac{\beta-\omega_0}{t-\omega_0}\right)q^{2(i+\alpha)}\right)} \times$$

$$\frac{\left(1-\left(\frac{\beta-\omega_0}{t-\omega_0}\right)q^{2\alpha}\right)-q^{2\alpha}\left(1-\left(\frac{\beta-\omega_0}{t-\omega_0}\right)\right)}{1-q^{2\alpha}}$$

$$= [\widetilde{\alpha}]_q (t-\omega_0)^{\alpha-1}q^{1-\alpha}\prod_{i=0}^\infty\left(\frac{1-\left(\frac{\beta-\omega_0}{t-\omega_0}\right)q^{2(i+1)}}{1-\left(\frac{\beta-\omega_0}{t-\omega_0}\right)q^{2(i+\alpha)}}\right)$$

$$= [\widetilde{\alpha}]_q (\rho_{q,\omega}(t)-\omega_0)^{\alpha-1}\prod_{i=0}^\infty\left(\frac{1-\left(\frac{\beta-\omega_0}{\rho_{q,\omega}(t)-\omega_0}\right)q^{2i+1}}{1-\left(\frac{\beta-\omega_0}{\rho_{q,\omega}(t)-\omega_0}\right)q^{2(i+\alpha-1)+1}}\right)$$

$$= [\widetilde{\alpha}]_q \widetilde{(\rho_{q,\omega}(t)-\beta)}_{q,\omega}^{\alpha-1}.$$

So, Lemma 4 (a) holds. Similarly to the above, we use Lemma 1 and Definition 1 to show that

$$\tilde{D}_{q,\omega}(\beta - \widetilde{\rho_{q,\omega}(t)})^{\widetilde{\alpha}}_{q,\omega} = -\widetilde{[\alpha]}_q(\widetilde{\beta - t})^{\widetilde{\alpha-1}}_{q,\omega}.$$

Then, Lemma 4 (b) holds. □

Definition 2 ([20]). *Let I be any closed interval of $\mathbb{R}$ containing a, b and ω_0 and $f : I \to \mathbb{R}$ be a given function. The symmetric Hahn integral of f from a to b is defined by*

$$\int_a^b f(t)\tilde{d}_{q,\omega}t := \int_{\omega_0}^b f(t)\tilde{d}_{q,\omega}t - \int_{\omega_0}^a f(t)\tilde{d}_{q,\omega}t,$$

where

$$\tilde{\mathcal{I}}_{q,\omega}f(t) = \int_{\omega_0}^x f(t)\tilde{d}_{q,\omega}t := (1-q^2)(x - \omega_0)\sum_{k=0}^{\infty} q^{2k} f\left(\sigma_{q,\omega}^{2k+1}(x)\right), \quad x \in I.$$

Providing that the above series converges at $x = a$ and $x = b$, f is called symmetric Hahn integrable on $[a, b]$. In addition, f is symmetric Hahn integrable on I if it is symmetric Hahn integrable on $[a, b]$ for all $a, b \in I$.

For $N \in \mathbb{N}$, we define an operator $\tilde{\mathcal{I}}_{q,\omega}^N$ by

[A] $\tilde{\mathcal{I}}_{q,\omega}^0 f(x) = f(x)$ and $\tilde{\mathcal{I}}_{q,\omega}^N f(x) = \tilde{\mathcal{I}}_{q,\omega}\tilde{\mathcal{I}}_{q,\omega}^{N-1}f(x), N \in \mathbb{N}.$

From the symmetric Hahn derivatives, we have

[B] $\tilde{D}_{q,\omega}\tilde{\mathcal{I}}_{q,\omega}f(x) = f(x)$ and $\tilde{\mathcal{I}}_{q,\omega}\tilde{D}_{q,\omega}f(x) = f(x) - f(\omega_0).$

Lemma 5 ([20]). *Properties of symmetric Hahn Integrals.*
Let $0 < q < 1$, $\omega > 0$, $a, b \in I_{q,\omega}^T$ and f, g be symmetric Hahn integrable on $I_{q,\omega}^T$. Then,

(a) $\int_a^a f(t)\tilde{d}_{q,\omega}t = 0,$
(b) $\int_a^b f(t)\tilde{d}_{q,\omega}t = -\int_b^a f(t)\tilde{d}_{q,\omega}t,$
(c) $\int_a^b f(t)\tilde{d}_{q,\omega}t = \int_c^b f(t)\tilde{d}_{q,\omega}t + \int_a^c f(t)\tilde{d}_{q,\omega}t, \; c \in I_{q,\omega}^T, \; a < c < b,$
(d) $\int_a^b [\alpha f(t) + \beta g(t)]\tilde{d}_{q,\omega}t = \alpha \int_a^b f(t)\tilde{d}_{q,\omega}t + \beta \int_a^b g(t)\tilde{d}_{q,\omega}t, \; \alpha, \beta \in \mathbb{R},$
(e) $\int_a^b \left[f(\rho_{q,\omega}(t))\tilde{D}_{q,\omega}g(t)\right]\tilde{d}_{q,\omega}t = [f(t)g(t)]_a^b - \int_a^b \left[g(\sigma_{q,\omega}(t))\tilde{D}_{q,\omega}f(t)\right]\tilde{d}_{q,\omega}t.$

We next introduce the fundamental theorem and Leibniz formula of symmetric Hahn calculus.

Lemma 6 ([20]). *The fundamental theorem of symmetric Hahn calculus*
Let $f : I \to \mathbb{R}$ be continuous at ω_0. Then

$$F(x) := \int_{\omega_0}^x f(t)\tilde{d}_{q,\omega}t, \quad x \in I$$

is continuous at ω_0 and $\tilde{D}_{q,\omega}F(x)$ exists for every $x \in \sigma_{q,\omega}(I) := \{qt + \omega : t \in I\}$ where

$$\tilde{D}_{q,\omega}F(x) = f(x).$$

In addition,

$$\int_a^b \tilde{D}_{q,\omega}f(t)\tilde{d}_{q,\omega}t = f(b) - f(a) \text{ for all } a, b \in I.$$

Lemma 7. *The Leibniz formula of symmetric Hahn calculus*
Let $f : I_{q,\omega}^T \times I_{q,\omega}^T \to \mathbb{R}$. Then,

$$\tilde{D}_{q,\omega}\left[\int_{\omega_0}^{t} f(t,s)\,\tilde{d}_{q,\omega}s\right] = \int_{\omega_0}^{\rho_{q,\omega}(t)} {}_t\tilde{D}_{q,\omega}f(t,s)\,\tilde{d}_{q,\omega}s + f\left(\sigma_{q,\omega}(t), t\right),$$

where ${}_t\tilde{D}_{q,\omega}$ is symmetric Hahn difference with respect to t.

Proof. For $t \in I_{q,\omega}^T$,

$$\tilde{D}_{q,\omega}\left[\int_{\omega_0}^{t} f(t,s)\,\tilde{d}_{q,\omega}s\right]$$

$$= \frac{1}{\sigma_{q,\omega}(t) - \rho_{q,\omega}(t)}\left\{\int_{\omega_0}^{\sigma_{q,\omega}(t)} f(\sigma_{q,\omega}(t))\,\tilde{d}_{q,\omega}s - \int_{\omega_0}^{\rho_{q,\omega}(t)} f(\rho_{q,\omega}(t),s)\,\tilde{d}_{q,\omega}s\right\}$$

$$= \frac{1}{\sigma_{q,\omega}(t) - \rho_{q,\omega}(t)}\left\{\left[\int_{\omega_0}^{\sigma_{q,\omega}(t)} f(\sigma_{q,\omega}(t),s)\,\tilde{d}_{q,\omega}s - \int_{\omega_0}^{\rho_{q,\omega}(t)} f(\sigma_{q,\omega}(t),s)\,\tilde{d}_{q,\omega}s\right]\right.$$

$$\left. + \left[\int_{\omega_0}^{\rho_{q,\omega}(t)} f(\sigma_{q,\omega}(t),s)\,\tilde{d}_{q,\omega}s - \int_{\omega_0}^{\rho_{q,\omega}(t)} f(\rho_{q,\omega}(t),s)\,\tilde{d}_{q,\omega}s\right]\right\}$$

$$= \frac{1}{\sigma_{q,\omega}(t) - \rho_{q,\omega}(t)}\int_{\omega_0}^{\rho_{q,\omega}(t)} \left[f(\sigma_{q,\omega}(t),s) - f(\rho_{q,\omega}(t),s)\right]\tilde{d}_{q,\omega}s$$

$$- \frac{q}{(1-q^2)(t-\omega_0)}\left\{(1-q^2)(\sigma_{q,\omega}(t)-\omega_0)\sum_{k=0}^{\infty} q^{2k} f\left(\sigma_{q,\omega}(t), \sigma_{q,\omega}^{2k+2}(t)\right)\right.$$

$$\left. - (1-q^2)(\rho_{q,\omega}(t)-\omega_0)\sum_{k=0}^{\infty} q^{2k} f\left(\sigma_{q,\omega}(t), \sigma_{q,\omega}^{2k}(t)\right)\right\}$$

$$= \int_{\omega_0}^{\rho_{q,\omega}(t)} {}_t\tilde{D}_{q,\omega}f(t,s)\,\tilde{d}_{q,\omega}s$$

$$- q\left\{\sum_{k=0}^{\infty} q^{2k+1} f(\sigma_{q,\omega}(t), \sigma_{q,\omega}^{2k+2}(t)) - \sum_{k=0}^{\infty} q^{2k-1} f(\sigma_{q,\omega}(t), \sigma_{q,\omega}^{2k}(t))\right\}$$

$$= \int_{\omega_0}^{\rho_{q,\omega}(t)} {}_t\tilde{D}_{q,\omega}f(t,s)\,\tilde{d}_{q,\omega}s + f\left(\sigma_{q,\omega}(t), t\right).$$

The proof is complete. □

Next, we give some auxiliary lemmas used for simplifying calculations.

Lemma 8. *Let $0 < q < 1$, $\omega > 0$ and $f : I \to \mathbb{R}$ be continuous at ω_0. Then,*

$$\int_{\omega_0}^{t}\int_{\omega_0}^{r} f(s)\,\tilde{d}_{q,\omega}s\,\tilde{d}_{q,\omega}r = q\int_{\omega_0}^{t}\int_{qs+\omega}^{t} f(qs+\omega)\,\tilde{d}_{q,\omega}r\,\tilde{d}_{q,\omega}s.$$

Proof. From Definition 2, we find that

$$\int_{\omega_0}^t \int_{\omega_0}^r f(s) \, \tilde{d}_{q,\omega} s \, \tilde{d}_{q,\omega} r$$

$$= \int_{\omega_0}^t \left[(1-q^2)(r-\omega_0) \sum_{k=0}^{\infty} q^{2k} f\left(\sigma_{q,\omega}^{2k+1}(r)\right) \right] \tilde{d}_{q,\omega} r$$

$$= \sum_{k=0}^{\infty} q^{2k}(1-q^2) \left[\int_{\omega_0}^t (r-\omega_0) f\left(\sigma_{q,\omega}^{2k+1}(r)\right) \tilde{d}_{q,\omega} r \right]$$

$$= q(1-q^2)^2 (t-\omega_0)^2 \sum_{k=0}^{\infty} \sum_{m=0}^{\infty} q^{4m+2k} f\left(\sigma_{q,\omega}^{2m+2k+2}(t)\right)$$

$$= q(1-q^2)^2 (t-\omega_0)^2 \sum_{k=0}^{\infty} \left[q^{4m} f\left(\sigma_{q,\omega}^{2m+2}(t)\right) + q^{4m+2} f\left(\sigma_{q,\omega}^{2m+4}(t)\right) + q^{4m+4} f\left(\sigma_{q,\omega}^{2m+6}(t)\right) + \dots \right]$$

$$= q(1-q^2)^2 (t-\omega_0)^2 \Big\{ \left[f\left(\sigma_{q,\omega}^2(t)\right) + q^2 f\left(\sigma_{q,\omega}^4(t)\right) + q^4 f\left(\sigma_{q,\omega}^6(t)\right) + \dots \right]$$

$$+ \left[q^4 f\left(\sigma_{q,\omega}^4(t)\right) + q^6 f\left(\sigma_{q,\omega}^6(t)\right) + q^8 f\left(\sigma_{q,\omega}^8(t)\right) + \dots \right]$$

$$+ \left[q^8 f\left(\sigma_{q,\omega}^6(t)\right) + q^{10} f\left(\sigma_{q,\omega}^8(t)\right) + q^{12} f\left(\sigma_{q,\omega}^{10}(t)\right) + \dots \right] + \dots \Big\}$$

$$= q(1-q^2)^2 (t-\omega_0)^2 \left\{ f\left(\sigma_{q,\omega}^2(t)\right) + q^2(1+q^2) f\left(\sigma_{q,\omega}^4(t)\right) + q^4(1+q^2+q^4) f\left(\sigma_{q,\omega}^6(t)\right) + \dots \right\}$$

$$= q(1-q^2)^2 (t-\omega_0)^2 \sum_{k=0}^{\infty} q^{2k} [\widetilde{k+1}]_q f\left(\sigma_{q,\omega}^{2k+2}(t)\right)$$

$$= q \int_{\omega_0}^t [t - \sigma_{q,\omega}(s)] f\left(\sigma_{q,\omega}(s)\right) \tilde{d}_{q,w} s$$

$$= q \int_{\omega_0}^t \left[\int_{\omega_0}^t f\left(\sigma_{q,\omega}(s)\right) \tilde{d}_{q,w} r - \int_{\omega_0}^{\sigma_{q,\omega}(s)} f\left(\sigma_{q,\omega}(s)\right) \tilde{d}_{q,w} r \right] \tilde{d}_{q,w} s$$

$$= q \int_{\omega_0}^t \int_{qs+\omega}^t f(qs+\omega) \, \tilde{d}_{q,\omega} r \, \tilde{d}_{q,\omega} s.$$

□

In the next theorem we evaluate the multiple symmetric Hahn integrals as follows.

Theorem 1. *For $f : I_{q,\omega}^T \to \mathbb{R}$, the multiple symmetric Hahn integral is given by*

$$\tilde{\mathcal{I}}_{q,\omega}^n f(x) := \int_{\omega_0}^x \int_{\omega_0}^{\tau_1} \dots \int_{\omega_0}^{\tau_{n-1}} f(\tau_n) \tilde{d}_{q,\omega} \tau_n \dots \tilde{d}_{q,\omega} \tau_2 \tilde{d}_{q,\omega} \tau_1$$

$$= \frac{1}{[n-1]_q!} q^{\binom{n}{2}} \int_{\omega_0}^t (\widetilde{t-\tau})_{q,\omega}^{n-1} f\left(\sigma_{q,\omega}^{n-1}(\tau)\right) \tilde{d}_{q,\omega} \tau, \quad (1)$$

where $n \in \mathbb{N}$ and $\binom{n}{k} = \frac{\Gamma(n+1)}{\Gamma(k+1)\Gamma(n-k+1)}$.

Proof. If $n = 1$, $\tilde{\mathcal{I}}_{q,\omega} f(x) = \int_{\omega_0}^x f(\tau) \tilde{d}_{q,\omega} \tau$.

If $n=2$, by using Lemma 8, we have

$$\begin{aligned}
\tilde{\mathcal{I}}^2_{q,\omega} f(x) &= \int_{\omega_0}^x \int_{\omega_0}^s f(\tau) \tilde{d}_{q,\omega} \tau \tilde{d}_{q,\omega} s = q \int_{\omega_0}^x \int_{\sigma_{q,\omega}(\tau)}^x f(\sigma_{q,\omega}(\tau)) \tilde{d}_{q,\omega} s \, \tilde{d}_{q,\omega} \tau \\
&= q \int_{\omega_0}^x [x - \sigma_{q,\omega}(\tau)] f(\sigma_{q,\omega}(\tau)) \tilde{d}_{q,w} \tau \\
&= q \int_{\omega_0}^x \widetilde{(x-\tau)}_{q,\omega}^{1} f(q\tau+\omega)) \tilde{d}_{q,\omega} \tau.
\end{aligned}$$

We suppose that Theorem 1 holds for $n=k$ and then prove that it is true for $n=k+1$ as follows:

$$\begin{aligned}
\tilde{\mathcal{I}}^{k+1}_{q,\omega} f(x) &= \tilde{\mathcal{I}}_{q,\omega} \left[\frac{1}{[k-1]_q!} q^{\binom{k}{2}} \int_{\omega_0}^x \widetilde{(x-\tau)}_{q,\omega}^{k-1} f\left(\sigma^{k-1}_{q,\omega}(\tau)\right) \tilde{d}_{q,\omega} \tau \right] \\
&= \frac{1}{[k-1]_q!} q^{\binom{k}{2}} \int_{\omega_0}^x \int_{\omega_0}^s \widetilde{(s-\tau)}_{q,\omega}^{k-1} f\left(\sigma^{k-1}_{q,\omega}(\tau)\right) \tilde{d}_{q,\omega} \tau \tilde{d}_{q,\omega} s \\
&= \frac{1}{[k-1]_q!} q^{\binom{k}{2}} (1-q^2)(x-\omega_0) \sum_{m=0}^\infty q^{2m}(1-q^2) \left[\sigma^{2m+1}_{q,\omega}(x) - \omega_0\right] \times \\
&\quad \sum_{l=0}^\infty q^{2l} \left(\sigma^{2m+1}_{q,\omega}(x) - \sigma^{2m+2l+2}_{q,\omega}(x)\right)_{q,\omega}^{k-1} f\left(\sigma^{k+2m+2l+1}_{q,\omega}(x)\right). \quad (2)
\end{aligned}$$

From (2), by using Lemma 1b, we obtain

$$\begin{aligned}
\tilde{\mathcal{I}}^{k+1}_{q,\omega} f(x) &= \frac{1}{[k-1]_q!} q^{\binom{k}{2}} (1-q^2)^2 (x-\omega_0) \sum_{m=0}^\infty \sum_{l=0}^\infty q^{4m+2l+1}(x-\omega_0) q^{(2m+1)(k-1)} \times \\
&\quad (x-\omega_0)^{k-1} \widetilde{(1-q^{2l+1})}_q^{k-1} f\left(\sigma^{k+2m+2l+1}_{q,\omega}(x)\right) \\
&= \frac{\widetilde{[k]_q}}{\widetilde{[k]_q!}} q^{-k} q^{\binom{k+1}{2}} (1-q^2)^2 (x-\omega_0)^{k+1} \sum_{l=0}^\infty \sum_{m=0}^l q^{2l} q^{(2m+1)k} \times \\
&\quad \widetilde{(1-q^{2l-2m+1})}_q^{k-1} f\left(\sigma^{k+2l+1}_{q,\omega}(x)\right). \quad (3)
\end{aligned}$$

From (2), by using Lemma 1a, we obtain

$$\begin{aligned}
\tilde{\mathcal{I}}^{k+1}_{q,\omega} f(x) &= \frac{1}{\widetilde{[k]_q!}} q^{\binom{k+1}{2}} \int_{\omega_0}^x \widetilde{(x-\tau)}_{q,\omega}^{k} f\left(\sigma^k_{q,\omega}(\tau)\right) \tilde{d}_{q,\omega} \tau \\
&= \frac{1}{\widetilde{[k]_q!}} q^{\binom{k+1}{2}} (1-q^2)(x-\omega_0) \sum_{l=0}^\infty q^{2l} \widetilde{\left(x - \sigma^{2l+1}_{q,\omega}(x)\right)}_{q,\omega}^k f\left(\sigma^{k+2l+1}_{q,\omega}(x)\right) \\
&= \frac{1}{\widetilde{[k]_q!}} q^{\binom{k+1}{2}} (1-q^2)(x-\omega_0)^{k+1} \sum_{l=0}^\infty q^{2l} \widetilde{(1-q^{2l+1})}_q^k f\left(\sigma^{k+2l+1}_{q,\omega}(x)\right). \quad (4)
\end{aligned}$$

Since

$$[\widetilde{k}]_q q^{-k}(1-q^2) \sum_{m=0}^{l} q^{(2m+1)k}\left(1-q^{\widetilde{2l-2m+1}}\right)_q^{k-1}$$

$$= (1-q^{2k}) \sum_{m=0}^{l} q^{2mk}\left(1-q^{\widetilde{2l-2m+1}}\right)_q^{k-1}$$

$$= (1-q^{2k})\left\{\prod_{i=0}^{k-2}\left[1-q^{2(l+i-1)}\right] + \ldots + q^{2(l-2)k}\prod_{i=0}^{k-2}\left[1-q^{2i+6}\right] + q^{2(l-1)k}\prod_{i=0}^{k-2}\left[1-q^{2i+4}\right]\right.$$
$$\left. + q^{2lk}\prod_{i=0}^{k-2}\left[1-q^{2i+2}\right]\right\}$$

$$= (1-q^{2k})\left\{\prod_{i=0}^{k-2}\left[1-q^{2(l+i-1)}\right] + \ldots + q^{2(l-2)k}\prod_{i=0}^{k-2}\left[1-q^{2i+6}\right]\right.$$
$$\left. + q^{2(l-1)k}\left(1-q^{2k+2}\right)\prod_{i=0}^{k-3}\left[1-q^{2i+4}\right]\right\}$$

$$= (1-q^{2k})\left\{\prod_{i=0}^{k-2}\left[1-q^{2(l+i-1)}\right] + \ldots + q^{2(l-3)k}\prod_{i=0}^{k-2}\left[1-q^{2i+8}\right]\right.$$
$$\left. + q^{2(l-2)k}\left[\prod_{i=0}^{k-2}\left[1-q^{2i+6}\right] + q^{2k}\left(1-q^{2k+2}\right)\prod_{i=0}^{k-3}\left[1-q^{2i+4}\right]\right]\right\}$$

$$= (1-q^{2k})\left\{\prod_{i=0}^{k-2}\left[1-q^{2(l+i-1)}\right] + \ldots + q^{2(l-3)k}\prod_{i=0}^{k-2}\left[1-q^{2i+8}\right]\right.$$
$$\left. + q^{2(l-2)k}\left(1-q^{2k+4}\right)\left(1-q^{2k+2}\right)\prod_{i=0}^{k-4}\left[1-q^{2i+6}\right]\right\}$$

$\vdots$

$$= (1-q^{2k})\left\{\prod_{i=0}^{k-2}\left[1-q^{2(l+i-1)}\right] + \ldots + q^{2k}\left(1-q^{2(l+k-1)}\right)\left(1-q^{2(l+k-2)}\right)\ldots \times \right.$$
$$\left. \left(1-q^{2k+2}\right)\prod_{i=0}^{k-(l+1)}\left[1-q^{2(l+i)}\right]\right\}$$

$$= \left(1-q^{2k+2l}\right)\left(1-q^{2k+2l-2}\right)\ldots\left(1-q^{2k+2}\right)\left(1-q^{2k}\right)\left(1-q^{2k-2}\right)\ldots\left(1-q^{2l+4}\right)\left(1-q^{2l+2}\right)$$

$$= \prod_{i=0}^{k-1}\left(1-q^{2(l+i+1)}\right)$$

$$= (1-q^{\widetilde{2l+1}}(x))_q^k.$$

We find that (1) holds when $n = k+1$.
Our proof is done using mathematical induction. □

3. Fractional Symmetric Hahn Integral

In Section 2, we have presented the multiple symmetric Hahn integral for integer order in the form (1). We next apply this result for fractional orders that can be used to further define fractional symmetric Hahn difference operators of Riemann–Liouville and Caputo types. We first introduce the fractional symmetric Hahn integral as follows.

Definition 3. Let $\alpha, \omega > 0$, $0 < q < 1$, and f be a function defined on $I_{q,\omega}^T$. The fractional symmetric Hahn integral is defined by

$$\widetilde{\mathcal{I}}_{q,\omega}^\alpha f(t) := \frac{q^{\binom{\alpha}{2}}}{\widetilde{\Gamma}_q(\alpha)} \int_{\omega_0}^t (\widetilde{t-s})_{q,\omega}^{\alpha-1} f\left(\sigma_{q,\omega}^{-1}(s)\right) \widetilde{d}_{q,\omega}s$$

$$= \frac{(1-q^2)q^{\binom{\alpha}{2}}(t-\omega_0)}{\widetilde{\Gamma}_q(\alpha)} \sum_{k=0}^\infty q^{2k} \left(t - \sigma_{q,\omega}^{2k+1}(t)\right)_{q,\omega}^{\alpha-1} f\left(\sigma_{q,\omega}^{2k+\alpha}(t)\right), \quad (5)$$

and $\widetilde{\mathcal{I}}_{q,\omega}^0 f)(t) = f(t)$.

By Lemma 1a, $(t - \widetilde{\sigma^{2k+1}}_{q,\omega}(t))_{q,\omega}^{\alpha-1} = (t-\omega_0)^{\alpha-1}(1-\widetilde{q^{2k+1}})_q^{\alpha-1}$. It implies that

$$\widetilde{\mathcal{I}}_{q,\omega}^\alpha f(t) = \frac{(1-q^2)q^{\binom{\alpha}{2}}(t-\omega_0)^\alpha}{\widetilde{\Gamma}_q(\alpha)} \sum_{k=0}^\infty q^{2k} \left(1 - \widetilde{q^{2k+1}}\right)_q^{\alpha-1} f\left(\sigma_{q,\omega}^{2k+\alpha}(t)\right). \quad (6)$$

Some properties of the fractional symmetric Hahn integral are given below.

Theorem 2. For $\alpha, \omega > 0$, $0 < q < 1$, and $f : I_{q,\omega}^T \to \mathbb{R}$,

$$\widetilde{\mathcal{I}}_{q,\omega}^\alpha f(t) = \widetilde{\mathcal{I}}_{q,\omega}^{\alpha+1}\left[\widetilde{D}_{q,\omega}f(t)\right] + \frac{f(\omega_0)}{\widetilde{\Gamma}_q(\alpha+1)} q^{\binom{\alpha}{2}} (t-\omega_0)^\alpha.$$

Proof. We apply Lemma 4b and Lemma 5e to (5). Then, we get

$$\widetilde{\mathcal{I}}_{q,\omega}^\alpha f(t) := \frac{q^{\binom{\alpha}{2}}}{\widetilde{\Gamma}_q(\alpha)} \int_{\omega_0}^t (\widetilde{t-s})_{q,\omega}^{\alpha-1} f\left(\sigma_{q,\omega}^{-1}(s)\right) \widetilde{d}_{q,\omega}s$$

$$= -\frac{q^{\binom{\alpha}{2}}}{\widetilde{\Gamma}_q(\alpha)[\alpha]_q} \int_{\omega_0}^t f\left(\rho_{q,\omega}\left(\sigma_{q,\omega}^\alpha(s)\right)\right) \widetilde{D}_{q,\omega}(t - \widetilde{\rho_{q,\omega}(s)})_{q,\omega}^\alpha \widetilde{d}_{q,\omega}s$$

$$= \frac{q^{\binom{\alpha}{2}}}{\widetilde{\Gamma}_q(\alpha+1)} \left\{ -\left[(t - \widetilde{\rho_{q,\omega}(s)})_{q,\omega}^\alpha f\left(\sigma_{q,\omega}^\alpha(s)\right)\right]_{\omega_0}^t + \right.$$

$$\left. q^\alpha \int_{\omega_0}^t \widetilde{D}_{q,\omega} f\left(\sigma_{q,\omega}^\alpha(s)\right) (\widetilde{t-s})_{q,\omega}^\alpha \widetilde{d}_{q,\omega}s \right\}$$

$$= \widetilde{\mathcal{I}}_{q,\omega}^{\alpha+1}\left[\widetilde{D}_{q,\omega}f(t)\right] + \frac{f(\omega_0)}{\widetilde{\Gamma}_q(\alpha+1)} q^{\binom{\alpha}{2}} (t-\omega_0)^\alpha.$$

□

Theorem 3. For $\alpha, \beta, \omega > 0$, $0 < q < 1$, $f : I_{q,\omega}^T \to \mathbb{R}$, and $a \in I_{q,\omega}^T$,

$$\int_{\omega_0}^a (\widetilde{t-s})_{q,\omega}^{\beta-1} \widetilde{\mathcal{I}}_{q,\omega}^\alpha f(s) \widetilde{d}_{q,\omega}s = 0.$$

Proof. From Definition 3, for $n \in \mathbb{N}_0$, we have

$$\widetilde{\mathcal{I}}_{q,\omega}^{\alpha} f\left(\sigma_{q,\omega}^{2n+1}(a)\right) = \frac{q^{\binom{\alpha}{2}}}{\widetilde{\Gamma}_q(\alpha)} \int_{\omega_0}^{\sigma_{q,\omega}^{2n+1}(a)} \left(\sigma_{q,\omega}^{2n+1}(a) - s\right)_{q,\omega}^{\widetilde{\alpha-1}} f\left(\sigma_{q,\omega}^{\alpha-1}(s)\right) \tilde{d}_{q,\omega} s$$

$$= \frac{(1-q^2)q^{\binom{\alpha}{2}}[\sigma_{q,\omega}^{2n+1}(a) - \omega_0]}{\widetilde{\Gamma}_q(\alpha)} \sum_{k=0}^{\infty} q^{2k} \left(\sigma_{q,\omega}^{2n+1}(a) - \sigma_{q,\omega}^{2k+2n+2}(a)\right)_{q,\omega}^{\widetilde{\alpha-1}} \times$$

$$f\left(\sigma_{q,\omega}^{2k+\alpha}(a)\right).$$

By using Lemma 2, we find that $\left(\sigma_{q,\omega}^{2n+1}(a) - \sigma_{q,\omega}^{2k+2n+2}(a)\right)_{q,\omega}^{\widetilde{\alpha-1}} = 0$. Therefore,

$$\widetilde{\mathcal{I}}_{q,\omega}^{\alpha} f\left(\sigma_{q,\omega}^{2n+1}(a)\right) = 0. \tag{7}$$

From Definition 2 and (7), we have

$$\int_{\omega_0}^{a} (\widetilde{t-s})_{q,\omega}^{\beta-1} \widetilde{\mathcal{I}}_{q,\omega}^{\alpha} f(s) \tilde{d}_{q,\omega} s$$

$$= (1-q^2)(a-\omega_0) \sum_{k=0}^{\infty} q^{2k} \left(t - \widetilde{\sigma_{q,\omega}^{2k+1}(a)}\right)_{q,\omega}^{\beta-1} \left[\widetilde{\mathcal{I}}_{q,\omega}^{\alpha} f\left(\sigma_{q,\omega}^{2k+1}(a)\right)\right] = 0.$$

□

Lemma 9 ([22]). *For $\mu, \alpha, \beta >\in \mathbb{R}^+$, the following identity is valid:*

$$\sum_{k=0}^{\infty} q^{\alpha k} \frac{(1-\mu q^{1-k})_q^{\alpha-1}(1-\mu q^{1+k})_q^{\beta-1}}{(1-q)_q^{\alpha-1}(1-q)_q^{\beta-1}} = \frac{(1-\mu q)_q^{\alpha+\beta-1}}{(1-q)_q^{\alpha+\beta-1}}.$$

Theorem 4. *For $\alpha, \beta, \omega > 0$, $0 < q < 1$, and $f : I_{q,\omega}^T \to \mathbb{R}$,*

$$\widetilde{\mathcal{I}}_{q,\omega}^{\alpha}\left[\widetilde{\mathcal{I}}_{q,\omega}^{\beta} f(t)\right] = \widetilde{\mathcal{I}}_{q,\omega}^{\beta}\left[\widetilde{\mathcal{I}}_{q,\omega}^{\alpha} f(t)\right] = \widetilde{\mathcal{I}}_{q,\omega}^{\alpha+\beta} f(t).$$

Proof. By Definition 3, for $t \in I_{q,\omega}^T$, we have

$$\tilde{\mathcal{I}}_{q,\omega}^\alpha \tilde{\mathcal{I}}_{q,\omega}^\beta f(t) = \tilde{\mathcal{I}}_{q,\omega}^\alpha \left[\frac{q^{\binom{\beta}{2}}}{\tilde{\Gamma}_q(\beta)} \int_{\omega_0}^t (t-s)_{q,\omega}^{\widetilde{\beta-1}} f\left(\sigma_{q,\omega}^{\beta-1}(s)\right) \tilde{d}_{q,\omega} s \right]$$

$$= \frac{q^{\binom{\alpha}{2}+\binom{\beta}{2}}}{\tilde{\Gamma}_q(\alpha)\tilde{\Gamma}_q(\beta)} \int_{\omega_0}^t (t-x)_{q,\omega}^{\widetilde{\alpha-1}} \int_{\omega_0}^{\sigma_{q,\omega}^{\alpha-1}(x)} (\sigma_{q,\omega}^{\alpha-1}(x) - s)_{q,\omega}^{\widetilde{\beta-1}} f\left(\sigma_{q,\omega}^{\beta-1}(s)\right) \tilde{d}_{q,\omega} s\, \tilde{d}_{q,\omega} x$$

$$= \frac{q^{\binom{\alpha}{2}+\binom{\beta}{2}+\alpha\beta}}{\tilde{\Gamma}_q(\alpha)\tilde{\Gamma}_q(\beta)} (1-q^2)^2 (t-\omega_0)^{\alpha+\beta} \times$$

$$\sum_{k=0}^\infty \sum_{h=0}^\infty q^{2k+2h+2k\beta} (1-\widetilde{q^{2k+1}})_q^{\widetilde{\alpha-1}} (1-\widetilde{q^{2h+1}})_q^{\widetilde{\beta-1}} f\left(\sigma_{q,\omega}^{2h+2k+\alpha+\beta}(t)\right)$$

$$= \frac{q^{\binom{\alpha+\beta}{2}}}{\tilde{\Gamma}_q(\alpha)\tilde{\Gamma}_q(\beta)} (1-q^2)^2 (t-\omega_0)^{\alpha+\beta} \times$$

$$\sum_{k=0}^\infty \sum_{h=k}^\infty q^{2h+2k\beta} (1-\widetilde{q^{2k+1}})_q^{\widetilde{\alpha-1}} (1-\widetilde{q^{2h-2k+1}})_q^{\widetilde{\beta-1}} f\left(\sigma_{q,\omega}^{2h+\alpha+\beta}(t)\right)$$

$$= \frac{q^{\binom{\alpha+\beta}{2}}}{\tilde{\Gamma}_q(\alpha)\tilde{\Gamma}_q(\beta)} (1-q^2)^2 (t-\omega_0)^{\alpha+\beta} \times$$

$$\sum_{h=0}^\infty q^{2h} \left[\sum_{k=0}^h q^{2k\beta} (1-\widetilde{q^{2k+1}})_q^{\widetilde{\alpha-1}} (1-\widetilde{q^{2h-2k+1}})_q^{\widetilde{\beta-1}} \right] f\left(\sigma_{q,\omega}^{2h+\alpha+\beta}(t)\right).$$

Using [21] (Theorem 2), Lemma 9, and $\tilde{\Gamma}_q(\alpha + \beta) = \frac{(1-q)_q^{\widetilde{\alpha+\beta-1}}}{(1-q^2)^{\alpha+\beta-1}}$, we obtain

$$\sum_{k=0}^h q^{2k\beta} (1-\widetilde{q^{2k+1}})_q^{\widetilde{\alpha-1}} (1-\widetilde{q^{2h-2k+1}})_q^{\widetilde{\beta-1}} = (1-\widetilde{q^2})_q^{\widetilde{\alpha-1}} (1-\widetilde{q^2})_q^{\widetilde{\beta-1}} \frac{(1-\widetilde{q^{2h+1}})_q^{\widetilde{\alpha+\beta-1}}}{(1-\widetilde{q^2})_q^{\widetilde{\alpha+\beta-1}}}$$

$$= \frac{\tilde{\Gamma}_q(\alpha)\tilde{\Gamma}_q(\beta)}{(1-q^2)\tilde{\Gamma}_q(\alpha+\beta)} (1-\widetilde{q^{2h+1}})_q^{\widetilde{\alpha+\beta-1}}.$$

Therefore,

$$\tilde{\mathcal{I}}_{q,\omega}^\alpha \tilde{\mathcal{I}}_{q,\omega}^\beta f(t) = \frac{q^{\binom{\alpha+\beta}{2}}}{\tilde{\Gamma}_q(\alpha+\beta)} (1-q^2)(t-\omega_0)^{\alpha+\beta} \sum_{h=0}^\infty q^{2h} (1-\widetilde{q^{2h+1}})_q^{\widetilde{\alpha+\beta-1}} f\left(\sigma_{q,\omega}^{2h+\alpha+\beta}(t)\right)$$

$$= \frac{q^{\binom{\alpha+\beta}{2}}}{\tilde{\Gamma}_q(\alpha+\beta)} (1-q^2)(t-\omega_0) \sum_{h=0}^\infty q^{2h} \left(t - \sigma_{q,\omega}^{2h+1}(t)\right)_{q,\omega}^{\widetilde{\alpha+\beta-1}} f\left(\sigma_{q,\omega}^{2h+\alpha+\beta}(t)\right)$$

$$= \frac{q^{\binom{\alpha+\beta}{2}}}{\tilde{\Gamma}_q(\alpha+\beta)} \int_0^t (t-s)_{q,\omega}^{\widetilde{\alpha+\beta-1}} f\left(\sigma_{q,\omega}^{\alpha+\beta-1}(s)\right) \tilde{d}_{q,\omega} s = \tilde{\mathcal{I}}_{q,\omega}^{\alpha+\beta} f(t).$$

Similarly to the above, by commuting the order of integrals, we have

$$\tilde{\mathcal{I}}_{q,\omega}^\beta \tilde{\mathcal{I}}_{q,\omega}^\alpha f(t) = \tilde{\mathcal{I}}_{q,\omega}^{\alpha+\beta} f(t).$$

□

4. The Fractional Symmetric Hahn Difference Operator of the Riemann–Liouville Type

In this section, we introduce the fractional symmetric Hahn difference operator of Riemann–Liouville as given in the following definition.

Definition 4. For $\alpha, \omega > 0$, $0 < q < 1$ and f defined on $I_{q,\omega}^T$, the fractional symmetric Hahn difference operator of Riemann–Liouville type of order α is defined by

$$\tilde{D}_{q,\omega}^\alpha f(t) := \tilde{D}_{q,\omega}^N \tilde{\mathcal{I}}_{q,\omega}^{N-\alpha} f(t),$$
$$\tilde{D}_{q,\omega}^0 f(t) = f(t)$$

where $N - 1 < \alpha < N$, $N \in \mathbb{N}$.

Next, we will establish some properties of fractional symmetric Hahn difference operators of the Riemann–Liouville type as follows.

Theorem 5. For $\alpha, \omega > 0$, $0 < q < 1$ and $f : I_{q,\omega}^T \to \mathbb{R}$,

$$\tilde{D}_{q,\omega}^\alpha \tilde{\mathcal{I}}_{q,\omega}^\alpha f(t) = f(t).$$

Proof. For some $N - 1 < \alpha < N$, $N \in \mathbb{N}$, we find that

$$\tilde{D}_{q,\omega}^\alpha \tilde{\mathcal{I}}_{q,\omega}^\alpha f(t) = \tilde{D}_{q,\omega}^N \tilde{\mathcal{I}}_{q,\omega}^{N-\alpha} \tilde{\mathcal{I}}_{q,\omega}^\alpha f(t) = \tilde{D}_{q,\omega}^N \tilde{\mathcal{I}}_{q,\omega}^N f(t) = f(t).$$

The proof is complete. □

Theorem 6. For $\alpha \in (0,1)$, $\omega > 0$, $0 < q < 1$ and $f : I_{q,\omega}^T \to \mathbb{R}$,

$$\tilde{\mathcal{I}}_{q,\omega}^\alpha \tilde{D}_{q,\omega}^\alpha f(t) = f(t) + C(t - \omega_0)^{\alpha - 1}, \quad C \in \mathbb{R}.$$

Proof. Let $C(t) = \tilde{\mathcal{I}}_{q,\omega}^\alpha \tilde{D}_{q,\omega}^\alpha f(t) - f(t)$. Taking $\tilde{D}_{q,\omega}^\alpha$ to both sides and using Theorem 5, we have

$$\tilde{D}_{q,\omega}^\alpha C(t) = \tilde{D}_{q,\omega}^\alpha \tilde{\mathcal{I}}_{q,\omega}^\alpha \tilde{D}_{q,\omega}^\alpha f(t) - \tilde{D}_{q,\omega}^\alpha f(t) = \tilde{D}_{q,\omega}^\alpha f(t) - \tilde{D}_{q,\omega}^\alpha f(t) = 0.$$

From

$$\int_{\omega_0}^t (\widetilde{t-s})_{q,\omega}^{-\alpha} (s - \omega_0)^{\alpha - 1} \tilde{d}_{q,\omega} s$$
$$= (1 - q^2)(t - \omega_0) \sum_{k=0}^\infty q^{2k} \left(t - \widetilde{\sigma_{q,\omega}^{2k+1}}(t) \right)_{q,\omega}^{-\alpha} \left(\sigma_{q,\omega}^{2k+1}(t) - \omega_0 \right)^{\alpha - 1}$$
$$= q^{\alpha - 1}(1 - q^2) \sum_{k=0}^\infty q^{2\alpha k} \left(1 - \widetilde{q^{2k+1}} \right)_q^{-\alpha},$$

and according to Definitions 3 and 4, we have

$$\tilde{D}_{q,\omega}^\alpha (t - \omega_0)^{\alpha - 1}$$
$$= \tilde{D}_{q,\omega} \tilde{\mathcal{I}}_{q,\omega}^{1-\alpha} (t - \omega_0)^{\alpha - 1}$$
$$= \tilde{D}_{q,\omega} \left[\frac{q^{\binom{1-\alpha}{2}}}{\tilde{\Gamma}_q(1-\alpha)} \int_{\omega_0}^t (\widetilde{t-s})_{q,\omega}^{-\alpha} \left(\sigma_{q,\omega}^{-\alpha}(s) - \omega_0 \right)^{\alpha - 1} \tilde{d}_{q,\omega} s \right]$$
$$= \tilde{D}_{q,\omega} \left[\frac{q^{\binom{1-\alpha}{2}}(1-q^2)(t-\omega_0)}{\tilde{\Gamma}_q(1-\alpha)} \sum_{k=0}^\infty q^{2k} \left(t - \widetilde{\sigma_{q,\omega}^{2k+1}}(t) \right)_{q,\omega}^{-\alpha} \left(\sigma_{q,\omega}^{2k-\alpha+1}(t) - \omega_0 \right)^{\alpha - 1} \right]$$
$$= \tilde{D}_{q,\omega} \left[\frac{q^{\binom{1-\alpha}{2} - (\alpha - 1)^2}(1-q^2)}{\tilde{\Gamma}_q(1-\alpha)} \sum_{k=0}^\infty q^{2k\alpha} \left(1 - \widetilde{\sigma_{q,\omega}^{2k+1}} \right)_{q,\omega}^{-\alpha} \right]$$
$$= 0.$$

Hence, $C(t) = C(t - \omega_0)^{\alpha-1}$. □

Theorem 7. Let $\alpha, \omega > 0$, $0 < q < 1$ and $f : I_{q,\omega}^T \to \mathbb{R}$. Then,

$$\tilde{\mathcal{I}}_{q,\omega}^\alpha \tilde{D}_{q,\omega}^\alpha f(t) = f(t) + C_1(t - \omega_0)^{\alpha-1} + C_2(t - \omega_0)^{\alpha-2} + \ldots + C_N(t - \omega_0)^{\alpha-N}$$

for some $C_i \in \mathbb{R}, i = 1, 2, \ldots, N$ and $N - 1 < \alpha < N$ for $N \in \mathbb{N}$.

Proof. By Theorem 2, we have

$$
\begin{aligned}
\tilde{\mathcal{I}}_{q,\omega}^\alpha \tilde{D}_{q,\omega}^\alpha f(t) &= \tilde{\mathcal{I}}_{q,\omega}^\alpha \tilde{D}_{q,\omega}^N \tilde{\mathcal{I}}_{q,\omega}^{N-\alpha} f(t) \\
&= \tilde{\mathcal{I}}_{q,\omega}^{\alpha-1} \tilde{D}_{q,\omega}^{N-1} \tilde{\mathcal{I}}_{q,\omega}^{N-\alpha} f(t) - \frac{q^{\binom{\alpha-1}{2}}}{\widetilde{\Gamma}_q(\alpha)} (t - \omega_0)^{\alpha-1} \tilde{D}_{q,\omega}^{N-1} \tilde{\mathcal{I}}_{q,\omega}^{N-\alpha} f(\omega_0) \\
&= \tilde{\mathcal{I}}_{q,\omega}^{\alpha-2} \tilde{D}_{q,\omega}^{N-2} \tilde{\mathcal{I}}_{q,\omega}^{N-\alpha} f(t) - \frac{q^{\binom{\alpha-2}{2}}}{\widetilde{\Gamma}_q(\alpha-1)} (t - \omega_0)^{\alpha-2} \tilde{D}_{q,\omega}^{N-2} \tilde{\mathcal{I}}_{q,\omega}^{N-\alpha} f(\omega_0) \\
&\quad - \frac{q^{\binom{\alpha-1}{2}}}{\widetilde{\Gamma}_q(\alpha)} (t - \omega_0)^{\alpha-1} \tilde{D}_{q,\omega}^{N-1} \tilde{\mathcal{I}}_{q,\omega}^{N-\alpha} f(\omega_0) \\
&\vdots \\
&= \tilde{\mathcal{I}}_{q,\omega}^{\alpha-N+1} \tilde{D}_{q,\omega}^{\alpha-N+1} f(t) - \frac{q^{\binom{\alpha-N+1}{2}}}{\widetilde{\Gamma}_q(\alpha-N+2)} (t - \omega_0)^{\alpha-N+1} \tilde{D}_{q,\omega} \tilde{\mathcal{I}}_{q,\omega}^{N-\alpha} f(\omega_0) \\
&\quad - \ldots - \frac{q^{\binom{\alpha-2}{2}}}{\widetilde{\Gamma}_q(\alpha-1)} (t - \omega_0)^{\alpha-2} \tilde{D}_{q,\omega}^{N-2} \tilde{\mathcal{I}}_{q,\omega}^{N-\alpha} f(\omega_0) \\
&\quad - \frac{q^{\binom{\alpha-1}{2}}}{\widetilde{\Gamma}_q(\alpha)} (t - \omega_0)^{\alpha-1} \tilde{D}_{q,\omega}^{N-1} \tilde{\mathcal{I}}_{q,\omega}^{N-\alpha} f(\omega_0).
\end{aligned}
$$

Using Theorem 6, we obtain

$$\tilde{\mathcal{I}}_{q,\omega}^\alpha \tilde{D}_{q,\omega}^\alpha f(t) = f(t) + C_1(t - \omega_0)^{\alpha-1} + C_2(t - \omega_0)^{\alpha-2} + \ldots + C_N(t - \omega_0)^{\alpha-N}.$$

The proof is complete. □

Corollary 1. Let $\alpha, \omega > 0$, $0 < q < 1$ and $f : I_{q,\omega}^T \to \mathbb{R}$. Then,

$$\tilde{\mathcal{I}}_{q,\omega}^\alpha \tilde{D}_{q,\omega}^\alpha f(t) = f(t) - \sum_{k=0}^{N-1} \frac{(t - \omega_0)^{\alpha-N+k} q^{\binom{\alpha-N+k}{2}}}{\widetilde{\Gamma}_q(\alpha - N + k + 1)} \left[\tilde{D}_{q,\omega}^{\alpha-N+k} f(\omega_0) \right]$$

where $N - 1 < \alpha < N$ for $N \in \mathbb{N}$.

5. The Fractional Symmetric Hahn Difference Operator of the Caputo type

Finally, we introduce the fractional symmetric Hahn difference operator of Caputo types as follows.

Definition 5. For $\alpha, \omega > 0$, $0 < q < 1$ and $f : I_{q,\omega}^T \to \mathbb{R}$, the fractional symmetric Hahn difference operator of Caputo type of order α is defined by

$$^C\tilde{D}_{q,\omega}^\alpha f(t) := \tilde{\mathcal{I}}_{q,\omega}^{N-\alpha} \tilde{D}_{q,\omega}^N f(t)$$

$$= \frac{q^{\binom{N-\alpha}{2}}}{\tilde{\Gamma}_q(N-\alpha)} \int_{\omega_0}^t (t\widetilde{-}s)_{q,\omega}^{N-\alpha-1} \tilde{D}_{q,\omega}^N f\left(\sigma_{q,\omega}^{N-\alpha-1}(s)\right) \tilde{d}_{q,\omega}s,$$

and $^C\tilde{D}_{q,\omega}^0 f(t) = f(t)$, where $N-1 < \alpha < N$, $N \in \mathbb{N}$.

Theorem 8. For $\alpha, \omega > 0$, $0 < q < 1$ and $f : I_{q,\omega}^T \to \mathbb{R}$,

$$^C\tilde{D}_{q,\omega}^\alpha f(t) = \frac{(1-q^2)q^{\binom{N-\alpha}{2}}}{\tilde{\Gamma}_q(N-\alpha)} (t-\omega_0)^{N-\alpha} \sum_{k=0}^\infty q^{2k} \left(1-\widetilde{q^{2k+1}}\right)_{q,\omega}^{N-\alpha-1} \tilde{D}_{q,\omega}^N f\left(\sigma_{q,\omega}^{2k+N-\alpha}(s)\right),$$

where $N-1 < \alpha < N$, $N \in \mathbb{N}$.

Proof. For $t \in I_{q,\omega}^T$ and by Definition 5, we have

$$^C\tilde{D}_{q,\omega}^\alpha f(t) = \frac{(1-q^2)q^{\binom{N-\alpha}{2}}}{\tilde{\Gamma}_q(N-\alpha)} (t-\omega_0) \sum_{k=0}^\infty q^{2k} \left(t - \widetilde{\sigma_{q,\omega}^{2k+1}(t)}\right)_{q,\omega}^{N-\alpha-1} \tilde{D}_{q,\omega}^N f\left(\sigma_{q,\omega}^{2k+N-\alpha}(s)\right)$$

$$= \frac{(1-q^2)q^{\binom{N-\alpha}{2}}}{\tilde{\Gamma}_q(N-\alpha)} (t-\omega_0)^{N-\alpha} \sum_{k=0}^\infty q^{2k} \left(1-\widetilde{q^{2k+1}}\right)_{q,\omega}^{N-\alpha-1} \tilde{D}_{q,\omega}^N f\left(\sigma_{q,\omega}^{2k+N-\alpha}(s)\right).$$

The proof is complete. □

Next, we present some properties of fractional symmetric Hahn difference operators of Caputo type as follows.

Theorem 9. For $\alpha, \omega > 0$, $0 < q < 1$ and $f : I_{q,\omega}^T \to \mathbb{R}$,

$$^C\tilde{D}_{q,\omega}^\alpha \tilde{\mathcal{I}}_{q,\omega}^\alpha f(t) = f(t).$$

Proof. For some $N-1 < \alpha < N$, $N \in \mathbb{N}$ and from Definition 5 and Corollary 1, we have

$$^C\tilde{D}_{q,\omega}^\alpha \tilde{\mathcal{I}}_{q,\omega}^\alpha f(t) = \tilde{\mathcal{I}}_{q,\omega}^{N-\alpha} \tilde{D}_{q,\omega}^N \tilde{\mathcal{I}}_{q,\omega}^\alpha f(t) = \tilde{\mathcal{I}}_{q,\omega}^{N-\alpha} \tilde{D}_{q,\omega}^{N-\alpha} f(t)$$

$$= f(t) - \sum_{k=0}^{N-1} \frac{q^{\binom{k-\alpha}{2}}}{\tilde{\Gamma}_q(k-\alpha+1)} (t-\omega_0)^{k-\alpha} \left[\tilde{D}_{q,\omega}^k \tilde{\mathcal{I}}_{q,\omega}^\alpha f(\omega_0)\right].$$

From (7), we have

$$\sum_{k=0}^{N-1} \frac{q^{\binom{k-\alpha}{2}}}{\tilde{\Gamma}_q(k-\alpha+1)} (t-\omega_0)^{k-\alpha} \left[\tilde{D}_{q,\omega}^k \tilde{\mathcal{I}}_{q,\omega}^\alpha f(\omega_0)\right] = 0.$$

It implies that

$$^C\tilde{D}_{q,\omega}^\alpha \tilde{\mathcal{I}}_{q,\omega}^\alpha f(t) = f(t).$$

The proof is complete. □

Theorem 10. For $\alpha, \omega > 0$, $0 < q < 1$ and $f : I_{q,\omega}^T \to \mathbb{R}$,

$$\mathcal{I}_{q,\omega}^\alpha {}^C D_{q,\omega}^\alpha f(t) = f(t) - \sum_{k=0}^{N-1} \frac{(t-\omega_0)^k}{[\tilde{k}]_q} \left[D_{q,\omega}^k f(\omega_0) \right],$$

where $N - 1 < \alpha < N$, $N \in \mathbb{N}$.

Proof. From Definition 5, Lemma 1a, and Corollary 1, we have

$$\begin{aligned}
\mathcal{I}_{q,\omega}^\alpha {}^C D_{q,\omega}^\alpha f(t) &= \mathcal{I}_{q,\omega}^\alpha \left[\mathcal{I}_{q,\omega}^{N-\alpha} D_{q,\omega}^N f(t) \right] = \mathcal{I}_{q,\omega}^N D_{q,\omega}^N f(t) \\
&= f(t) - \sum_{k=0}^{N-1} \frac{q^{\binom{k}{2}}}{\tilde{\Gamma}_q(k+1)} \left[\tilde{D}_{q,\omega}^k f(\omega_0) \right] (t-\omega_0)^k \\
&= f(t) - \sum_{k=0}^{N-1} \frac{q^{\binom{k}{2}}}{[\tilde{k}]_q} \left[\tilde{D}_{q,\omega}^k f(\omega_0) \right] (t-\omega_0)^k.
\end{aligned}$$

The proof is complete. □

Corollary 2. Let $\alpha, \omega > 0$, $0 < q < 1$ and $f : I_{q,\omega}^T \to \mathbb{R}$. Then,

$$\tilde{\mathcal{I}}_{q,\omega}^\alpha {}^C \tilde{D}_{q,\omega}^\alpha f(t) = f(t) + C_0 + C_1(t-\omega_0) + \ldots + C_{N-1}(t-\omega_0)^{N-1},$$

for some $C_i \in \mathbb{R}, i = 0, 1, \ldots, N-1$ and $N - 1 < \alpha < N, N \in \mathbb{N}$.

6. Conclusions

Throughout the paper, fractional symmetric Hahn integral, Riemann–Liouville and Caputo fractional symmetric Hahn difference operators have been introduced. In addition, the properties of these fractional symmetric Hahn operators have been proven. This work might be able to used as a basis for related research, such as defining the Laplace transform for fractional symmetric Hanh calculus or investigating the fractional symmetric Hahn-convolution product and computing its fractional symmetric Hahn–Laplace transform. Finally, we hope to employ these properties to solve symmetric Hahn difference problems in future works.

Author Contributions: Conceptualization, N.P. and T.S.; Methodology, N.P. and T.S.; Validation, N.P. and T.S.; Formal Analysis, N.P. and T.S.; Investigation, N.P. and T.S.; Writing—Original Draft Preparation, N.P. and T.S.; Writing—Review & Editing, N.P. and T.S.; Funding Acquisition, N.P.

Funding: This research was funded by King Mongkut's University of Technology North Bangkok. Contract no. KMUTNB-61-KNOW-027.

Acknowledgments: The last author of this research was supported by Suan Dusit University.

Conflicts of Interest: The authors declare no conflicts of interest regarding the publication of this paper.

References

1. Costas-Santos, R.S.; Marcellán, F.; Second structure Relation for *q*-semiclassical polynomials of the Hahn Tableau. *J. Math. Anal. Appl.* **2007**, *329*, 206–228. [CrossRef]
2. Kwon, K.H.; Lee, D.W.; Park, S.B.; Yoo, B.H. Hahn class orthogonal polynomials. *Kyungpook Math. J.* **1998**, *38*, 259–281.
3. Foupouagnigni, M. Laguerre-Hahn Orthogonal Polynomials with Respect to the Hahn Operator: Fourth-Order Difference Equation for the rth Associated and the Laguerre-Freud Equations Recurrence Coefficients. Ph.D. Thesis, National University of Benin, Proto Novo, Benin, 1998.
4. Hahn, W. Über Orthogonalpolynome, die *q*-Differenzenlgleichungen genügen. *Math. Nachr.* **1949**, *2*, 4–34. [CrossRef]

5. Aldwoah, K.A. Generalized Time Scales and Associated Difference Equations. Ph.D. Thesis, Cairo University, Cairo, Egypt, 2009.
6. Annaby, M.H.; Hamza, A.E.; Aldwoah, K.A. Hahn difference operator and associated Jackson-Nörlund integrals. *J. Optim. Theory Appl.* **2012**, *154*, 133–153. [CrossRef]
7. Malinowska, A.B.; Torres, D.F.M. The Hahn quantum variational calculus. *J. Optim. Theory Appl.* **2010**, *147*, 419–442. [CrossRef]
8. Malinowska, A.B.; Torres, D.F.M. Quantum Variational Calculus. In *Spinger Briefs in Electrical and Computer Engineering-Control, Automation and Robotics*; Springer: Berlin/Heidelberg, Germany, 2014.
9. Malinowska, A.B.; Martins, N. Generalized transversality conditions for the Hahn quantum variational calculus. *Optim. J. Math. Program. Oper. Res.* **2013**, *62*, 323–344. [CrossRef]
10. Hamza, A.E.; Ahmed, S.M. Theory of linear Hahn difference equations. *J. Adv. Math.* **2013**, *4*, 441–461.
11. Hamza, A.E.; Ahmed, S.M. Existence and uniqueness of solutions of Hahn difference equations. *Adv. Differ. Equ.* **2013**, *2013*, 316. [CrossRef]
12. Hamza, A.E.; Makharesh, S.D. Leibniz' rule and Fubinis theorem associated with Hahn difference operator. *J. Adv. Math.* **2016**, *12*, 6335–6345.
13. Sitthiwirattham, T. On a nonlocal boundary value problem for nonlinear second-order Hahn difference equation with two different q, ω-derivatives. *Adv. Differ. Equ.* **2016**, *2016*, 116. [CrossRef]
14. Sriphanomwan, U.; Tariboon, J.; Patanarapeelert, N.; Ntouyas, S.K.; Sitthiwirattham, T. Nonlocal boundary value problems for second-order nonlinear Hahn integro-difference equations with integral boundary conditions. *Adv. Differ. Equ.* **2017**, *2017*, 170. [CrossRef]
15. Brikshavana, T.; Sitthiwirattham, T. On fractional Hahn calculus. *Adv. Differ. Equ.* **2017**, *2017*, 354. [CrossRef]
16. Patanarapeelert, N.; Sitthiwirattham, T. Existence Results for Fractional Hahn Difference and Fractional Hahn Integral Boundary Value Problems. *Discrete Dyn. Nat. Soc.* **2017**, *2017*, 7895186. [CrossRef]
17. Patanarapeelert, N.; Brikshavana, T.; Sitthiwirattham, T. On nonlocal Dirichlet boundary value problem for sequential Caputo fractional Hahn integrodifference equations. *Bound. Value Probl.* **2018**, *2018*, 6. [CrossRef]
18. Patanarapeelert, N.; Sitthiwirattham, T. On Nonlocal Robin Boundary Value Problems for Riemann-Liouville Fractional Hahn Integrodifference Equation. *Bound. Value Probl.* **2018**, *2018*, 46. [CrossRef]
19. Dumrongpokaphan, T.; Patanarapeelert, N.; Sitthiwirattham, T. Existence Results of a Coupled System of Caputo Fractional Hahn Difference Equations with Nonlocal Fractional Hahn Integral Boundary Value Conditions. *Mathematics* **2019**, *7*, 15. [CrossRef]
20. Artur, M.C.; Cruz, B.; Martins, N.; Torres, D.F.M. Hahn's symmetric quantum variational calculus. *Numer. Algebra Control Optim.* **2013**, *3*, 77–94.
21. Sun, M.; Jin, Y.; Hou, C. Certain fractional q-symmetric integrals and q-symmetric derivatives and their application. *Adv. Differ. Equ.* **2016**, *2016*, 222. [CrossRef]
22. Rajkovic P.M.; Marinković S.D. ; Stanković M.S. Fractional integrals and derivatives in q-calculus. *Appl. Anal. Discrete Math.* **2007**, *1*, 311–323.

© 2019 by the authors. Licensee MDPI, Basel, Switzerland. This article is an open access article distributed under the terms and conditions of the Creative Commons Attribution (CC BY) license (http://creativecommons.org/licenses/by/4.0/).

Article

Approximate Controllability of Infinite-Dimensional Degenerate Fractional Order Systems in the Sectorial Case

Dumitru Baleanu [1,2], Vladimir E. Fedorov [3,4,*], Dmitriy M. Gordievskikh [5] and Kenan Taş [1]

[1] Department of Mathematics, Faculty of Arts and Sciences, Çankaya University, TR-06530 Ankara, Turkey
[2] Institute of Space Science, R-077125 Măgurle-Bucharest, Romania
[3] Department of Mathematical Analysis, Chelyabinsk State University, 454001 Chelyabinsk, Russia
[4] Laboratory of Functional Materials, South Ural State University, 454080 Chelayabinsk, Russia
[5] Department of Physics, Mathematics and Information Technology Education,
Shadrinsk State Pedagogical University, 641870 Shadrinsk, Russia
* Correspondence: kar@csu.ru; Tel.: +7-952-514-1719

Received: 1 June 2019; Accepted: 9 August 2019; Published: 12 August 2019

Abstract: We consider a class of linear inhomogeneous equations in a Banach space not solvable with respect to the fractional Caputo derivative. Such equations are called degenerate. We study the case of the existence of a resolving operators family for the respective homogeneous equation, which is an analytic in a sector. The existence of a unique solution of the Cauchy problem and of the Showalter—Sidorov problem to the inhomogeneous degenerate equation is proved. We also derive the form of the solution. The approximate controllability of infinite-dimensional control systems, described by the equations of the considered class, is researched. An approximate controllability criterion for the degenerate fractional order control system is obtained. The criterion is illustrated by the application to a system, which is described by an initial-boundary value problem for a partial differential equation, not solvable with respect to the time-fractional derivative. As a corollary of general results, an approximate controllability criterion is obtained for the degenerate fractional order control system with a finite-dimensional input.

Keywords: approximate controllability; degenerate evolution equation; fractional Caputo derivative; sectorial operator

MSC: 93B05; 35R11; 34G99

1. Introduction

Infinite-dimensional systems with distributed control, whose dynamics are described by the fractional order equation of the form

$$D_t^\alpha Lx(t) = Mx(t) + f(t), \quad t \in (0,T], \tag{1}$$

are studied. Here $\mathcal{X}$ and $\mathcal{Y}$ are reflexive Banach spaces, $L, M : \mathcal{X} \to \mathcal{Y}$ are linear closed operators, defined on dense in $\mathcal{X}$ linear subspaces $\mathcal{D}_L$ and $\mathcal{D}_M$ respectively, $m-1 < \alpha \leq m \in \mathbb{N}$, D_t^α is the Caputo derivative, $f \in C^\gamma([0,T];\mathcal{Y})$, $\gamma \in (0,1]$, where $C^\gamma([0,T];\mathcal{Y})$ is the space of Hölder functions (see the definition before Theorem 1). Equation (1) is supposed to be degenerate, that is, ker $L \neq \{0\}$, and the pair (L, M) generates an analytic in a sector resolving operators family of the homogeneous ($f \equiv 0$) Equation (1). The existence of a unique solution of the Cauchy problem and of the Showalter—Sidorov problem

$$(Lx)^{(k)}(0) = y_k \in \mathcal{Y}, \quad k = 0, 1, \ldots, m-1, \qquad (2)$$

to the inhomogeneous degenerate Equation (1) is proved and the form of the solution is also derived. The approximate controllability is investigated for distributed systems of control of the form

$$D_t^\alpha Lx(t) = Mx(t) + B(t)u(t) + g(t), \quad t \in (0, T], \qquad (3)$$

with $g \in C^\gamma([0,T]; \mathcal{Y})$, $\gamma \in (0,1]$, $B \in C^\gamma([0,T]; \mathcal{L}(\mathcal{U}; \mathcal{Y}))$, where $\mathcal{U}$ is a Banach space, $u \in C^\gamma([0,T]; \mathcal{U})$ is a control function. Taking into account the obtained results on the initial problems to the degenerate equation, the initial state is determined by the Showalter—Sidorov conditions, not by the Cauchy conditions. The equivalence of the approximate controllability of the original degenerate system and of two its subsystems on the degeneration subspace and its complement is proved. Based on this result, the obtained criteria of the approximate controllability of the subsystems are used to get a criterion for the whole degenerate control system. The criterion is illustrated by the application to an initial-boundary value problem for a partial differential equation with a degenerate spatial differential operator at the Caputo time derivative. As a corollary of the general result, an approximate controllability criterion is obtained for the degenerate fractional order control system (3) with a finite-dimensional input, that is, when $\mathcal{U} = \mathbb{R}^n$.

In the case of $\mathcal{X} = \mathcal{Y}$, $L = I$, $\alpha = 1$ controllability and approximate controllability issues have been studied in classical papers [1–5], and in many other works (see the surveys in References [6,7]). For fractional α see References [8,9] and others.

For various classes of degenerate ($\ker L \neq \{0\}$) systems (3) of the order $\alpha = 1$ the controllability and the approximate controllability were researched in References [10–14]. In References [15–17] the approximate controllability issues are studied for system (3) of fractional order α under the condition of (L, p)-boundedness of the operator M, it is a more restrictive condition on the pair of operators L, M than in this work.

The solvability of various optimal control problems for systems, described by Equation (3) with (L, p)-bounded operator M and respective semilinear equations, is studied in References [18,19] and others.

2. Nondegenerate System Solvability

To study the approximate controllability of fractional order control systems, we formulate the existence and uniqueness theorems for the equations, which describe their dynamics. Firstly, we consider the equation, which is resolved with respect to the fractional derivative.

Denote $g_\beta(t) = t^{\beta-1}/\Gamma(\beta)$ at $t > 0$, $\beta > 0$, where $\Gamma(\cdot)$ is the Euler Gamma function,

$$J_t^\beta h(t) := (g_\beta * h)(t) := \frac{1}{\Gamma(\beta)} \int_0^t (t-s)^{\beta-1} h(s) ds.$$

Let $m - 1 < \alpha \leq m \in \mathbb{N}$, D_t^α is the fractional Caputo derivative, that is,

$$D_t^\alpha h(t) := D_t^m J_t^{m-\alpha} \left(h(t) - \sum_{k=0}^{m-1} h^{(k)}(0) g_{k+1}(t) \right).$$

Let $\overline{\mathbb{R}}_+ = \mathbb{R}_+ \cup \{0\}$, $\mathcal{Z}$ be a Banach space, $\mathcal{L}(\mathcal{Z})$ be the Banach space of all linear bounded operators on $\mathcal{Z}$, $Cl(\mathcal{Z})$ be the set of all linear closed operators, densely defined in $\mathcal{Z}$, acting into $\mathcal{Z}$.

We shall write $A \in \mathcal{A}_\alpha(\theta_0, a_0)$ for some $\alpha > 0$, $\theta_0 \in (\pi/2, \pi)$, $a_0 \geq 0$, if an operator $A \in Cl(\mathcal{Z})$ satisfies the following conditions:

(i) for every $\lambda \in S_{\theta_0, a_0} := \{\mu \in \mathbb{C} : |\arg(\mu - a_0)| < \theta_0, \mu \neq a_0\}$ we have $\lambda^\alpha \in \rho(A) := \{\mu \in \mathbb{C} : (\mu I - A)^{-1} \in \mathcal{L}(\mathcal{Z})\}$;

(ii) for any $a > a_0$, $\theta \in (\pi/2, \theta_0)$ there exists $K = K(\theta, a) > 0$, such that at all $\lambda \in S_{\theta, a}$

$$\|(\lambda^\alpha I - A)^{-1}\|_{\mathcal{L}(\mathcal{Z})} \leq \frac{K(\theta, a)}{|\lambda^{\alpha-1}(\lambda - a)|}.$$

Remark 1. *It is known that at $\alpha \in (0, 2)$ an operator $A \in Cl(\mathcal{Z})$ satisfies conditions (i) and (ii), if and only if there exists a resolving family of operators for the linear homogeneous equation $D_t^\alpha z(t) = Az(t)$ (see Theorem 2.14 [20], and more general Theorem I.2.1 [21]). Moreover, $A \in \mathcal{A}_1(\theta_0, a_0)$, if and only if it generates an analytic in a sector operator semigroup. In this case it is often called a sectorial operator.*

Denote by $\partial S_{a,\theta}$ the boundary of $S_{a,\theta} := \{\mu \in \mathbb{C} : |\arg(\mu - a)| < \theta, \mu \neq a\}$ at some $\theta \in (\pi/2, \theta_0)$, $a > a_0$.

Lemma 1 ([22]). *Let $\alpha > 0$, $A \in \mathcal{A}_\alpha(\theta_0, a_0)$, $\theta \in (\pi/2, \theta_0)$, $a > a_0$. Then the families of operators*

$$\left\{ Z_\beta(t) = \frac{1}{2\pi i} \int_{\partial S_{a,\theta}} \mu^{\alpha-1-\beta}(\mu^\alpha I - A)^{-1} e^{\mu t} d\mu \in \mathcal{L}(\mathcal{Z}) : t \in \mathbb{R}_+ \right\}, \quad \beta \in \mathbb{R},$$

admit analytic extensions to $\Sigma_{\theta_0} := \{\tau \in \mathbb{C} : |\arg \tau| < \theta_0 - \pi/2, \tau \neq 0\}$.

Remark 2. *It can be shown that for a bounded operator $A \in \mathcal{L}(\mathcal{Z})$ we have $Z_\beta(t) = t^\beta E_{\alpha, \beta+1}(t^\alpha A)$, where $E_{\alpha, \beta+1}$ is the Mittag-Leffler function.*

Consider the Cauchy problem

$$z^{(k)}(0) = z_k, \quad k = 0, 1, \ldots, m - 1, \qquad (4)$$

for the inhomogeneous equation

$$D_t^\alpha z(t) = Az(t) + f(t), \quad t \in (0, T], \qquad (5)$$

where $A \in \mathcal{A}_\alpha(\theta_0, a_0)$, $T > 0$, $f : [0, T] \to \mathcal{Z}$. A solution of problem (4) and (5) is a function $z \in C((0, T]; \mathcal{D}_A) \cap C^{m-1}([0, T]; \mathcal{Z})$, such that

$$g_{m-\alpha} * \left(z - \sum_{k=0}^{m-1} z^{(k)}(0) g_{k+1} \right) \in C^m((0, T]; \mathcal{Z})$$

and Equalities (4) and (5) for all $t \in (0, T]$ are satisfied.

Remark 3. *It is known [20] that the resolving operators family for the homogeneous ($f \equiv 0$) Equation (5) is $\{Z_0(t) : t \in \mathbb{R}_+\}$, where $Z_0(0) = I$.*

A mapping $f \in C([0, T]; \mathcal{Z})$ is called Hölder function with a power $\gamma \in (0, 1]$, if there exists a constant $C > 0$, such that for all $t, s \in [0, T]$ we have $\|f(t) - f(s)\|_{\mathcal{Z}} \leq C|t - s|^\gamma$. Denote the linear space of such functions with a fixed γ by $C^\gamma([0, T]; \mathcal{Z})$.

Theorem 1 ([23]). *Let $\alpha > 0$, $A \in \mathcal{A}_\alpha(\theta_0, a_0)$, $f \in C^\gamma([0, T]; \mathcal{Z})$ for some $\gamma \in (0, 1]$. Then for any $z_k \in \mathcal{D}_A$, $k = 0, 1, \ldots, m - 1$, there exists a unique solution of problem (4) and (5). It has the form*

$$z(t) = \sum_{k=0}^{m-1} Z_k(t) z_k + \int_0^t Z_{\alpha-1}(t-s) f(s) ds.$$

Remark 4. *Analogous result with $f \in C([0, T]; \mathcal{D}_A)$ is obtained in [24]. The case of a bounded operator A and $f \in C([0, T]; \mathcal{Z})$ is studied in [25].*

3. Degenerate System Solvability

We now obtain an existence and uniqueness theorem for the degenerate equation, which describes the dynamics of fractional order degenerate systems.

Let $\mathcal{X}$, $\mathcal{Y}$ be Banach spaces, $\mathcal{L}(\mathcal{X}; \mathcal{Y})$ be the Banach space of all linear bounded operators from $\mathcal{X}$ into $\mathcal{Y}$, $Cl(\mathcal{X}; \mathcal{Y})$ be the set of all linear closed densely defined in $\mathcal{X}$ operators, acting into the space $\mathcal{Y}$. Let $L, M \in Cl(\mathcal{X}; \mathcal{Y})$, $\ker L \neq \{0\}$. The set of points $\mu \in \mathbb{C}$, such that the operator $\mu L - M : \mathcal{D}_L \cap \mathcal{D}_M \to \mathcal{Y}$ is injective, and $(\mu L - M)^{-1} L \in \mathcal{L}(\mathcal{X})$, $L(\mu L - M)^{-1} \in \mathcal{L}(\mathcal{Y})$, is called L-resolvent set $\rho^L(M)$ of the operator M. Introduce denotations $R_\mu^L(M) := (\mu L - M)^{-1} L$, $L_\mu^L(M) := L(\mu L - M)^{-1}$.

Definition 1. *Let $\alpha > 0$, $L, M \in Cl(\mathcal{X}; \mathcal{Y})$. We say that a pair of operators (L, M) belongs to the class $\mathcal{H}_\alpha(\theta_0, a_0)$, if*

(i) *there exist $\theta_0 \in (\pi/2, \pi)$ and $a_0 \geq 0$, such that for all $\lambda \in S_{\theta_0, a_0}$ inclusion $\lambda^\alpha \in \rho^L(M)$ is valid;*

(ii) *for any $\theta \in (\pi/2, \theta_0)$, $a > a_0$ there exists a constant $K = K(\theta, a) > 0$, such that for all $\lambda \in S_{\theta, a}$*

$$\max \left\{ \|R_{\lambda^\alpha}^L(M)\|_{\mathcal{L}(\mathcal{X})}, \|L_{\lambda^\alpha}^L(M)\|_{\mathcal{L}(\mathcal{Y})} \right\} \leq \frac{K(\theta, a)}{|\lambda^{\alpha-1}(\lambda - a)|}.$$

Remark 5. *If there exists an inverse operator $L^{-1} \in \mathcal{L}(\mathcal{Y}; \mathcal{X})$, then $(L, M) \in \mathcal{H}_\alpha(\theta_0, a_0)$, if and only if $L^{-1}M \in \mathcal{A}_\alpha(\theta_0, a_0)$ and $ML^{-1} \in \mathcal{A}_\alpha(\theta_0, a_0)$.*

It is not difficult to show that the subspaces $\ker R_\mu^L(M) = \ker L$, $\mathrm{im} R_\mu^L(M)$, $\ker L_\mu^L(M)$, $\mathrm{im} L_\mu^L(M)$ do not depend on $\mu \in \rho^L(M)$. Introduce the denotations $\ker R_\mu^L(M) := \mathcal{X}^0$, $\ker L_\mu^L(M) := \mathcal{Y}^0$. By $\mathcal{X}^1$ ($\mathcal{Y}^1$) we denote the closure of $\mathrm{im} R_\mu^L(M)$ ($\mathrm{im} L_\mu^L(M)$) in the norm of the space $\mathcal{X}$ ($\mathcal{Y}$). By L_k (M_k) the restriction of the operator L (M) on $\mathcal{D}_{L_k} := \mathcal{D}_L \cap \mathcal{X}^k$ ($\mathcal{D}_{M_k} := \mathcal{D}_M \cap \mathcal{X}^k$) is denoted, $k = 0, 1$.

Theorem 2 ([22]). *Let Banach spaces $\mathcal{X}$ and $\mathcal{Y}$ be reflexive, $(L, M) \in \mathcal{H}_\alpha(\theta_0, a_0)$. Then*

(i) $\mathcal{X} = \mathcal{X}^0 \oplus \mathcal{X}^1$, $\mathcal{Y} = \mathcal{Y}^0 \oplus \mathcal{Y}^1$;

(ii) *the projector P (Q) on the subspace $\mathcal{X}^1$ ($\mathcal{Y}^1$) along $\mathcal{X}^0$ ($\mathcal{Y}^0$) has the form $P = \text{s-}\lim_{n\to\infty} nR_n^L(M)$ ($Q = \text{s-}\lim_{n\to\infty} nL_n^L(M)$);*

(iii) $L_0 = 0$, $M_0 \in Cl(\mathcal{X}^0; \mathcal{Y}^0)$, $L_1, M_1 \in Cl(\mathcal{X}^1; \mathcal{Y}^1)$;

(iv) *there exist inverse operators $L_1^{-1} \in Cl(\mathcal{Y}^1; \mathcal{X}^1)$, $M_0^{-1} \in \mathcal{L}(\mathcal{Y}^0; \mathcal{X}^0)$;*

(v) $\forall x \in \mathcal{D}_L$ $Px \in \mathcal{D}_L$ and $LPx = QLx$;

(vi) $\forall x \in \mathcal{D}_M$ $Px \in \mathcal{D}_M$ and $MPx = QMx$;

(vii) *let $S := L_1^{-1}M_1 : \mathcal{D}_S \to \mathcal{X}^1$, then $\mathcal{D}_S := \{x \in \mathcal{D}_{M_1} : M_1 x \in \mathrm{im} L_1\}$ is dense in $\mathcal{X}$;*

(viii) *let $V := M_1 L_1^{-1} : \mathcal{D}_V \to \mathcal{Y}^1$, then $\mathcal{D}_V := \{y \in \mathrm{im} L_1 : L_1^{-1} y \in \mathcal{D}_{M_1}\}$ is dense in $\mathcal{Y}$;*

(ix) *if $L_1 \in \mathcal{L}(\mathcal{X}^1; \mathcal{Y}^1)$, or $M_1 \in \mathcal{L}(\mathcal{X}^1; \mathcal{Y}^1)$, then $S \in \mathcal{A}_\alpha(\theta_0, a_0)$;*

(x) *if $L_1^{-1} \in \mathcal{L}(\mathcal{Y}^1; \mathcal{X}^1)$, or $M_1^{-1} \in \mathcal{L}(\mathcal{Y}^1; \mathcal{X}^1)$, then $V \in \mathcal{A}_\alpha(\theta_0, a_0)$;*

(xi) the families of operators

$$\left\{ X_\beta(t) = \frac{1}{2\pi i} \int_{\partial S_{a,\theta}} \mu^{\alpha-1-\beta} R^L_{\mu^\alpha}(M) e^{\mu t} d\mu \in \mathcal{L}(\mathcal{X}) : t \in \mathbb{R}_+ \right\}, \quad \beta \in \mathbb{R},$$

$$\left\{ Y_\beta(t) = \frac{1}{2\pi i} \int_{\partial S_{a,\theta}} \mu^{\alpha-1-\beta} L^L_{\mu^\alpha}(M) e^{\mu t} d\mu \in \mathcal{L}(\mathcal{Y}) : t \in \mathbb{R}_+ \right\}, \quad \beta \in \mathbb{R},$$

admit analytic extensions to $\Sigma_{\theta_0} := \{t \in \mathbb{C} : |\arg t| < \theta_0 - \pi/2, t \neq 0\}$. For any $\theta \in (\pi/2, \theta_0)$, $a > a_0$ there exists such $C_\beta = C_\beta(\theta, a)$, that for each $t \in \Sigma_\theta$

$$\max\{\|X_\beta(t)\|_{\mathcal{L}(\mathcal{X})}, \|Y_\beta(t)\|_{\mathcal{L}(\mathcal{Y})}\} \leq C_\beta(\theta, a) e^{a \operatorname{Re} t}(|t|^{-1} + a)^{-\beta}, \quad \beta \leq 0, \tag{6}$$

$$\max\{\|X_\beta(t)\|_{\mathcal{L}(\mathcal{X})}, \|Y_\beta(t)\|_{\mathcal{L}(\mathcal{Y})}\} \leq C_\beta(\theta, a) e^{a \operatorname{Re} t} |t|^\beta, \quad \beta > 0. \tag{7}$$

Consider the degenerate ($\ker L \neq \{0\}$) inhomogeneous equation

$$D^\alpha_t L x(t) = M x(t) + f(t), \quad t \in (0, T], \tag{8}$$

with a given $f : [0, T] \to \mathcal{Y}$. Its solution is a function $x \in C((0, T]; \mathcal{D}_M)$, such that $Lx \in C^{m-1}([0, T]; \mathcal{Y})$, $g_{m-\alpha} * \left(Lx - \sum_{k=0}^{m-1} (Lx)^{(k)}(0) g_{k+1} \right) \in C^m((0, T]; \mathcal{Y})$, and for all $t \in (0, T]$ equality (8) is fulfilled. A solution of the Cauchy problem

$$x^{(k)}(0) = x_k, \quad k = 0, 1, \ldots, m-1, \tag{9}$$

for Equation (8) is a solution of the equation, such that $x \in C^{m-1}([0, T]; \mathcal{X})$ and conditions (9) are satisfied.

Theorem 3. *Let $\alpha > 0$, Banach spaces $\mathcal{X}$, $\mathcal{Y}$ be reflexive, $(L, M) \in \mathcal{H}_\alpha(\theta_0, a_0)$, $L_1 \in \mathcal{L}(\mathcal{X}^1; \mathcal{Y}^1)$ or $M_1 \in \mathcal{L}(\mathcal{X}^1; \mathcal{Y}^1)$, $f : [0, T] \to \mathcal{Y}^0 + \operatorname{im} L_1$, at some $\gamma \in (0, 1]$ $L_1^{-1} Qf \in C^\gamma([0, T]; \mathcal{X})$, $(I - Q)f \in C^{m-1}([0, T]; \mathcal{Y})$, $x_k \in \mathcal{D}_M$, $Px_k \in \mathcal{D}_S$, $k = 0, 1, \ldots, m-1$, equalities*

$$D^k_t \big|_{t=0} M_0^{-1}(I - Q) f(t) = -(I - P) x_k, \quad k = 0, 1, \ldots, m-1, \tag{10}$$

are valid. Then there exists a unique solution of problem (8) and (9), moreover, it has the form

$$x(t) = \sum_{k=0}^{m-1} X_k(t) x_k + \int_0^t X_{\alpha-1}(t-s) L_1^{-1} Q f(s) ds - M_0^{-1}(I - Q) f(t). \tag{11}$$

Proof. Put $x^0(t) := (I - P)x(t)$, $x^1(t) := Px(t)$. By virtue of Theorem 2 Equation (8) can be reduced to the system of the two equations

$$0 = x^0(t) + M_0^{-1}(I - Q) f(t),$$

$$D^\alpha_t x^1(t) = S x^1(t) + g(t), \quad S := L_1^{-1} M_1, \quad g(t) := L_1^{-1} Q f(t). \tag{12}$$

Therefore, $x^0(t) = -M_0^{-1}(I - Q) f(t)$, and for the satisfying of Cauchy conditions (9) it is necessarry the fulfillment of (10). Due to Theorem 2 $S \in \mathcal{A}_\alpha(\theta_0, a_0)$, hence Theorem 1 implies the existence of a unique solution of the Cauchy problem $x^{1(k)}(0) = Px_k$, $k = 0, 1, \ldots, m-1$, to Equation (12). Besides,

$$x^1(t) = \frac{1}{2\pi i} \sum_{k=0}^{m-1} \int_{\partial S_{a,\theta}} \mu^{\alpha-k-1}(\mu^\alpha I - S)^{-1} e^{\mu t} d\mu Px_k + \frac{1}{2\pi i} \int_0^t \int_{\partial S_{a,\theta}} (\mu^\alpha I - S)^{-1} e^{\mu(t-s)} d\mu g(s) ds =$$

$$= \frac{1}{2\pi i} \sum_{k=0}^{m-1} \int_{\partial S_{a,\theta}} \mu^{\alpha-k-1}(\mu^\alpha L - M)^{-1} L e^{\mu t} d\mu x_k + \frac{1}{2\pi i} \int_0^t \int_{\partial S_{a,\theta}} (\mu^\alpha L - M)^{-1} L e^{\mu(t-s)} d\mu g(s) ds,$$

since $L(I - P) = 0$, the operator $(\lambda L_0 - M_0)^{-1} = -M_0^{-1}$ exists for every $\lambda \in \mathbb{C}$. □

Theorem 4. *Let $\alpha > 0$, Banach spaces $\mathcal{X}$, $\mathcal{Y}$ be reflexive, $(L, M) \in \mathcal{H}_\alpha(\theta_0, a_0)$, $L_1^{-1} \in \mathcal{L}(\mathcal{Y}^1; \mathcal{X}^1)$ or $M_1^{-1} \in \mathcal{L}(\mathcal{Y}^1; \mathcal{X}^1)$, $f \in C([0, T]; \mathcal{Y})$, $Qf \in C^\gamma([0, T]; \mathcal{Y})$ at some $\gamma \in (0, 1]$, $(I - Q)f \in C^{m-1}([0, T]; \mathcal{Y})$, $x_k \in \mathcal{D}_M$, $Px_k \in \mathcal{D}_L$, $k = 0, 1, \ldots, m-1$, equalities (10) are valid. Then there exists a unique solution of problem (8) and (9), and it has form (11).*

Proof. In this case, instead of Equation (12) we obtain the equation

$$D_t^\alpha y(t) = Vy(t) + h(t), \quad V := M_1 L_1^{-1}, \quad h(t) := Qf(t), \tag{13}$$

where $y(t) := L_1 x^1(t) = L_1 Px(t)$. Theorem 2 implies, that $V \in \mathcal{A}_\alpha(\theta_0, a_0)$, and due to Theorem 1 there exists a unique solution of the Cauchy problem $y^{(k)}(0) = L_1 Px_k \in \mathcal{D}_V$, $k = 0, 1, \ldots, m-1$, for Equation (13). The solution has the form

$$y(t) = \frac{1}{2\pi i} \sum_{k=0}^{m-1} \int_{\partial S_{a,\theta}} \mu^{\alpha-k-1}(\mu^\alpha I - V)^{-1} e^{\mu t} d\mu L_1 Px_k + \frac{1}{2\pi i} \int_0^t \int_{\partial S_{a,\theta}} (\mu^\alpha I - V)^{-1} e^{\mu(t-s)} d\mu h(s) ds =$$

$$= \frac{1}{2\pi i} \sum_{k=0}^{m-1} Y_k(t) L_1 Px_k + \int_0^t Y_{\alpha-1}(t-s) Qf(s) ds, \tag{14}$$

therefore, $x^1(t) = L_1^{-1} y(t)$ has form (11). The function $x^0(t)$ is the same as in the previous proof. □

So, the Cauchy problem for degenerate Equation (8) is overdetermined due to the necessity of conditions (10). Consider the so-called Showalter—Sidorov problem

$$(Lx)^{(k)}(0) = y_k, \quad k = 0, 1, \ldots, m-1, \tag{15}$$

which is natural for weakly degenerate evolution equations, when the degeneration subspace $\mathcal{X}^0$ coincides with $\ker L$. A solution of this problem to Equation (8) is a solution of the equation, such that conditions (15) are satisfied.

Reasoning as before, we can prove the next assertions.

Theorem 5. *Let $\alpha > 0$, Banach spaces $\mathcal{X}$, $\mathcal{Y}$ be reflexive, $(L, M) \in \mathcal{H}_\alpha(\theta_0, a_0)$, $L_1 \in \mathcal{L}(\mathcal{X}^1; \mathcal{Y}^1)$ or $M_1 \in \mathcal{L}(\mathcal{X}^1; \mathcal{Y}^1)$, $f \in C([0, T]; \mathcal{Y})$, $Qf(t) \in \mathrm{im} L$ for all $t \in [0, T]$, $L_1^{-1} Qf \in C^\gamma([0, T]; \mathcal{X})$ at some $\gamma \in (0, 1]$, $y_k \in L[\mathcal{D}_L \cap \mathcal{D}_M]$, $L_1^{-1} y_k \in \mathcal{D}_S$, $k = 0, 1, \ldots, m-1$. Then there exists a unique solution of problem (8) and (15), and it has form*

$$x(t) = \sum_{k=0}^{m-1} X_k(t) L_1^{-1} y_k + \int_0^t X_{\alpha-1}(t-s) L_1^{-1} Qf(s) ds - M_0^{-1}(I - Q)f(t). \tag{16}$$

Theorem 6. *Let $\alpha > 0$, Banach spaces $\mathcal{X}$, $\mathcal{Y}$ be reflexive, $(L, M) \in \mathcal{H}_\alpha(\theta_0, a_0)$, $L_1^{-1} \in \mathcal{L}(\mathcal{Y}^1; \mathcal{X}^1)$ or $M_1^{-1} \in \mathcal{L}(\mathcal{Y}^1; \mathcal{X}^1)$, $f \in C([0, T]; \mathcal{Y})$, $Qf \in C^\gamma([0, T]; \mathcal{Y})$ for some $\gamma \in (0, 1]$, $y_k \in L[\mathcal{D}_L \cap \mathcal{D}_M]$, $k = 0, 1, \ldots, m - 1$. Then there exists a unique solution of problem (8) and (15), and it has form (16).*

Here, in contrast to the proofs of Theorems 3 and 4 we have no initial conditions for $(I - Q)x(t)$ and there is not condition $(I - Q)f \in C^{m-1}([0, T]; \mathcal{Y})$ nor condition (10) of the matching of initial data with the right-hand side of Equation (8).

Remark 6. *Note that, due to Theorem 2 $L = L_1 P + 0(I - P) = L_1 P$, therefore, $\operatorname{im} L = \operatorname{im} L_1 \subset \mathcal{Y}^1$. Thus, $y_k \in L[\mathcal{D}_L \cap \mathcal{D}_M] = L_1[\mathcal{D}_{L_1} \cap \mathcal{D}_{M_1}]$, if and only if $y_k \in \mathcal{D}_V$. So, under the conditions of Theorem 6 the set $L[\mathcal{D}_L \cap \mathcal{D}_M] = \mathcal{D}_V$ is dense in $\mathcal{Y}^1$.*

Remark 7. *Study of the degenerate system controllability will be carried out in the next sections on the basis of Theorem 6, since its conditions on f and y_k are less restrictive than those in Theorem 5.*

Remark 8. *It can be shown that in the case of reflexive Banach spaces $\mathcal{X}$ and $\mathcal{Y}$ for $(L, M) \in \mathcal{H}_\alpha(\theta_0, a_0)$ conditions (15) are equivalent to the conditions $(Px)^{(k)}(0) = L_1^{-1} y_k$, $k = 0, 1, \ldots, m - 1$. Recall that $\operatorname{im} L \subset \mathcal{X}^1$.*

4. Approximate Controllability of Subsystems

Here, we reduce the degenerate control system to two subsystems on mutually complement subspaces.

Let $\mathcal{X}$, $\mathcal{Y}$ be reflexive Banach spaces, $\mathcal{U}$ be a Banach space, $L, M \in \mathcal{C}l(\mathcal{X}; \mathcal{Y})$, $(L, M) \in \mathcal{H}_\alpha(\theta_0, a_0)$. Denote by $C_Q^\gamma([0, T]; \mathcal{L}(\mathcal{U}; \mathcal{Y}))$ for some $\gamma \in (0, 1]$ the linear space of all operator-valued functions $B \in C([0, T]; \mathcal{L}(\mathcal{U}; \mathcal{Y}))$, such that $QB \in C^\gamma([0, T]; \mathcal{L}(\mathcal{U}; \mathcal{Y}))$. Analogously, $C_Q^\gamma([0, T]; \mathcal{Y})$ is the set of all vector-valued functions $g \in C([0, T]; \mathcal{Y})$, such that $Qg \in C^\gamma([0, T]; \mathcal{Y})$.

Further, we shall assume that $B \in C_Q^\gamma([0, T]; \mathcal{L}(\mathcal{U}; \mathcal{Y}))$, $g \in C_Q^\gamma([0, T]; \mathcal{Y})$ for some $\gamma \in (0, 1]$. Control functions $u(\cdot)$ for the system, which is described by the Showalter—Sidorov problem

$$(Lx)^{(k)}(0) = y_k, \quad k = 0, 1, \ldots, m - 1, \tag{17}$$

$$D_t^\alpha Lx(t) = Mx(t) + B(t)u(t) + g(t), \tag{18}$$

will be choosen from the space $C^\gamma([0, T]; \mathcal{U})$, hence $Bu \in C_Q^\gamma([0, T]; \mathcal{Y})$. By means of Theorem 2 problem (17) and (18) can be reduced to the initial value problem

$$y^{(k)}(0) = y_k, \quad k = 0, 1, \ldots, m - 1, \tag{19}$$

for the system of equations

$$D_t^\alpha y(t) = Vy(t) + QB(t)u(t) + Qg(t), \tag{20}$$

$$x^0(t) = -M_0^{-1}(I - Q)(B(t)u(t) + g(t)) \tag{21}$$

on the subspaces $\mathcal{Y}^1$ and $\mathcal{X}^0$, respectively. Here $V = M_1 L_1^{-1} \in \mathcal{C}l(\mathcal{Y}^1)$, $y(t) = L_1 Px(t)$, $x^0(t) = (I - P)x(t)$. Note that due to Theorem 1 the solution of problem (19) and (20) has the form

$$y(t) = \sum_{k=0}^{m-1} Y_k(t) y_k + \int_0^t Y_{\alpha-1}(t - s) Q(B(s)u(s) + g(s)) ds. \tag{22}$$

Denoted by $x(T; \bar{y}; u)$, the value at the time moment T of the solution to problem (17) and (18) with the initial data $\bar{y} = (y_0, y_1, \ldots, y_{m-1})$ in (17) and with a control function u. Denoted by $y(T; \bar{y}; u)$, the value at the time T of the solution for the subsystem, described by (19), (20). And by $x^0(T;u)$ denotes the value at $t = T$ of function (21).

System (18) is called approximately controllable in time $T > 0$, if, for every $\varepsilon > 0$, $\hat{x} \in \mathcal{X}$, $\overline{y} = (y_0, y_1, \ldots, y_{m-1}) \in (L[\mathcal{D}_L \cap \mathcal{D}_M])^m$ in (17) there exists a control function $u \in C^\gamma([0,T]; \mathcal{U})$, such that $\|x(T;\overline{y};u) - \hat{x}\|_\mathcal{X} \leq \varepsilon$.

System (20) is called approximately controllable in time $T > 0$, if for all $\varepsilon > 0$, $\hat{y} \in \mathcal{Y}^1$, $\overline{y} = (y_0, y_1, \ldots, y_{m-1}) \in (\mathcal{D}_V)^m$ in (19) there exists a control function $u \in C^\gamma([0,T]; \mathcal{U})$, such that $\|y(T;\overline{y};u) - \hat{y}\|_{\mathcal{Y}^1} \leq \varepsilon$.

System (21) is called approximately controllable in time $T > 0$, if for every $\varepsilon > 0$, $\hat{x}^0 \in \mathcal{X}^0$ there exists $u \in C([0,T]; \mathcal{U})$, such that $\|x^0(T;u) - \hat{x}^0\|_{\mathcal{X}^0} \leq \varepsilon$.

Remark 9. *We take u not from $C^\gamma([0,T]; \mathcal{U})$ in the last definition, since due to the definition of problem (8) and (15) solution, the continuity of u is sufficient for the existence of the subsystem (21) solution, since $x^0(t) \in \ker L$ for all t.*

The following result shows that, while controlling two systems (20) and (21) by the same function $u(\cdot)$, we can, nevertheless, simultaneously lead the trajectories of the both systems into the ε-neighborhood of respective given points $\hat{y} \in \mathcal{Y}^1$, $\hat{x}^0 \in \mathcal{X}^0$.

Theorem 7. *Let Banach spaces $\mathcal{X}$, $\mathcal{Y}$ be reflexive, $(L,M) \in \mathcal{H}_\alpha(\theta_0, a_0)$, $L_1^{-1} \in \mathcal{L}(\mathcal{Y}^1; \mathcal{X}^1)$, $B \in C_Q^\gamma([0,T]; \mathcal{L}(\mathcal{U}; \mathcal{Y}))$, $g \in C_Q^\gamma([0,T]; \mathcal{Y})$ for some $\gamma \in (0,1]$. Then system (18) is approximately controllable in time T, if and only if its subsystems (20) and (21) are approximately controllable in time T.*

Proof. The direct assertion of Theorem 7 is obvious, since system (18) splits into two mutually complementary subsystems (20) and (21). Consider the inverse assertion of Theorem 7. Let for all $\hat{x}^0 \in \mathcal{X}^0$, $\varepsilon > 0$ there exists a function $u^0 \in C([0,T]; \mathcal{U})$, such that

$$\left\| -M_0^{-1}(I-Q)(B(T)u^0(T) + g(T)) - \hat{x}^0 \right\|_\mathcal{X} \leq \varepsilon/3,$$

and

$$\forall \overline{y} \in (\mathcal{D}_T)^m \ \forall \hat{y} \in \mathcal{Y}^1 \ \forall \varepsilon > 0 \ \exists u^1 \in C^\gamma([0,T]; \mathcal{U}) \ \|y(T;\overline{y};u^1) - \hat{y}\|_\mathcal{Y} \leq \varepsilon/\left(3\|L_1^{-1}\|_{\mathcal{L}(\mathcal{Y}^1; \mathcal{X}^1)}\right).$$

Then choose the new control function u, such that $u(t) = u^1(t)$ at $t \in [0,\delta]$ for some $\delta \in (T/2, T)$, and $u(t) = u^1(\delta) + \gamma(t-\delta) + b(t-\delta)^2$ at $t \in (\delta, T]$ with

$$\gamma = \frac{du^1}{dt}(\delta), \quad b = \frac{u^0(T) - u^1(\delta) - \gamma(T-\delta)}{(T-\delta)^2}.$$

Then

$$u(t) = u^1(\delta) + \gamma(t-\delta) + \frac{(t-\delta)^2}{(T-\delta)^2}\left(u^0(T) - u^1(\delta) - \gamma(T-\delta)\right) \in C^\gamma([0,T]; \mathcal{U}), \quad u(T) = u^0(T).$$

Note that for any $\delta \in (T/2, T)$

$$\|u(t)\|_\mathcal{U} \leq C := 2 \max_{t \in [0,T]} \|u^1(t)\|_\mathcal{U} + \|\gamma\|_\mathcal{U} T + \|u^0(T)\|_\mathcal{U}, \quad t \in [0,T],$$

where C is independent of δ.

For arbitrary $\hat{x} \in \mathcal{X}$ take the control function, constructed as it was explained before with $\hat{x}^0 = (I - P)\hat{x}$ and $\hat{y} = L\hat{x}$, then for sufficiently small $T - \delta > 0$,

$$\|x(T; \overline{y}; u) - \hat{x}\|_\mathcal{X} \leq \|x^0(T; u^0) - (I - P)\hat{x}\|_\mathcal{X} + \|L_1^{-1}y(T; \overline{y}; u^1) - L_1^{-1}L\hat{x}\|_\mathcal{X} +$$

$$+ \|L_1^{-1}y(T; \overline{y}; u) - L_1^{-1}y(T; \overline{y}; u^1)\|_\mathcal{X} \leq 2\varepsilon/3 + 2C\|L_1^{-1}\|_{\mathcal{L}(\mathcal{Y}^1; \mathcal{X}^1)} \int_\delta^T \|Y_{\alpha-1}(T - s)QB(s)\|_{\mathcal{L}(\mathcal{U}; \mathcal{Y})} ds \leq \varepsilon.$$

Here, we take into account estimate (7) for $\alpha > 1$. At $\alpha \in (0, 1]$ due to (6) we also have

$$\int_\delta^T \|Y_{\alpha-1}(T - s)QB(s)\|_{\mathcal{L}(\mathcal{U}; \mathcal{Y})} ds \leq C_1(1 + aT)^{1-\alpha}(T - \delta)^\alpha \to 0 \text{ as } \delta \to T-.$$

□

Analogously, the notion of the approximate controllability in free time can be defined. For example, system (18) is called approximately controllable in free time, if for every $\varepsilon > 0$, $\hat{x} \in \mathcal{X}$, $\overline{y} = (y_0, y_1, \ldots, y_{m-1}) \in (L[\mathcal{D}_L \cap \mathcal{D}_M])^m$ in (17) there exists $T > 0$ and a control function $u \in C^\gamma([0, T]; \mathcal{U})$, $\gamma \in (0, 1]$, such that $\|x(T; \overline{y}; u) - \hat{x}\|_\mathcal{X} \leq \varepsilon$.

Theorem 8. *Let Banach spaces $\mathcal{X}$, $\mathcal{Y}$ be reflexive, $(L, M) \in \mathcal{H}_\alpha(\theta_0, a_0)$, $L_1^{-1} \in \mathcal{L}(\mathcal{Y}^1; \mathcal{X}^1)$, for every $T > 0$ $B \in C_Q^{\gamma(T)}([0, T]; \mathcal{L}(\mathcal{U}; \mathcal{Y}))$, $g \in C_Q^{\gamma(T)}([0, T]; \mathcal{Y})$, $\gamma(T) \in (0, 1]$. Then system (18) is approximately controllable in free time, if and only if its subsystems (20) and (21) are approximately controllable in free time.*

Proof. This statement can be proved as Theorem 7. Let us prove the inverse assertion. Let $\varepsilon > 0$, $\hat{x} \in \mathcal{X}$, $\overline{y} = (y_0, y_1, \ldots, y_{m-1}) \in (L[\mathcal{D}_L \cap \mathcal{D}_M])^m$ and there exist $T_1 > 0$, $u^1 \in C^{\gamma(T_1)}([0, T_1]; \mathcal{U})$, such that $\|y(T_1; \overline{y}; u^1) - L\hat{x}\|_{\mathcal{Y}^1} \leq \varepsilon/3$, and $T_0 > 0$, $u^0 \in C([0, T_0]; \mathcal{U})$, such that $\|x^0(T_0; u^0) - (I - P)\hat{x}\|_{\mathcal{X}^0} \leq \varepsilon/3$. Take the control function u as in the proof of Theorem 7 with $T = T_1$, then $\|x(T_1; \overline{y}; u) - \hat{x}\|_{\mathcal{Y}^1} \leq \varepsilon$. □

5. Criterion of Approximate Controllability

Now let us obtain a criterion of the fractional order degenerate control system approximate controllability in terms of the operators from the respective equation.

Let $\mathcal{Z}$ be a Banach space, $\mathcal{A}$ be some set of indices, $\alpha \in \mathcal{A}$, $\mathcal{D}_\alpha \subset \mathcal{Z}$. By $\text{span}\{\mathcal{D}_\alpha : \alpha \in \mathcal{A}\}$ we denote the linear span of the sets $\mathcal{D}_\alpha$ union, $\alpha \in \mathcal{A}$, and by $\overline{\text{span}}\{\mathcal{D}_\alpha : \alpha \in \mathcal{A}\}$ its closure in the space $\mathcal{Z}$ is denoted. We denote by $\overline{\text{im}}A$ the closure of the image $\text{im}A$ of an operator $A : \mathcal{D}_A \to \mathcal{Z}$.

Lemma 2. *Let Banach spaces $\mathcal{X}$, $\mathcal{Y}$ be reflexive, $(L, M) \in \mathcal{H}_\alpha(\theta_0, a_0)$, $L_1^{-1} \in \mathcal{L}(\mathcal{Y}^1; \mathcal{X}^1)$, $QB \in C^\gamma([0, T]; \mathcal{L}(\mathcal{U}; \mathcal{Y}))$, $Qg \in C^\gamma([0, T]; \mathcal{Y})$ for some $\gamma \in (0, 1]$. Then system (20) is approximately controllable in time T, if and only if*

$$\overline{\text{span}}\{\text{im}Y_{\alpha-1}(T - s)QB(s) : 0 < s < T\} = \mathcal{Y}^1. \tag{23}$$

Proof. Form (22) of the Cauchy problem solution implies that it is sufficient to consider only the approximate controllability of system (20) from zero ($\overline{y} = 0$). Suppose that the system is not approximately controllable from zero. Then the set of vectors of the form

$$\int_0^T Y_{\alpha-1}(T - s)QB(s)u(s)ds, \quad u \in C^\gamma([0, T]; \mathcal{U}),$$

is not dense in the space $\mathcal{Y}^1$. By the Hahn—Banach Theorem, in this case there exists $f \in \mathcal{Y}^{1*} \setminus \{0\}$, such that

$$f\left(\int_0^T Y_{\alpha-1}(T-s)QB(s)u(s)ds\right) = \int_0^T f\left(Y_{\alpha-1}(T-s)QB(s)u(s)\right) ds = 0 \qquad (24)$$

for all $u \in C^\gamma([0,T];\mathcal{U})$.

For every v from the Lebesgue—Bochner space $L_p(0,T;\mathcal{U})$, $\max\{1,1/\alpha\} < p < \infty$, there exists a sequence $\{u_n\} \subset C^\gamma([0,T];\mathcal{U})$, such that $\lim_{n\to\infty} u_n = v$ in $L_p(0,T;\mathcal{U})$. Therefore, using reasoning as in the end of Theorem 7 proof, i.e., applying inequalities (6) and (7), obtain

$$\left|\int_0^T f(Y_{\alpha-1}(T-s)QB(s)(u_n(s)-v(s)))ds\right| \leq C\|f\|_{\mathcal{Y}^{1*}} \int_0^T s^{(\alpha-1)p'} ds \int_0^T \|u_n(s)-v(s)\|_{\mathcal{U}}^p ds \to 0$$

as $n \to \infty$. Here we take into account, that inequality $p > 1/\alpha$ implies that $(\alpha-1)p' + 1 > 0$, where $p' = p/(p-1)$. Consequently, equality (24) is valid for all $u \in L_p(0,T;\mathcal{U})$.

Take $t_0 \in (0,T)$ and small $\delta > 0$, $u_\delta(t) = w \in \mathcal{U}$ at $t \in [t_0-\delta, t_0+\delta]$, $u_\delta(t) = 0$ for $t \in [0,T] \setminus [t_0-\delta, t_0+\delta]$. Then $u_\delta \in L_p(0,T;\mathcal{U})$, and by the continuity of the integrand

$$0 = \frac{1}{2\delta} \int_{t_0-\delta}^{t_0+\delta} f\left(Y_{\alpha-1}(T-s)QB(s)w\right) ds = f\left(Y_{\alpha-1}(T-\xi)QB(\xi)w\right)$$

for some $\xi \in (t_0-\delta, t_0+\delta)$. We pass to the limit as $\delta \to 0+$ and obtain the equality $f\left(Y_{\alpha-1}(T-t_0)QB(t_0)w\right) = 0$ for all $t_0 \in (0,T)$, $w \in \mathcal{U}$. Hence condition (23) is not satisfied.

The inverse statement is obvious due to the integral form (22) of the solution of Equation (20) with zero initial data. □

This assertion can be formulated in terms of Section 2 in the next form.

Theorem 9. *Let $A \in \mathcal{A}_\alpha(\theta_0, a_0)$, $B \in C^\gamma([0,T];\mathcal{L}(\mathcal{U};\mathcal{Z}))$, $g \in C^\gamma([0,T];\mathcal{Z})$ for some $\gamma \in (0,1]$. Then the system $D_t^\alpha z(t) = Az(t) + B(t)u(t) + g(t)$ is approximately controllable in time T, if and only if*

$$\overline{\text{span}}\{\text{im} Z_{\alpha-1}(T-s)B(s) : 0 < s < T\} = \mathcal{Z}.$$

Remark 10. *If $QB(t)$ does not depend on t, then the approximate controllability of system (20) in time T implies its approximate controllability in any time $T_1 > T$, since*

$$\overline{\text{span}}\{\text{im} Y_{\alpha-1}(s)QB : 0 < s < T\} \subset \overline{\text{span}}\{\text{im} Y_{\alpha-1}(s)QB : 0 < s < T_1\}.$$

The criterion of system (21) approximate controllability is obvious.

Lemma 3. *Let Banach spaces $\mathcal{X}$, $\mathcal{Y}$ be reflexive, $(L,M) \in \mathcal{H}_\alpha(\theta_0, a_0)$, moreover, $(I-Q)B \in C([0,T];\mathcal{L}(\mathcal{U};\mathcal{Y}))$ and $(I-Q)g \in C([0,T];\mathcal{Y})$. Then system (21) is approximately controllable in time T, if and only if $\overline{\text{im}} M_0^{-1}(I-Q)B(T) = \mathcal{X}^0$.*

Remark 11. *If $(I-Q)B(t)$ does not depend on t, then the approximate controllability of system (21) in time T implies its approximate controllability at any time $T_1 > 0$.*

Theorem 10. *Let Banach spaces $\mathcal{X}$, $\mathcal{Y}$ be reflexive, $(L,M) \in \mathcal{H}_\alpha(\theta_0, a_0)$, $L_1^{-1} \in \mathcal{L}(\mathcal{Y}^1; \mathcal{X}^1)$, $B \in C_Q^\gamma([0,T];\mathcal{L}(\mathcal{U};\mathcal{Y}))$, $g \in C_Q^\gamma([0,T];\mathcal{Y})$ for some $\gamma \in (0,1]$. Then system (18) is approximately controllable in time T, if and only if $\overline{\text{im}} M_0^{-1}(I-Q)B(T) = \mathcal{X}^0$, $\overline{\text{span}}\{\text{im} Y_{\alpha-1}(T-s)QB(s) : 0 < s < T\} = \mathcal{Y}^1$.*

Proof. The required result follows from Theorem 7, Lemmas 2 and 3. □

Remark 12. *By Remarks 10, 11 and Theorem 10, if $B(t)$ does not depend on t, then the approximate controllability of system (18) in time T implies its approximate controllability in any greater time $T_1 > T$.*

Similar result for the controllability in free time can be obtained analogously.

Theorem 11. *Let Banach spaces $\mathcal{X}$, $\mathcal{Y}$ be reflexive, $(L, M) \in \mathcal{H}_\alpha(\theta_0, a_0)$, $L_1^{-1} \in \mathcal{L}(\mathcal{Y}^1; \mathcal{X}^1)$, for all $T > 0$ $B \in C_Q^{\gamma(T)}([0,T]; \mathcal{L}(\mathcal{U}; \mathcal{Y}))$, $g \in C_Q^{\gamma(T)}([0,T]; \mathcal{Y})$, $\gamma(T) \in (0,1]$. Then system (18) is approximately controllable in free time, if and only if $\overline{\text{span}}\{\text{im} M_0^{-1}(I - Q)B(T) : T \in \mathbb{R}_+\} = \mathcal{X}^0$,*

$$\overline{\text{span}}\{\text{im} Y_{\alpha-1}(T-s)QB(s) : 0 < s < T, T \in \mathbb{R}_+\} = \mathcal{Y}^1.$$

6. Application to an Initial-Boundary Value Problem

We shall apply the obtained criterion to the control system, which is described by an initial-boundary value problem for a partial differential equation, not solvable with respect to the time fractional derivative.

Let $\alpha \in (1,2)$, $a_k \in C([0,T]; \mathbb{R})$, $k \in \mathbb{N}$, $\sup_{k \in \mathbb{N}} |a_k(t)| < \infty$ for every $t \in [0,T]$, $v_0, v_1 \in H_0^2(0, \pi) := \{x \in H^2(0,\pi) : x(0) = x(\pi) = 0\}$. Consider the initial-boundary value problem

$$v(0,t) = v(\pi, t) = v_{\xi\xi}(0, t) = v_{\xi\xi}(\pi, t) = 0, \quad t \in (0, T], \tag{25}$$

$$v(\xi, 0) + v_{\xi\xi}(\xi, 0) = v_0(\xi), \quad \xi \in (0, \pi), \tag{26}$$

$$v_t(\xi, 0) + v_{\xi\xi t}(\xi, 0) = v_1(\xi), \quad \xi \in (0, \pi), \tag{27}$$

to the equation

$$D_t^\alpha (v + v_{\xi\xi}) = v_{\xi\xi} + 2 v_{\xi\xi\xi\xi} + \sum_{k=1}^\infty a_k(t) \langle u(\eta, t), \sin k\eta \rangle_{L_2(0,\pi)} \sin k\xi, \quad (\xi, t) \in (0, \pi) \times (0, T]. \tag{28}$$

Choose $\mathcal{X} = H_0^2(0, \pi)$, $\mathcal{Y} = \mathcal{U} = L_2(0, \pi)$, $L = 1 + \frac{\partial^2}{\partial \xi^2} \in \mathcal{L}(\mathcal{X}; \mathcal{Y})$,

$$\mathcal{D}_M = H_0^4(0, \pi) := \left\{ x \in H^4(0, \pi) : x(0) = x(\pi) = x''(0) = x''(\pi) = 0 \right\},$$

$$M = \frac{\partial^2}{\partial \xi^2} + 2 \frac{\partial^4}{\partial \xi^4} \in \mathcal{C}l(\mathcal{X}; \mathcal{Y}), \quad B(t) = \sum_{k=1}^\infty a_k(t) \langle \cdot, \sin k\eta \rangle_{L_2(0,\pi)} \sin k\xi \in \mathcal{L}(\mathcal{U}; \mathcal{Y}), \ t \in [0,T].$$

Thus, problem (25)–(28) has form (17) and (18) with $g \equiv 0$. Here we have $\ker L = \text{span}\{\sin \xi\} \neq \{0\}$, hence Equation (28) is degenerate.

It is known that the set $\{\sqrt{2/\pi} \sin k\xi : k \in \mathbb{N}\}$ is the orthonormal basis in $L_2(0, \pi)$ of eigenfunctions of the operator $\frac{\partial^2}{\partial \xi^2}$ with domain $H_0^2(0, \pi)$, which correspond to the eigenvalues $\{-k^2 : k \in \mathbb{N}\}$. Since the polynomials $1 + \lambda$ and $\lambda + 2\lambda^2$ have no common roots, by Theorem 7 [22] the operator $L_1 : \mathcal{X}^1 \to \mathcal{Y}^1$ is a homeomorphism and for $\alpha \in [1,2)$ there exist $\theta_0 \in (\pi/2, \pi)$, $a_0 \geq 0$, such that $(L, M) \in \mathcal{H}_\alpha(\theta_0, a_0)$. Besides, from Theorem 7 [22] it follows, that $\mathcal{X}^0 = \mathcal{Y}^0 = \text{span}\{\sin \xi\}$, $\mathcal{Y}^1$ is the closure of span$\{\sin k\xi : k = 2, 3, \dots\}$ in $L_2(0, \pi)$, $\mathcal{X}^1$ is the closure of the same set in $H_0^2(0, \pi)$.

By Lemma 3 subsystem (21) is controllable in time T, if and only if $a_1(T) \neq 0$. Besides, it is controllable in free time if and only if $a_1 \not\equiv 0$ on $\mathbb{R}_+$. In the both cases we can say about the exact controllability on the one-dimensional space $\mathcal{X}^0$.

For $y \in L_2(0, \pi)$ we have

$$Y_{\alpha-1}(t)y = \sum_{k=2}^{\infty} y_k \sin k\xi \frac{1}{2\pi i} \int_{\Gamma} \frac{e^{\mu t} d\mu}{\mu^{\alpha} - \mu_k} = \sum_{k=2}^{\infty} y_k \sin k\xi \frac{1}{2\pi i} \int_{t\Gamma} \frac{t^{\alpha-1} e^{\lambda} d\lambda}{\lambda^{\alpha} - t^{\alpha} \mu_k} =$$

$$= \sum_{k=2}^{\infty} y_k \sin k\xi \sum_{n=0}^{\infty} t^{\alpha(n+1)-1} \mu_k^n \frac{1}{2\pi i} \int_{t\Gamma} e^{\lambda} \lambda^{-\alpha(n+1)} d\lambda =$$

$$= \sum_{k=2}^{\infty} y_k \sin k\xi \sum_{n=0}^{\infty} \frac{t^{\alpha(n+1)-1} \mu_k^n}{\Gamma(\alpha(n+1))} = \sum_{k=2}^{\infty} t^{\alpha-1} E_{\alpha,\alpha}(\mu_k t^{\alpha}) y_k \sin k\xi,$$

where

$$y_k = \langle y(\eta), \sin k\eta \rangle_{L_2(0,\pi)}, \quad \mu_k = \frac{2k^4 - k^2}{1 - k^2}, \quad E_{\alpha,\beta}(z) = \sum_{n=0}^{\infty} \frac{z^n}{\Gamma(\alpha n + \beta)}$$

is the Mittag-Leffler function. So,

$$Y_{\alpha-1}(T-s)QB(s) = \sum_{k=2}^{\infty} a_k(s)(T-s)^{\alpha-1} E_{\alpha,\alpha}(\mu_k(T-s)^{\alpha}) \langle \cdot, \sin k\eta \rangle_{L_2(0,\pi)} \sin k\xi,$$

therefore, subsystem (20) is approximately controllable in time T, if and only if for every $k \in \mathbb{N} \setminus \{1\}$ there exists $s_k \in (0, T)$, such that

$$a_k(s_k) E_{\alpha,\alpha} \left(\frac{2k^4 - k^2}{1 - k^2} (T - s_k)^{\alpha} \right) \neq 0.$$

Since $E_{\alpha,\alpha}$ is the entire function and has isolated zeros only, such a condition is equivalent to the condition: $a_k \not\equiv 0$ on $[0, T]$ for every $k \in \mathbb{N} \setminus \{1\}$.

Analogously, subsystem (20) is approximately controllable in free time if and only if $a_k \not\equiv 0$ on $\mathbb{R}_+$ for all $k \in \mathbb{N} \setminus \{1\}$.

Moreover, it is easy to check that

$$\sup_{k=2,3,\ldots} \frac{2k^4 - k^2}{1 - k^2} \leq 0,$$

therefore, $(L, M) \in \mathcal{H}_{\alpha}(\theta_0, a_0)$ for some $\theta_0 \in (\pi/2, \pi)$, $a_0 \geq 0$ in the case $\alpha \in (0, 1]$ (see Theorem 7 [22]). Hence we can study problem (25), (26) and (28) with $\alpha \in (0, 1]$ analogously.

Proposition 1. *Let $\alpha \in (0, 2)$. System (25) and (28) is approximately controllable in time T if and only if $a_1(T) \neq 0$ and for every $k \in \mathbb{N} \setminus \{1\}$ $a_k \not\equiv 0$ on $[0, T]$.*

Analogously, we can obtain the next assertion by the obvious way.

Proposition 2. *Let $\alpha \in (0, 2)$. System (25) and (28) is approximately controllable in free time, if and only if $a_k \not\equiv 0$ on $\mathbb{R}_+$ for all $k \in \mathbb{N}$.*

7. Approximate Controllability of Systems with Finite-Dimensional Input

Let $g : [0, T] \to \mathcal{Y}$, $b_i \in \mathcal{Y}$, $i = 1, 2, \ldots, n$, be given. Consider the control system

$$D_t^{\alpha} Lx(t) = Mx(t) + \sum_{i=1}^{n} b_i u_i(t) + g(t), \qquad (29)$$

where $u_i : [0, T] \to \mathbb{R}$, $i = 1, 2, \ldots, n$. It is a partial case of system (18). Indeed, we can take $\mathcal{U} = \mathbb{R}^n$, $u = (u_1, u_2, \ldots, u_n)$, $Bu(t) = \sum_{i=1}^{n} b_i u_i(t)$. Such a control system is called a system with finite-dimensional input. It is evident that $B \in \mathcal{L}(\mathbb{R}^n; \mathcal{Y})$. Control function $u = (u_1, \ldots, u_n)$ will be chosen from the space $C^\gamma([0, T]; \mathbb{R}^n)$. Theorem 10 and Theorem 11 implies the next assertion.

Corollary 1. *Let Banach spaces $\mathcal{X}$, $\mathcal{Y}$ be reflexive, $(L, M) \in \mathcal{H}_\alpha(\theta_0, a_0)$, $L_1^{-1} \in \mathcal{L}(\mathcal{Y}^1; \mathcal{X}^1)$, $b_i \in \mathcal{Y}$, $i = 1, 2, \ldots, n$, $g \in C_Q^\gamma([0, T]; \mathcal{Y})$ for some $\gamma \in (0, 1]$. Then*

(i) *system (29) is approximately controllable in time T if and only if*

$$\text{span}\{(I - Q)b_i, i = 1, 2, \ldots, n\} = \mathcal{Y}^0, \quad \overline{\text{span}}\{Y_{\alpha-1}(s)Qb_i, 0 < s < T, i = 1, 2, \ldots, n\} = \mathcal{Y}^1.$$

(ii) *system (29) is approximately controllable in free time if and only if*

$$\text{span}\{(I - Q)b_i, i = 1, 2, \ldots, n\} = \mathcal{Y}^0, \quad \overline{\text{span}}\{Y_{\alpha-1}(s)Qb_i, s \in \mathbb{R}_+, i = 1, 2, \ldots, n\} = \mathcal{Y}^1.$$

Proof. By Theorem 10 the condition $\mathcal{X}^0 = \overline{\text{im}} M_0^{-1}(I - Q)B = \overline{\text{span}}\left\{M_0^{-1}(I - Q)b_i, i = 1, 2, \ldots, n\right\}$ is necessary and sufficient for the approximate controllability in time T of the subsystem on the subspace $\mathcal{X}^0$. This set is finite-dimensional, and the operator M_0 is densely defined, therefore

$$\mathcal{X}^0 = \text{span}\left\{M_0^{-1}(I - Q)b_i, i = 1, 2, \ldots, n\right\} = \mathcal{D}_{M_0},$$

it is equivalent to the equality $\mathcal{Y}^0 = M[\mathcal{D}_{M_0}] = \text{span}\{(I - Q)b_i, i = 1, 2, \ldots, n\}$. Other equalities follow from Theorems 10 and 11 in an obvious way. □

Remark 13. *So, we see that under the conditions of Corollary 1 from the approximate controllability of system (29) it follows that $\dim \mathcal{X}^0 = \dim \mathcal{Y}^0 \leq n$.*

Remark 14. *In the conditions of Corollary 1 from the approximate controllability of system (29) it follows that $M_0 \in \mathcal{L}(\mathcal{X}^0; \mathcal{Y}^0)$, since $\mathcal{D}_{M_0} = \mathcal{X}^0$ and the operator M_0 is closed.*

Let $\triangle$ be the Laplace operator and the system with one-dimensional input be described by the equation

$$D_t^\alpha (5v + \triangle v) = \triangle v + 2\triangle^2 v + b(\xi, \eta) u(t), \quad (\xi, \eta, t) \in (0, \pi) \times (0, \pi) \times (0, T], \quad (30)$$

with initial conditions of form (26) at $\alpha \in (0, 1]$, or of form (26), (27) at $\alpha \in (1, 2)$, defined on $(0, \pi) \times (0, \pi)$ and with boundary conditions of the form

$$v(0, \eta, t) = v(\pi, \eta, t) = v(\xi, 0, t) = v(\xi, \pi, t) = 0, \quad \xi, \eta \in (0, \pi), \, t \in (0, T], \quad (31)$$

$$\triangle v(0, \eta, t) = \triangle v(\pi, \eta, t) = \triangle v(\xi, 0, t) = \triangle v(\xi, \pi, t) = 0, \quad \xi, \eta \in (0, \pi), \, t \in (0, T]. \quad (32)$$

Here $b \in L_2((0, \pi) \times (0, \pi))$. Reasoning as in Section 6, we see that system (30)–(32) is not controllable in free time even, since the subspace $\mathcal{Y}^0 = \text{span}\{\sin \xi \sin 2\eta, \sin 2\xi \sin \eta\}$ is two-dimensional, and the condition $\text{span}\{(I - Q)b\} = \mathcal{Y}^0$ can not be satisfied.

8. Conclusions

Thus, the work obtained the necessary and sufficient conditions for approximate controllability for a class of degenerate fractional order evolution equations in terms of operators from the equation. The cases of infinite-dimensional and finite-dimensional input were studied. Using the concrete control systems described by the initial-boundary value problems for the partial differential equations, the applications of the obtained abstract results were demonstrated.

Author Contributions: Conceptualization, D.B.; methodology, D.B. and V.E.F.; validation, D.M.G. and K.T.; formal analysis, V.E.F., D.M.G. and K.T.; investigation, D.M.G. and K.T.; Writing—Original Draft preparation, V.E.F.; Writing—Review and Editing, D.B. and V.E.F.; supervision, D.B. and V.E.F.; project administration, V.E.F. and D.M.G.

Funding: The reported study was funded by the Russian Foundation for Basic Research, project number 19-41-450001; by Act 211 of Government of the Russian Federation, contract 02.A03.21.0011; and by Ministry of Science and Higher Education of the Russian Federation, task number 1.6462.2017/BCh.

Conflicts of Interest: The authors declare no conflict of interest.

References

1. Kalman, R.E.; Ho, Y.S.; Narendra, K.S. Controllability of linear dynamical systems. *Contrib. Differ. Equ.* **1963**, *1*, 189–213.
2. Krasovskii, N.N. On the theory of controllability and observability of linear dynamic systems. *J. Appl. Math. Mech.* **1964**, *28*, 1–14. [CrossRef]
3. Fattorini, H.O. On complete controllability of linear systems. *J. Differ. Equ.* **1967**, *3*, 391–402. [CrossRef]
4. Kurzhanskiy, A.B. Towards controllability in Banach spaces. *Differ. Equ.* **1969**, *5*, 1715–1718. (In Russian)
5. Triggiani, R. Controllability and observability in Banach space with bounded operators. *SIAM J. Control* **1975**, *13*, 462–491. [CrossRef]
6. Curtain, R.F. The Salamon—Weiss class of well-posed infinite dimensional linear systems: A survey. *IMA J. Math. Control Inf.* **1997**, *14*, 207–223. [CrossRef]
7. Sholokhovich, F.A. On controllability of linear dynamical systems. *News Ural State Univ.* **1998**, *10*, 103–126. (In Russian)
8. Debbouche, A.; Baleanu, D. Controllability of fractional evolution nonlocal impulsive quasilinear delay integro-differential systems. *Comput. Math. Appl.* **2011**, *62*, 1442–1450. [CrossRef]
9. Chalishajar, D.N.; Malar, K.; Karthikeyan, K. Approximate controllability of abstract impulsive fractional neutral evolution equations with infinite delay in Banach spaces. *Electron. J. Differ. Equ.* **2013**, *2013*, 1–21.
10. Fedorov, V.E.; Ruzakova, O.A. Controllability of linear Sobolev type equations with relatively p-radial operators. *Russ. Math.* **2002**, *46*, 54–57.
11. Fedorov, V.E.; Ruzakova, O.A. Controllability in dimensions of one and two of Sobolev-type equations in Banach spaces. *Math. Notes* **2003**, *74*, 583–592. [CrossRef]
12. Fedorov, V.E.; Shklyar, B. Exact null controllability of degenerate evolution equations with scalar control. *Sbornik Math.* **2012**, *203*, 1817–1836. [CrossRef]
13. Plekhanova, M.V.; Fedorov, V.E. *Optimal Control for Degenerate Distributed Systems*; Publishing Center of South Ural State University: Chelyabinsk, Russia, 2013. (In Russian)
14. Plekhanova, M.V.; Fedorov, V.E. On controllability of degenerate distributed systems. *Ufa Math. J.* **2014**, *6*, 77–96. [CrossRef]
15. Fedorov, V.E.; Gordievskikh, D.M.; Baybulatova, G.D. Controllability of a class of weakly degenerate fractional order evolution equations. *AIP Conf. Proc.* **2017**, *1907*, 020009-1–020009-14.
16. Fedorov, V.E.; Gordievskikh, D.M.; Turov, M.M. Infinite-dimensional and finite-dimensional ε-controllability for a class of fractional order degenerate evolution equations. *Chelyabinsk Phys. Math. J.* **2018**, *3*, 5–26. (In Russian)
17. Fedorov, V.E.; Gordievskikh, D.M. Approximate controllability of strongly degenerate fractional order system of distributed control. In Proceedings of the 17th IFAC Workshop on Control Applications of Optimization CAO 2018, Yekaterinburg, Russia, 15–19 October 2018; Volume 51, pp. 675–680.
18. Plekhanova, M.V. Distributed control problems for a class of degenerate semilinear evolution equations. *J. Comput. Appl. Math.* **2017**, *312*, 39–46. [CrossRef]
19. Plekhanova, M.V. Optimal control for quasilinear degenerate distributed systems of higher order. *J. Math. Sci.* **2016**, *19*, 236–244. [CrossRef]
20. Bajlekova, E.G. Fractional Evolution Equations in Banach Spaces. Ph.D. Thesis, University Press Facilities, Eindhoven University of Technology, Eindhoven, The Netherlands, 2001.

21. Prüss, J. *Evolutionary Integral Equations and Applications*; Springer: Basel, Switzerland, 1993.
22. Fedorov, V.E.; Romanova, E.A.; Debbouche, A. Analytic in a sector resolving families of operators for degenerate evolution fractional equations. *J. Math. Sci.* **2018**, *228*, 380–394. [CrossRef]
23. Fedorov, V.E. A class of fractional order semilinear evolutions in Banach spaces. In *Integral Equations and Their Applications, Proceedings of the University Network Seminar on the Occasion of the Third Mongolia—Russia—Vietnam Workshop on NSIDE 2018, Hung Yen, Vietnam, 27–28 October 2018*; Hanoi Mathematical Society, Hung Yen University of Technology and Education: Hung Yen, Vietnam, 2011; pp. 11–20.
24. Fedorov, V.E.; Romanova, E.A. Inhomogeneous evolution equations of fractional order in the sectorial case. *Itogi Nauki i Tekhniki. Contemp. Math. and its Appl. Thematic Reviews.* **2018**, *149*, 103–112. (In Russian)
25. Fedorov, V.E.; Gordievskikh, D.M.; Plekhanova, M.V. Equations in Banach spaces with a degenerate operator under a fractional derivative. *Differ. Equ.* **2015**, *51*, 1360–1368. [CrossRef]

© 2019 by the authors. Licensee MDPI, Basel, Switzerland. This article is an open access article distributed under the terms and conditions of the Creative Commons Attribution (CC BY) license (http://creativecommons.org/licenses/by/4.0/).

Article

On the Solution of an Imprecisely Defined Nonlinear Time-Fractional Dynamical Model of Marriage

Rajarama Mohan Jena [1], Snehashish Chakraverty [1] and Dumitru Baleanu [2,3,*]

1 Department of Mathematics, National Institute of Technology Rourkela, Rourkela 769008, India
2 Department of Mathematics, Faculty of Art and Sciences, Cankaya University, Balgat, Ankara 06530, Turkey
3 Institute of Space Sciences, 077125 Magurele-Bucharest, Romania
* Correspondence: dumitru@cankaya.edu.tr

Received: 27 June 2019; Accepted: 25 July 2019; Published: 1 August 2019

Abstract: The present paper investigates the numerical solution of an imprecisely defined nonlinear coupled time-fractional dynamical model of marriage (FDMM). Uncertainties are assumed to exist in the dynamical system parameters, as well as in the initial conditions that are formulated by triangular normalized fuzzy sets. The corresponding fractional dynamical system has first been converted to an interval-based fuzzy nonlinear coupled system with the help of a single-parametric gamma-cut form. Further, the double-parametric form (DPF) of fuzzy numbers has been used to handle the uncertainty. The fractional reduced differential transform method (FRDTM) has been applied to this transformed DPF system for obtaining the approximate solution of the FDMM. Validation of this method was ensured by comparing it with other methods taking the gamma-cut as being equal to one.

Keywords: fractional calculus; triangular fuzzy number; double-parametric form; FRDTM; fractional dynamical model of marriage

1. Introduction

In the present era, fractional-order derivatives have become widespread due to their wide interdisciplinary applications and implementation in various fields of science and technology, such as solid mechanics, fluid dynamics, financial mathematics, social sciences, and other areas of science and engineering (see References [1–5]). As the solutions of non-integer order differential equations are more complicated than integer-order differential equations, computationally efficient and reliable numerical methods need to be developed to handle these. Authors have written different books (see References [6–10]) in which various studies and analyses on fractional calculus may be found that will support the authors for better understanding of the concepts of fractional calculus.

The hypothesis of entropy has been connected formerly with thermodynamics only; however, in present-day, it has additionally been utilized in different areas like data hypothesis, psychodynamics, biophysical financial aspects, human relations, etc. The second law of thermodynamics expresses that entropy increases with time. It demonstrates the unpredictability of a structure over some time if there is nothing to balance out it. Likewise, in human interactions, every day, various associations lead to some turmoil. Recently, the discussion of the titled model has been attaining recognition throughout the past few years. Relational relations emerge from numerous points of view, for instance, marriage, blood relations, close attachments, work, clubs [11,12], and so forth. Many authors have studied various research related to FDMM. The nonlinear coupled fractional FDMM was first investigated by Ozalp and Koca [13]. In that paper, they performed a balance situation for equilibrium points. Khader and Alqahtani [14] applied the Bernstein collocation method for obtaining the solution of a nonlinear FDMM, and they also compared their results with the Runge–Kutta fourth-order method. They defined the fractional derivative in the Riemann–Liouville sense, and via the utilization of Bernstein polynomials, they converted the FDMM to a system of nonlinear algebraic equations, which were solved using Newton's iterative method. Khader et al. [15] also solved the same model by implementing the Legendre spectral collocation method and affirmed the natural behavior of the present system. Singh et al. [16] implemented a q-homotopy analysis method coupled with Sumudu transform and Adomian decomposition method to solve FDMM and comparison results with the existing literature are also included. Goyal et al. [17] studied the FDMM utilizing a variation iteration method and a homotopy perturbation transform method.

Few authors have scientifically investigated the causes of extramarital interactions in marriage. It is essential and challenging to find out why some wedded couples separate, while a few couples do not. Moreover, among wedded couples, a few are fulfilled, while some are not fulfilled with each other. As such, the number of divorce cases are increasing every day all over the globe. A survey inside the U.S. uncovered that inside a forty-year interval, the probability of a first marriage finishing in separation are roughly 50 to 67 percent. The record is 10 percent higher for a second marriage. Around the world, the U.S. has the highest divorce rate. In this regard, experiments may be tough to conduct and may also be restricted for personal concerns, and so a mathematical model happens to be advantageous. As such, recently, researchers are investigating different dynamical models for interpersonal relations.

The most recent model of marriage is the Romeo and Juliet model [18]. Assume that at any moment t, we need to determine Romeo's adoration or loathing for Juliet, $R(t)$ and Juliet's affection or hate for Romeo, $J(t)$. Positive estimations of these propose love, and negative values specify hate.

The presumption about this model is that the change in Romeo's adoration for Juliet is a small amount of his present love in addition to a small amount of her present love. Also, Juliet's affection for Romeo will change by a small amount of her present love for Romeo and a small amount of Romeo's adoration for her. This presumption prompts the model as given below [18,19]:

$$\begin{aligned}\frac{dR}{dt} &= aR(t) + bJ(t).\\ \frac{dJ}{dt} &= cR(t) + dJ(t).\end{aligned} \quad (1)$$

where a, b, c, and d are constants.

Gottman et al. [20] studied the discrete dynamical model to characterize the connection between them. Since the layouts of research in those fields are cumbersome and restrained through the moral reflections, mathematical models may furthermore have a fundamental influence in considering the elements of relations and conduct highlights. A few models are present for describing the romantic relationship; however, they may be limited to integer-order differential equations.

An integer order mathematical model of love is given as follows:

$$\begin{aligned}\frac{d\psi}{dt} &= -a_1\psi + b_1\xi\left(1 - \delta\xi^2\right) + c_1.\\ \frac{d\xi}{dt} &= -a_2\xi + b_2\psi\left(1 - \delta\psi^2\right) + c_2.\end{aligned} \quad (2)$$

Here, variables ψ and ξ measure the adoration of a man or woman for his/her partner. The parameters $a_i, b_i, c_i (1 \leq i \leq 2)$ denote the oblivion, reaction, and attraction constants. We have measured that the decay of the feelings for one's partner occurs exponentially quickly within the absence of a partner. The parameter a_i indicates the degree to which one is stimulated by way of one's personal feeling. It is used as a level of dependency along with fretfulness regarding other's affirmation in relationships. The parameter b_i represents the level to which one is supported by one's partner and additionally anticipates him/her to be useful. It measures the tendency to keep away from or seek closeness in a relationship. The term $-a_i\psi$ and $-a_i\xi$ state that one's adoration measure decays exponentially without one's partner, $1/a_i$ suggests the time needed for love to diminish and δ is a compensatory constant.

In the present study, a time-fractional order dynamical system has been considered instead of its integer order system because fractional order equations are generalizations of integer order differential equations and fractional order models hold memory. Interpersonal relationships are influenced by memory, which makes the modeling more appropriate than the integer one for this kind of dynamical system. This fact confirms that fractional modeling is best suited for this kind of system. Hence, the investigation of the time-fractional systems is significant. The FDMM is given as:

$$\begin{aligned} \frac{d^\alpha \psi}{dt^\alpha} &= -a_1\psi + b_1\xi\left(1 - \delta\xi^2\right) + c_1. \\ \frac{d^\alpha \xi}{dt^\alpha} &= -a_2\xi + b_2\psi\left(1 - \delta\psi^2\right) + c_2. \end{aligned} \quad (3)$$

where $0 < \alpha \le 1$ $a_i \ge 0$

with initial conditions (ICs):

$$\psi(0) = 0 = \xi(0) \quad (4)$$

It is observed that all the authors mentioned above have considered the parameters and variables involved in FDMM as crisp or precise. However, in real life, it may not always be possible to take crisp values due to errors in experiments, observations, and many other errors. Therefore, the parameters and variables may be considered as uncertain. Here, the uncertainties are considered as intervals/fuzzy. The parameters $a_i, b_i, c_i (1 \le i \le 2)$ and δ denote the oblivion, reaction, attraction, and compensatory constants, respectively. As these parameters are related to attractions and reactions of the model, its values may not always be fixed. As such, the main targets of the authors are to consider these parameters as fuzzy and then solve this fuzzy fractional model using an efficient method.

Let us consider the coupled fuzzy FDMM as given below:

$$\begin{aligned} \frac{d^\alpha \widetilde{\psi}}{dt^\alpha} &= -(a_1 - 0.02, a_1, a_1 + 0.02)\widetilde{\psi} + (b_1 - 0.02, b_1, b_1 + 0.02)\widetilde{\xi} \\ &\quad \left\{1 - (\delta - 0.01, \delta, \delta + 0.01)\widetilde{\xi}^2\right\} + (c_1 - 0.2, c_1, c_1 + 0.2). \\ \frac{d^\alpha \widetilde{\xi}}{dt^\alpha} &= -(a_2 - 0.02, a_2, a_2 + 0.02)\widetilde{\xi} + (b_2 - 0.02, b_2, b_2 + 0.02)\widetilde{\psi} \\ &\quad \left\{1 - (\delta - 0.01, \delta, \delta + 0.01)\widetilde{\psi}^2\right\} + (c_2 - 0.2, c_2, c_2 + 0.2). \end{aligned} \quad (5)$$

with fuzzy ICs

$$\widetilde{\psi}(0) = \widetilde{\xi}(0) = (-0.1, 0, 0.1) \quad (6)$$

where variables $\widetilde{\psi}$ and $\widetilde{\xi}$ describe the uncertain adoration of a man or woman for his/her partner.

The basic concepts of fuzzy variables were first presented by Chang and Zadeh [21], where they suggested the theory of a fuzzy derivative. The extensive analysis in Chang and Zadeh [21] was well-defined and studied by Dubois and Prade [22]. Kaleva [23] and Seikkala [24] studied the fuzzy differential equations (FDEs) and initial value problems. Various problems related to the differential FDEs are broadly studied by Chakraverty et al. (see References [25–27]). As fuzzy fractional differential equations (FFDEs) are quite challenging to solve as compared to fractional differential equations, computationally efficient numerical methods should be developed. In this research, we have applied a fractional reduced differential transform method (FRDTM) along with imprecisely defined parameters involved in the FDMM in order to study this dynamical system. Also, the convergence analysis of the present solution has been discussed with an increasing number of terms of the solution. The double-parametric form of a fuzzy number is applied to find the solution of the fractional fuzzy dynamical model of marriage. This model has not yet been studied using FRDTM. The main benefit of using this technique are: First, this procedure achieves the expansions of the solutions. Second, this technique does not require any discretization, perturbations, or modification of the ICs. Also, this technique needs fewer computations with high precision, as well as less time compared to other techniques. In view of the above literature, FFDEs are first changed to a differential equation using a double-parametric form (DPF). Then, the equivalent equation is solved using FRDTM to have an interval/fuzzy solution in terms of the DPF.

The remaining parts of the manuscript are arranged as follows. In the "Preliminaries" section, we give essential information related to fuzzy arithmetic, triangular fuzzy number, and double-parametric form of a fuzzy number. In section "Fractional Reduced Differential Transform Method," we discuss methodology and important theorems related to this technique. The double-parametric form-based solution of FDMM is given in section "Double-Parametric-Based Solution of Uncertainty FDMM Using FRDTM." Next, numerical outcomes and deliberations are given in the "Results and Discussions" section. Finally, conclusions are drawn.

2. Preliminaries

In this segment, some basic definitions, and notations of fuzzy variables are discussed (see References [25–27]).

Definition 1. **(Fuzzy Number)** *A fuzzy number $\widetilde{\psi}$ is a convex normalized fuzzy set $\widetilde{\psi}$ of the real line $\Re$ such that:*

$$\left\{\mu_{\widetilde{\psi}}(x) : \Re \to [0, 1], \forall x \in \Re\right\}$$

where $\mu_{\widetilde{\psi}}$ is a membership function and is piecewise continuous.

Definition 2. **(Triangular Fuzzy Number)** *A triangular fuzzy number $\widetilde{\psi}$ is a convex normalized fuzzy set $\widetilde{\psi}$ of the real line $\Re$ such that:*

(a) *There exists exactly one $x_0 \in \Re$ with $\mu_{\widetilde{\psi}}(x_0)$ (x_0 is called the mean value of $\widetilde{\psi}$), where $\mu_{\widetilde{\psi}}$ is called the membership function of the fuzzy set.*
(b) *$\mu_{\widetilde{\psi}}(x)$ is piecewise continuous.*

The membership function $\mu_{\widetilde{\psi}}$ of a triangular fuzzy number $\widetilde{\psi} = (a_1, b_1, c_1)$ is defined as:

$$\mu_{\widetilde{\psi}}(x) = \begin{cases} 0, & x \leq a_1, \\ \frac{x-a_1}{b_1-a_1}, & a_1 \leq x \leq b_1, \\ \frac{c_1-x}{c_1-b_1}, & b_1 \leq x \leq c_1, \\ 0, & x \geq c_1. \end{cases}$$

Definition 3. **(Single-Parametric Form of Fuzzy Numbers)** *The triangular fuzzy number $\widetilde{\psi} = (a_1, b_1, c_1)$ can be characterized by an ordered pair of functions through the γ-cut approach $\lfloor \underline{\psi}(\gamma), \overline{\psi}(\gamma) \rfloor = [(b_1 - a_1)\gamma + a_1, -(c_1 - b_1)\gamma + c_1]$, where $\gamma \in [0, 1]$. The γ-cut form is well-known as the single-parametric form of fuzzy numbers. It is observed that the lower and upper bounds of the fuzzy numbers satisfy the below statements:*

(i) *$\underline{\psi}(\gamma)$ is a left-bounded nondecreasing continuous function over $[0, 1]$.*
(ii) *$\overline{\psi}(\gamma)$ is a right-bounded nonincreasing continuous function over $[0, 1]$.*
(iii) *$\underline{\psi}(\gamma) \leq \overline{\psi}(\gamma)$, where $0 \leq \gamma \leq 1$.*

Definition 4. **(Double-Parametric Form of Fuzzy Number)** *Using the single-parametric form, as discussed in Definition 3, we have $\widetilde{\psi} = [\underline{\psi}(\gamma), \overline{\psi}(\gamma)]$.*

Now we can write this as crisp with DPF as:

$$\widetilde{\psi}(\gamma, \beta) = \beta\bigl(\overline{\psi}(\gamma) - \underline{\psi}(\gamma)\bigr) + \underline{\psi}(\gamma)$$

where γ and $\beta \in [0, 1]$.

Definition 5. **(Fuzzy Arithmetic)** *For arbitrary fuzzy numbers $\widetilde{x} = \lfloor \underline{x}(\gamma), \overline{x}(\gamma) \rfloor$, $\widetilde{y} = \lfloor \underline{y}(\gamma), \overline{y}(\gamma) \rfloor$ and scalar m, fuzzy arithmetics are well-defined as below:*

(i) *$\widetilde{x} = \widetilde{y}$ if and only if $\underline{x}(\gamma) = \underline{y}(\gamma)$ and $\overline{x}(\gamma) = \overline{y}(\gamma)$.*

(ii) $\tilde{x} + \tilde{y} = \lfloor \underline{x}(\gamma) + \underline{y}(\gamma), \overline{x}(\gamma) + \overline{y}(\gamma) \rfloor$.

(iii) $\tilde{x} \times \tilde{y} = \begin{bmatrix} \min\bigl(\underline{x}(\gamma) \times \underline{y}(\gamma),\ \underline{x}(\gamma) \times \overline{y}(\gamma), \overline{x}(\gamma) \times \underline{y}(\gamma),\ \overline{x}(\gamma) \times \overline{y}(\gamma)\bigr), \\ \max\bigl(\underline{x}(\gamma) \times \underline{y}(\gamma),\ \underline{x}(\gamma) \times \overline{y}(\gamma), \overline{x}(\gamma) \times \underline{y}(\gamma),\ \overline{x}(\gamma) \times \overline{y}(\gamma)\bigr) \end{bmatrix}$.

(iv) $k\tilde{x} = \begin{cases} [k\overline{x}(\gamma), k\underline{x}(\gamma)], & k < 0 \\ [k\underline{x}(\gamma), k\overline{x}(\gamma)], & k \geq 0 \end{cases}$.

3. Fractional Reduced Differential Transform Method

Let us take an analytic and k-times continuously differentiable function $\psi(x,t)$. Assume that $\psi(x,t)$ is denoted as a product of two functions as $\psi(x,t) = a(x)b(t)$. From Momani and Odibat [28], this function is written as follows

$$\psi(x,t) = \left(\sum_{m=0}^{\infty} A(m)x^m\right)\left(\sum_{n=0}^{\infty} B(n)t^n\right) = \sum_{m=0}^{\infty}\sum_{n=0}^{\infty} F(m,n)\, x^m t^n \tag{7}$$

where $F(m,n) = A(m)B(n)$ is named as the spectrum of $\psi(x,t)$.

Lemma 1. *The fractional reduced differential transform (FRDT) of an analytic function $\psi(x,t)$ is defined as:*

$$\psi_k(x) = \frac{1}{\Gamma(1+\alpha k)}[D_t^{\alpha k}\psi(x,t)]_{t=t_0} \text{ for } k = 0, 1, 2, \ldots \tag{8}$$

The inverse transform of $\psi_k(x)$ is well-defined as

$$\psi(x,t) = \sum_{k=0}^{\infty} \psi_k(x)(t-t_0)^{\alpha k} \tag{9}$$

From Equations (8) and (9), we obtain:

$$\psi(x,t) = \sum_{k=0}^{\infty} \frac{1}{\Gamma(\alpha k+1)}[D_t^{\alpha k}\psi(x,t)]_{t=t_0}(t-t_0)^{\alpha k} \tag{10}$$

In particular, at $t_0 = 0$, we have:

$$\psi(x,t) = \sum_{k=0}^{\infty} \frac{1}{\Gamma(\alpha k+1)}[D_t^{\alpha k}\psi(x,t)]_{t=0}\, t^{\alpha k} \tag{11}$$

Theorem 1. *Let $\psi(x,t)$, $\xi(x,t)$, and $\zeta(x,t)$ be three analytical functions such that $\psi(x,t) = R_D^{-1}[\psi_k(x)]$, $\xi(x,t) = R_D^{-1}[\xi_k(x)]$, and $\zeta(x,t) = R_D^{-1}[\zeta_k(x)]$. Hence from References [29–32]):*

(i) If $\psi(x,t) = c_1\xi(x,t) \pm c_2\zeta(x,t)$, then $\psi_k(x) = c_1\xi_k(x) \pm c_2\zeta_k(x)$, where c_1 and c_2 are constants.

(ii) If $\psi(x,t) = a\,\xi(x,t)$, then $\psi_k(x) = a\,\xi_k(x)$.

(iii) If $\psi(x,t) = x^m t^n$, then $\psi_k(x) = x^m \delta(k-n)$ where $\delta(k) = \begin{cases} 1, & k = 0 \\ 0, & k \neq 0 \end{cases}$.

(iv) If $\psi(x,t) = x^m t^n \xi(x,t)$, then $\psi_k(x) = x^m \xi_{k-n}(x)$.

(v) If $\psi(x,t) = \xi(x,t)\zeta(x,t)$, then $\psi_k(x) = \sum_{i=0}^{j} \xi_i(x)\zeta_{j-i}(x) = \sum_{i=0}^{j} \zeta_i(x)\xi_{j-i}(x)$.

(vi) If $\psi(x,t) = \xi(x,t)\zeta(x,t)\varsigma(x,t)$, then $\psi_k(x) = \sum_{j=0}^{k}\sum_{i=0}^{j} \xi_i(x)\zeta_{j-i}(x)\varsigma_{k-j}(x)$.

Theorem 2. Let $\psi(x,t)$ and $\xi(x,t)$ are two analytical functions such that $\psi(x,t) = R_D^{-1}[\psi_k(x)]$, and $\xi(x,t) = R_D^{-1}[\xi_k(x)]$. Hence:

(i) If $\psi(x,t) = \frac{\partial^m}{\partial x^m}\xi(x,t)$, then $\psi_k(x) = \frac{\partial^m}{\partial x^m}\xi_k(x)$.

(ii) If $\psi(x,t) = \frac{\partial^{na}}{\partial t^{na}}\xi(x,t)$, then $\psi_k(x) = \frac{\Gamma(1+(k+n)\alpha)}{(1+k\alpha)}\xi_{k+n}(x)$.

Corollary 1. If $\zeta(x,t) = e^{\delta t + \theta x}$, then $\zeta_k(x) = \frac{\delta^k}{k!}e^{\theta x}$.

Corollary 2. If $\xi(x,t) = \sin(\theta t + \mu x)$ and $\zeta(x,t) = \cos(\theta t + \mu x)$, then $\xi_k(x) = \frac{\theta^k}{k!}\sin\left(\frac{k\pi}{2} + \mu x\right)$ and $\zeta_k(x) = \frac{\theta^k}{k!}\cos\left(\frac{k\pi}{2} + \mu x\right)$.

In order to explain the concept of FRDTM, let us consider the following equation in the operator form as:

$$L\psi(x,t) + R\psi(x,t) + N\psi(x,t) = h(x,t) \tag{12}$$

with IC:

$$\psi(x,0) = g(x) \tag{13}$$

where $L = \frac{\partial^\alpha}{\partial t^\alpha}$; R, N are linear, nonlinear operators; and $h(x,t)$ is an inhomogeneous source term.
Using Theorem 2 and Equations (8) and (12), this reduces to:

$$\frac{\Gamma(1+\alpha k + \alpha)}{\Gamma(1+\alpha k)}\psi_{k+1}(x) = H_k(x) - R\psi_k(x) - N\psi_k(x) \text{ for } k = 0,1,2\ldots \tag{14}$$

where $\psi_k(x)$ and $H_k(x)$ are the transformed form of $\psi(x,t)$ and $h(x,t)$, respectively.
Applying FRDTM on the IC, we obtain:

$$\psi_0(x) = g(x) \tag{15}$$

Using Equations (14) and (15), $\psi_k(x)$ for $k = 1, 2, 3, \ldots$ can be determined.
Then, taking the inverse transformation of $\{\psi_k(x)\}_{k=0}^n$ gives the n-term approximate solution as:

$$\psi_n(x,t) = \sum_{k=0}^n \psi_k(x) t^{\alpha k} \tag{16}$$

Therefore, the analytical result of Equation (12) is written as $\psi(x,t) = \lim_{n \to \infty}\psi_n(x,t)$.

4. Double-Parametric-Based Solution of an Uncertain FDMM Using FRDTM

To begin with, by applying the single parametric form, the FDMM is changed to an interval-based FDE. At that moment, by applying the DPF, the interval-based FDE is transformed into an FDMM having two parameters that may control the uncertainty. Finally, FRDTM is then applied to solve the corresponding double parametrized FDMM for obtaining the needed solution in terms of intervals/fuzzy variables.

Equations (5) and (6) can now be modified in single-parametric form as:

$$\left[\tfrac{d^\alpha}{dt^\alpha}\underline{\psi}(t;\gamma), \tfrac{d^\alpha}{dt^\alpha}\overline{\psi}(t;\gamma)\right] = -[(0.02\gamma + a_1 - 0.02), (-0.02\gamma + a_1 + 0.02)]\left[\underline{\psi}(t;\gamma), \overline{\psi}(t;\gamma)\right]$$
$$+ [(0.02\gamma + b_1 - 0.02), (-0.02\gamma + b_1 + 0.02)]\left[\underline{\xi}(t;\gamma), \overline{\xi}(t;\gamma)\right]$$
$$\left[1 - \{(0.01\gamma + \delta - 0.01), (-0.01\gamma + \delta + 0.01)\}\{\underline{\xi}(t;\gamma), \overline{\xi}(t;\gamma)\}^2\right]$$
$$+ [(0.2\gamma + c_1 - 0.2), (-0.2\gamma + c_1 + 0.2)]$$

$$\left[\tfrac{d^\alpha}{dt^\alpha}\underline{\xi}(t;\gamma), \tfrac{d^\alpha}{dt^\alpha}\overline{\xi}(t;\gamma)\right] = -[(0.02\gamma + a_2 - 0.02), (-0.02\gamma + a_2 + 0.02)]\left[\underline{\xi}(t;\gamma), \overline{\xi}(t;\gamma)\right] \quad (17)$$
$$+ [(0.02\gamma + b_2 - 0.02), (-0.02\gamma + b_2 + 0.02)]\left[\underline{\psi}(t;\gamma), \overline{\psi}(t;\gamma)\right]$$
$$\left[1 - \{(0.01\gamma + \delta - 0.01), (-0.01\gamma + \delta + 0.01)\}\{\underline{\psi}(t;\gamma), \overline{\psi}(t;\gamma)\}^2\right]$$
$$+ [(0.2\gamma + c_2 - 0.2), (-0.2\gamma + c_2 + 0.2)].$$

with fuzzy ICs:

$$[\underline{\psi}(0;\gamma), \overline{\psi}(0;\gamma)] = [\underline{\xi}(0;\gamma), \overline{\xi}(0;\gamma)] = [0.1\gamma - 0.1, -0.1\gamma + 0.1] \quad (18)$$

where Equations (17) and (18) are in interval form. One can find out the solution of this interval equations, but sometimes it is complicated to handle such types of interval equations. Therefore, one may require the double-parametric form to handle this interval computation. Applying double-parametric form to Equations (17) and (18), we have:

$$\left\{\beta\left(\tfrac{d^\alpha}{dt^\alpha}\overline{\psi}(t;\gamma) - \tfrac{d^\alpha}{dt^\alpha}\underline{\psi}(t;\gamma)\right) + \tfrac{d^\alpha}{dt^\alpha}\underline{\psi}(t;\gamma)\right\} = \{\beta(0.04 - 0.04\gamma) + 0.02\gamma - 0.02 + a_1\}$$
$$\left\{\beta(\overline{\psi}(t;\gamma) - \underline{\psi}(t;\gamma)) + \underline{\psi}(t;\gamma)\right\} + \{\beta(0.04 - 0.04\gamma) + 0.02\gamma + b_1 - 0.02\}$$
$$\left\{\beta(\overline{\xi}(t;\gamma) - \underline{\xi}(t;\gamma)) + \underline{\xi}(t;\gamma)\right\}\left[\begin{array}{c} 1 - \{\beta(0.02 - 0.02\gamma) + 0.01\gamma + \delta - 0.01\} \\ \{\beta(\overline{\xi}(t;\gamma) - \underline{\xi}(t;\gamma)) + \underline{\xi}(t;\gamma)\}^2 \end{array}\right] \quad (19)$$
$$+ \{\beta(0.4 - 0.4\gamma) + 0.2\gamma + c_1 - 0.2\}.$$

$$\left\{\beta\left(\tfrac{d^\alpha}{dt^\alpha}\overline{\xi}(t;\gamma) - \tfrac{d^\alpha}{dt^\alpha}\underline{\xi}(t;\gamma)\right) + \tfrac{d^\alpha}{dt^\alpha}\underline{\xi}(t;\gamma)\right\} = \{\beta(0.04 - 0.04\gamma) + 0.02\gamma - 0.02 + a_2\}$$
$$\left\{\beta(\overline{\xi}(t;\gamma) - \underline{\xi}(t;\gamma)) + \underline{\xi}(t;\gamma)\right\} + \{\beta(0.04 - 0.04\gamma) + 0.02\gamma + b_2 - 0.02\}$$
$$\left\{\beta(\overline{\psi}(t;\gamma) - \underline{\psi}(t;\gamma)) + \underline{\psi}(t;\gamma)\right\}\left[\begin{array}{c} 1 - \{\beta(0.02 - 0.02\gamma) + 0.01\gamma + \delta - 0.01\} \\ \{\beta(\overline{\psi}(t;\gamma) - \underline{\psi}(t;\gamma)) + \underline{\psi}(t;\gamma)\}^2 \end{array}\right] \quad (20)$$
$$+ \{\beta(0.4 - 0.4\gamma) + 0.2\gamma + c_2 - 0.2\}.$$

with fuzzy ICs:

$$\left\{\beta(\overline{\psi}(0;\gamma) - \underline{\psi}(0;\gamma)) + \underline{\psi}(0;\gamma)\right\} = \left\{\beta(\overline{\xi}(0;\gamma) - \underline{\xi}(0;\gamma)) + \underline{\xi}(0;\gamma)\right\} \quad (21)$$
$$= \beta(-0.2\gamma + 0.2) + (0.1\gamma - 0.1)$$

Let us take:

$$\left\{\beta\left(\tfrac{d^\alpha}{dt^\alpha}\overline{\xi}(t;\gamma) - \tfrac{d^\alpha}{dt^\alpha}\underline{\xi}(t;\gamma)\right) + \tfrac{d^\alpha}{dt^\alpha}\underline{\xi}(t;\gamma)\right\} = \tfrac{d^\alpha}{dt^\alpha}\widetilde{\xi}(t;\gamma,\beta)$$
$$\left\{\beta\left(\tfrac{d^\alpha}{dt^\alpha}\overline{\psi}(t;\gamma) - \tfrac{d^\alpha}{dt^\alpha}\underline{\psi}(t;\gamma)\right) + \tfrac{d^\alpha}{dt^\alpha}\underline{\psi}(t;\gamma)\right\} = \tfrac{d^\alpha}{dt^\alpha}\widetilde{\psi}(t;\gamma,\beta)$$
$$\left\{\beta(\overline{\psi}(t;\gamma) - \underline{\psi}(t;\gamma)) + \underline{\psi}(t;\gamma)\right\} = \widetilde{\psi}(t;\gamma,\beta)$$
$$\left\{\beta(\overline{\xi}(t;\gamma) - \underline{\xi}(t;\gamma)) + \underline{\xi}(t;\gamma)\right\} = \widetilde{\xi}(t;\gamma,\beta).$$
$$\{\beta(0.04 - 0.04\gamma) + 0.02\gamma - 0.02 + a_1\} = \widetilde{a}_1.$$
$$\{\beta(0.04 - 0.04\gamma) + 0.02\gamma - 0.02 + a_2\} = \widetilde{a}_2.$$

$$\{\beta(0.04 - 0.04\gamma) + 0.02\gamma + b_1 - 0.02\} = \widetilde{b}_1.$$
$$\{\beta(0.04 - 0.04\gamma) + 0.02\gamma + b_2 - 0.02\} = \widetilde{b}_2.$$
$$\{\beta(0.4 - 0.4\gamma) + 0.2\gamma + c_1 - 0.2\} = \widetilde{c}_1.$$
$$\{\beta(0.4 - 0.4\gamma) + 0.2\gamma + c_2 - 0.2\} = \widetilde{c}_2.$$
$$\{\beta(0.02 - 0.02\gamma) + 0.01\gamma + \delta - 0.01\} = \widetilde{\delta}.$$
$$\{\beta(\overline{\psi}(0;\gamma) - \underline{\psi}(0;\gamma)) + \underline{\psi}(0;\gamma)\} = \widetilde{\psi}(0;\gamma,\beta)$$

and

$$\{\beta(\overline{\xi}(0;\gamma) - \underline{\xi}(0;\gamma)) + \underline{\xi}(0;\gamma)\} = \widetilde{\xi}(0;\gamma,\beta)$$

Substituting all the above equations in Equations (19)–(21), we get:

$$\begin{aligned}\frac{d^\alpha \widetilde{\psi}(t;\gamma,\beta)}{dt^\alpha} &= -\widetilde{a}_1 \widetilde{\psi}(t;\gamma,\beta) + \widetilde{b}_1 \widetilde{\xi}(t;\gamma,\beta)\left(1 - \widetilde{\delta}\widetilde{\xi}^2(t;\gamma,\beta)\right) + \widetilde{c}_1.\\ \frac{d^\alpha \widetilde{\xi}(t;\gamma,\beta)}{dt^\alpha} &= -\widetilde{a}_2 \widetilde{\xi}(t;\gamma,\beta) + \widetilde{b}_2 \widetilde{\psi}(t;\gamma,\beta)\left(1 - \widetilde{\delta}\widetilde{\psi}^2(t;\gamma,\beta)\right) + \widetilde{c}_2.\end{aligned} \quad (22)$$

with ICs:

$$\widetilde{\psi}(0;\gamma,\beta) = \widetilde{\xi}(0;\gamma,\beta) = \beta(-0.2\gamma + 0.2) + (0.1\gamma - 0.1) = \eta \quad (23)$$

Solving Equation (22) with the ICs in Equation (23), we have $\widetilde{\psi}_1(t;\gamma,\beta)$ and $\widetilde{\psi}_2(t;\gamma,\beta)$ in terms of γ and β. To find the lower and upper bounds of the solutions in single parametric form, we have to substitute $\beta = 0$ and $\beta = 1$, respectively. Mathematically these are written as:

$$\widetilde{\psi}(t;\gamma,0) = \underline{\psi}(t;\gamma), \widetilde{\xi}(t;\gamma,0) = \underline{\xi}(t;\gamma) \text{ and } \widetilde{\psi}(t;\gamma,1) = \overline{\psi}(t;\gamma), \widetilde{\xi}(t;\gamma,1) = \overline{\xi}(t;\gamma)$$

Applying FRDTM to both sides of Equation (22), and using Theorems 1 and 2, we have:

$$\begin{aligned}\frac{\Gamma(1+k\alpha+\alpha)}{\Gamma(1+k\alpha)} \widetilde{\psi}_{k+1}(\gamma,\beta) &= -\widetilde{a}_1 \widetilde{\psi}_k(\gamma,\beta) + \widetilde{b}_1 \widetilde{\xi}_k(\gamma,\beta) - \widetilde{b}_1\widetilde{\delta}\left(\sum_{i=0}^{k}\sum_{j=0}^{i} \widetilde{\xi}_{i-j}\widetilde{\xi}_j\widetilde{\xi}_{k-i}\right) + \widetilde{c}_1 \delta(k)\\ \frac{\Gamma(1+k\alpha+\alpha)}{\Gamma(1+k\alpha)} \widetilde{\xi}_{k+1}(\gamma,\beta) &= -\widetilde{a}_2 \widetilde{\xi}_k(\gamma,\beta) + \widetilde{b}_2 \widetilde{\psi}_k(\gamma,\beta) - \widetilde{b}_2\widetilde{\delta}\left(\sum_{i=0}^{k}\sum_{j=0}^{i} \widetilde{\psi}_{i-j}\widetilde{\psi}_j\widetilde{\psi}_{k-i}\right) + \widetilde{c}_2 \delta(k)\end{aligned} \quad (24)$$

Where:

$$\delta(k) = \begin{cases} 1, & k = 0 \\ 0, & k \neq 0 \end{cases}$$

Using FRDTM on the IC, we get:

$$\widetilde{\psi}_0(\gamma,\beta) = \widetilde{\xi}_0(\gamma,\beta) = \eta \quad (25)$$

Using Equation (25) in Equation (24), the following values of $\widetilde{\psi}_k$ and $\widetilde{\xi}_k$ for $k = 1, 2, \ldots$ are obtained:

$$\begin{aligned}\widetilde{\psi}_1 &= -\widetilde{a}_1 \eta + \widetilde{b}_1 \eta - \widetilde{b}_1 \widetilde{\delta} \eta^3 + \widetilde{c}_1.\\ \widetilde{\xi}_1 &= -\widetilde{a}_2 \eta + \widetilde{b}_2 \eta - \widetilde{b}_2 \widetilde{\delta} \eta^3 + \widetilde{c}_2.\end{aligned} \quad (26)$$

$$\begin{aligned}\widetilde{\psi}_2 &= -\widetilde{a}_1\left(-\widetilde{a}_1\eta + \widetilde{b}_1\eta - \widetilde{b}_1\widetilde{\delta}\eta^3 + \widetilde{c}_1\right) + \widetilde{b}_1\left(-\widetilde{a}_2\eta + \widetilde{b}_2\eta - \widetilde{b}_2\widetilde{\delta}\eta^3 + \widetilde{c}_2\right)\\ &\quad - 3\widetilde{b}_1\widetilde{\delta}\eta^2\left(-\widetilde{a}_2\eta + \widetilde{b}_2\eta - \widetilde{b}_2\widetilde{\delta}\eta^3 + \widetilde{c}_2\right).\\ \widetilde{\xi}_2 &= -\widetilde{a}_2\left(-\widetilde{a}_2\eta + \widetilde{b}_2\eta - \widetilde{b}_2\widetilde{\delta}\eta^3 + \widetilde{c}_2\right) + \widetilde{b}_2\left(-\widetilde{a}_1\eta + \widetilde{b}_1\eta - \widetilde{b}_1\widetilde{\delta}\eta^3 + \widetilde{c}_1\right)\\ &\quad - 3\widetilde{b}_1\widetilde{\delta}\eta^2\left(-\widetilde{a}_1\eta + \widetilde{b}_1\eta - \widetilde{b}_1\widetilde{\delta}\eta^3 + \widetilde{c}_1\right).\end{aligned} \quad (27)$$

Continuing the above procedure, all the values of $\{\widetilde{\psi}_i, \widetilde{\xi}_i\}_{i=3}^{\infty}$ can be calculated. Therefore, according to FRDTM, the n-term solutions of Equation (22) with Equation (23) are written as:

$$\widetilde{\psi}(t;\gamma,\beta) = \sum_{k=0}^{n} \widetilde{\psi}_k(\gamma,\beta) t^{\alpha k}.$$
$$\widetilde{\xi}(t;\gamma,\beta) = \sum_{k=0}^{n} \widetilde{\xi}_k(\gamma,\beta) t^{\alpha k}. \tag{28}$$

Substituting $\beta = 0$ and $\beta = 1$, the lower and upper bounds of the solution can be calculated, which, respectively, are as follows:

$$\widetilde{\psi}(t;\gamma,0) = \sum_{k=0}^{n} \widetilde{\psi}_k(\gamma,0) t^{\alpha k}.$$
$$\widetilde{\xi}(t;\gamma,0) = \sum_{k=0}^{n} \widetilde{\xi}_k(\gamma,0) t^{\alpha k}. \tag{29}$$

and

$$\widetilde{\psi}(t;\gamma,1) = \sum_{k=0}^{n} \widetilde{\psi}_k(\gamma,1) t^{\alpha k}.$$
$$\widetilde{\xi}(t;\gamma,1) = \sum_{k=0}^{n} \widetilde{\xi}_k(\gamma,1) t^{\alpha k}. \tag{30}$$

5. Results and Discussion

In this section, an approximate solution of a fuzzy FDMM using FRDTM has been studied. Various numerical computations have been carried out by taking different values of parameters involved in the equation and ICs. In this article, all the figures and tables are included by considering the values of the parameters as $a_1 = 0.05, b_1 = 0.04, c_1 = 0.2, a_2 = 0.07, b_2 = 0.06, c_2 = 0.3$ and $\delta = 0.01$ (see References [16,17]). The achieved outcomes are compared with the solution of Singh et al. [16] and Goyal et al. [17], which show the validation of the present study. Calculated results are displayed in terms of plots.

Here, all the numerical calculations have been computed by truncating the infinite series to a finite number of terms ($n = 5$). Fuzzy solutions of FDMM are portrayed in Figures 1–6 by changing time t from 0 to 1 and for different values of α. Next, interval solutions for different values of α have been illustrated in Figures 7–12 by considering $\gamma - cut$ 0.4, 0.8 and 1, and varying time t from 0 to 10. From these Figures 7–12, one may see that the line at $\gamma = 1$ is the central line, and all other solutions are present on both sides of the $\gamma = 1$ line.

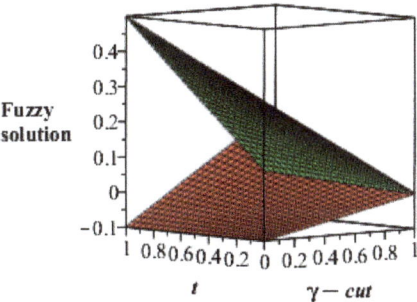

Figure 1. Lower and upper bounds fuzzy solutions of $\psi(t)$ at $\alpha = 1$.

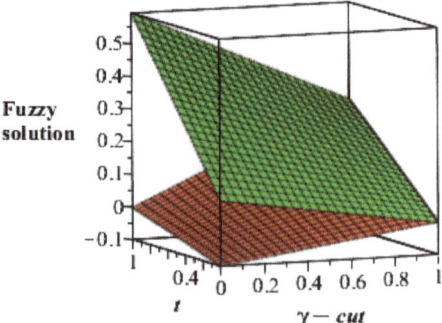

Figure 2. Lower and upper bounds fuzzy solutions of $\xi(t)$ at $\alpha = 1$.

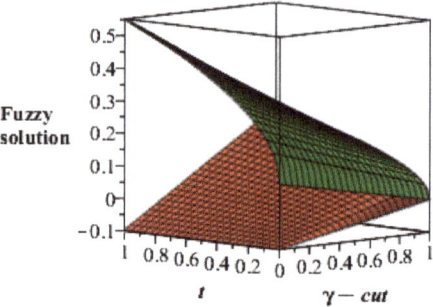

Figure 3. Lower and upper bounds fuzzy solutions of $\psi(t)$ at $\alpha = 0.5$.

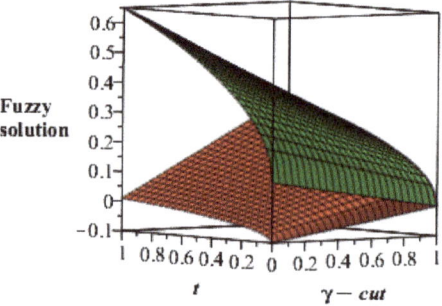

Figure 4. Lower and upper bounds fuzzy solutions of $\xi(t)$ at $\alpha = 0.5$.

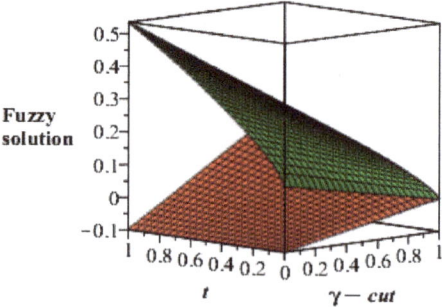

Figure 5. Lower and upper bounds fuzzy solutions of $\psi(t)$ at $\alpha = 0.75$.

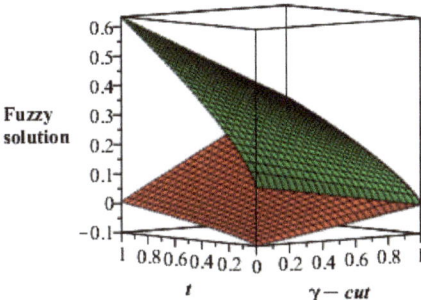

Figure 6. Lower and upper bounds fuzzy solutions of $\xi(t)$ at $\alpha = 0.75$.

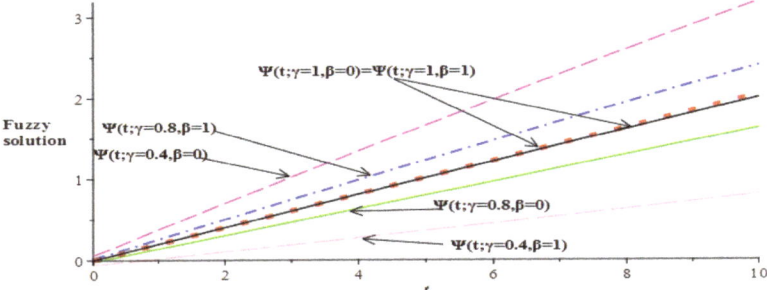

Figure 7. Lower and upper bounds interval solutions of $\psi(t)$ at $\alpha = 1$.

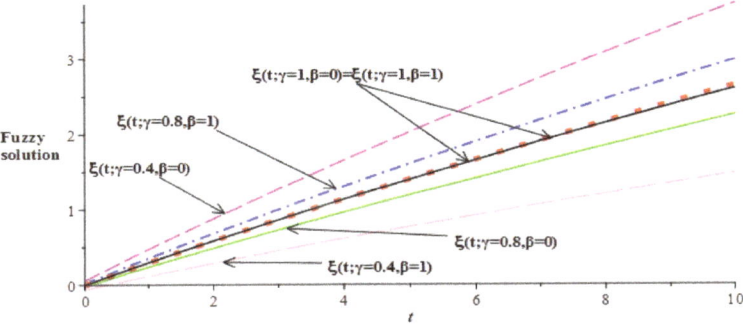

Figure 8. Lower and upper bounds interval solutions of $\xi(t)$ at $\alpha = 1$.

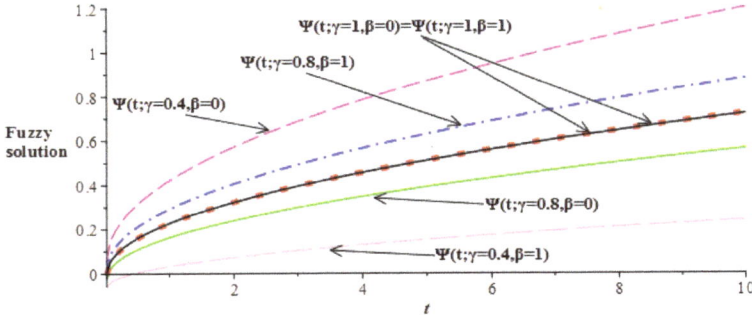

Figure 9. Lower and upper bounds interval solutions of $\psi(t)$ at $\alpha = 0.5$.

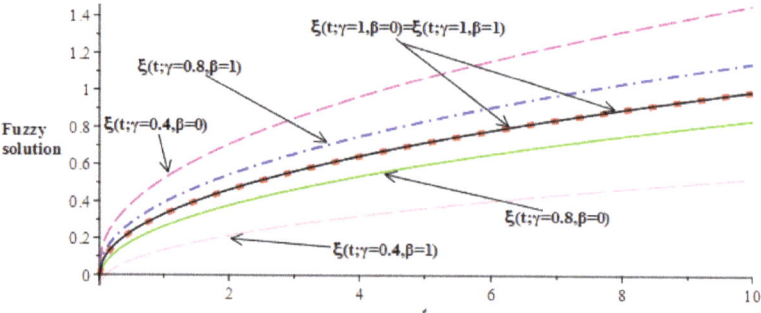

Figure 10. Lower and upper bounds interval solutions of $\xi(t)$ at $\alpha = 0.5$.

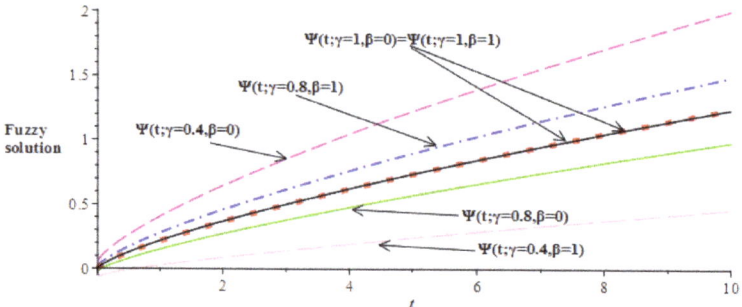

Figure 11. Lower and upper bounds interval solutions of $\psi(t)$ at $\alpha = 0.75$.

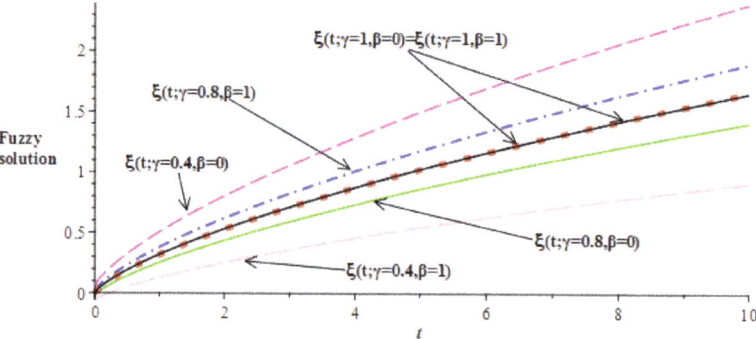

Figure 12. Lower and upper bounds interval solutions of $\xi(t)$ at $\alpha = 0.75$.

Also, it can be seen that the central line (crisp result, i.e., at $\gamma - cut = 1$) of Figures 7–12 gradually decreased with a decrease in α. Alternatively, we may say that a decrease in the values of α decreased the adoration of man or woman for his/her partner. From Tables 1–6, it is clear that the lower and upper bounds at different values of α were the same at $\gamma = 1$ and the obtained results matched with the solutions of Singh et al. [16] and Goyal et al. [17].

Table 1. Fuzzy and crisp solution of $\psi(t)$ at $\alpha = 1$ and $\gamma = 1$.

$t \rightarrow$ No of Approximation $\downarrow$	0	0.2	0.4	0.6	0.8	1
$n=1$ $[\underline{\psi}, \overline{\psi}]$	[0,0]	[0.04,0.03999]	[0.08,0.07999]	[0.12,0.119999]	[0.16,0.159999]	[0.20,0.199999]
$n=2$ $[\underline{\psi}, \overline{\psi}]$	[0,0]	[0.04,0.04003]	[0.0801,0.0802]	[0.1203,0.120359]	[0.1606,0.16064]	[0.2009,0.20099]
$n=3$ $[\underline{\psi}, \overline{\psi}]$	[0,0]	[0.04,0.04003]	[0.0801,0.08015]	[0.1203,0.120342]	[0.1606,160599]	[0.2009,0.20092]
$n=4$ $[\underline{\psi}, \overline{\psi}]$	[0,0]	[0.04,0.04004]	[0.0801,0.08015]	[0.1203,0.120342]	[0.1606,160597]	[0.2009,0.20092]
$n=5$ $[\underline{\psi}, \overline{\psi}]$	[0,0]	[0.04,0.04004]	[0.0801,0.08015]	[0.1203,0.120342]	[0.1606,160597]	[0.2009,0.20092]
Refs. [16,17] Crisp value at $\gamma=1$	0	0.04	0.0801	0.1203	0.1606	0.2009

Table 2. Fuzzy and crisp solution of $\xi(t)$ at $\alpha = 1$ and $\gamma = 1$.

$t \rightarrow$ No of Approximation $\downarrow$	0	0.2	0.4	0.6	0.8	1
$n=1$ $[\underline{\xi}, \overline{\xi}]$	[0,0]	[0.06,0.059999]	[0.12,0.1199]	[0.18,0.179999]	[0.24,0.239999]	[0.30,0.29999]
$n=2$ $[\underline{\xi}, \overline{\xi}]$	[0,0]	[0.0598,0.0598]	[0.1192,0.1197]	[0.1783,0.178389]	[0.2371,0.237189]	[0.2954,0.29549]
$n=3$ $[\underline{\xi}, \overline{\xi}]$	[0,0]	[0.0598,0.0598]	[0.1192,0.1192]	[0.1784,0.178406]	[0.2371,0.237182]	[0.2956,0.29562]
$n=4$ $[\underline{\xi}, \overline{\xi}]$	[0,0]	[0.0598,0.0598]	[0.1192,0.1192]	[0.1784,0.178405]	[0.2371,0.237179]	[0.2956,0.29561]
$n=5$ $[\underline{\xi}, \overline{\xi}]$	[0,0]	[0.0598,0.0598]	[0.1192,0.1192]	[0.1784,0.178405]	[0.2371,0.237179]	[0.2956,0.29561]
Refs. [16,17] Crisp value at $\gamma=1$	0	0.0598	0.1192	0.1784	0.2371	0.2956

Table 3. Fuzzy and crisp solution of $\psi(t)$ at $\alpha = 0.5$.

$t \rightarrow$ γ-cut $\downarrow$	0	0.2	0.4	0.6	0.8	1
$\gamma=0$ $[\underline{\psi}, \overline{\psi}]$	[−0.1,0.1]	[−0.991,0.3016]	[−0.0985,0.3853]	[−0.0980,0.4496]	[−0.0975,0.5038]	[−0.0970,0.5515]
$\gamma=0.2$ $[\underline{\psi}, \overline{\psi}]$	[−0.08,0.08]	[−0.059,0.2616]	[−0.0501,0.3370]	[−0.0432,0.3949]	[−0.0373,0.4437]	[−0.0321,0.4867]
$\gamma=0.4$ $[\underline{\psi}, \overline{\psi}]$	[−0.06,0.06]	[−0.0189,0.2215]	[−0.0017,0.2886]	[0.0115,0.3401]	[0.0228,0.3836]	[0.0328,0.4219]
$\gamma=0.6$ $[\underline{\psi}, \overline{\psi}]$	[−0.04,0.04]	[0.0211,0.1814]	[0.0467,0.2402]	[0.0663,0.2854]	[0.0829, 0.3234]	[0.0976,0.3570]
$\gamma=0.8$ $[\underline{\psi}, \overline{\psi}]$	[−0.02,0.02]	[0.0613,0.1414]	[0.0950,0.1918]	[0.1211,0.2306]	[0.1431,0.2633]	[0.1625, 0.2922]
$\gamma=1$ $[\underline{\psi}, \overline{\psi}]$	[0,0]	[0.1013,0.1013]	[0.1434, 0.1434]	[0.1759, 0.1758]	[0.2032, 0.2032]	[0.2273, 0.2273]
Refs. [16,17] Crisp value at $\gamma=1$	0	0.1013	0.1434	0.1759	0.2032	0.2273

Table 4. Fuzzy and crisp solution of $\xi(t)$ at $\alpha = 0.5$.

$t \rightarrow$ γ-cut $\downarrow$	0	0.2	0.4	0.6	0.8	1
$\gamma=0$ $[\underline{\xi}, \overline{\xi}]$	[−0.1,0.1]	[−0.0500,0.3493]	[−0.0299,0.4511]	[−0.0146,0.5288]	[−0.0019,0.5938]	[0.0092,0.6509]
$\gamma=0.2$ $[\underline{\xi}, \overline{\xi}]$	[−0.08,0.08]	[−0.0101,0.3094]	[0.0182,0.4030]	[0.0397,0.4744]	[0.0577, 0.5343]	[0.0734,0.5867]
$\gamma=0.4$ $[\underline{\xi}, \overline{\xi}]$	[−0.06,0.06]	[0.0298,0.2694]	[0.0663,0.3549]	[0.0940,0.4200]	[0.1172,0.4747]	[0.1375,0.5226]
$\gamma=0.6$ $[\underline{\xi}, \overline{\xi}]$	[−0.04,0.04]	[0.0698,0.2295]	[0.1144,0.3068]	[0.1484,0.3657]	[0.1768,0.4151]	[0.2017, 0.4584]
$\gamma=0.8$ $[\underline{\xi}, \overline{\xi}]$	[−0.02,0.02]	[0.1097,0.1896]	[0.1625,0.2587]	[0.2027,0.3114]	[0.2364,0.355]	[0.2659, 0.3942]
$\gamma=1$ $[\underline{\xi}, \overline{\xi}]$	[0,0]	[0.1496,0.1496]	[0.2106,0.2106]	[0.2570,0.2570]	[0.2960,0.2960]	[0.3300,0.3300]
Refs. [16,17] Crisp value at $\gamma=1$	0	0.1496	0.2106	0.2570	0.2960	0.3300

Table 5. Fuzzy and crisp solution of $\psi(t)$ at $\alpha = 0.75$.

γ-cut	0	0.2	0.4	0.6	0.8	1
$\gamma = 0 \ [\underline{\psi}, \overline{\psi}]$	[−0.1,0.1]	[−0.0995,0.2210]	[−0.0991,0.3187]	[−0.0986,0.3965]	[−0.0980,0.4681]	[−0.0974,0.5353]
$\gamma = 0.2 \ [\underline{\psi}, \overline{\psi}]$	[−0.08,0.08]	[−0.0666,0.1970]	[−0.0573,0.2769]	[−0.0491,0.3470]	[−0.0414,0.4115]	[−0.0342,0.4720]
$\gamma = 0.4 \ [\underline{\psi}, \overline{\psi}]$	[−0.06,0.06]	[−0.0336,0.1640]	[−0.0155,0.2351]	[0.0005,0.2975]	[0.0152,0.3549]	[0.0291,0.4087]
$\gamma = 0.6 \ [\underline{\psi}, \overline{\psi}]$	[−0.04,0.04]	[−0.0007,0.1311]	[0.0263,0.1933]	[0.0410,0.2480]	[0.0718,0.2983]	[0.0924,0.3455]
$\gamma = 0.8 \ [\underline{\psi}, \overline{\psi}]$	[−0.02,0.02]	[0.0323,0.0982]	[0.0680,0.1516]	[0.0995,0.1985]	[0.12843,0.2417]	[0.1557,0.2822]
$\gamma = 1 \ [\underline{\psi}, \overline{\psi}]$	[0,0]	[0.0652,0.0652]	[0.1098,0.1098]	[0.1490,0.1490]	[0.1850,0.1850]	[0.2189,0.2189]
Refs. [16,17] Crisp value at $\gamma = 1$	0	0.0652	0.1098	0.1490	0.1850	0.2189

Table 6. Fuzzy and crisp solution of $\xi(t)$ at $\alpha = 0.75$.

γ-cut	0	0.2	0.4	0.6	0.8	1
$\gamma = 0 \ [\underline{\xi}, \overline{\xi}]$	[−0.1,0.1]	[−0.0675,0.2615]	[−0.0457,0.3707]	[−0.0268,0.4658]	[−0.0097,0.5526]	[0.0063,0.6336]
$\gamma = 0.2 \ [\underline{\xi}, \overline{\xi}]$	[−0.08,0.08]	[−0.0346,0.2286]	[−0.0040,0.3290]	[0.0225,0.4165]	[0.0466,0.4963]	[0.0690,0.5709]
$\gamma = 0.4 \ [\underline{\xi}, \overline{\xi}]$	[−0.06,0.06]	[−0.0017,0.1957]	[0.0376,0.2874]	[0.0717,0.3672]	[0.1028,0.4401]	[0.1317,0.5081]
$\gamma = 0.6 \ [\underline{\xi}, \overline{\xi}]$	[−0.04,0.04]	[0.0312,0.1628]	[0.0792,0.2457]	[0.1210,0.3179]	[0.1590,0.3839]	[0.1945,0.4454]
$\gamma = 0.8 \ [\underline{\xi}, \overline{\xi}]$	[−0.02,0.02]	[0.0641,0.1299]	[0.1209,0.2041]	[0.1702,0.2687]	[0.2152,0.3276]	[0.2572,0.3827]
$\gamma = 1 \ [\underline{\xi}, \overline{\xi}]$	[0,0]	[0.0970,0.0970]	[0.1625,0.1625]	[0.2195, 0.2195]	[0.2714,0.2714]	[0.3199,0.3199]
Refs. [16,17] Crisp value at $\gamma = 1$	0	0.097	0.1625	0.2195	0.2714	0.3199

6. Conclusions

In this paper, approximate solutions of a fuzzy FDMM were found with the help of an efficient method, namely FRDTM. In the procedure, the DPF of fuzzy number was applied. This methodology was found to be straight forward as it converted FDEs to an advantageous form involving two parameters that controlled the uncertainty. Attained outcomes were compared with the existing results and were found to be in agreement. The main benefit of applying this method is that it does not require any assumption, perturbation, or discretization for solving the governing time-fractional dynamical model. Also, the computation time was less compared to other techniques. From this study, it is concluded that the decrease in the values of α decreased romantic relations between the couple.

Author Contributions: Each author has contributed equally towards preparing and finalizing the whole research work of the present paper.

Funding: This research received no external funding.

Acknowledgments: The first-named author acknowledges the Department of Science and Technology of the Government of India for providing INSPIRE Fellowship (IF170207) in order to carry out the present research.

Conflicts of Interest: The authors declare that they have no competing interests.

References

1. Jena, R.M.; Chakraverty, S.; Jena, S.K. Dynamic Response Analysis of Fractionally Damped Beams Subjected to External Loads using Homotopy Analysis Method. *J. Appl. Comput. Mech.* **2019**, *5*, 355–366.
2. Jena, R.M.; Chakraverty, S. Solving time-fractional Navier–Stokes equations using homotopy perturbation Elzaki transform. *SN Appl. Sci.* **2019**, *1*, 16. [CrossRef]
3. Jena, R.M.; Chakraverty, S. Residual Power Series Method for Solving Time-fractional Model of Vibration Equation of Large Membranes. *J. Appl. Comput. Mech.* **2019**, *5*, 603–615.

4. Jena, R.M.; Chakraverty, S. A new iterative method based solution for fractional Black–Scholes option pricing equations (BSOPE). *SN Appl. Sci.* **2019**, *1*, 95–105. [CrossRef]
5. Edeki, S.O.; Motsepa, T.; Khalique, C.M.; Akinlabi, G.O. The Greek parameters of a continuous arithmetic Asian option pricing model via Laplace Adomian decomposition method. *Open Phys.* **2018**, *16*, 780–785. [CrossRef]
6. Podlubny, I. *Fractional Differential Equations*; Academic Press: San Diego, CA, USA, 1999.
7. Kilbas, A.A.; Srivastava, H.M.; Trujillo, J.J. *Theory and Applications of Fractional Differential Equations*; North-Holland Mathematical Studies; Elsevier Science Publishers: Amsterdam, The Netherlands; London, UK; New York, NY, USA, 2006; Volume 204.
8. Baleanu, D.; Diethelm, K.; Scalas, E.; Trujillo, J.J. *Fractional Calculus: Models and Numerical Methods*; World Scientific: Boston, MA, USA: 2012.
9. Baleanu, D.; Machado, J.A.T.; Luo, A.C. *Fractional Dynamics and Control*; Springer: Berlin, Germany, 2012.
10. Miller, K.S.; Ross, B. *An Introduction to the Fractional Calculus and Fractional Differential Equations*; A Wiley-Interscience Publication; John Wiley and Sons: New York, NY, USA; Chichester, UK; Brisbane, Australian; Toronto, ON, Canada; Singapore, 1993.
11. Barley, K.; Cherif, A. Stochastic nonlinear dynamics of interpersonal and romantic relationships. *Appl. Math. Comput.* **2011**, *217*, 6273–6281. [CrossRef]
12. Rinaldi, S. Love dynamics: The case of linear couples. *Appl. Math. Comput.* **1998**, *95*, 181–192. [CrossRef]
13. Ozalp, N.; Koca, I. A fractional order nonlinear dynamical model of interpersonal relationships. *Adv. Differ. Equ.* **2012**, *189*, 1–7. [CrossRef]
14. Khader, M.M.; Alqahtani, R. Approximate solution for system of fractional non-linear dynamical marriage model using Bernstein polynomials. *J. Nonlinear Sci. Appl.* **2017**, *10*, 865–873. [CrossRef]
15. Khader, M.M.; Shloof, A.; Ali, H. On the numerical simulation and convergence study for system of non-linear fractional dynamical model of marriage. *NTMSCI* **2017**, *5*, 130–141. [CrossRef]
16. Singh, J.; Kumar, D.; Qurashi, M.A.; Baleanu, D. A Novel Numerical Approach for a Nonlinear Fractional Dynamical Model of Interpersonal and Romantic Relationships. *Entropy* **2017**, *19*, 375. [CrossRef]
17. Goyal, M.; Prakash, A.; Gupta, S. Numerical simulation for time-fractional nonlinear coupled dynamical model of romantic and interpersonal relationships. *Pramana J. Phys.* **2019**, *92*, 82. [CrossRef]
18. Martin, M.T.C.; Bumpass, B.L. Recent trends in marital disruption. *Demography* **1989**, *26*, 37–51. [CrossRef] [PubMed]
19. Strogatz, S.H. *Nonlinear Dynamics and Chaos: With Applications to Physics, Biology, Chemistry and Engineering*; Reading, M.A., Ed.; Addison-Wesley: Boston, MA, USA, 1994.
20. Gottman, J.M.; Murray, J.D.; Swanson, C.C.; Tyson, R.; Swanson, K.R. *The Mathematics of Marriage*; MIT Press: Cambridge, MA, USA, 2002.
21. Chang, S.L.; Zadeh, L.A. On fuzzy mapping and control. *IEEE Trans. Syst. Man Cybern.* **1972**, *2*, 30–34. [CrossRef]
22. Dubois, D.; Prade, H. Towards fuzzy differential calculus part 3: Differentiation. *Fuzzy Sets Syst.* **1982**, *8*, 225–233. [CrossRef]
23. Kaleva, O. The Cauchy problem for fuzzy differential equations. *Fuzzy Sets Syst.* **1990**, *35*, 389–396. [CrossRef]
24. Seikkala, S. On the fuzzy initial value problem. *Fuzzy Sets Syst.* **1987**, *24*, 319–330. [CrossRef]
25. Chakraverty, S.; Tapaswini, S.; Behera, D. *Fuzzy Arbitrary Order System: Fuzzy Fractional Differential Equations and Applications*; John Wiley & Sons Inc.: Hoboken, NJ, USA, 2016.
26. Chakraverty, S.; Tapaswini, S.; Behera, D. *Fuzzy Differential Equations and Applications for Engineers and Scientists*; Taylor and Francis Group: Boca Raton, FL, USA, 2016.
27. Chakraverty, S.; Sahoo, D.M.; Mahato, N.R. *Concepts of Soft Computing: Fuzzy and ANN with Programming*; Springer: Singapore, 2019.
28. Momani, S.; Odibat, Z. A generalized differential transform method for linear partial differential equations of fractional order. *Appl. Math. Lett.* **2008**, *21*, 194–199.
29. Singh, B.K.; Kumar, P. FRDTM for numerical simulation of multi-dimensional, time-fractional model of Navier–Stokes equation. *Ain Shams Eng. J.* **2018**, *9*, 827–834. [CrossRef]
30. Singh, J.; Kumar, D.; Swroop, R.; Kumar, S. An efficient computational approach for time-fractional Rosenau–Hyman equation. *Neural Comput. Appl.* **2018**, *30*, 3063–3070. [CrossRef]

31. Rawashdeh, M.S. An Efficient Approach for Time–Fractional Damped Burger and Time—Sharma—Tasso—Olver Equations Using the FRDTM. *Appl. Math. Inf. Sci.* **2015**, *9*, 1239–1246.
32. Keskin, Y.; Oturan, G. Reduced Differential Transform Method for Partial Differential Equations. *Int. J. Nonlinear Sci. Numer. Simul.* **2009**, *10*, 741–749. [CrossRef]

© 2019 by the authors. Licensee MDPI, Basel, Switzerland. This article is an open access article distributed under the terms and conditions of the Creative Commons Attribution (CC BY) license (http://creativecommons.org/licenses/by/4.0/).

Article

Structure of Non-Oscillatory Solutions for Second Order Dynamic Equations on Time Scales

Yong Zhou [1,2,*], Bashir Ahmad [2] and Ahmed Alsaedi [2]

1. Faculty of Mathematics and Computational Science, Xiangtan University, Xiangtan 411105, China
2. Nonlinear Analysis and Applied Mathematics (NAAM) Research Group, Faculty of Science, King Abdulaziz University, Jeddah 21589, Saudi Arabia
* Correspondence: yzhou@xtu.edu.cn

Received: 28 June 2019; Accepted: 29 July 2019; Published: 30 July 2019

Abstract: In this paper, we make a detailed analysis of the structure of non-oscillatory solutions for second order superlinear and sublinear dynamic equations on time scales. The sufficient and necessary conditions for existence of non-oscillatory solutions are presented.

Keywords: dynamic equations; time scales; classification; existence; necessary and sufficient conditions

MSC: 34C10; 34N05

1. Introduction

During the past few decades, an active worldwide research on the oscillation and nonoscillation for dynamic equations on time scales has been carried out by many mathematicians. Some interesting monographs [1–5] contain many important works in this area. In particular, many researchers have studied oscillation of second order dynamic equations. For some recent results on the topic, we refer the reader to the works [6–20] and the references cited therein.

Consider the second order dynamic equation on time scales

$$[p(t)x^\Delta(t)]^\Delta + g(t, x(\eta(t))) = 0, \quad t \in \mathbb{T}_0 \subseteq \mathbb{T}, \tag{1}$$

where $p \in C_{rd}(\mathbb{T}_0, \mathbb{R}^+)$, $\eta \in C_{rd}(\mathbb{T}_0, \mathbb{T})$, $g : \mathbb{T}_0 \times \mathbb{R} \to \mathbb{R}$ is continuous and $\operatorname{sgn} g(t, x) = \operatorname{sgn} x$ for $t \in \mathbb{T}_0$, $\lim_{t \to \infty} \eta(t) = \infty$.

Oscillation of the Equation (1) has been studied by Došlý and Hilger [6], Grace, Agarwal, Bohner and O'Regan [7], Zhou, Ahmad and Alsaedi [20]. A non-oscillatory of Equation (1) is also considered by Graef and Hill [21], Erbe, Baoguo and Peterson [22]. For more details, we refer the reader to see the references cited therein. However, to the authors' knowledge, there are no papers dealing with the analysis of structure of non-oscillatory solutions and sufficient and necessary conditions for existence of all kinds of non-oscillatory solutions for dynamic equations on time scales.

Our aim is to give a classification of non-oscillatory solutions to second order superlinear and sublinear dynamic equations on time scales, which is presented in Section 2. Then, we obtain the sufficient and necessary conditions for existence of some kinds of non-oscillatory solutions in Section 3.

2. Classification of Non-Oscillatory Solutions

Let $\mathbb{T}$ be a time scale (i.e., a closed subset of the real numbers $\mathbb{R}$) with $\sup \mathbb{T} = \infty$. We assume throughout that $\mathbb{T}$ has the topology that it inherits from the standard topology on the real numbers $\mathbb{R}$. For $t \in \mathbb{T}$, we define the forward jump operator $\sigma : \mathbb{T} \to \mathbb{T}$ by $\sigma(t) := \inf\{s \in \mathbb{T} : s > t\}$. Denote by $C_{rd}(\mathbb{T}, \mathbb{R})$ the space consisting of all functions which are right-dense points in $\mathbb{T}$ and its left-sided limits exist (finite) at left-dense points in $\mathbb{T}$. Furthermore, let us put $[t_0, \infty) := \mathbb{T}_0 = \{t \in \mathbb{T} : t_0 \leq t < \infty\}$.

Definition 1. If
$$\frac{f(t,x)}{x} \geq \frac{f(t,y)}{y} \quad \text{for } x \geq y > 0 \text{ or } x \leq y < 0, \ t \in \mathbb{T}_0,$$
then f is said to be superlinear. If
$$\frac{f(t,x)}{x} \leq \frac{f(t,y)}{y} \quad \text{for } x \geq y > 0 \text{ or } x \leq y < 0, \ t \in \mathbb{T}_0,$$
then f is said to be sublinear.

Next, for convenience, we set
$$P(t) = \int_{t_0}^{t} \frac{1}{p(s)} \Delta s, \quad \hat{P}(t) = \int_{t}^{\infty} \frac{1}{p(s)} \Delta s.$$

Lemma 1. Assume that $\int_{t_0}^{\infty} \frac{1}{p(s)} \Delta s = \infty$ and $x(t)$ is an eventually positive solution of Equation (1). Then, there exist $c_1 > 0, c_2 > 0$ and $t_1 \in \mathbb{T}_0$ such that
$$x^{\Delta}(t) > 0, \ c_1 \leq x(t) \leq c_2 P(t), \ t \geq t_1.$$

Proof. Choose $t \geq t_0$ sufficiently large such that $x(\eta(t)) > 0$. Suppose that there exists $t_1 > t_0$ such that $x^{\Delta}(t_1) \leq 0$. Integrating Equation (1) from t_1 to t, we get
$$p(t)x^{\Delta}(t) - p(t_1)x^{\Delta}(t_1) + \int_{t_1}^{t} g(s, x(\eta(s))) \Delta s = 0. \tag{2}$$

Dividing Equation (2) by $p(t)$, and then integrating from $t_2 (> t_1)$ to t, we have
$$x(t) - x(t_2) - p(t_1)x^{\Delta}(t_1) \int_{t_2}^{t} \frac{1}{p(s)} \Delta s + \int_{t_2}^{t} \left[\frac{1}{p(s)} \int_{t_1}^{s} g(\theta, x(\eta(\theta))) \Delta \theta \right] \Delta s = 0. \tag{3}$$

Noting that $\operatorname{sgn} g(\theta, x) = \operatorname{sgn} x$ and $x^{\Delta}(t_1) \leq 0$, after the transposition of terms, letting $t \to \infty$, we get $x(t) \to -\infty$, which is a contradiction. Hence, $x^{\Delta}(t) > 0$. Therefore, there exists $c_1 > 0$ such that $x(t) \geq c_1$. By Equation (3), there exists $c_2 > 0$ such that $x(t) \leq c_2 P(t)$. The proof is complete. □

Lemma 2. Assume that $\int_{t_0}^{\infty} \frac{1}{p(s)} \Delta s < \infty$ and $x(t)$ is an eventually positive solution of Equation (1). Then, there exist $c_1 > 0, c_2 > 0$ and $t_1 \in \mathbb{T}_0$ such that
$$x(t) \geq -p(t)x^{\Delta}(t)\hat{P}(t), \ c_1 \hat{P}(t) \leq x(t) \leq c_2, \ t \geq t_1.$$

Proof. Let $t \geq t_0$ be sufficiently large so that $x(\eta(t)) > 0$. Then, it follows from Equation (1) and $\operatorname{sgn} g(t,x) = \operatorname{sgn} x$ that $(p(t)x^{\Delta}(t))^{\Delta} < 0$, for $t \geq t_1$. Hence,
$$p(s)x^{\Delta}(s) \leq p(t)x^{\Delta}(t) \quad s > t \geq t_1. \tag{4}$$

Dividing Equation (4) by $p(s)$, and then integrating from t to $t_2 (> t_1)$, we have
$$x(t_2) - x(t) \leq p(t)x^{\Delta}(t) \int_{t}^{t_2} \frac{1}{p(s)} \Delta s.$$

We now show that $\lim_{t \to \infty} x(t) < \infty$. If not, let $\lim_{t \to \infty} x(t) = \infty$. Integrating Equation (1) from t_0 to t, we get
$$p(t)x^{\Delta}(t) - p(t_0)x^{\Delta}(t_0) + \int_{t_0}^{t} g(s, x(\eta(s))) \Delta s = 0,$$

Dividing the above equation by $p(t)$, and then integrating from t_0 to t yields

$$x(t) = x(t_0) + \hat{P}(t_0)p(t_0)x^\Delta(t_0) - \hat{P}(t)p(t_0)x^\Delta(t_0) - \int_{t_0}^{t}\left[\frac{1}{p(s)}\int_{t_0}^{s}g(\theta,x(\eta(\theta)))\Delta\theta\right]\Delta s.$$

Hence, we obtain that $\hat{P}(t) \to \infty$ as $t \to \infty$, which is a contradiction. Consequently, we get

$$x(t) \geq -p(t)x^\Delta(t)\int_{t}^{\infty}\frac{1}{p(s)}\Delta s.$$

Thus, the first part of the lemma holds. On the other hand, dividing Equation (4) by $p(s)$, and then integrating from t_1 to t, we have

$$x(t) \leq x(t_1) + p(t_1)x^\Delta(t_1)\int_{t_1}^{t}\frac{1}{p(s)}\Delta s \leq x(t_1) + |p(t_1)x^\Delta(t_1)|\hat{P}(t_1) \triangleq c_2.$$

Since $p(t)x^\Delta(t)$ is decreasing, we get

$$x(t) + p(t_1)x^\Delta(t_1)\int_{t}^{\infty}\frac{1}{p(s)}\Delta s \geq x(t) + p(t)x^\Delta(t)\int_{t}^{\infty}\frac{1}{p(s)}\Delta s \geq 0.$$

If $x^\Delta(t_1) < 0$, then $x(t) \geq |p(t_1)x^\Delta(t_1)|\hat{P}(t) = c_1\hat{P}(t)$. If $x^\Delta(t_1) \geq 0$, then we can assume that $x^\Delta(t) \geq 0$, for $t \geq t_1$. Otherwise, by choosing $t_2 = t_1$, we repeat the above process. Thus, $x(t)$ is nondecreasing for $t \geq t_1$. Therefore,

$$x(t) \geq x(t_1) = \frac{x(t_1)}{\hat{P}(t_1)}\hat{P}(t_1) \geq \frac{x(t_1)}{\hat{P}(t_1)}\hat{P}(t) \triangleq c_1\hat{P}(t).$$

The proof is complete. □

Remark 1. *If $x(t)$ is an eventually negative solution of Equation (1), then there are analogous conclusions to Lemma 1 and Lemma 2, in which we just need to change the sign of constants c_1, c_2 into negative values and inverse the sign of inequalities.*

Theorem 1. *Let S denote the set of all non-oscillatory solutions of Equation (1). Assume that $\int_{t_0}^{\infty}\frac{1}{p(s)}\Delta s = \infty$. Then, any non-oscillatory solutions of Equation (1) must belong to one of the following classes:*

$$A_c^0 = \{x(t) \in S : \lim_{t\to\infty} x(t) = c \neq 0, \lim_{t\to\infty} p(t)x^\Delta(t) = 0\},$$
$$A_\infty^c = \{x(t) \in S : \lim_{t\to\infty} x(t) = \infty, \lim_{t\to\infty} p(t)x^\Delta(t) = c \neq 0\},$$
$$A_\infty^0 = \{x(t) \in S : \lim_{t\to\infty} x(t) = \infty, \lim_{t\to\infty} p(t)x^\Delta(t) = 0\}.$$

Proof. Without loss of generality, let $x(t)$ be an eventually positive solution of Equation (1). By Lemma 1, it is easy to see that either $\lim_{t\to\infty} x(t) = c > 0$, or $\lim_{t\to\infty} x(t) = +\infty$.

(i) If $\lim_{t\to\infty} x(t) = c > 0$, then $x(t)$ and $p(t)x^\Delta(t)$ are eventually positive. From Equation (1), since $(p(t)x^\Delta(t))^\Delta \leq 0$, that is, $p(t)x^\Delta(t)$ is nonincreasing, so $\lim_{t\to\infty} p(t)x^\Delta(t)$ exists. Now, we will show that $\lim_{t\to\infty} p(t)x^\Delta(t) = 0$. On the contrary, suppose that $\lim_{t\to\infty} p(t)x^\Delta(t) = c' > 0$. Furthermore, $\int_{t_0}^{t} x^\Delta(s)\Delta s \geq \int_{t_0}^{t}\frac{c'}{p(s)}\Delta s$. Thus, $\lim_{t\to\infty} x(t) = \infty$, which leads to a contradiction.

(ii) If $\lim_{t\to\infty} x(t) = +\infty$, then, in view of the fact that $\lim_{t\to\infty} p(t)x^\Delta(t)$ exists, it follows by L'Hôpital's rule that

$$\lim_{t\to\infty}\frac{x(t)}{P(t)} = \lim_{t\to\infty} p(t)x^\Delta(t).$$

327

On the other hand, by Lemma 1, we get

$$0 \leq \frac{x(t)}{P(t)} \leq c_2.$$

Therefore, either $\lim_{t\to\infty} p(t)x^\Delta(t) = c \neq 0$ or $\lim_{t\to\infty} p(t)x^\Delta(t) = 0$. □

Theorem 2. *Let S denote the set of all non-oscillatory solutions of Equation (1). Assume that $\int_{t_0}^\infty \frac{1}{p(s)} \Delta s < \infty$. Then, any non-oscillatory solutions of Equation (1) must belong to one of the following classes:*

$$A_c = \{x(t) \in S : \lim_{t\to\infty} x(t) = c \neq 0\},$$
$$A_0^c = \{x(t) \in S : \lim_{t\to\infty} x(t) = 0, \lim_{t\to\infty} p(t)x^\Delta(t) = c \neq 0\},$$
$$A_0^\infty = \{x(t) \in S : \lim_{t\to\infty} x(t) = 0, \lim_{t\to\infty} p(t)x^\Delta(t) = \infty\}.$$

Proof. Without loss of generality, let $x(t)$ be an eventually positive solution of Equation (1). By Equation (1), for sufficiently large t, we have that $(p(t)x^\Delta(t))^\Delta < 0$. Then, $p(t)x^\Delta(t)$ and $x^\Delta(t)$ are monotone and have eventually the same sign (either positive or negative). Firstly, we show that $\lim_{t\to\infty} x(t) = \infty$ does not hold. Indeed, if $\lim_{t\to\infty} x(t) = \infty$, then we get by integrating Equation (1) from t_0 to t that

$$p(t)x^\Delta(t) - p(t_0)x^\Delta(t_0) + \int_{t_0}^t g(s, x(\eta(s)))\Delta s = 0.$$

Dividing the above equation by $p(t)$, and then integrating from t_0 to t, we obtain

$$\begin{aligned}
x(t) &= x(t_0) + p(t_0)x^\Delta(t_0)\int_{t_0}^t \frac{1}{p(\theta)}\Delta\theta \\
&\quad - \int_{t_0}^t \frac{1}{p(\theta)}\int_{t_0}^\theta g(s, x(\eta(s)))\Delta s \Delta\theta \\
&= x(t_0) + p(t_0)x^\Delta(t_0)\left[\int_{t_0}^\infty \frac{1}{p(\theta)}\Delta\theta - \int_t^\infty \frac{1}{p(\theta)}\Delta\theta\right] \\
&\quad - \int_{t_0}^t \frac{1}{p(\theta)}\int_{t_0}^\theta g(s, x(\eta(s)))\Delta s \Delta\theta.
\end{aligned}$$

Therefore, $\lim_{t\to\infty}\int_t^\infty \frac{1}{p(\theta)}\Delta\theta = -\infty$, which is a contradiction. Hence, either $\lim_{t\to\infty} x(t) = c \neq 0$ or $\lim_{t\to\infty} x(t) = 0$. Since $p(t)x^\Delta(t)$ has a deterministic sign and it is monotone, this means that either $\lim_{t\to\infty} p(t)x^\Delta(t)$ exists or $\lim_{t\to\infty} p(t)x^\Delta(t) = \infty$. If $\lim_{t\to\infty} x(t) = 0$ and $\lim_{t\to\infty} p(t)x^\Delta(t)$ exists, then, by L'Hôpital's rule,

$$\lim_{t\to\infty} \frac{x(t)}{\hat{P}(t)} = -\lim_{t\to\infty} p(t)x^\Delta(t).$$

By Lemma 2, either $\lim_{t\to\infty} p(t)x^\Delta(t) = c \neq 0$, or $\lim_{t\to\infty} p(t)x^\Delta(t) = \infty$. □

3. Existence of Non-Oscillatory Solutions

In this section, we establish sufficient and necessary conditions for existence of some kinds of non-oscillatory solutions for Equation (1).

Theorem 3. *Assume that*
(i) $\int_{t_0}^\infty \frac{1}{p(s)}\Delta s = \infty$;
(ii) $g(t, x)$ *is superlinear or sublinear.*

Then, Equation (1) has a non-oscillatory solution $x(t) \in A_c^0$ if and only if

$$\int_{t_0}^{\infty} P(\sigma(s))|g(s,k)|\Delta s < \infty, \quad \text{for some } k \neq 0. \tag{5}$$

Proof. *Necessity.* Without loss of generality, let $x(t) \in A_c^0$ be eventually positive. By Lemma 1, there exist $c_1 > 0, c_2 > 0$ and $t_1 \in \mathbb{T}_0$ such that

$$x^{\Delta}(t) > 0, \quad c_1 \leq x(\eta(t)) \leq c_2, \quad t \geq t_1.$$

Multiplying Equation (1) by $P(\sigma(t))$, and then integrating from t_1 to t, we have

$$P(t)p(t)x^{\Delta}(t) - P(t_1)p(t_1)x^{\Delta}(t_1) - x(t) + x(t_1) + \int_{t_1}^{t} P(\sigma(s))g(s, x(\eta(s)))\Delta s = 0.$$

Since $\text{sgn } g(s,x) = \text{sgn } x$, it follows that the first term of above identity is finite as t tends to infinity. Therefore,

$$\int_{t_1}^{\infty} P(\sigma(s))g(s, x(\eta(s)))\Delta s < \infty.$$

If g is superlinear, then

$$g(t, c_1) \leq \frac{c_1 g(t, x(\eta(s)))}{x(\eta(s))} \leq g(t, x(\eta(s))),$$

which implies that

$$\int_{t_1}^{\infty} P(\sigma(s))g(s, c_1)\Delta s < \infty.$$

Similarly, if g is sublinear, then

$$\int_{t_1}^{\infty} P(\sigma(s))g(s, c_2)\Delta s < \infty.$$

Sufficiency. Without loss of generality, we let $k > 0$. If g is superlinear, then let $c = k/2$; if g is sublinear, then let $c = k$.

Choose $T > t_0$ so large that

$$\int_{T}^{\infty} P(\sigma(s))|g(s,c)|\Delta s < \frac{c}{2}.$$

Let

$$X = \left\{ x \mid x \in C_{rd}(\mathbb{T}_0, \mathbb{R}), \sup_{t \in \mathbb{T}_0} |x(t)| < \infty \right\}.$$

Endowed on X with the norm $\|x\| = \sup_{t \in \mathbb{T}_0} |x(t)|$, X is a Banach space. We define the set

$$\Omega = \{x = x(t) : x \in X, c \leq x(t) \leq 2c, t \in \mathbb{T}_0\}.$$

Clearly, Ω is a bounded, closed and convex subset of X. Define the map $\mathcal{S}$ on Ω as follows:

$$(\mathcal{S}x)(t) = \begin{cases} c + \int_{T}^{t} P(\sigma(s))g(s, x(\eta(s)))\Delta s + P(t)\int_{t}^{\infty} g(s, x(\eta(s)))\Delta s, & t \geq T, \\ (\mathcal{S}x)(T), & t_0 \leq t \leq T. \end{cases}$$

Step I. $\mathcal{S}$ maps Ω into Ω. Obviously, letting $x = x(t) \in \Omega$, we have $c \leq x(t) \leq 2c$ for $t \geq T$. Then,

$$c \leq (\mathcal{S}x)(t) \leq c + \int_{T}^{\infty} P(\sigma(s))g(s, x(\eta(s)))\Delta s < c + 2\int_{T}^{\infty} P(\sigma(s))g(s,c)\Delta s \leq 2c.$$

329

Hence, $c \leq (\mathcal{S}x)(t) \leq 2c$, for $t \in \mathbb{T}_0$. Therefore, $\mathcal{S}\Omega \subseteq \Omega$.

Step II. $\mathcal{S}$ is completely continuous.

We first claim that $\mathcal{S}$ is continuous. Let $x_n \in \Omega$ and $\|x_n - x\| \to 0$ as $n \to \infty$. Since Ω is a closed set, $x \in \Omega$. For $t \geq T$, we get

$$|(\mathcal{S}x_n)(t) - (\mathcal{S}x)(t)| \leq \int_T^\infty P(\sigma(s))|g(s,x_n(s)) - g(s,x(s))|\Delta s.$$

Since $|g(s,x_n(s)) - g(s,x(s))| \to 0$ as $n \to \infty$, so, by using the Lebesgue dominated convergence theorem, we conclude that $\lim_{n\to\infty} \|\mathcal{S}x_n - \mathcal{S}x\| = 0$, which implies that $\mathcal{S}$ is continuous in Ω.

Next, we show that $\mathcal{S}\Omega$ is relatively compact. It suffices to prove that the family of functions $\{\mathcal{S}x : x \in \Omega\}$ is bounded and uniformly Cauchy, and $\{\mathcal{S}x : x \in \Omega\}$ is equi-continuous on $[t_0, T_1]$ for any $T_1 \in [t_0, \infty)$. The boundedness is obvious. By Equation (5), for any $\varepsilon > 0$, let $T^* \geq T$ be so large that

$$\int_{T^*}^\infty P(\sigma(s))g(s,c)\Delta s < \frac{\varepsilon}{6}.$$

Then, for $x \in \Omega$, $t_2 > t_1 \geq T^*$, we have

$$|(\mathcal{S}x)(t_2) - (\mathcal{S}x)(t_1)|$$
$$\leq \left|\int_{t_1}^{t_2} P(\sigma(s))g(s,x(\eta(s)))\Delta s\right|$$
$$+ \left|P(t_2)\int_{t_2}^\infty g(s,x(\eta(s)))\Delta s\right| + \left|P(t_1)\int_{t_1}^\infty g(s,x(\eta(s)))\Delta s\right|$$
$$\leq \left|\int_{t_1}^{t_2} P(\sigma(s))g(s,x(\eta(s)))\Delta s\right|$$
$$+ \left|\int_{t_2}^\infty P(\sigma(s))g(s,x(\eta(s)))\Delta s\right| + \left|\int_{t_1}^\infty P(\sigma(s))g(s,x(\eta(s)))\Delta s\right|$$
$$\leq 3\left|\int_{T^*}^\infty P(\sigma(s))g(s,x(\eta(s)))\Delta s\right|$$
$$\leq 6\left|\int_{T^*}^\infty P(\sigma(s))g(s,c)\Delta s\right| < \varepsilon.$$

Hence, $\{\mathcal{S}x : x \in \Omega\}$ is uniformly Cauchy.

Furthermore, for any $T_1 \in [t_0, \infty)$ and $x \in \Omega$ with $T \leq t_1 < t_2 \leq T_1$, we get

$$|(\mathcal{S}x)(t_2) - (\mathcal{S}x)(t_1)|$$
$$\leq \left|\int_{t_1}^{t_2} P(\sigma(s))g(s,x(\eta(s)))\Delta s\right.$$
$$\left. + [P(t_2) - P(t_1)]\int_{t_1}^\infty g(s,x(\eta(s)))\Delta s - P(t_2)\int_{t_1}^{t_2} g(s,x(\eta(s)))\Delta s\right|$$
$$\leq \frac{c|P(t_2) - P(t_1)|}{P(T)} + P(T_1)\int_{t_1}^{t_2} g(s,c)\Delta s.$$

Hence, there exists a $\delta > 0$ such that

$$|(\mathcal{S}x)(t_2) - (\mathcal{S}x)(t_1)| < \varepsilon, \text{ when } 0 < t_2 - t_1 < \delta.$$

From the definition of operator $\mathcal{S}$, clearly, we have

$$|(\mathcal{S}x)(t_2) - (\mathcal{S}x)(t_1)| = 0 < \varepsilon, \text{ when } t_0 \leq t_1 < t_2 \leq T.$$

Thus, it follows that $\{Sx : x \in \Omega\}$ is equi-continuous on $[t_0, T_1]$. Hence, S is completely continuous. By Schauder's fixed point theorem, we deduce that there exists a $x_0 \in \Omega$ such that $Sx_0 = x_0$, which is a non-oscillatory solution of Equation (1) with $x_0 \in A_c^0$. The proof is completed. □

Theorem 4. *Assume that*
(i) $\int_{t_0}^{\infty} \frac{1}{p(s)} \Delta s = \infty$;
(ii) $g(t, x)$ is superlinear or sublinear.
Then, Equation (1) has a non-oscillatory solution $x(t) \in A_\infty^c$ if and only if

$$\int_{t_0}^{\infty} |g(s, kP(\eta(s))| \Delta s < \infty, \text{ for some } k \neq 0. \tag{6}$$

Proof. *Necessity.* Let $x(t) \in A_\infty^c$ be eventually positive. By Lemma 1 and $p(t)x^\Delta(t) \to c$ as $t \to \infty$, there exist $c_1 > 0, c_2 > 0$ and $t_1 \in \mathbb{T}_0$ such that

$$x^\Delta(t) > 0, \ c_1 P(\eta(t)) \leq x(\eta(t)) \leq c_2 P(\eta(t)), \ t \geq t_1.$$

Integrating Equation (1) from t_1 to t, we have

$$p(t)x^\Delta(t) - p(t_1)x^\Delta(t_1) + \int_{t_1}^{t} g(s, x(\eta(s))) \Delta s = 0,$$

which implies that

$$\int_{t_1}^{\infty} g(s, x(\eta(s))) \Delta s < \infty.$$

If g is superlinear, then

$$g(s, c_1 P(\eta(s))) \leq \frac{c_1 P(\eta(s))}{x(\eta(s))} g(s, x(\eta(s))) \leq g(s, x(\eta(s))),$$

which implies that

$$\int_{t_1}^{\infty} g(s, c_1 P(\eta(s))) \Delta s < \infty.$$

Similarly, if g is sublinear, then

$$\int_{t_1}^{\infty} g(s, c_2 P(\eta(s))) \Delta s < \infty.$$

Sufficiency. Without loss of generality, let $k > 0$. If g is superlinear, then let $c = k/2$; if g is sublinear, then let $c = k$.
Choose $T > t_0$ so large that

$$\int_{T}^{\infty} |g(s, cP(\eta(s))| \Delta s < \frac{c}{2}.$$

Let

$$X = \left\{ x \mid x \in C_{rd}(\mathbb{T}_0, \mathbb{R}), \sup_{t \in \mathbb{T}_0} \frac{|x(t)|}{P(t)} < \infty \right\}.$$

Endowed on X with the norm $\|x\| = \sup_{t \in \mathbb{T}_0} \frac{|x(t)|}{P(t)}$, X is a Banach space. Introduce a set

$$\Omega = \{x = x(t) : x \in X, \ cP(t) \leq x(t) \leq 2cP(t), \ t \in \mathbb{T}_0\}.$$

Clearly, Ω is a bounded, closed and convex subset of X. Define a map S on Ω by

$$(\mathcal{S}x)(t) = \begin{cases} cP(t) + \int_T^t P(\sigma(s))g(s,x(\eta(s)))\Delta s + P(t)\int_t^\infty g(s,x(\eta(s)))\Delta s, & t \geq T, \\ (\mathcal{S}x)(T), & t_0 \leq t \leq T. \end{cases}$$

Step I. $\mathcal{S}$ maps Ω into Ω. Let $x = x(t) \in \Omega$. Then, $cP(t) \leq x(t) \leq 2cP(t)$ for $t \geq T$, and

$$cP(t) \leq (\mathcal{S}x)(t) \leq cP(t) + P(t)\int_T^\infty g(s,x(\eta(s)))\Delta s < P(t)\left[c + 2\int_T^\infty g(s,c)\Delta s\right] \leq 2cP(t).$$

Hence, $\mathcal{S}\Omega \subseteq \Omega$.

Step II. $\mathcal{S}$ is completely continuous.

We first prove that $\mathcal{S}$ is continuous. Let $x_n \in \Omega$ and $\|x_n - x\| \to 0$ as $n \to \infty$. Since Ω is a closed set, $x \in \Omega$. For $t \geq T$, we get

$$|(\mathcal{S}x_n)(t) - (\mathcal{S}x)(t)| \leq P(t)\int_T^\infty |g(s,x_n(s)) - g(s,x(s))|\Delta s.$$

Since $|g(s,x_n(s)) - g(s,x(s))| \to 0$ as $n \to \infty$, $\lim_{n\to\infty}\|\mathcal{S}x_n - \mathcal{S}x\| = 0$, which implies that $\mathcal{S}$ is continuous in Ω.

Next, we show $\mathcal{S}\Omega$ is relatively compact. By Equation (6), for any $\varepsilon > 0$, let $t^* \geq T$ be sufficiently large such that

$$\int_{t^*}^\infty |g(s,cP(\eta(s)))|\Delta s < \frac{\varepsilon}{8}.$$

Since $\lim_{t\to\infty} P(t) = \infty$, there exists a $T^* \geq t^*$ such that

$$\frac{1}{P(t)}\left|\int_T^{t^*} P(\sigma(s))g(s,x(\eta(s)))\Delta s\right| < \frac{\varepsilon}{8}, \quad \text{for } t \geq T^*.$$

Hence, for $x \in \Omega$, $t_2 > t_1 \geq T^*$, we have

$$|(P^{-1}\mathcal{S}x)(t_2) - (P^{-1}\mathcal{S}x)(t_1)|$$
$$\leq \frac{1}{P(t_2)}\left|\int_T^{t^*} P(\sigma(s))g(s,x(\eta(s)))\Delta s\right| + \frac{1}{P(t_1)}\left|\int_T^{t^*} P(\sigma(s))g(s,x(\eta(s)))\Delta s\right|$$
$$+ \left|\frac{1}{P(t_2)}\int_{t^*}^{t_2} P(\sigma(s))g(s,x(\eta(s)))\Delta s\right| + \left|\frac{1}{P(t_1)}\int_{t^*}^{t_1} P(\sigma(s))g(s,x(\eta(s)))\Delta s\right|$$
$$+ \left|\int_{t_1}^{t_2} g(s,x(\eta(s)))\Delta s\right|$$
$$\leq \frac{\varepsilon}{4} + 3\int_{t^*}^\infty |g(s,x(\eta(s)))|\Delta s$$
$$\leq \frac{\varepsilon}{4} + 6\int_{t^*}^\infty |g(s,cP(\eta(s)))|\Delta s < \varepsilon.$$

Hence, $\{\mathcal{S}x : x \in \Omega\}$ is uniformly Cauchy. Furthermore, for any $T_1 \in [t_0,\infty)$ and $x \in \Omega$, if $T \leq t_1 < t_2 \leq T_1$, then

$$|(P^{-1}\mathcal{S}x)(t_2) - (P^{-1}\mathcal{S}x)(t_1)|$$
$$= \left|\left[\frac{1}{P(t_2)} - \frac{1}{P(t_1)}\right]\int_T^{t_1} P(\sigma(s))g(s,x(\eta(s)))\Delta s\right.$$
$$\left. + \frac{1}{P(t_2)}\int_{t_1}^{t_2} P(\sigma(s))g(s,x(\eta(s)))\Delta s + \int_{t_1}^{t_2} g(s,x(\eta(s)))\Delta s\right|.$$

Hence, there exists a $\delta > 0$ such that

$$|(P^{-1}Sx)(t_2) - (P^{-1}Sx)(t_1)| < \varepsilon, \text{ if } 0 < t_2 - t_1 < \delta.$$

From the definition of operator S, it is clear that

$$|(P^{-1}Sx)(t_2) - (P^{-1}Sx)(t_1)| = 0 < \varepsilon, \text{ if } t_0 \le t_1 < t_2 \le T.$$

From the foregoing arguments, we deduce that $\{Sx : x \in \Omega\}$ is equi-continuous on $[t_0, T_1]$. Hence, S is completely continuous. Consequently, by Schauder's fixed point theorem, there exists a $x_0 \in \Omega$ such that $Sx_0 = x_0$, which is a non-oscillatory solution of Equation (1) with $x_0 \in A_\infty^c$. The proof is completed. □

Theorem 5. *Assume that*
(i) $\int_{t_0}^\infty \frac{1}{p(s)} \Delta s < \infty$;
(ii) $g(t, x)$ *is superlinear or sublinear.*
Then, Equation (1) has a non-oscillatory solution $x(t) \in A_c$ if and only if

$$\int_{t_0}^\infty \hat{P}(s)|g(s,k)|\Delta s < \infty, \text{ for some } k \ne 0.$$

Proof. *Necessity.* Let $x(t) \in A_c$ be eventually positive. Firstly, we show that $\int_{t_0}^\infty \hat{P}(s)|g(s,x(\eta(t)))|\Delta s < \infty$. If not, multiplying Equation (1) by $\hat{P}(t)$, and then integrating from t_0 to t, we have

$$\hat{P}(t)p(t)x^\Delta(t) - \hat{P}(t_0)p(t_0)x^\Delta(t_0) - x(t) + x(t_0) + \int_{t_0}^t \hat{P}(s)g(s,x(\eta(s)))\Delta s = 0.$$

Then, $\lim_{t\to\infty} \hat{P}(t)p(t)x^\Delta(t) = -\infty$. Therefore, there exist $t_1 \ge t_0$, $M > 0$ such that $\hat{P}(t)p(t)x^\Delta(t) \le -M$, for $t \ge t_1$. Hence,

$$x(t) - x(t_1) \le M \ln\left(\frac{\hat{P}(t)}{\hat{P}(t_1)}\right).$$

Since $\lim_{t\to\infty} \hat{P}(t) = 0$, $\lim_{t\to\infty} \ln\left(\frac{\hat{P}(t)}{\hat{P}(t_1)}\right) = -\infty$. This is a contradiction.
By Lemma 2, there exist $c_1 > 0, c_2 > 0$ and $t_1 \in \mathbb{T}_0$ such that $c_1 \le x(\eta(t)) \le c_2$, $t \ge t_1$. Therefore, if g is superlinear, then

$$\int_{t_1}^\infty \hat{P}(t)g(s,c_1)\Delta s < \infty.$$

If g is sublinear, then

$$\int_{t_1}^\infty \hat{P}(t)g(s,c_2)\Delta s < \infty.$$

Sufficiency. Without loss of generality, we let $k > 0$. If g is superlinear, then let $c = k/2$; if g is sublinear, then let $c = k$.
Let $T > t_0$ be large so that

$$\int_T^\infty \hat{P}(\eta(s))|g(s,c)|\Delta s < \frac{c}{2}.$$

Let

$$X = \left\{x \mid x \in C_{rd}(\mathbb{T}_0, \mathbb{R}), \sup_{t \in \mathbb{T}_0} |x(t)| < \infty\right\}.$$

Endowed on X with the norm $\|x\| = \sup_{t \in \mathbb{T}_0} |x(t)|$, X is a Banach space. Define the set

$$\Omega = \{x = x(t) : x \in X, c \le x(t) \le 2c, t \in \mathbb{T}_0\}.$$

Define a map $\mathcal{S}$ on Ω as follows:

$$(\mathcal{S}x)(t) = \begin{cases} c + \hat{P}(t)\int_T^t g(s, x(\eta(s)))\Delta s + \int_t^\infty \hat{P}(s)g(s, x(\eta(s)))\Delta s, & t \geq T, \\ (\mathcal{S}x)(T), & t_0 \leq t \leq T. \end{cases}$$

Similarly to the previous process, we can show that $\mathcal{S}$ has a fixed point x_0, which is a non-oscillatory solution of Equation (1) with $x_0 \in A_c$. The proof is complete. □

Theorem 6. *Assume that*
(i) $\int_{t_0}^\infty \frac{1}{p(s)}\Delta s < \infty$;
(ii) $g(t, x)$ *is superlinear or sublinear.*
Then, Equation (1) has a non-oscillatory solution $x(t) \in A_0^c$ if and only if

$$\int_{t_0}^\infty |g(s, k\hat{P}(\eta(s)))|\Delta s < \infty, \text{ for some } k \neq 0.$$

Proof. *Necessity.* Let $x(t) \in A_0^c$ be eventually positive. Then, $\lim_{t\to\infty} x(t) = 0$, $\lim_{t\to\infty} \frac{x(t)}{\hat{P}(t)} = c \neq 0$. Assume that $c > 0$. Then, there exist $c_1 > 0$, $c_2 > 0$ and $t_1 \in \mathbb{T}_0$ such that $c_1\hat{P}(\eta(t)) \leq x(\eta(t)) \leq c_2\hat{P}(\eta(t))$, $t \geq t_1$. By Lemma 2, we have that $x(t) \geq -p(t)x^\Delta(t)\hat{P}(t)$, $t \geq t_1$. Thus, $-p(t)x^\Delta(t) \leq c_2$, $t \geq t_1$. On the other hand,

$$\int_{t_1}^t g(s, x(\eta(s)))\Delta s = p(t_1)x^\Delta(t_1) - p(t)x^\Delta(t) \leq |p(t_1)x^\Delta(t_1)| + c_2,$$

therefore

$$\int_{t_1}^t g(s, x(\eta(s)))\Delta s < \infty.$$

If g is superlinear, then

$$\int_{t_1}^\infty g(s, c_1\hat{P}(\eta(s)))\Delta s < \infty.$$

If g is sublinear, then

$$\int_{t_1}^\infty g(s, c_2\hat{P}(\eta(s)))\Delta s < \infty.$$

Sufficiency. Without loss of generality, we let $k > 0$. If g is superlinear, then let $c = k/2$; if g is sublinear, then let $c = k$.
Let $T > t_0$ be large so that

$$\int_T^\infty \hat{P}(\eta(s))|g(s, c)|\Delta s < \frac{c}{2}.$$

Let

$$X = \left\{ x \mid x \in C_{rd}(\mathbb{T}_0, \mathbb{R}), \sup_{t \in \mathbb{T}_0} \frac{|x(t)|}{\hat{P}(t)} < \infty \right\}.$$

Endowed on X with the norm $\|x\| = \sup_{t \in \mathbb{T}_0} \frac{|x(t)|}{\hat{P}(t)}$, X is a Banach space. Define the set

$$\Omega = \left\{ x = x(t) : x \in X, c\hat{P}(t) \leq x(t) \leq 2c\hat{P}(t), t \in \mathbb{T}_0 \right\}.$$

Define a map $\mathcal{S}$ on Ω as follows

$$(\mathcal{S}x)(t) = \begin{cases} c\hat{P}(t) + \hat{P}(t)\int_T^t g(s, x(\eta(s)))\Delta s + \int_t^\infty \hat{P}(s)g(s, x(\eta(s)))\Delta s, & t \geq T, \\ (\mathcal{S}x)(T), & t_0 \leq t \leq T. \end{cases}$$

Using the earlier arguments, one can show that Equation (1) has a non-oscillatory solution in A_0^c. The proof is complete. □

4. Conclusions

In the current paper, the structure of non-oscillatory solutions for a class of second order dynamic equations on time scales is considered. Under the differentiable assumptions, we first establish two classifications of non-oscillatory solutions. Forthermore, by using the assumptions of superlinear and sublinear function, we obtain four sufficient and necessary conditions for existence of some kinds of non-oscillatory solutions.

Author Contributions: All authors have read and approved the final manuscript. Y.Z. finished the manuscript and B.A. and A.A. made the content correction and English language checking.

Funding: This research was funded by the National Natural Science Foundation of P.R. China (11671339).

Conflicts of Interest: The authors declare no conflict of interest.

References

1. Bohner, M.; Georgiev, S.G. *Multivariable Dynamic Calculus on Time Scales*; Springer: Berlin/Heidelberg, Germany, 2017.
2. Georgiev, S.G. *Fractional Dynamic Calculus and Fractional Dynamic Equations on Time Scales*; Springer: Berlin/Heidelberg, Germany, 2018.
3. Martynyuk, A.A. *Stability Theory for Dynamic Equations on Time Scales*; Springer: Berlin/Heidelberg, Germany, 2016.
4. Georgiev, S.G. *Integral Equations on Time Scales*; Atlantis Press: Paris, France, 2016.
5. Saker, S. *Oscillation Theory of Dynamic Equations on Time Scales: Second and Third Orders*; Lap Lambert, Academic Publishing: Saarland, Germany, 2010.
6. Došlý, O.; Hilger, S. A necessary and sufficient condition for oscillation of the Sturm-Liouville dynamic equation on time scales. *J. Comput. Appl. Math.* **2002**, *141*, 147–158. [CrossRef]
7. Grace, S.R.; Agarwal, R.P.; Bohner, M.; O'Regan, D. Oscillation of second-order strongly superlinear and strongly sublinear dynamic equations. *Commun. Nonlinear Sci. Numer. Simul.* **2009**, *14*, 3463–3471. [CrossRef]
8. Zhou, Y.; Lan, Y. Classification and existence of non-oscillatory solutions of second-order neutral delay dynamic equations on time scales. *Nonlinear Oscil.* **2013**, *16*, 191–206.
9. Agarwal, R.P.; Bohner, M.; Li, T.; Zhang, C. Oscillation criteria for second-order dynamic equations on time scales. *Appl. Math. Lett.* **2014**, *31*, 34–40. [CrossRef]
10. Deng, X.H.; Wang, Q.R.; Zhou, Z. Oscillation criteria for second order nonlinear delay dynamic equations on time scales. *Appl. Math. Comput.* **2015**, *269*, 834–840. [CrossRef]
11. Deng, X.H.; Wang, Q.R.; Zhou, Z. Generalized Philos-type oscillation criteria for second order nonlinear neutral delay dynamic equations on time scales. *Appl. Math. Lett.* **2016**, *57*, 69–76. [CrossRef]
12. Senel, M.T.; Utku, N.; El-Sheikh, M.M.A.; Li, T. Kamenev-type criteria for nonlinear second-order delay dynamic equations. *Hacet. J. Math. Stat.* **2018**, *47*, 339–345.
13. Bohner, M.; Hassan, T.S.; Li, T. Fite-Hille-Wintner-type oscillation criteria for second-order half-linear dynamic equations with deviating arguments. *Indag. Math.* **2018**, *29*, 548–560. [CrossRef]
14. Hasil, P.; Veselý, M. Oscillation and non-oscillation results for solutions of perturbed half-linear equations. *Math. Meth. Appl. Sci.* **2018**, *41*, 3246–3269. [CrossRef]
15. Negi, S.S.; Abbas, S.; Malik, M. Oscillation criteria of singular initial-value problem for second order nonlinear dynamic equation on time scales. *Nonauton. Dyn. Syst.* **2018**, *5*, 102–112. [CrossRef]
16. Negi, S.S.; Abbas, S.; Malik, M.; Xia, Y.-H. New oscillation criteria of special type second-order nonlinear dynamic equations on time scales. *Math. Sci.* **2018**, *12*, 25–39. [CrossRef]
17. Agarwal, R.P.; Grace, S.R.; O'Regan, D. Non-oscillatory solutions for higher order dynamic equations. *J. Lond. Math. Soc.* **2003**, *67*, 165–179. [CrossRef]
18. Zhu, Z.Q.; Wang, Q.R. Existence of nonoscillatory solutions to neutral dynamic equations on time scales. *J. Math. Anal. Appl.* **2007**, *335*, 751–762. [CrossRef]

19. Zhou, Y. Nonoscillation of higher order neutral dynamic equations on time scales. *Appl. Math. Lett.* **2019**, *94*, 204–209. [CrossRef]
20. Zhou, Y.; Ahmad, B.; Alsaedi, A. Necessary and sufficient conditions for oscillation of second-order dynamic equations on time scales. *Math. Meth. Appl. Sci.* **2019**. [CrossRef]
21. Graef, J.R.; Hill, M. Nonoscillation of all solutions of a higher order nonlinear delay dynamic equation on time scales. *J. Math. Anal. Appl.* **2015**, *423*, 1693–1703. [CrossRef]
22. Erbe, L.; Baoguo, J.; Peterson, A. Nonoscillation for second order sublinear dynamic equations on time scales. *J. Comput. Appl. Math.* **2009**, *232*, 594–599. [CrossRef]

© 2019 by the authors. Licensee MDPI, Basel, Switzerland. This article is an open access article distributed under the terms and conditions of the Creative Commons Attribution (CC BY) license (http://creativecommons.org/licenses/by/4.0/).

MDPI
St. Alban-Anlage 66
4052 Basel
Switzerland
Tel. +41 61 683 77 34
Fax +41 61 302 89 18
www.mdpi.com

Mathematics Editorial Office
E-mail: mathematics@mdpi.com
www.mdpi.com/journal/mathematics

www.ingramcontent.com/pod-product-compliance
Lightning Source LLC
LaVergne TN
LVHW070224100526
838202LV00015B/2089